ANNUAL PLANT REVIEWS VOLUME 45

ANNUAL PLANT REVIEWS VOLUME 45

The Evolution of Plant Form

Edited by

Barbara A. Ambrose

Cullman Assistant Curator in Plant Genomics
The New York Botanical Garden
Bronx, NY 10458, USA

Michael Purugganan

Dorothy Schiff Professor of Genomics
Department of Biology
Center for Genomics and Systems Biology
New York University
New York, NY 10011, USA

and

Center for Genomics and Systems Biology
NYU Abu Dhabi Institute
New York University Abu Dhabi
Abu Dhabi, United Arab Emirates

WILEY-BLACKWELL

A John Wiley & Sons, Ltd., Publication

Annual Plant Reviews

Blackwell Publishing was acquired by John Wiley & Sons in February 2007. Blackwell's publishing program has been merged with Wiley's global Scientific, Technical and Medical business to form Wiley-Blackwell.

Registered office: John Wiley & Sons, Ltd, The Atrium, Southern Gate, Chichester, West Sussex, PO19 8SQ, UK

Editorial offices: 9600 Garsington Road, Oxford, OX4 2DQ, UK
2121 State Avenue, Ames, Iowa 50014-8300, USA

For details of our global editorial offices, for customer services and for information about how to apply for permission to reuse the copyright material in this book please see our website at www.wiley.com/wiley-blackwell.

Library of Congress Cataloging-in-Publication Data

The evolution of plant form / edited by Barbara A. Ambrose, Michael Purugganan.
p. cm. – (Annual plant reviews ; v. 45)
Includes bibliographical references and index.
ISBN 978-1-4443-3001-4 (hardback : alk. paper) – ISBN 978-1-118-25388-5 (epdf/ebook) – ISBN 978-1-118-25393-9 (emobi) – ISBN 978-1-118-25385-4 (epub) – ISBN 978-1-118-30588-1 (obook) 1. Plants–Evolution. 2. Plant morphology. I. Ambrose, Barbara A. II. Purugganan, Michael. III. Series: Annual plant reviews ; v. 45.
QK980.E87 2012
571.3′2–dc23

2012024775

A catalogue record for this book is available from the British Library.

Wiley also publishes its books in a variety of electronic formats. Some content that appears in print may not be available in electronic books.

Cover image: A close up of the female lobe of the liverwort *Marchantia polymorpha*. See chapter 2 for more details. Photograph by Bernard Goffinet.
Cover design by www.hisandhersdesign.co.uk

Set in 10/12pt Palatino by Aptara® Inc., New Delhi, India
Printed and bound in Malaysia by Vivar Printing Sdn Bhd

1 2013

Annual Plant Reviews

A series for researchers and postgraduates in the plant sciences. Each volume in this series focuses on a theme of topical importance and emphasis is placed on rapid publication.

Titles in the series:

1. **Arabidopsis**
   Edited by M. Anderson and J.A. Roberts
2. **Biochemistry of Plant Secondary Metabolism**
   Edited by M. Wink
3. **Functions of Plant Secondary Metabolites and their Exploitation in Biotechnology**
   Edited by M. Wink
4. **Molecular Plant Pathology**
   Edited by M. Dickinson and J. Beynon
5. **Vacuolar Compartments**
   Edited by D.G. Robinson and J.C. Rogers
6. **Plant Reproduction**
   Edited by S.D. O'Neill and J.A. Roberts
7. **Protein–Protein Interactions in Plant Biology**
   Edited by M.T. McManus, W.A. Laing and A.C. Allan
8. **The Plant CellWall**
   Edited by J.K.C. Rose
9. **The Golgi Apparatus and the Plant Secretory Pathway**
   Edited by D.G. Robinson
10. **The Plant Cytoskeleton in Cell Differentiation and Development**
    Edited by P.J. Hussey
11. **Plant–Pathogen Interactions**
    Edited by N.J. Talbot
12. **Polarity in Plants**
    Edited by K. Lindsey
13. **Plastids**
    Edited by S.G. Moller
14. **Plant Pigments and their Manipulation**
    Edited by K.M. Davies
15. **Membrane Transport in Plants**
    Edited by M.R. Blatt
16. **Intercellular Communication in Plants**
    Edited by A.J. Fleming
17. **Plant Architecture and Its Manipulation**
    Edited by C.G.N. Turnbull
18. **Plasmodeomata**
    Edited by K.J. Oparka

19. **Plant Epigenetics**
    Edited by P. Meyer
20. **Flowering and Its Manipulation**
    Edited by C. Ainsworth
21. **Endogenous Plant Rhythms**
    Edited by A. Hall and H. McWatters
22. **Control of Primary Metabolism in Plants**
    Edited by W.C. Plaxton and M.T. McManus
23. **Biology of the Plant Cuticle**
    Edited by M. Riederer
24. **Plant Hormone Signaling**
    Edited by P. Hadden and S.G. Thomas
25. **Plant Cell Separation and Adhesion**
    Edited by J.R. Roberts and Z. Gonzalez-Carranza
26. **Senescence Processes in Plants**
    Edited by S. Gan
27. **Seed Development, Dormancy and Germination**
    Edited by K.J. Bradford and H. Nonogaki
28. **Plant Proteomics**
    Edited by C. Finnie
29. **Regulation of Transcription in Plants**
    Edited by K. Grasser
30. **Light and Plant Development**
    Edited by G. Whitelam
31. **Plant Mitochondria**
    Edited by D.C. Logan
32. **Cell Cycle Control and Plant Development**
    Edited by D. Inzé
33. **Intracellular Signaling in Plants**
    Edited by Z. Yang
34. **Molecular Aspects of Plant Disease Resistance**
    Edited by J. Parker
35. **Plant Systems Biology**
    Edited by G.M. Coruzzi and R.A. Gutiérrez
36. **The Moss *Physcomitrella patens***
    Edited by C.D. Knight, P-F. Perroud and D.J. Cove
37. **Root Development**
    Edited by T. Beeckman
38. **Fruit Development and Seed Dispersal**
    Edited by L. Østergaard
39. **Function and Biotechnology of Plant Secondary Metabolites**
    Edited by M. Wink
40. **Biochemistry of Plant Secondary Metabolism**
    Edited by M. Wink
41. **Plant Polysaccharides**
    Edited by P. Ulvskov
42. **Nitrogen Metabolism in Plants in the Post-genomic Era**
    Edited by C. Foyer and H. Zhang
43. **Biology of Plant Metabolomics**
    Edited by R.D. Hall
44. **The Plant Hormone Ethylene**
    Edited by M.T. McManus
45. **The Evolution of Plant Form**
    Edited by B.A. Ambrose and M. Purugganan

CONTENTS

A color plate section falls between pages 62 and 63

LIST OF CONTRIBUTORS

Barbara A. Ambrose
The New York Botanical Garden
Bronx, NY 10458, USA

William R. Buck
The New York Botanical Garden
Bronx, NY 10458, USA

Kathleen Donohue
Biology Department
Duke University
Durham, NC 27708, USA

James A. Doyle
Department of Evolution and Ecology
University of California
Davis, CA 95616, USA

Bernard Goffinet
Department of Ecology and Evolutionary Biology
University of Connecticut
Storrs, CT 06269, USA

Amy Litt
The New York Botanical Garden
Bronx, NY 10458, USA

Cristina Ferrándiz
Instituto de Biología Molecular y Celular de Plantas (IBMCP),
Consejo Superior de Investigaciones Cientificas-Universidad Politecnica de Valencia,
46022 Valencia, Spain

Michael Purugganan
Department of Biology
Center for Genomics and Systems Biology
New York University
New York, NY 10011, USA

and

Center for Genomics and Systems Biology
NYU Abu Dhabi Institute,
New York University Abu Dhabi
Abu Dhabi, United Arab Emirates

Paula J. Rudall
Royal Botanic Gardens, Kew, Richmond,
Surrey TW9 3AB, UK

Harald Schneider
Department of Botany
Natural History Museum
London SW7 5BD, UK

Kristen Shepard
Department of Biological Sciences
Barnard College
New York, NY 10027, USA

Dennis Wm. Stevenson
The New York Botanical Garden
Bronx, NY 10458, USA

PREFACE

From Linnaeus to Mendel to Darwin, studies of land plants have been crucial for elucidating basic biological principles. Plant species have been important, in particular, in advancing the study of evolutionary principles, both from classical and modern molecular studies. Despite nearly three centuries of study, there remain several gaps in our understanding of the evolution of morphology in land plants, associated with their prolific diversification over the last ~500 million years.

Current scientific knowledge of land plant morphology, phylogeny, evolutionary mechanisms, and genomic tools can now provide us with a more comprehensive theory for the evolution of plant form. Plant morphological evolution has been studied through paleobotany and phylogenetic analyses, and major morphological innovations in the land plants have been mapped onto phylogenetic trees. Questions about morphological evolution in land plants are being addressed, but there is still a divide between the classical morphologists (who know the questions) and the molecular geneticist (who has new tools to address these questions). Some of these tools include whole genome sequences, functional plant model systems, and knowledge of the transcription factors and regulatory networks underlying plant morphology. Population genetic studies in plants can also elegantly demonstrate how changes at the microevolutionary level lead to differentiation in form, separating plant species. However, to understand morphological innovations in the land plants as well as the molecular basis of morphological evolution in land plants, it is time for the integration of studies in population genetics, adaptation, phenotypic plasticity, ecology, modularity, and development.

Previous books on the evolution of plant form have either focused on morphological evolution in the angiosperms and/or on developmental genetics or population genetics as sole mechanisms of evolutionary change. This volume aims to bring together a range of plant biologists to provide a reference and inspiration for future studies in the evolution of plant form.

This volume covers land plant morphology, mechanisms of evolutionary change, as well as the tools available to help address the evolution of plant form. The first half of the book presents the classic questions on the evolution of plant form. Chapter 1 provides a phylogenetic reference for morphological evolution in the land plants as well as indicates the origins of major morphological innovations in the land plants. Chapters 2–6 provide a more in-depth analysis of morphological evolution within the major land plant groups: bryophytes, ferns, gymnosperms, and angiosperms.

Chapters 7–10 present a range of studies that address mechanisms of evolutionary change and the tools available to address mechanisms of evolutionary change. Chapter 11 provides a synthesis of plant morphology and evolutionary mechanisms to present a more inclusive theory for plant morphological evolution.

Barbara A. Ambrose and Michael Purugganan

ACKNOWLEDGMENTS

We are indebted to the authors of this volume, not only for their contributions but also for their patience and perseverance throughout the making of this book. We would like to thank M.T. McManus for initially suggesting this topic for a volume in *Annual Plant Reviews* and to Wiley-Blackwell for their support. BAA is eternally grateful to her husband, Emilio, with whom everything is possible, and to their son, Santiago, a little marvel. Michael would not have been able to work on this without the support, love, and encouragement of his wife, Alessandra, who saw this project from the start to its completion.

Annual Plant Reviews (2013) **45**, 1–50
doi: 10.1002/9781118305881.ch1

http://onlinelibrary.wiley.com

Chapter 1

PHYLOGENETIC ANALYSES AND MORPHOLOGICAL INNOVATIONS IN LAND PLANTS

James A. Doyle

Department of Evolution and Ecology, University of California, Davis, CA, USA

Abstract: An increasingly robust phylogenetic framework based on molecular and fossil data clarifies the sequence of evolutionary innovations in land plants. Oogamy and cellular novelties (phragmoplast, plasmodesmata, incipient meristems) evolved in aquatic streptophytes. Invasion of the land entailed interpolation of the sporophyte, jacketed gametangia and sporangia, and air-dispersed spores, followed by stomata. Origin of vascular plants involved branching of the sporophyte and stepwise evolution of vascular tissue. Leaves originated independently in lycophytes and euphyllophytes; in some euphyllophytes leaves were derived from single dichotomous branches, in others from whole branch systems. In seed plants, secondary growth evolved before the seed. Pinnately compound leaves were replaced by simple leaves in coniferophytes. The origin of the angiosperm flower remains unresolved, but bitegmic ovules may be derived from cupules, and the ancestral carpel can be reconstructed as ascidiate. Evolution of double fertilization was a stepwise process that continued within angiosperms; vessels also evolved within the group. Monocots show major reorganization tied to loss of secondary growth, while pentamerous flowers evolved from dimerous within eudicots.

Keywords: evolution; phylogeny; morphology; innovations; land plants; angiosperms.

Annual Plant Reviews Volume 45: The Evolution of Plant Form, First Edition.
Edited by Barbara A. Ambrose and Michael Purugganan.

1.1 Introduction

As in other groups of terrestrial organisms, the evolution of land plants involved a series of radiations linked with major evolutionary innovations, many of them clearly adaptations that allowed progressively more efficient and varied occupation of the land environment. The morphology of familiar plants such as the model system *Arabidopsis thaliana* therefore represents a hierarchical accumulation of structural features that arose at different points on the line from their distant aquatic ancestors, with older advances shared with a successively wider range of relatives. This chapter attempts to summarize the present picture of the sequence of evolutionary innovations in the latest phylogenetic framework, as well as outstanding unresolved issues.

Some of the main events in the evolutionary history of land plants have been recognized since the late nineteenth century. Key insights were recognition of the alternation of haploid and diploid generations, seen in its most basic and obvious form in "bryophytes" such as mosses and "lower vascular plants" such as ferns, and realization that this life cycle persists in modified form into seed plants (Hofmeister 1862; Bower 1890, 1908; Strasburger 1894). However, the details have become much clearer over the past century as a result of many factors, including fossil discoveries that show intermediate stages and character combinations no longer preserved in the living flora, technical advances that revealed new suites of characters at the microscopic and ultrastructural level, development of more explicit methods of analysis of phylogenetic relationships, and the application of these methods to molecular sequence data. Methods of phylogeny reconstruction, many derived from earlier partial insights (notably Zimmermann 1931; Donoghue & Kadereit 1992) but first clearly synthesized in English by Hennig (1966), were elaborated under the rubric of "cladistics" in the 1970s and 1980s and used in analyses of morphological characters. These methods used the principle of parsimony to search for the phylogenetic tree involving the fewest character state changes, on the assumption that this is the tree most consistent with the totality of characters recognized.

Whereas some precladistic discussions assumed that phylogeny could only be approached by consideration of fossils and identification of direct ancestors, which implied that the phylogeny of groups such as angiosperms with a supposedly poor fossil record could not be understood, cladistic methods could be applied to both living and fossil organisms. These methods also made it possible to draw conclusions on the origin of groups and their ancestral states by recognition of closest outgroups without identification of actual direct ancestors. At the level of land plants, many analyses included both fossil and living taxa. There was considerable discussion of the relative importance of the two sorts of data, some arguing that the main relationships among living organisms could be reconstructed without fossil data, which necessarily have far fewer characters due to lack of preservation of parts (Patterson 1981). Others argued that inclusion of fossils was necessary to obtain correct relationships, as in amniote vertebrates (Gauthier *et al.* 1988), and even when fossils were not required to infer the correct topology of the tree of living organisms, they could be needed to reconstruct the evolutionary steps leading to living clades, which are often separated from their closest relatives by large numbers of morphological changes (Donoghue *et al.* 1989). In molecular hindsight, morphological cladistic analyses correctly resolved many contentious problems that had plagued earlier intuitive approaches, such as the monophyly of land plants and angiosperms. On other questions, however, such as rooting of the angiosperm phylogenetic tree and relationships among vascular plant and seed plant lines, the results varied from one analysis to another, presumably due to homoplasy (evolutionary convergence and reversal), different interpretations of characters, and variable sampling of both extant and fossil taxa.

This picture has improved dramatically in the past two decades with the accumulation of vast quantities of molecular sequence data from more and more species, which has led to increasingly complete, consistent, and statistically well-supported trees of living organisms. The first analyses of sequences of single genes showed many of the same sorts of inconsistencies seen in morphological analyses. However, as more genes have been sequenced and combined into multigene and even whole-genome analyses, many tentative early results have stabilized and become statistically robust, and with some conspicuous exceptions, most early conflicts between genes have been firmly resolved. These studies started with parsimony analysis, but newer maximum likelihood and Bayesian methods take a more statistical approach to changes on branches. This has led to a role reversal—whereas formerly ideas on the evolution of morphological characters were used to reconstruct phylogenies, phylogenies based on molecular data are now used to reconstruct the evolution of morphological characters, by plotting (optimization) of character states on trees derived from molecular data, using parsimony or likelihood-based methods, thus avoiding dangers of circular reasoning.

For understanding of major evolutionary innovations, a major weakness of the increasingly exclusive reliance on molecular data for phylogeny reconstruction is that purely molecular analyses cannot include fossil taxa, since with the exception of very recent fossils (such as human relatives) no DNA remains. There is much discussion of whether morphological data have any role at all in reconstruction of phylogenies of living organisms (Scotland *et al.* 2003; Wiens 2004). Because of the vastly greater number of DNA characters, combined analyses of morphological and molecular data tend to be dominated by molecular data. Morphology may still have some role in resolving parts of the phylogeny where molecular analyses of different genes give conflicting or poorly supported results, such as among seed plants and major groups above the base of the angiosperms, but it may be argued that this represents only a brief intermediate phase before all relationships are cemented by genomic data.

Even if molecular analyses completely resolve all relationships among living taxa, other approaches will be needed to address those cases where fossils provide evidence on transitions that are not preserved in the living flora. Molecular analyses consider only relationships among crown groups—where a crown group includes the most recent common ancestor of the living members of a clade and all its derivatives—and not fossils on or attached to the stem lineages leading to these clades, known as stem relatives (Doyle & Donoghue 1993). Integration of these fossils will still require compilation and analysis of morphological data from both the fossils and all relevant living taxa. The ideal approach may be a "total evidence" analysis that combines both molecular and morphological data (Hermsen & Hendricks 2008), but this may not be so easy because of problems in choice of molecular data sets and differences in taxon sampling: single species in molecular data sets, but often higher taxa for which ancestral states have been reconstructed in morphological data sets. A few studies of land plants have used a total evidence approach to integrate morphological data from fossils with molecular data (e.g., Rothwell & Nixon 2006), but in the meantime others have used a "molecular scaffold" approach (Springer *et al.* 2001; Manos *et al.* 2007; Doyle & Endress 2010) that integrates fossils by analyzing a morphological data set for both living and fossil taxa with the relationships among living taxa fixed to a "constraint tree" based on molecular data. Such analyses essentially ask what additional insights fossils provide if the relationships inferred from molecular data are correct.

This chapter does not pretend to be a comprehensive review of the literature, but rather a selective though hopefully balanced survey of current ideas and evidence. Currently understood phylogenetic relationships and the placement of morphological innovations are summarized in Figures 1.1 and 1.2. A good general review of land plant phylogeny is provided by Judd *et al.* (2008). Except where noted, most information on morphological characters of the living and fossil taxa discussed here can be found in standard plant morphology and paleobotany texts (Smith 1955; Gifford & Foster 1989; Stewart & Rothwell 1993; Crum 2001; Taylor *et al.* 2009) and Graham (1993) for algal outgroups. Not all the innovations mentioned are equally "major" in an evolutionary sense, but some more obscure ones are of interest as providing morphological support for relationships. In most cases, I refer to clades above the ordinal level with anglicized versions of names in the phylogenetic nomenclature of Cantino *et al.* (2007), and I use quotes to mark traditional paraphyletic groups when these are first mentioned (e.g., "bryophytes"). Hopefully this summary will be useful as a framework for investigations on the developmental-genetic and functional bases of the evolutionary changes inferred.

1.2 Basic innovations in cell structure and life cycle: aquatic streptophytes

Both morphological and molecular phylogenetic analyses confirm the long-standing view that land plants are members of the clade of green organisms,

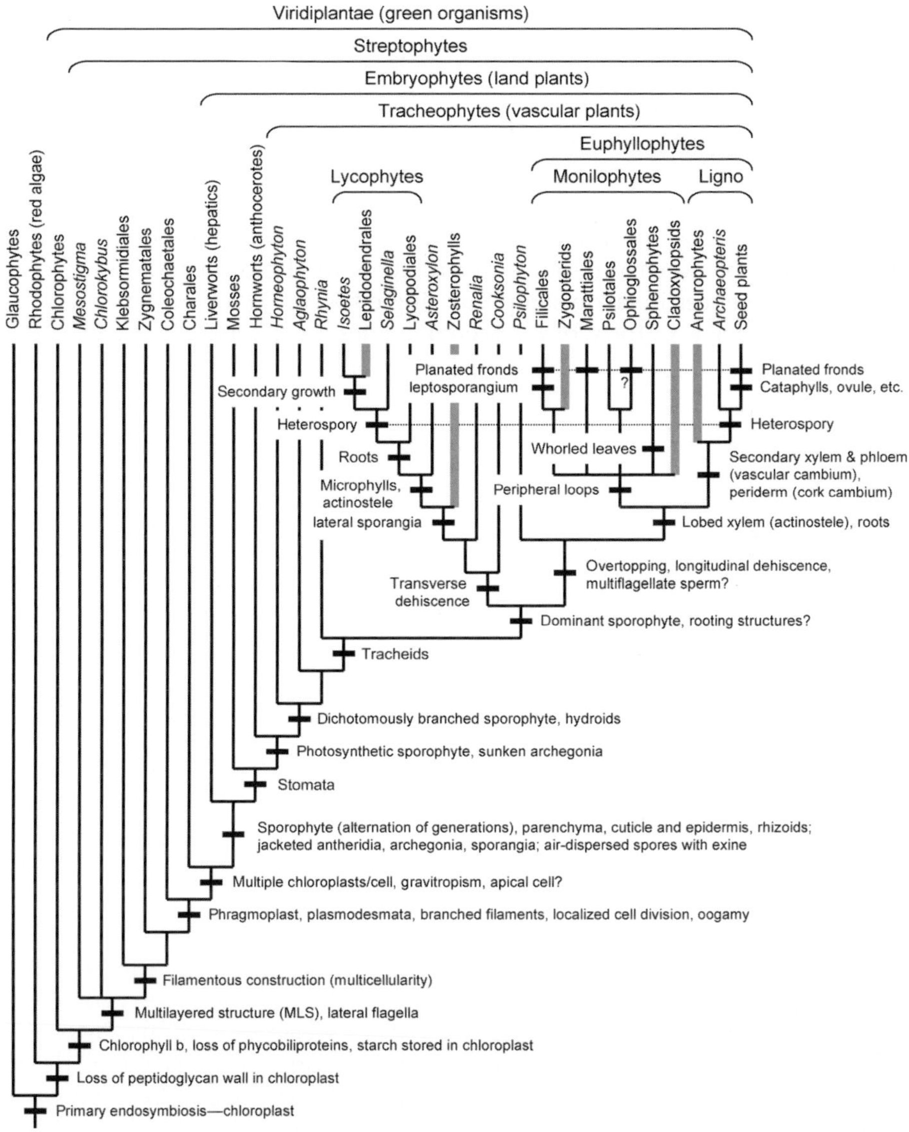

Figure 1.1 Phylogenetic relationships and evolutionary innovations from the origin of plastids to seed plants. See the text for sources of phylogenetic information. Wide gray lines indicate presumed paraphyletic groups. Question marks indicate cases where the exact placement of an innovation is uncertain (e.g., because of lack of information in fossils). Ligno, lignophytes.

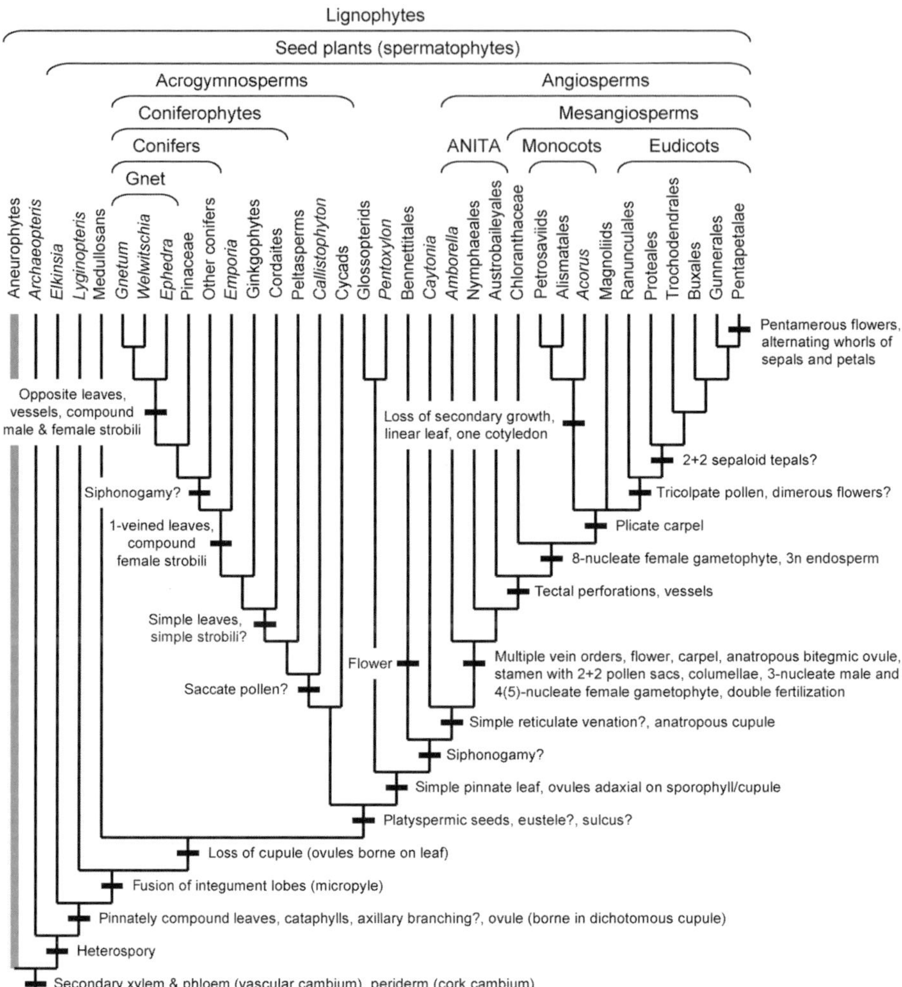

Figure 1.2 Phylogenetic relationships and evolutionary innovations in lignophytes ("progymnosperms" and seed plants). Phylogenetic relationships based largely on Doyle (2008) and Endress and Doyle (2009). Gnet, Gnetales. See Figure 1.1 for other conventions.

or Viridiplantae, in which other members have been traditionally called green algae. The most conspicuous innovations that unite this clade are the origin of chlorophyll *b* (in addition to chlorophyll *a*) and storage of starch in the chloroplasts. Green organisms are in turn linked by molecular data to red algae (rhodophytes) and the unicellular glaucophytes. Molecular phylogenies indicate that these three groups were derived from the line in which the chloroplast first originated by primary endosymbiosis with a cyanobacterium, from which chloroplasts of all other photosynthetic eukaryotes were derived by secondary endosymbiosis (incorporation of a red or green alga) or tertiary endosymbiosis (Delwiche & Palmer 1997; Keeling 2004). The chloroplasts of glaucophytes retain a remnant of the peptidoglycan cell wall seen in free-living cyanobacteria and other eubacteria, whereas both glaucophytes and red algae retain phycobiliproteins, the characteristic photosynthetic accessory pigments of cyanobacteria, which were lost in green plants.

Within green organisms, studies of cell ultrastructure in the 1960s and 1970s led to the view that certain groups of "green algae" are more closely related to land plants than others (Pickett-Heaps 1969, 1972, 1975, 1979; Stewart & Mattox 1975; Graham 1993), such as Charales (complex filaments with whorled branches), Coleochaetales (*Chaetosphaeridium*, with branched filaments, and *Coleochaete*, with disks made up of radiating branched filaments or a solid sheet of cells), and Zygnematales (including the unbranched filamentous alga *Spirogyra* and single-celled desmids). This conclusion has been confirmed and extended by molecular data (Karol *et al.* 2001; Lewis & McCourt 2004; Qiu 2008; Becker & Marin 2009). The clade of land plants and their aquatic algal relatives is known as the streptophytes, while the sister clade including most green algae (such as the familiar genera *Chlamydomonas*, *Acetabularia*, and *Ulva*, as well as basal unicellular lines known as "prasinophytes") is called the chlorophytes.

Aquatic streptophytic algae form a paraphyletic grade, that is, a series of successive branches diverging from the line leading to land plants. Molecular analyses are inconsistent on which of these groups is most closely related to land plants (their sister group), but most recent studies (Karol *et al.* 2001; Lewis & McCourt 2004; Qiu *et al.* 2006b; Qiu 2008; Becker & Marin 2009) identify the sister group as Charales and the next outgroup as Coleochaetales. This view has been challenged by analyses that linked Zygnematales with land plants (Turmel *et al.* 2006; Chang & Graham 2011; Wodniok *et al.* 2011), although Qiu (2008) argued that this result may be a function of rearrangements in the chloroplast genome of Zygnematales and sparse taxon sampling. Zygnematales are morphologically simpler than *Coleochaete* and Charales; if they are sister to land plants, their simplicity may be due to reversal (Wodniok *et al.* 2011). The unicellular flagellate *Mesostigma*, *Chlorokybus* (with small clusters of cells), and Klebsormidiales (unbranched filaments) appear to be lower on the tree, with *Mesostigma* alone or *Mesostigma* plus *Chlorokybus* (Lemieux *et al.* 2007) most basal. All these organisms are freshwater or soil algae, indicating that plants invaded the land not from the sea but from freshwater, a less radical step from a physiological point of view. In fact, adaptation to freshwater may have been a preadaptation for invasion of the land (Becker & Marin 2009).

Aquatic streptophytes show varying numbers of innovations retained in land plants, which were the morphological evidence (presumed synapomorphies) that originally led to their recognition as land plant relatives (Pickett-Heaps 1969, 1979; Stewart & Mattox 1975; Graham 1993). Even *Mesostigma*, which was formerly placed in the prasinophytes with unicellular chlorophytes, has two laterally inserted flagella attached to a distinctive multilayered structure (MLS) of microtubules (Rogers *et al.* 1981; Melkonian 1989), features retained in the motile sperm cells well into land plants, until flagella were lost within seed plants.

More important for the origin of land plants, the most derived aquatic streptophyte lines, Coleochaetales and Charales, show the appearance of the characteristic land plant mode of cell division. Cell division in these groups involves a phragmoplast, where the spindle fibers persist in their original orientation between the two daughter nuclei and cell wall material is deposited centrifugally to form the cell plate, leading to a transverse wall rigidly connected to the sidewalls. In these algae, perforations left in the cell plate by the spindle fibers become the sites of plasmodesmata, or connections between the cytoplasms of adjacent cells, a major feature of multicellularity in land plants (Lucas *et al.* 1993; Qiu 2008). This contrasts with the predominant mode of cell division in chlorophytes, with a phycoplast (Pickett-Heaps 1972), where the spindle fibers collapse to the plane between the two nuclei, and cell separation occurs by centripetal furrowing of the cell membrane. More basal streptophytes have a persistent spindle but the cells divide by furrowing, with late formation of a rudimentary phragmoplast in some Zygnematales, indicating that the phragmoplast evolved within streptophytes (Pickett-Heaps 1969, 1979; Stewart & Mattox 1975; Graham 1993).

The land plant mode of cell division, which is so familiar that it may be taken for granted, ultimately formed the basis for the characteristic solid, three-dimensional tissue construction of land plants (Hagemann 1999). It originated, however, in algae with a simpler, basically filamentous construction. Although Zygnematales grow by intercalary divisions of cells along the filament, *Coleochaete* and Charales resemble land plants in having incipient meristems. Cell divisions are localized at the margin of the disk or the ends of the radiating filaments in *Coleochaete* and in a distinct apical cell in Charales (Graham 1993). An additional innovation that Charales share with land plants is gravitropism, a feature that took on greater importance in the land environment (Qiu 2008).

A significant consideration for the origin of land plants is the fact that aquatic streptophytes have a haploid life cycle, like most though not all members of the chlorophyte clade. This means that all mitotic divisions occur in the haploid phase and the only diploid cell is the zygote, which undergoes meiosis directly (usually after forming a resting zygospore). Within the aquatic streptophyte grade, the most important reproductive innovation was the origin of oogamy, seen in Coleochaetales and Charales, with a sessile egg and motile sperm rather than undifferentiated, usually motile isogametes. The exact point of origin of oogamy is uncertain, since sexual reproduction has not been described in the most basal streptophytes (Graham 1993). In Zygnematales filaments or cells conjugate and a protoplast migrates from one cell into another as a nonflagellated gamete. This is presumably autapomorphic, but without better evidence on relationships and data on life cycles in basal streptophytes it is uncertain whether it was derived from an oogamous or an isogamous condition. From a phylogenetic point of view, the oogamous haploid life cycle of "higher" aquatic streptophytes forms the starting point for the alternation of generations of land plants.

Coleochaete and Charales resemble land plants in additional reproductive innovations: retention of the zygote in the haploid thallus of *Coleochaete*, its investment by haploid cells in Charales, and presence of cells around the egg in *Coleochaete* that resemble placental transfer cells in the archegonium of land plants (Graham & Wilcox 1983). Which, if any, of these features are synapomorphies shared with land plants depends in part on which line is more closely related to land plants (Haig 2008).

1.3 Invasion of the land: "bryophytes"

The evolution of land plants (embryophytes) from their freshwater ancestors involved many major innovations, most of which are obvious adaptations to the radically different requirements for structural support, uptake of water, prevention of desiccation, and gas exchange in a terrestrial environment. The sequence of origin of these and subsequent land adaptations has been clarified by improved resolution of phylogenetic relationships, particularly where the land plant tree is rooted.

Since discovery of the basic life cycle of land plants, with alternating haploid (gametophyte) and diploid (sporophyte) generations (Hofmeister 1862; Strasburger 1894), it has been recognized that "bryophytes" and "lower" vascular plants, which have free-swimming sperm cells, represent a primitive "amphibious" stage in land plant evolution (Bower 1890, 1908). However, the relationship between bryophytes, with a dominant gametophyte, and vascular plants, with a dominant sporophyte, remained unsettled until the advent of morphological cladistic analyses, which made much use of data on cellular characters in aquatic streptophytes (Mishler & Churchill 1984, 1985; Kenrick & Crane 1997).

Although it had been widely thought that bryophytes and vascular plants were divergent sister groups, or even separate lines of evolution from aquatic ancestors, these analyses indicated that bryophytes are a paraphyletic series of three lines diverging below vascular plants, a result confirmed by molecular studies (Qiu *et al.* 1998, 2006b; Karol *et al.* 2001; Qiu 2008). This indicates that the bryophytic life cycle is ancestral, as is also implied by the fact that it is less removed from the life cycle in aquatic streptophytes, where the diploid phase consists of only one cell, the zygote (Mishler & Churchill 1984; Haig 2008). The morphological analysis of Mishler and Churchill (1984) indicated that liverworts (hepatics) were basal (sister to all other land plants), followed by hornworts (anthocerotes), with mosses sister to vascular plants. Although earlier molecular studies gave inconsistent rootings, multigene analyses appear to be stabilizing on the basal position for liverworts, but with hornworts rather than mosses as the sister group of vascular plants (Qiu *et al.* 2006b; Qiu 2008; Chang & Graham 2011).

A key innovation in land plants was origin of the alternation of generations, or specifically, given the haploid life cycle of the outgroups, origin of a multicellular diploid sporophyte, a topic of debate for more than a century (Haig 2008). The present view corresponds to the classic antithetic or interpolation theory (Bower 1890, 1908, 1935): that the sporophyte was a new phase interpolated into a haploid life cycle by a delay of meiosis, so that the zygote underwent mitotic divisions that produced a mass of diploid cells, some of which underwent meiosis. Compared to an alga with zygotic meiosis, this would result in production of vastly more numerous and genetically more varied haploid spores per fertilization event (Bower 1908; Becker & Marin 2009). This contrasts with the opposing homologous or transformation theory (Scott 1895; Zimmermann 1952; Remy 1982): that the ancestral life cycle was already an alternation of generations, but with isomorphic haploid and diploid phases that were subsequently modified in opposite ways.

In vegetative morphology, a major advance of land plants was evolution of more bulky three-dimensional parenchymatous construction, which would have retarded desiccation by decreasing the surface to volume ratio. The closest approach to parenchyma in aquatic streptophytes is in some species of *Coleochaete*, where areas around the zygote become several cells thick. Other antidesiccation devices were origin of a waxy cuticle and differentiation of epidermal cells at the surface of the plant body. Another innovation was origin of hair-like rhizoid cells on parts of the gametophyte in contact with the soil, which function for anchoring and water uptake (Kenrick 2002). Fragmentary Silurian fossils may represent bryophytic cuticles with attached rhizoids (Graham *et al.* 2004). The symbiotic association of land plants with mycorrhizal fungi, which has been considered an important key to their success (Pirozynski & Malloch 1975), presumably also originated near this point, since it occurs in liverworts and hornworts. Genes required for mycorrhiza formation also occur in mosses, although mycorrhizae themselves are known only in the basal genus *Takakia* (Wang *et al.* 2010).

Reproductive innovations of land plants include multicellular gametangia and sporangia with a jacket of sterile cells: antheridia producing numerous motile sperm cells, archegonia with neck canal cells and a basal egg, and sporangia with numerous sporogenous cells that undergo meiosis to produce the spores. The sterile jacket is functionally comparable to the epidermis, as a layer protecting the inner cells from desiccation. Although the initial stages differ, development of all these structures involves periclinal division of one or more surface cells, followed by derivation of the jacket from the outer cell(s) and the fertile cells from the inner cell(s) (Smith 1955; Crum 2001). This is an innovation relative to *Coleochaete* and Charales, where the egg is surrounded by sterile cells, but these are not derived by periclinal division of the same initial. There are closer approaches to land plant antheridia in the male structures of algal outgroups. In some species of *Coleochaete*, asymmetrical divisions produce both sterile and spermatogenous cells, while in some Charales periclinal divisions produce outer shield cells and inner cells that give rise to spermatogenous filaments (Pickett-Heaps 1975; Graham 1993).

One of the most important adaptations to the land environment was evolution of spores with a highly resistant outer wall or exine, which allowed dispersal of the meiotic products by air rather than water (Becker & Marin 2009). Being produced by meiosis, these spores are formed in tetrads, but in most living groups they separate before being shed from the sporangium. In the fossil spore record, however, there is an initial phase in which spores remained united in tetrads, extending back to the Middle Ordovician and probably the Cambrian (Taylor & Strother 2008). Discovery of masses of such tetrads in a fragmentary sporangium (Wellman *et al.* 2003) demonstrated that they were produced by a multicellular sporophyte. Tetrads are joined in the Late Ordovician (Steemans *et al.* 2009) by single spores with a prominent trilete (triradiate) tetrad scar, which represents the junction of the three contact faces with other spores in the tetrad. Steemans *et al.* (2009) suggested that these spores were produced by vascular plants, but this is unwarranted; although a trilete scar is absent or poorly developed in most liverworts and mosses (Gray 1985), it is well developed in hornworts (Shaw & Renzaglia 2004) and presumably originated in their common ancestor with vascular plants.

The bryophytic lines show the stepwise origin of additional new terrestrial adaptations. The most striking was the origin of stomata for gas exchange, seen in the sporophytes of mosses and hornworts. The sporophyte is largest, longest-lived, photosynthetic, and almost independent in hornworts, a possible synapomorphy supporting their sister group relationship to vascular plants (Campbell 1924; Qiu *et al.* 2006b). Tubular conducting cells occur in one or both generations of many mosses (Hébant 1977), but the water-conducting cells are hydroids that lack the internal secondary wall thickenings of the tracheids of living vascular plants. The fact that hydroids are absent in basal groups such as *Takakia*, *Sphagnum*, and Andraeales implies that they evolved within mosses. In the reproductive sphere, there was a noteworthy change in the archegonia: from long, narrow, and stalked in liverworts and mosses, to basally sunken with a short neck in hornworts and vascular plants.

1.4 Origin of vascular plants: the importance of fossils

The two most important innovations of vascular plants (tracheophytes) are the origin of (1) a branched sporophyte that produces numerous sporangia, and (2) vascular tissue consisting of xylem, made up of dead cells that conduct water and minerals from the soil to aerial parts, and phloem, with living cells that conduct sugars and other products of photosynthesis to the rest of the plant. Present data indicate that living vascular plants consist of three main clades: two clades of spore-bearing plants, lycophytes and monilophytes, which include ferns, *Equisetum* (the only living representative of sphenophytes), and Psilotales; and seed plants, including "gymnosperms" and angiosperms. All of these groups have a dominant sporophyte consisting of leaves, stems, and roots (except in Psilotales, apparently as a result of secondary loss). However, fossil data indicate that these were preceded by more primitive plants consisting entirely of dichotomously branched stems, with terminal sporangia. Cladistic analyses by Kenrick and Crane (1997) confirmed that these "rhyniophytes" were stem relatives of living (crown group) vascular plants. Such plants are known in best anatomical detail in the remarkably preserved Early Devonian Rhynie Chert (Kidston & Lang 1917, 1920) but extend back into the Middle Silurian.

Rhyniophytes provide unique evidence that the origin of vascular tissue, or xylem and phloem, was a stepwise process. Two Rhynie plants, *Horneophyton* and *Aglaophyton*, which represent the first two branches in the phylogeny of Kenrick and Crane (1997), had xylem consisting not of tracheids, with secondary wall thickenings laid down on the inside of the primary cell wall, but rather of hydroids, with no secondary thickenings, as in mosses (Hébant 1977; Edwards 1986). Mishler and Churchill (1984, 1985) considered vascular tissue a synapomorphy of mosses and vascular plants, which were sister groups in their analysis. However, with hornworts sister to vascular plants and the absence of hydroids in basal mosses this now appears to be a convergence. *Rhynia*, in the third branch of Kenrick and Crane (1997), had true tracheids but secondary thickenings of a different structure than those of living vascular plants (Edwards 1980; Kenrick & Crane 1991, 1997).

A major breakthrough in understanding early vascular plants was the discovery that some axes in the Rhynie Chert bore antheridia and archegonia. Moreover, it was recognized that similar plants had already been described from other Devonian floras under the name *Sciadophyton*, which had a rosette of radiating axes and flared tips bearing gametangia (Remy & Remy 1980; Remy 1982; Remy *et al.* 1993; Kenrick 1994). The Rhynie Chert axes had vascular tissue similar to that of the co-occurring sporophytes.

Remy (1982) and other fairly recent authors (e.g., Stewart & Rothwell 1993) have cited the similarity of the sporophytes and gametophytes in rhyniophytes as evidence for the homologous or transformation theory—that the common ancestor of land plants had an isomorphic alternation of generations, which was modified by reduction of the sporophyte in bryophytes but elaboration of the sporophyte and reduction of the gametophyte in vascular plants. This has been refuted by morphological and molecular phylogenetic evidence that bryophytes are a paraphyletic group consisting of three successive branches, all with a simple, unbranched sporophyte. Maintaining the homologous theory would require that the sporophyte was independently reduced in each of these lines, a much less parsimonious scenario. The existence of two comparable phases in rhyniophytes is entirely consistent with the interpolation theory, since at some point in its elaboration the sporophyte must have "passed" the gametophyte in complexity. Furthermore, it would not be surprising if genes involved in development of the gametophyte were co-opted by the sporophyte as it became independent (Haig 2008).

The rhyniophytic stem relatives of vascular plants also furnish insights on evolution of the vegetative architecture of the vascular plant sporophyte that could not be gained directly from living plants. Rhyniophytes are famous for their dichotomously branched creeping rhizomes and erect aerial stems, in which each successive dichotomy occurred at right angles to the last, giving a bushy, three-dimensional structure, with terminal sporangia at the tips of some axes. The rhizomes bore rhizoids that are presumably homologous with the rhizoids of bryophytes and the root hairs of more derived vascular plants. This architecture served as the basis for the telome theory of Zimmermann (1930, 1952), where the term "telome" refers to the free tips above the last dichotomy. According to the telome theory, leaves were derived by a series of elementary processes, each representing a simple change in developmental processes. These were (1) overtopping, a shift from equal to unequal branching of the apical meristem; (2) planation, restriction of branching from three dimensions to one plane; and (3) webbing, transformation of an open dichotomous system, where the apical meristem splits into two distinct meristems, into a sheet, by formation of a continuous marginal meristem. The telome theory has been much criticized for being overly simplistic and even untestable, but in explicitly relating evolutionary changes to changes in development it anticipated current evo-devo thinking (Stein & Boyer 2006).

1.5 Early innovations within vascular plants: leaves, roots, and heterospory

Among living vascular plants, most morphological and molecular phylogenetic analyses indicate that lycophytes (club mosses and relatives) are the sister group of the remaining vascular plants, or euphyllophytes, which in turn consist of monilophytes and seed plants (Raubeson & Jansen 1992; Kenrick & Crane 1997; Pryer *et al.* 2001; Qiu *et al.* 2007). Although the sporophyte and gametophyte were similar in complexity in the rhyniophytic stem relatives of vascular plants, the sporophyte is dominant in all members of the crown group and presumably had become so in their most recent common ancestor. Based on a tree of living taxa only, Schneider *et al.* (2002) inferred that leaves and roots originated once in the common ancestor of living vascular plants, but as argued by Friedman *et al.* (2004) inclusion of fossil outgroups of the living clades indicates that both organs arose more than once.

Typical roots, with positive geotropism and a root cap, appear to have originated at least twice, in lycophytes and euphyllophytes (Kenrick 2002; Friedman *et al.* 2004). However, stem relatives of both clades, such as *Zosterophyllum*, *Bathurstia*, and *Asteroxylon* on the line leading to lycophytes

(Gensel *et al.* 2001) and *Psilophyton* on the line to euphyllophytes (Banks *et al.* 1975; Doran 1980), had downward-growing dichotomous "rooting structures" transitional between rhizomes and roots, suggesting that such structures may have arisen below the crown node of vascular plants. Their independent modification into typical roots may be reflected in differences between roots of living lycophytes and other vascular plants, notably apical dichotomous branching rather than endogenous lateral branching from the pericycle.

An early anatomical innovation in vascular plants was origin of periderm (cork, consisting of rows of suberized cells). The oldest known example of this tissue is in *Psilophyton* (Banks 1981; Banks & Colthart 1993), where it acted as a mechanism for repairing wounds in the epidermis, doubtless its original function. Since periderm of this sort also occurs in living lycophytes (Lu 1996), it presumably evolved before the crown node of vascular plants, but it is not known in rhyniophytes (Banks 1981).

In lycophytes, the most conspicuous innovations are simple one-veined leaves, known as microphylls, and the position of the sporangia in the axils or on the adaxial surface of fertile leaves, or sporophylls. Another is lobing of the xylem in the stele (actinostele). The origin of the leaves and the sporangial position pose special problems, which are closely linked.

The sporangia of lycophytes are more derived than those of rhyniophytes in being globose or reniform and having transverse dehiscence (with two valves opening horizontally relative to the stem), rather than elongate with no visible structural modification for dehiscence. In both characters they resemble sporangia of several Late Silurian and Early Devonian fossil taxa: *Cooksonia*, with dichotomously branched, leafless stems bearing terminal sporangia; *Renalia*, with small dichotomous lateral branches with terminal sporangia; *Zosterophyllum* and other "zosterophylls," with sporangia borne laterally on leafless stems; and the Rhynie Chert plant *Asteroxylon*, which had leaves but lateral sporangia borne directly on the stem. According to the cladistic analysis of Kenrick and Crane (1997), all these fossils are stem relatives of lycophytes. Their arrangement is consistent with a scenario in which the sporangia became lateral by overtopping and then reduction of fertile branches.

Since Bower (1935), it has been widely assumed that the one-veined leaves of lycophytes were derived from nonvascularized outgrowths or enations rather than overtopped branches (e.g., Stewart & Rothwell 1993). This view seemed to be supported by the occurrence of enations in Early Devonian "spiny zosterophylls" such as *Sawdonia* and the fact that the leaves of *Asteroxylon* were intermediate in having a vascular strand that went only to the base. However, the analysis of Kenrick and Crane (1997) separated *Asteroxylon* and typical lycophytes from the spiny zosterophylls and nested them among zosterophylls that lack enations, implying that enations and microphylls were not homologous. As an alternative hypothesis, Kenrick and Crane (1997) proposed that microphylls originated by sterilization of lateral sporangia in a nonspiny zosterophyll, which would be consistent with the fact that sporangia and microphylls are intermixed in *Asteroxylon*. The association of sporangia with sporophylls would presumably be a later event.

Within lycophytes, an important innovation was the origin of heterospory, a synapomorphy of the living genera *Selaginella* and *Isoetes* and numerous Late Devonian and Carboniferous fossils, where instead of producing spores of one size that develop into bisexual gametophytes (homospory), sporophytes produce microspores and much larger megaspores that develop into male and female gametophytes, respectively. Because the gametophytes are retained inside the spore wall (endospory), this represents another step in reduction of the gametophyte generation. The shift from homospory to heterospory was repeated in several other fossil and living groups of vascular plants, most notably seed plants, making it one of the most conspicuous iterative trends in plant evolution (Bateman & DiMichele 1994). Proposed adaptive causes for this trend range from primarily genetic, such as assuring outcrossing and genetic diversity or allowing sporophytic control over sex expression, to nutritional and ecological, as a means of parental investment in the next sporophyte generation, by provisioning the megaspore and thus the female gametophyte with nutrients (Chaloner & Sheerin 1981; Bateman & DiMichele 1994).

Another major innovation in Late Paleozoic lycophytes (Lepidodendrales, or Isoetales: Bateman *et al.* 1992; Kenrick & Crane 1997) was the origin of secondary growth and the tree habit. This secondary growth differed from that of seed plants in involving a unifacial cambium that produced secondary xylem but no secondary phloem. There was also a cork cambium that produced periderm (so-called secondary cortex) near the outside of the trunk, which differed from the periderm of seed plants in functioning as the main support tissue. These innovations allowed Lepidodendrales to dominate Carboniferous coal swamp vegetation, but they have little significance for modern plants. They persist only in vestigial form in the radically reduced aquatic genus *Isoetes*, which has a single cambium that produces a mixture of xylem and phloem to the inside and parenchymatous cortical tissue to the outside. Some Lepidodendrales also showed a convergent origin of functionally seed-like structures (*Lepidocarpon*), derived by envelopment of the megasporangium by the sporophyll.

The most important innovation of the euphyllophyte clade, seen in stem relatives such as the Early Devonian genus *Psilophyton*, was differentiation of one or more orders of main axes with dichotomous lateral branches, as a result of unequal dichotomy of the apical meristem (Banks *et al.* 1975)—a prime example of the telome process of overtopping. Another innovation was longitudinal dehiscence of the sporangia, that is, opening along a slit down one side, a feature retained in most later euphyllophytes, as illustrated by the pollen sacs of angiosperms. A synapomorphy seen in living euphyllophytes is a shift from biflagellate to multiflagellate sperm. *Psilophyton* had a round xylem cylinder, but this had become lobed (giving an actinostele) in *Pertica* (Gensel 1984) and basal crown euphyllophytes.

The dichotomous lateral branches of *Psilophyton* have often been interpreted as illustrating a stage in the origin of leaves, the next steps being planation and establishment of dorsiventral (adaxial-abaxial) polarity. However, later euphyllophytes suggest that leaves were derived in more than one way from structures in a *Psilophyton*-like ancestor, such that leaves in different lines have different homologies (Beck 1970; Doyle 1998; Galtier 2010). Either individual ultimate dichotomous branches could become simple, dichotomously organized leaves, or whole branch systems could be modified into pinnately compound leaves, with leaflets derived from the dichotomous ultimate branches. Leaves of the first sort might be homologous with leaflets of the second sort (setting aside later fusion of leaflets and other complications). Both types of leaves have been traditionally described as megaphylls, an unfortunate term that glosses over their presumably different origins (Tomescu 2009). Evolution of a continuous lamina with dichotomous venation, as a result of origin of a marginal meristem, occurred many times in leaves of both types (Boyce & Knoll 2002).

The finding that the living members of the monilophyte clade included not only ferns but also sphenophytes (with whorled simple leaves) and Psilotales (with one-veined or nonvascularized simple leaves and no roots) has been regarded as an unexpected result of molecular studies (Pryer *et al.* 2001). However, it had been anticipated by the morphological cladistic analysis of Kenrick and Crane (1997) and earlier suggestions that both ferns and sphenophytes were derived from Middle and Late Devonian "cladoxylopsids," which had branch systems with dichotomous ultimate appendages and a stele with lobed or subdivided xylem (Scheckler 1974; Stein *et al.* 1984). Ironically, the clearest morphological synapomorphy of monilophytes is an obscure feature not seen in living ferns but characteristic of cladoxylopsids and some Paleozoic plants thought to be early ferns, including the Late Devonian genus *Rhacophyton*, Carboniferous "zygopterids," and *Ankyropteris*, a probable stem relative of the main living fern order Filicales—an area of parenchymatous protoxylem near the tip of each lobe of the xylem, surrounded by a "peripheral loop" of metaxylem. If these fossils are related to living ferns, this character was lost or modified beyond recognition in the latter. However, it may be represented by the protoxylem canals of *Equisetum* and related Paleozoic sphenophytes (Equisetales), modified by stretching of the internodes due to intercalary meristematic growth.

A relationship of cladoxylopsids and zygopterids to living ferns was rejected by Rothwell and Nixon (2006), based on a morphological and molecular analysis, in which molecular data prevailed in uniting living members of the monilophyte clade, but cladoxylopsids and zygopterids formed a basal grade below all living vascular plants. However, this may be a result of incorrect rooting of vascular plants: lycophytes were linked with seed plants, whereas most other analyses place lycophytes at the base of vascular plants and unite monilophytes and seed plants. This may be a result of insufficient taxon sampling, especially the lack of fossils such as *Cooksonia* and zosterophylls as outgroups of lycophytes. If all groups in the tree of Rothwell and Nixon (2006) except *Aglaophyton* are rerooted between lycophytes and the remaining groups, cladoxylopsids and zygopterids are linked with crown monilophytes, roughly consistent with the view of Kenrick and Crane (1997).

The most important evolutionary innovation in monilophytes was evolution of the compound leaf (frond) of ferns. Another innovation was the whorled phyllotaxis of sphenophytes, which also occurred in some members of the cladoxylopsid grade (Berry & Stein 2000; Soria & Meyer-Berthaud 2003; Cordi & Stein 2005). In sphenophytes, available data support the view that leaves were derived from single dichotomous branches like those of *Psilophyton*, or the more leaf-like ultimate appendages of cladoxylopsids. Although *Equisetum* and its closest fossil relatives, including the Carboniferous tree genus *Calamites*, have one-veined leaves (typologically microphylls), other sphenophytes had leaves that were more like dichotomous branches. Examples include the wedge-shaped, dichotomously veined leaves of *Sphenophyllum* and the unwebbed dichotomous leaves of *Archaeocalamites*, a basal member of the Equisetales. However, Ophioglossales, Marattiales, and Filicales have basically compound leaves more plausibly derived from whole branch systems bearing dichotomous appendages.

A bizarre but potentially significant feature of some cladoxylopsids (e.g., *Arachnoxylon*), *Rhacophyton*, Carboniferous zygopterids, and the Carboniferous genus *Stauropteris* is quadriseriate branching, with secondary axes or pinnae produced in alternating pairs perpendicular to the rachis (Cornet *et al.* 1976; Stein 1981; Phillips & Galtier 2005; Galtier 2010). In *Rhacophyton*, Cornet *et al.* (1976) showed various degrees of reduction of one pinna per pair, and they suggested that reduction of this sort led to the biseriate (pinnate) pinna arrangement of living ferns. A connection of such fossils with Filicales in particular is supported by *Ankyropteris*, which had normal biseriate pinnae but resembled zygopterids in having an H-shaped vascular strand with peripheral loops in the petiole. These considerations suggest that planation proceeded from higher to lower orders (Galtier 2010). First, the ultimate appendages became biseriate, resulting in pinnae with two rows of pinnules, but pinnae were still borne in a quadriseriate arrangement. Second, the whole frond became planated by a shift from quadriseriate to biseriate pinnae. Such a scenario might never be suspected without fossil evidence. This hypothesis requires further phylogenetic testing, since the only analysis to include relevant taxa is that of Rothwell and Nixon (2006), which linked *Ankyropteris* and Filicales but separated them from groups with quadriseriate fronds. There is no evidence on the mode of origin of the leaves of Marattiales and Ophioglossales, since these groups had planated fronds as far back as they are known (Late Carboniferous for Marattiales, earliest Tertiary for Ophioglossales: Rothwell & Stockey 1989), and they have not been associated with more primitive fossil relatives.

Psilotales, traditionally associated with rhyniophytes because of their dichotomous shoot organization and lack of roots, are strongly linked with

Ophioglossales by molecular data (Manhart 1995; Pryer *et al.* 2001; Rothwell & Nixon 2006). Possible morphological synapomorphies are axial mycotrophic gametophytes and fertile appendages with an adaxial sporangium-bearing portion (Doyle 1998). The fact that young appendages of *Psilotum* resemble primordia of fern fronds (Kaplan 1977, 2001) is consistent with the view that they were reduced from fronds. Although the lack of roots in Psilotales has been considered primitive (e.g., Bremer 1985), current phylogenies indicate that it is a result of loss.

In contrast to living monilophytes, some Paleozoic forms had secondary growth, or possible precursor conditions. Tree sphenophytes (*Calamites*) had extensive secondary wood, and a bifacial cambium is well documented in *Sphenophyllum* (Eggert & Gaunt 1973; Cichan 1985). Secondary xylem has also been reported in some cladoxylopsids, *Rhacophyton*, and zygopterid ferns, but it is unusual in being limited in amount. In cladoxylopsids this tissue has been interpreted as aligned metaxylem, as in extant Ophioglossales (Rothwell & Karrfalt 2008), because it lacks rays (Scheckler 1974). However, rays are known in *Rhacophyton* and *Zygopteris* (Dittrich *et al.* 1983; Phillips & Galtier 2005). Banks *et al.* (1975) reported aligned metaxylem in larger stems of *Psilophyton*. Expanding on a suggestion of Kenrick and Crane (1997), this might mean that a first step toward secondary growth, namely, periclinal divisions of tracheid initials, occurred on the stem lineage of euphyllophytes. This is of minor consequence for living monilophytes, since even if their ancestors had secondary growth it must have been subsequently lost.

In the main fern clade, Filicales, the most conspicuous innovation is the leptosporangium (hence their name "leptosporangiate ferns"). This differs from the eusporangium of other groups in its smaller size, fewer spores, narrow stalk, thin wall, and a row of thick-walled cells, the annulus, which corresponds to a patch of cells in *Ankyropteris* and other Carboniferous stem relatives. Current phylogenies (Pryer *et al.* 1995) confirm the traditional view that the sporangium underwent a shift from longitudinal to transverse dehiscence within Filicales. This trend culminated in the famous snapping dehiscence of Polypodiaceae *sensu lato*, which contrary to older views form a clade. Another innovation within Filicales was reduction of the number of jacket cells in the antheridium, which may have occurred once or twice (Pryer *et al.* 1995). Heterospory evolved in water ferns (Marsileaceae and Salviniaceae), which were previously interpreted as two unrelated lines but now appear to form a clade.

1.6 Innovations on the line to seed plants: "progymnosperms" and "seed ferns"

Two major paleobotanical advances in the last century shed new light on the origin of seed plants (spermatophytes) and their evolutionary innovations, which include not only the seed but also secondary growth. First was the recognition of "seed ferns" or "pteridosperms" (Oliver & Scott 1903), based on the association of fern-like pinnately compound leaves with woody stems and seeds in the Late Carboniferous genus *Lyginopteris*. This discovery showed that early seed plants had a leaf type not retained in any living members, while later studies of more primitive seeds from the Late Devonian and Early Carboniferous (Long 1966, 1975; Pettitt & Beck 1968; Serbet & Rothwell 1992) revealed apparent steps in the origin of the seed.

Second was recognition of the "progymnosperms," based on association of woody trunks and leafy branch systems with sporangium-bearing appendages in the Late Devonian tree *Archaeopteris* and Middle and Late Devonian "aneurophytes" (Beck 1960, 1970, 1971; Scheckler & Banks 1971a). If these fossils are stem relatives of seed plants, as confirmed by morphological cladistic analyses (Doyle & Donoghue 1986; Rothwell & Serbet 1994; Hilton & Bateman 2006), they show that the first major seed plant innovation was not the seed. Rather it was secondary growth, with secondary xylem and phloem produced indefinitely by a bifacial vascular cambium and periderm produced by a cork cambium. The resulting origin of large trees with a woody trunk marked another step in the trend for elaboration of the sporophyte generation. The clade including both progymnosperms and seed plants has been called lignophytes, after the secondary wood. Origin of a cork cambium was an extension of the original role of periderm as a wound-repair device and was clearly an adaptation to distension and splitting of the epidermis and cortex due to growth of the vascular cylinder.

Progymnosperms also clarify the origin of the original fern-like leaf of seed plants. The more primitive aneurophytes had three-dimensional branch systems, with spiral or opposite-decussate secondary branches bearing dichotomous appendages. In *Archaeopteris*, however, the secondary axes were in one plane, resulting in a branch system that was originally misinterpreted as a fern-like frond. In all these groups the ultimate appendages, usually called leaves, were spiral or decussate. Such branch systems could be transformed into fronds of the seed fern type by planation, with the primary axis becoming the rachis, secondary axes becoming pinna rachises, and ultimate appendages becoming leaflets (Beck 1970; Doyle 1998; Galtier 2010).

In contrast to ferns, there is no sign that planation occurred first in higher order branches—in fact, the combination of spiral leaves and biseriate secondary axes in *Archaeopteris* might suggest the opposite sequence. However, the related genus *Svalbardia* had three-dimensional branch systems, suggesting that planation occurred independently in *Archaeopteris* and seed plants (the alternative, that seed plants are more closely related to *Archaeopteris* than to *Svalbardia*, is less likely because the two genera share anatomical advances not seen in basal seed plants, such as grouped pitting). This reasoning implies that seed plants are from an ancestor with branch systems like those of aneurophytes rather than *Archaeopteris*. Late Devonian and Carboniferous seed ferns differed from both ferns and *Archaeopteris* in showing more or less extensive dichotomy of the frond rachis (Walton 1931; Serbet & Rothwell

1992; Galtier 2010), but dichotomy of main axes is known in at least one aneurophyte, *Proteokalon* (Scheckler & Banks 1971b).

This scenario implies that origin of the seed plant leaf proceeded first by overtopping of simple dichotomous appendages, at the level of basal euphyllophytes, and then by planation of a whole branch system bearing such leaves into a compound leaf, between progymnosperms and seed ferns. It is suggestive that two gene families that specify adaxial-abaxial polarity in angiosperms, KANADI and Class III HD-Zip, occur throughout vascular plants, whereas the YABBY family is known across seed plants but has not been identified in ferns or more basal groups (Floyd & Bowman 2007). This raises the possibility that KANADI and Class III HD-Zip genes were involved in the initial overtopping seen in basal euphyllophytes, but YABBY genes were involved in the subsequent planation of branch systems into fronds in seed plants, which was independent of that in ferns (Sarojam *et al.* 2010).

Another major innovation of seed plants was axillary branching, a fundamental aspect of the architecture of living seed plants. Its exact point of origin is unclear, partly because early seed plants rarely branched, but it is known in *Lyginopteris* (Brenchley 1913) and the Early Carboniferous genus *Calamopitys* (Galtier & Holmes 1982; Galtier 1988). A less commonly noted synapomorphy of seed plants is the presence of pointed cataphylls (scale leaves) at the bases of shoots and around buds, in addition to fronds. These may represent leaf primordia whose development was arrested before production of pinnae.

Inside the stem, the most conspicuous innovation of living seed plants is the eustele, with a ring of primary vascular bundles around a pith. A series of intermediates in basal seed ferns show steps in origin of a eustele from a lobed protostele (actinostele), as in aneurophytes, by differentiation of the central tissue into parenchyma rather than xylem, leaving the lobes as separate vascular strands (Beck 1970). *Archaeopteris* also had a eustele, but the fact that basal seed ferns had actinosteles implies that this was a convergence, like planation of the branch systems. A eustele was firmly established in crown group seed plants, but its number of origins is unclear. *Lyginopteris* had a eustele, but medullosans, which all analyses place nearer the crown group, had either an actinostele (*Quaestora*) or a special type of eustele with tangentially elongate primary vascular strands and internal as well as external secondary wood (Basinger *et al.* 1974).

The first major reproductive innovation in lignophytes was heterospory, an essential step toward the seed. This step is seen in *Archaeopteris*, where it is the main evidence that this group was more closely related to seed plants than were aneurophytes (Doyle & Donoghue 1986; Hilton & Bateman 2006). In *Archaeopteris*, there was no external difference between either the microsporangia and megasporangia or the fertile appendages bearing them, which were dichotomous structures arranged like leaves on secondary axes of a branch system.

The key process in origin of the seed itself, or more precisely its immature stage, the ovule, was formation of an integument (future seed coat) around the megasporangium (nucellus), with a micropyle at the apex for pollen capture. Evidence on this process comes from Late Devonian (*Elkinsia*) and Early Carboniferous seed ferns (Long 1966, 1975; Rothwell & Scheckler 1988; Serbet & Rothwell 1992), which appear to attach to the seed plant stem lineage (Doyle & Donoghue 1986; Rothwell & Serbet 1994; Hilton & Bateman 2006). In these fossils the megasporangium was surrounded by a ring of integument lobes that show every degree of fusion from the base up, culminating in a complete integument with a typical micropyle. Reduction to one functional megaspore had occurred by this point. The lobes have been interpreted as outer telomes of a dichotomous branch (Andrews 1963) or sterilized outer sporangia of a synangium (Kenrick & Crane 1997), but these alternatives are not as different as they may seem, since synangia would presumably represent modified fertile telomic branches. The ovules were borne in so-called cupules, which were dichotomous structures resembling two facing hands with ovules on the palms. These cupules have been homologized with the similarly constructed fertile appendages of progymnosperms; in at least one aneurophyte, *Tetraxylopteris* (Bonamo & Banks 1967), the sporangia were in clusters that could be transformed into ovules (Kenrick & Crane 1997). In the Late Carboniferous genus *Lyginopteris*, the cupule had been simplified to several lobes around a single ovule.

Subsequent evolution of the ovule presents a striking story of transfer of function. In living gymnospermous seed plants, the pollen is caught by a pollination drop exuded from micropyle, but this was impossible before fusion of the integument lobes. Instead the pollination drop was exuded by an extension of the nucellus, called the lagenostome or salpinx, in which the epidermis separated from the inner tissue to form a central column surrounded by a pollen chamber (hydrasperman reproduction: Rothwell & Scheckler 1988). At first the pollen did not differ externally from the trilete spores of "lower" plants, with a trilete scar that presumably opened to release motile sperm that swam in liquid in the pollen chamber to archegonia at the apex of the female gametophyte (Chaloner 1970). The female gametophyte, inside the megaspore membrane, was relatively large, but the archegonia had been further simplified by reduction of the number of neck cells. The distinctive male gametophyte of living seed plants, with two sperm cells and a row of sterile cells, presumably evolved from a gametophyte with a larger number of cells. Its point of origin is unclear, but two sperm cells have been described in the pollen of medullosan seed ferns (Stewart 1951). In medullosans and crown group seed plants, the cupule was apparently lost and the ovule came to be borne directly on a leaf.

Another innovation of crown group seed plants that has been less discussed is the typical embryo, in which the root apex is located at the opposite pole from the shoot apex, rather than in a lateral position, as in other vascular plants. Understanding the timing and mode of its origin is prevented by the absence of mature embryos in Carboniferous seeds (Rothwell 1988). Embryos of a modern type are known in conifers near the Carboniferous–Permian boundary (Mapes *et al.* 1989), but these were already in the crown group.

1.7 Innovations within seed plants, especially conifers

Tracing evolutionary innovations within seed plants is hampered by uncertain relationships among clades, especially angiosperms and Gnetales, the two most derived groups. Prior to cladistic analyses (e.g., Chamberlain 1935), seed plants were widely assumed to be diphyletic, with the seed originating independently in "cycadophytes" (cycads, seed ferns, and fossils such as Bennettitales) and "coniferophytes" (conifers, ginkgophytes, and Paleozoic cordaites). This was thought to be reflected in the symmetry of the seeds: radial in cycadophytes (radiospermic), and bilateral or biradial in coniferophytes (platyspermic). After the recognition of progymnosperms, Beck (1966) argued that cycadophytes were related to aneurophytes, while coniferophytes were related to *Archaeopteris*. In this view, the fronds of seed ferns were derived from whole branch systems but the simple leaves of coniferophytes were derived from dichotomous ultimate appendages of the *Archaeopteris* type. However, these schemes were contradicted by morphological cladistic analyses, which nested coniferophytes within seed plants (Crane 1985; Doyle & Donoghue 1986; Nixon *et al.* 1994; Rothwell & Serbet 1994; Doyle 1996, 2006; Hilton & Bateman 2006; Rothwell *et al.* 2009). Furthermore, although molecular data do not directly address relationships between living taxa and progymnosperms, they refute the classic diphyletic hypothesis by failing to split seed plants into cycadophytes and coniferophytes (except Mathews *et al.* 2010).

Most morphological cladistic analyses have linked angiosperms and Gnetales, along with Mesozoic Bennettitales and *Pentoxylon*, in a clade called anthophytes because its members have more or less flower-like reproductive structures—an updated version of the anthostrobilus hypothesis of Arber & Parkin (1907). Some analyses placed anthophytes among "seed ferns" such as Permian glossopterids and Mesozoic corystosperms and *Caytonia* (Crane 1985; Doyle & Donoghue 1986; Hilton & Bateman 2006), which had often been proposed as angiosperm relatives (Gaussen 1946; Stebbins 1974; Doyle 1978; Retallack & Dilcher 1981). Others, however, related the anthophytes to coniferophytes (Nixon *et al.* 1994; Rothwell & Serbet 1994; Rothwell *et al.* 2009).

In contrast, only a few molecular analyses, of ribosomal genes, have linked angiosperms and Gnetales, with weak support (Hamby & Zimmer 1992; Stefanovic *et al.* 1998; Rydin *et al.* 2002). Most multigene analyses (reviewed in Mathews 2009) have either nested Gnetales within conifers, with Pinaceae (the "gnepine" hypothesis: Qiu *et al.* 1999, 2007; Bowe *et al.* 2000; Chaw *et al.* 2000; Hajibabaei *et al.* 2006), or placed them at the base of seed plants (e.g., Rai *et al.* 2008). The latter result is implausible from a geological point of view, since Gnetales and angiosperms were the last major groups to appear in the fossil record. There is evidence that trees with Gnetales basal are a result of long branch attraction, where homoplastic changes on lines with large amounts of molecular evolution lead to incorrect inferences on relationships. Those data sets that give Gnetales-basal trees when analyzed with parsimony place Gnetales within conifers when analyzed with likelihood methods, which are thought to correct for long branch effects (Sanderson *et al.* 2000; Magallón & Sanderson 2002; Soltis *et al.* 2002; Burleigh & Mathews 2004). The conflict with morphology may be less severe than is often thought. Views that Gnetales are related to conifers rather than angiosperms were common before cladistics (Bailey 1944, 1949; Eames 1952; Bierhorst 1971; Doyle 1978), and in the morphological analysis of Doyle (2008) trees with the two positions of Gnetales became equally parsimonious after inclusion of recently described conifer-like features in Gnetales (e.g., Carlquist 1996).

In most molecular analyses that place Gnetales in conifers, angiosperms are the sister group of living gymnosperms, with the result that gymnosperms, long assumed to be paraphyletic, form a monophyletic group. However, this conclusion applies only to living gymnosperms. All morphological analyses that include fossils, including those with living taxa constrained to a molecular arrangement (Doyle 2006, 2008), place a series of seed fern taxa, which are typologically gymnosperms, below the crown seed plant clade. With this in mind, the name acrogymnosperms has been proposed for the clade of living gymnosperms (Cantino *et al.* 2007). The greatest uncertainty concerns cycads, which different analyses have placed at the base of acrogymnosperms, with *Ginkgo*, or with angiosperms (Mathews *et al.* 2010).

Setting aside angiosperms, the most striking innovations within crown group seed plants occur in coniferophytes. Coniferophytes differ markedly from seed ferns in their simple leaves: fan-shaped or strap-shaped with dichotomous venation in ginkgophytes and cordaites, scale-like or needle-like with one vein in most conifers (conifers with several veins, such as some Araucariaceae and Podocarpaceae, appear to be derived, judging from their nested positions). The idea that coniferophytes were derived from a seed fern prototype was proposed by Rothwell (1982), based on recognition of conifer-like platyspermic seeds and saccate pollen (as in cordaites, Paleozoic conifers, and living Pinaceae and Podocarpaceae) in the Late Carboniferous seed fern *Callistophyton*. Given that seed ferns had compound fronds, this would require a radical change in leaf morphology, which Rothwell suggested could have been a result of heterochrony, a change in the timing of developmental events. Like other seed ferns, *Callistophyton* had not only fronds but also pointed cataphylls—bud scales at the base of the axillary branches. Rothwell suggested that the leaves of conifers might be derived not from fronds but from cataphylls, which were already essentially like conifer leaves, if the plant continued to produce cataphylls throughout its life, without ever shifting to fronds. This hypothesis explains the leaves of conifers better than the dichotomously veined leaves of cordaites and ginkgophytes, although it might be a smaller step to derive leaves of the latter sort from cataphylls than from large fronds.

Origin of coniferophyte reproductive structures also involved a change from fertile fronds, like those of *Callistophyton*, which had microsporangia or ovules on the abaxial surface, to simple sporophylls. In ginkgophytes these sporophylls were apparently grouped into simple male and female strobili. The female strobili were reduced to a stalk with two sessile ovules in *Ginkgo* but still bore several ovules in Mesozoic fossils (Zhou & Zhang 1992; Zhou 2009). In contrast, cordaites had compound strobili consisting of an axis bearing bracts and axillary short shoots (simple strobili) with scale leaves and simple sporophylls. In conifers the male cones are simple strobili with scale-like sporophylls, but the female cones have woody cone scales bearing one or more ovules. Because each cone scale is subtended by a bract, there had been suggestions since the 1800s that the cone scales are axillary branches and the whole cone is derived from a compound strobilus, as in cordaites (Worsdell 1900). This was confirmed by studies of Late Carboniferous and Permian conifer stem relatives by Florin (1951, 1954), who showed that the female cones had bracts and obvious axillary short shoots bearing scale leaves and sporophylls. Modern cone scales would be derived by transformation of the fertile short shoot into a woody scale; Florin described intermediate conditions in Permian and Triassic conifers, although some of the details have required modification (Mapes & Rothwell 1984; Clement-Westerhof 1988).

These comparisons have often been taken as evidence for derivation of conifers from cordaites. The simple organization of the conifer male cone, however, suggests that the two groups were derived from a common ancestor that had both simple male and simple female strobili. Both types of strobili would be grouped into compound strobili in cordaites, but only the female strobili in conifers. However, this picture is complicated by recognition that there were Late Paleozoic conifers with compound male strobili (*Thucydia*: Hernandez-Castillo *et al.* 2001).

Platyspermic seeds occur not only in *Callistophyton* and coniferophytes (best seen in cordaites) but also in Permian and Mesozoic "seed ferns," including peltasperms, corystosperms, glossopterids, and *Caytonia*. Rothwell and Serbet (1994) questioned the distinction between platyspermy and radiospermy, but it can be made fairly consistently if defined in terms of anatomy (Doyle 1996). The shift from radiospermic to platyspermic seeds marks a clade called platysperms, which may be equivalent to crown group seed plants, depending on the position of *Callistophyton* and cycads. Many analyses that include fossils have nested cycads within platysperms, which would imply that their classic radiospermic seeds are a reversal, a conclusion consistent with the bilateral anatomy of seeds of *Cycas* (Stevenson 1990).

Other innovations of crown group seed plants involved modifications of the original spore-like pollen of more basal seed plants. One was origin of saccate pollen, with a single tire-like air sac or two sacs, formed by separation of the inner and outer layers of the exine. Today sacs are restricted to the conifer families Pinaceae and Podocarpaceae, but in the past they also occurred in *Callistophyton*, cordaites, Paleozoic conifers, some peltasperms (*Autunia*), corystosperms, glossopterids, and *Caytonia*. Whether sacs evolved once and were lost in the many nonsaccate seed plant taxa or arose several times is unclear because of uncertainty on the position of various fossils near the crown group node. However, the phylogenetic trees of Doyle (2008) imply that sacs are homologous in *Callistophyton* and other members of the acrogymnosperm line and were lost in *Ginkgo*, Araucariaceae, other conifers, and Gnetales (Doyle 2010). Experimental studies show that the sacs function to float the pollen upward in the liquid in the micropylar canal to the nucellus of the ovule, which is oriented downward, while their loss is correlated with a shift to upward-oriented ovules (Doyle 1945; Leslie 2010).

Another pollen innovation, also seen in *Callistophyton*, was the origin of a sulcus, an elongate thin area that serves as the site of germination of a pollen tube (actually known in *Callistophyton*: Rothwell 1972), correlated with loss of the proximal tetrad scar. Some cordaites and Paleozoic conifers had a tetrad scar and no sulcus, interpreted by Poort *et al.* (1996) as representing a stage before origin of a pollen tube. However, the nested position of these taxa in most analyses makes it more parsimonious to assume that the sulcus was lost. This may seem implausible, especially if one assumes that loss of the sulcus would imply loss of the pollen tube. This could mean that the sulcus was lost but the tube was retained, that the tube evolved once but a sulcus originated later in several lines (Doyle & Donoghue 1986; Friedman 1993), or that cordaites and conifers are more basal than current analyses indicate.

A final related innovation was a shift from swimming to nonmotile sperm, correlated with a change in function of the pollen tube: from anchoring the male gametophyte and absorption of nutrients (haustorial pollen tube, as in cycads and *Ginkgo*) to transfer of sperm to the archegonia (siphonogamy, as in living conifers, Gnetales, and angiosperms). The long, slender pollen tube of *Callistophyton* has been taken as evidence for siphonogamy (Rothwell 1981; Nishida *et al.* 2004), but Taylor (1988) and Friedman (1993) considered the function of the tube to be unknown. In acrogymnosperms, if ginkgophytes are sister to fossil and living conifers (Doyle 2008) and Paleozoic conifers were zooidogamous (Poort *et al.* 1996), siphonogamy originated on the line leading to living conifers and Gnetales.

Additional innovations evolved in Gnetales, many of them convergences with angiosperms if Gnetales are nested in conifers, such as presence of an outer tunica layer in the apical meristem and cellular embryogeny. Some would be interpreted as convergences with angiosperms even if the two groups were related, as noted in morphological cladistic studies (Doyle & Donoghue 1986; Doyle 1996), such as the angiosperm-like leaves of *Gnetum*, considering the scale-like and strap-shaped leaves of *Ephedra* and *Welwitschia*, which are consistent with a position in conifers, and vessels in the wood, which apparently arose within angiosperms. Although the reproductive units making up the strobili have been compared with flowers, they can also be interpreted as axillary shoots of a compound strobilus, like that of cordaites and Paleozoic conifers, with the "perianth" of the male "flower" and the outer integument around the ovule derived from sterile scale leaves on the axillary fertile shoot (cf. Eames 1952; Doyle 1994).

The details need further examination in light of developmental evidence that the male structures are more complex than previously assumed (Mundry & Stützel 2004).

1.8 Origin of angiosperms and their innovations

Phylogenetic analyses show a marked contrast between strong evidence on relationships within angiosperms and great uncertainty on their closest relatives. Early morphological analyses appeared to narrow the list of outgroups to Gnetales, Bennettitales, and *Pentoxylon*, but some linked the resulting anthophyte clade with corystosperms, glossopterids, and *Caytonia* (Crane 1985; Doyle & Donoghue 1986), while others nested the clade in coniferophytes (Nixon *et al.* 1994; Rothwell & Serbet 1994; Rothwell *et al.* 2009). Some later studies diluted the anthophyte concept by moving *Caytonia* up to a position as the sister group of angiosperms (Doyle 1996, 2006, 2008; Hilton & Bateman 2006). However, molecular analyses that place Gnetales in the conifers break up the anthophytes still more, and if angiosperms and acrogymnosperms are sister groups, no living gymnosperm taxon is closer to the angiosperms than any other, so any evidence on the origin of angiosperm innovations must come from fossils. Determining that cycads are sister to angiosperms (Mathews *et al.* 2010) would not help much, as cycads share few innovations with angiosperms. In either case, the angiosperm line would have diverged in the Carboniferous, since the oldest known acrogymnosperms are Late Carboniferous (cordaites, early conifers) and the oldest known cycads are Early Permian.

This picture offers interesting parallels and contrasts with the situation in vertebrates, where the lines leading to living mammals and reptiles (including birds) also diverged in Carboniferous. However, there is a long series of uncontested fossil stem relatives attached to the mammalian stem lineage, the so-called mammal-like reptiles (Gauthier *et al.* 1988). In contrast, there is no consensus among paleobotanists that any known fossils are angiosperm stem relatives. For example, glossopterids, *Pentoxylon*, Bennettitales, and *Caytonia* were identified as stem relatives in the morphological analysis of Doyle (2008) when living taxa were constrained to a molecular arrangement, with Gnetales in conifers, and in some trees obtained without constraints. However, glossopterids and *Caytonia* were far removed from angiosperms in morphological analyses of Rothwell *et al.* (2009), in which angiosperms were sister to Gnetales and nested in conifers (a result strongly contradicted by molecular data). Both studies associated Bennettitales and *Pentoxylon* with angiosperms, but reproductive structures of these taxa are highly modified, so even if they are related to angiosperms they shed little light on the origin of angiosperm reproductive features.

In contrast, although morphological cladistic analyses varied greatly on rooting of the angiosperms (Dahlgren & Bremer 1985; Donoghue & Doyle

1989; Loconte & Stevenson 1991; Nixon *et al.* 1994; Doyle 1996, 2006), molecular analyses of many separate and combined genes have given remarkably consistent results (Mathews & Donoghue 1999; Parkinson *et al.* 1999; Qiu *et al.* 1999, 2006a, 2010; Soltis *et al.* 1999, 2000; Barkman *et al.* 2000; Graham & Olmstead 2000; Zanis *et al.* 2002; Jansen *et al.* 2007; Moore *et al.* 2007). The so-called ANITA lines, namely, the New Caledonian endemic *Amborella*, Nymphaeales (water lilies), and Austrobaileyales, form a grade at the base. The remaining groups, called mesangiosperms, form five major clades: Chloranthaceae (notable for their highly simplified flowers), the reduced aquatic genus *Ceratophyllum*, magnoliids, monocots, and eudicots (united by tricolpate pollen). Rootings on *Ceratophyllum* (based on *rbcL*: Chase *et al.* 1993) and grasses (Goremykin *et al.* 2003) appear to be effects of long branch attraction and inadequate taxon sampling (e.g., Qiu *et al.* 2001; Degtjareva *et al.* 2004; Soltis *et al.* 2004). The main uncertainty concerns whether *Amborella* and Nymphaeales form two successive basal lines or a clade (Barkman *et al.* 2000; Qiu *et al.* 2006a, 2010; Goremykin *et al.* 2009), with the latter arrangement supported especially by mitochondrial genes. Relationships among the mesangiosperm clades remain poorly resolved, but relationships within them are largely stable. These results make it possible to reconstruct many features of the ancestral angiosperms by character optimization, independent of assumptions about outgroups (e.g., any state shared by all three ANITA lines can be reconstructed as ancestral).

Besides the flower and other reproductive advances, angiosperms share major vegetative innovations, the most obvious being in the leaves. These can be reconstructed as originally simple, with pinnate major venation and a hierarchy of reticulate higher vein orders (Doyle 2007), so their evolution must have involved several changes from the ancestral seed fern leaf, which had leaflets with a midrib and one order of dichotomous fine venation. If glossopterids, *Pentoxylon*, Bennettitales, and *Caytonia* are angiosperm stem relatives, the first step toward the angiosperm leaf would be a marked simplification to the "simple pinnate" type (Doyle & Donoghue 1986) seen in the first three of these taxa. This type includes both simple leaves with a midrib and one order of secondary veins, as in *Pentoxylon* and glossopterids, and once-compound leaves with a rachis bearing leaflets with only one order of dichotomous or parallel venation, as in modern cycads, both of which occurred in Bennettitales and fossil cycads (*Caytonia* had four leaflets that resemble glossopterid leaves; if it is nested within this clade, its leaves are presumably secondarily compound). This shift would occur earlier if cycads are located below these fossils (Mathews *et al.* 2010). In addition, glossopterids and *Caytonia* had simple reticulate venation, with a network consisting of one vein order, suggesting that this may have been a first step toward the complex reticulate venation of angiosperms. If so, it must have been followed by elaboration of several vein orders, a change ascribed to a shift from marginal to diffuse meristematic activity (Doyle & Hickey 1976; Boyce & Knoll 2002; Boyce 2005; Doyle 2006).

Another innovation that had great physiological and ecological consequences was the origin of vessels from tracheids, with perforations formed by loss of the primary cell wall between vessel members, rather than pits. This apparently occurred within angiosperms, since vessels are lacking not only in putative outgroups (assuming Gnetales are in conifers) but also in *Amborella* and Nymphaeales (the latter have transitional conducting cells with porose pit membranes that have been called vessels but may be better considered tracheids: Carlquist & Schneider 2009). The famous cases of vessel-less wood in Winteraceae (magnoliids) and Trochodendraceae (eudicots), which are deeply nested within angiosperms, are most parsimoniously interpreted as due to reversal (Doyle & Endress 2000; Feild *et al.* 2002).

The angiosperm flower is often regarded as the signature innovation of the group, but it is not easy to define how it differs from the strobili of other seed plants (Doyle 2008; Rudall & Bateman 2010). It is unusual in having a much shorter axis relative to the length of the appendages, as also seen in so-called flowers of Bennettitales, presumably an adaptation for attraction of pollinators. The presence of a sterile perianth (reconstructed as present at the crown group node: Endress & Doyle 2009) is not a radical advance, since there are sterile appendages at the base of the strobili in many other seed plants (Doyle 2008), although their elaboration for attraction may have been new. More distinctive is bisexual organization, which is unique among living plants but did occur in some Bennettitales. However, it is not certain whether bisexuality was ancestral in angiosperms. Because *Amborella* and other members of the basal angiosperm grade (Hydatellaceae, Trimeniaceae, Schisandraceae) are unisexual and others are bisexual, the ancestral state is equivocal on parsimony grounds, although the presence of nonfunctional stamens in the female flowers of *Amborella* suggests it was bisexual (Endress & Doyle 2009). Possible developmental genetic mechanisms for the transition from separate male and female strobili to a bisexual strobilus have been the subject of much discussion (Frohlich & Chase 2007).

Both the anthostrobilus hypothesis of Arber and Parkin (1907) and the anthophyte hypothesis postulated that the flower originated well before the angiosperms, in their common ancestor with Bennettitales and Gnetales. However, if *Caytonia* is the sister group of angiosperms and Bennettitales are the second outgroup, flowers either arose separately in angiosperms and Bennettitales or were lost in *Caytonia*, since *Caytonia* had relatively large sporophylls that are unlikely to have been closely aggregated (Doyle 2008). Of these, the former scenario may be more plausible on functional grounds.

In the fertile parts of flower, the best-known innovation is enclosure of the ovules in the carpel, or angiospermy. However, the ovules themselves differ from those of other seed plants in being bitegmic (having two integuments), except in some derived taxa (e.g., most asterids), and usually anatropous (reflexed). Presumably the nucellus and the inner integument represent the original seed plant ovule, while the outer integument had a different origin.

Hypotheses on the origin of the carpel and its homologies with structures in other groups should also explain the bitegmic ovule.

Molecular data on relationships within angiosperms have led to important new insights on the ancestral carpel. Formerly it was widely thought that the most primitive carpels were plicate (conduplicate), like a leaf folded down the middle, as in magnoliids such as *Degeneria* and Winteraceae (Bailey & Swamy 1951). Such carpels were often described as unsealed, but actually they are closed by postgenital fusion of the margins of the U-shaped carpel primordium (Igersheim & Endress 1997; Endress & Igersheim 2000). However, the basal ANITA groups have carpels of the ascidiate type, considered primitive by Leinfellner (1969) and van Heel (1981). Here there is a cross-zone of meristematic tissue between the margins of the primordium, and the whole structure grows up like a tube. The carpel is sealed by secretion in the narrow canal leading from the stigma to the ovary (Endress & Igersheim 2000)—what might be considered an incompletely angiospermous condition. The number and position of ovules vary among near-basal groups, but optimization of characters on molecular trees (Endress & Doyle 2009) indicates that the ancestral carpel had a single pendent ovule attached to the cross-zone, as in *Amborella*, Hydatellaceae (Nymphaeales), and *Trimenia* (Austrobaileyales).

During the development of anatropous bitegmic ovules in the ANITA grade and magnoliids, the inner integument grows up from a ring around the nucellus, but the outer integument grows over from one side like a hood (Umeda *et al.* 1994; Imaichi *et al.* 1995; Igersheim & Endress 1997; Yamada *et al.* 2001a, 2003). The ovule of *Amborella* is orthotropous (erect rather than reflexed) but dorsiventral early in development, suggesting derivation from an anatropous ancestor (Yamada *et al.* 2001b). There is evidence that the outer integument is a leaf-like structure with the nucellus and inner integument on its adaxial surface. First, when there are vascular bundles in the outer integument, the xylem is to the inside, implying that this side is adaxial (Frohlich & Chase 2007). Second, in *Arabidopsis* the gene *INO*, a member of the YABBY family, which specifies abaxial identity and was implicated above in origin of the seed plant leaf from a progymnosperm branch system, is expressed in the outer epidermis of the outer integument (Balasubramanian & Schneitz 2000; Meister *et al.* 2002; Skinner *et al.* 2004) but not in the inner integument (McAbee *et al.* 2006). This is consistent with the view that the two integuments have different origins and the outer integument is leaf-like (Skinner *et al.* 2004; McAbee *et al.* 2006; Doyle 2008).

More specific ideas on homologies of the outer integument have centered on the cupules of various "seed fern" groups, which appear to have different homologies in different taxa. As discussed above, the cupules of the oldest seed ferns were dichotomous structures that were probably derived from the fertile appendages of progymnosperms and lost in more derived seed ferns, such as medullosans and *Callistophyton*. In contrast, younger peltasperms, corystosperms, glossopterids, and *Caytonia* had dorsiventral cupules that are more likely modified leaves (sporophylls) or leaflets, with ovules on one surface. Among these, anatomical and positional evidence (reviewed in Doyle 2008) indicates that the ovules were on the abaxial surface of the cupule in peltasperms and corystosperms but on the adaxial surface in glossopterids. Glossopterids had one or more cupules attached to the midrib of a leaf, most simply interpreted as sporophylls on an adnate axillary branch (Retallack & Dilcher 1981). *Caytonia* had reflexed cupules that have been interpreted as leaflets borne along the rachis of a sporophyll, with adaxial ovules (Harris 1940; Gaussen 1946; Harris 1951; Doyle 1978). However, as emphasized by Rothwell *et al.* (2009), this has not been confirmed by anatomical data, and there are putatively related fossils with cupules arranged in a spiral, suggesting that they were whole sporophylls borne on a branch (Schweitzer & Kirchner 1998; Wang 2010).

These observations are consistent with precladistic hypotheses that the angiosperm outer integument is homologous with the cupule of glossopterids and/or *Caytonia* (Gaussen 1946; Stebbins 1974; Doyle 1978; Retallack & Dilcher 1981). They are also consistent with analyses in which glossopterids and *Caytonia* are angiosperm stem relatives, although the fact that the same analyses also associate angiosperms with *Pentoxylon* and Bennettitales poses problems, since these taxa had ovule-bearing structures that are difficult to interpret in these terms (or any others; Doyle 2008). *Caytonia* is more like angiosperms in having anatropous cupules, which could be transformed into anatropous bitegmic ovules by reduction to one ovule per cupule.

Gaussen (1946) and later authors (Doyle 1978; Crane 1985) proposed that the carpel was derived by broadening and folding of a cupule-bearing rachis of the *Caytonia* type. However, others have homologized the carpel with a glossopterid leaf-cupule complex, with the carpel wall derived from the subtending leaf and the bitegmic ovule from a cupule on an axillary branch (Stebbins 1974; Retallack & Dilcher 1981; Doyle 1996; Doyle 2008). This would be consistent with the inference that the ancestral carpel had one bitegmic ovule on the cross-zone, corresponding to an axillary position, and with hints that the carpel wall and placenta in *Arabidopsis* are under separate genetic control (Skinner *et al.* 2004). However, ovule position appears to have been highly labile in early angiosperms (Frohlich & Chase 2007; Endress & Doyle 2009), suggesting that positional arguments for homology should be used with caution. A problem in relating angiosperms to both glossopterids and *Caytonia* is the fact that it is difficult to homologize the cupule-bearing structures of these fossils with each other; this might be easier if the supposed *Caytonia* sporophyll turns out to be a sporophyll-bearing branch (Doyle 2008).

Although discussions on the origin of the flower tend to emphasize the carpel, the stamens too show major innovations and present similar problems. They are unique in having four microsporangia fused into two thecae, one on either side of the medial plane. Thomas (1925) compared angiosperm stamens with microsynangia of *Caytonia*, which consisted of four sporangia, but as noted by Harris (1937) the four sporangia were not separated into two pairs and were borne on a branched structure interpreted as a sporophyll. If angiosperms are related to *Caytonia*, it might be better to homologize each theca with a synangium and the stamen with the whole sporophyll, drastically reduced to an unbranched structure with two synangia (Gaussen 1946; Doyle 2008). Character optimization on molecular trees (Endress & Doyle 2009) indicates that the microsporangia were originally lateral or adaxial (introrse). This could be compared to the situation in glossopterids, where microsporangia were borne on a branch from the adaxial side of a leaf, or in Bennettitales, where microsynangia were borne on the adaxial side of simple or pinnate structures interpreted as sporophylls. However, in both fossil groups the microsporangia were perpendicular to the leaf-like structure, not parallel to its surface as in angiosperms, and like ovule position, microsporangial position was highly labile in early angiosperms (Endress & Doyle 2009).

Molecular phylogenetic results confirm the standard view that the first angiosperms had monosulcate pollen. If angiosperms had ancestors with saccate pollen, as in glossopterids and *Caytonia*, the sacs must have been lost below the crown group. Rothwell *et al.* (2009) argued that the saccate pollen of *Caytonia* was evidence against a relationship with angiosperms. However, even if *Caytonia* is sister to angiosperms, it is not clear that their common ancestor had sacs, and as already noted loss of sacs was a common phenomenon in seed plants, tied to shifts away from capture of pollen by a pollination drop secreted by a downward-oriented ovule (Doyle 2010; Leslie 2010).

An important innovation in angiosperm pollen was columellar exine structure, with radial rods connecting the inner nexine and outer tectum layers. In potential outgroups, the infratectal structure was alveolar (glossopterids, *Caytonia*) or granular (Bennettitales, *Pentoxylon*, Gnetales), whereas in angiosperms it is granular or columellar. Earlier workers (Van Campo & Lugardon 1973; Doyle *et al.* 1975; Walker & Skvarla 1975; Walker 1976) argued that granular (including "atectate") exine structure was ancestral, since it occurs both in gymnosperms and in Magnoliales, which were assumed to be primitive. However, the ANITA lines have columellar exines, or in *Amborella* an apparently related type with an undulating tectum (Sampson 1993; Hesse 2001), indicating that columellae evolved on the angiosperm stem lineage (Doyle 2005, 2009). Columellar exines often have perforations in the tectum, resulting in reticulate sculpture, but the tectum is continuous in *Amborella* and Nymphaeales, implying that perforations arose within angiosperms, below the Austrobaileyales node.

Another advance on the line leading to angiosperms was siphonogamy, which appears to have originated independently from siphonogamy in conifers and Gnetales, based on the placement of zooidogamous cycads and *Ginkgo* lower in the acrogymnosperms. Multiflagellated sperm cells have been described in glossopterids (Nishida *et al.* 2004), which would imply that siphonogamy originated after this group diverged from angiosperms, if the two are related. Based on similarities in the pattern of pollen tube growth in ovules of Bennettitales and araucariaceous conifers, Stockey and Rothwell (2003) argued that Bennettitales were siphonogamous.

Some of the most remarkable innovations in angiosperms are related to radical reduction of the gametophyte generation. In the male gametophyte, loss of sterile cells left only two sperm cells and the tube nucleus. The ancestral female gametophyte (embryo sac) was long assumed to be of the eight-nucleate *Polygonum* type, with no recognizable archegonium but an egg and two synergid cells, two free polar nuclei, and three antipodal cells. This serves as the setup for double fertilization, where one of the two sperm cells fuses with the egg to produce the embryo, the other with the polar nuclei to produce the triploid endosperm, the nourishing tissue for the embryo in most angiosperms. However, recent studies confirmed earlier reports (e.g., Yoshida 1960; Batygina *et al.* 1982) that two of the ANITA lines, Nymphaeales and Austrobaileyales, have a four-nucleate female gametophyte with only an egg, two synergids, and one polar nucleus, which fuses with the second sperm to produce diploid endosperm (Williams & Friedman 2002; Friedman *et al.* 2003; Williams & Friedman 2004). Friedman *et al.* (2003) proposed that this type was ancestral and the *Polygonum* type was derived by duplication of the four-nucleate module, with the antipodals representing a sterilized egg and synergids.

A problem is that *Amborella* has a female gametophyte resembling the *Polygonum* type (with an exception discussed below), so it seems equally parsimonious to assume that the four-nucleate type was either ancestral or derived twice from eight-nucleate. However, Williams and Friedman (2004) showed that dissection of this character into developmental changes favors a scenario in which the first step, on the angiosperm stem lineage, was reduction from many nuclei to a four-nucleate module, which was duplicated once in *Amborella* and once after the divergence of Austrobaileyales, that is, at the base of the mesangiosperms. If so, endosperm originated as a diploid tissue and later become triploid. Its triploid state would be a byproduct of duplication of the polar nucleus, but advantages of triploidy in endosperm tissue may have been the selective factor that favored this change (Friedman & Williams 2004) and a key innovation that contributed to the vast diversity of the mesangiosperm clade (Williams & Friedman 2004), which includes around 99.9% of angiosperm species.

This scenario is complicated by the discovery that the female gametophyte of *Amborella* has nine nuclei rather than eight (Friedman 2006). The extra nucleus is derived from division of one of the micropylar cells to produce the egg and a outer cell. This could be the sole remnant in angiosperms of the periclinal division of the archegonial initial cell, established in the first land plants, or it could be an autapomorphy (Friedman & Ryerson 2009). In the former case, it could mean that the ancestral female gametophyte had five nuclei rather than four.

There is little evidence on the timing of these changes in the life cycle, because relevant details are rarely preserved in fossils. In the heyday of the anthophyte hypothesis, when a form of double fertilization was confirmed in Gnetales (where fusion of the second sperm with another nucleus of the female gametophyte produces an extra embryo), this was proposed as a step toward the angiosperm condition (Friedman 1990, 1994; Friedman & Carmichael 1996). However, in the light of molecular phylogenies this now appears to be a convergence (Friedman & Floyd 2001). Glossopterids and Bennettitales had relatively large female gametophytes, so if they are related to angiosperms the main reduction must have occurred between their divergence and the base of the angiosperms. Unfortunately none of these characters are preserved in *Caytonia*.

Another issue concerns embryogeny, which involves only cellular divisions in angiosperms but an initial free-nuclear phase in other seed plants, except Gnetales (*Ephedra* is a complex case discussed in Doyle 2006). An exception in angiosperms is *Paeonia*, which was considered primitive by Stebbins (1974) but is clearly derived based on its position in the eudicot order Saxifragales. However, if acrogymnosperms and angiosperms are sister groups, it is equivocal whether cellular or nuclear embryogenesis is ancestral for crown group seed plants. This is a case where a position of cycads on the line leading to angiosperms (Mathews *et al.* 2010) would have an effect on scenarios, since it would imply that the angiosperm condition was derived, but the point at which cellular embryogeny evolved on the angiosperm stem lineage would remain unknown.

1.9 Innovations within angiosperms: monocots and eudicots

Evolutionary innovations within angiosperms are too numerous to treat thoroughly, but a few deserve special attention. One already mentioned was derivation of the plicate carpel, with margins sealed by postgenital fusion of the young epidermises, from the ancestral ascidiate carpel (Endress & Igersheim 2000). In trees based on combined morphological and molecular data, this occurred above the base of mesangiosperms, after divergence of the ascidiate Chloranthaceae and *Ceratophyllum* lines, but in trees with other arrangements of mesangiosperms it occurred either several times or once at the base of the clade, followed by reversals (Doyle & Endress 2000; Endress & Doyle 2009). Syncarpy, which allows fertilization of ovules in all carpels by pollen landing on any stigma, arose many times within angiosperms, usually by congenital fusion of the carpels (Endress 1994).

Monocots underwent a major reorganization of vegetative morphology tied with loss of secondary growth, including scattered primary vascular bundles (an atactostele), early abortion of the primary root, and production of adventitious roots from rhizomes (Arber 1925; Takhtajan 1969; Dahlgren *et al.* 1985). Similar changes also occurred in Nymphaeales, where it is clear that they are functionally related to an aquatic habit. This and the assumption that the most primitive monocots were aquatic members of the Alismatales (Helobiae) led to the hypothesis that monocots and Nymphaeales were derived from an aquatic common ancestor (e.g., Cronquist 1968; Takhtajan 1969), which seemed to be supported by morphological cladistic analyses that linked the two groups (Dahlgren & Bremer 1985; Donoghue & Doyle 1989; Loconte & Stevenson 1991). However, the molecular separation of monocots from Nymphaeales indicates that their similar derived features are purely convergent. Molecular data have further clarified the situation by showing that *Acorus* and Alismatales are basal to the remaining monocot groups (core monocots, or petrosaviids). *Acorus* is a marsh plant but less aquatic than classic Helobiae, which are nested in Alismatales among Araceae and Tofieldiaceae, many of which are also marsh or bog plants. This supports a more nuanced view that loss of secondary growth and related features evolved in wet but not fully aquatic habitats.

Because *Acorus*, Tofieldiaceae, and many basal Araceae are like other monocots in having linear leaves with "parallel" venation (actually joining at the apex: Kaplan 1973; Doyle *et al.* 2008), such leaves presumably evolved on the stem lineage of monocots. In most monocots the leaf blade develops from the lower zone of the leaf primordium, not the upper zone as in other groups, which implies that it corresponds to the leaf base of other angiosperms (Kaplan 1973). However, in *Acorus*, Alismatales, and many Liliales, the blade develops from the upper leaf zone (Kaplan 1973; Bharathan 1996; Rudall & Buzgo 2002), indicating that the change in blade morphology occurred first and the new mode of development later.

The signature innovation of monocots is the single cotyledon of the embryo, in contrast to the two cotyledons of other angiosperms and most gymnosperms. Whether the monocot condition evolved by loss of one cotyledon or fusion of two is a long-standing topic of discussion (Arber 1925; Takhtajan 1969). There are no intermediate conditions in taxa that might be phylogenetically relevant. In Hydatellaceae, minute aquatics once considered monocots but recently found to be related to Nymphaeales (Saarela *et al.* 2007), Sokoloff *et al.* (2008) described fusion of the cotyledons into a bilobed structure and suggested that this might be a step toward a single cotyledon, but in the current phylogenetic context this cannot be more than an analogy.

In precladistic classifications angiosperms were divided into monocots and dicots, but "dicots" have been abandoned as a grossly paraphyletic group. However, some 95% of dicot species form a clade, the eudicots, which include such familiar plants as the model system *Arabidopsis*. Their most definite synapomorphy is tricolpate pollen, with three germination furrows (colpi) running along lines of longitude relative to the polar axis (defined by the center of the meiotic tetrad and the center of the grain), rather than one furrow (sulcus) at the distal pole. Tricolpate pollen was ancestral to many other types, such as tricolporate and triporate. This innovation may have allowed more rapid penetration of the stigma by the pollen tube, regardless of the orientation of the pollen. Precladistic workers assumed that tricolpate pollen evolved independently in several lines (e.g., Cronquist 1968), but molecular studies confirmed preliminary indications from morphological analyses (Dahlgren & Bremer 1985; Donoghue & Doyle 1989) that eudicots are monophyletic and tricolpate pollen originated only once (except for a superficial convergence in *Illicium* and *Schisandra*, in the Austrobaileyales, where the three furrows are oriented differently: Huynh 1976; Doyle 2005). Phylogenetic analyses provide little evidence on the mode of origin of tricolpate pollen, since the potential outgroups are basically monosulcate (Doyle 2005), but some of the oldest Early Cretaceous tricolpate pollen has obliquely oriented and interconnected colpi, suggesting an origin by elongation, spiralization, and fragmentation of a sulcus (Doyle & Hotton 1991; Blackmore & Crane 1998).

Another change that may have occurred on the line to eudicots but was not recognized until recently is a shift from trimerous to dimerous flowers, with parts in whorls of two (Drinnan *et al.* 1994; Endress & Doyle 2009). It is equivocal whether dimery arose once at the base of eudicots or more than once within them. In the basal order Ranunculales the Papaveraceae are dimerous but most of the other families are trimerous, and in the other line the basal order Proteales includes both the dimerous family Proteaceae and taxa with other conditions (*Nelumbo*, *Platanus*). However, dimery was established in Trochodendraceae, Buxaceae, and Gunnerales, the lines closest to the remaining eudicots. In this part of the phylogeny the flower can be reconstructed as having two pairs of sepaloid tepals and two pairs of stamens (Ronse De Craene 2008; Endress & Doyle 2009). Given that most of these taxa (except *Nelumbo* and Proteaceae) are wind pollinated, whereas more basal groups are largely insect pollinated, this floral architecture appears to reflect a shift from insect to wind pollination.

Within eudicots, a major innovation was origin of the typical pentamerous (less commonly tetramerous) "dicot" flower of textbooks, with alternating whorls of five protective sepals, five attractive petals, and five or ten stamens. This appears to be a synapomorphy of a major clade, called core eudicots or Pentapetalae, which is the sister group of Gunnerales (*Gunnera* and *Myrothamnus*). The fact that Pentapetalae are nested among the dimerous groups suggests that their typically insect pollinated flowers were derived from simpler wind pollinated flowers, a reversal of the earlier trend. Wanntorp and Ronse De Craene (2005) argued that the flowers of *Gunnera* are not relevant for the origin of Pentapetalae because they are reduced for wind pollination, but this does not follow, since parsimony optimization indicates that the whole line went through a reduced phase (Ronse De Craene 2004; Endress & Doyle 2009). Based on the spiral perianth of *Berberidopsis*, near the base of the Pentapetalae, Ronse De Craene (2004) suggested that the pentamerous flower was derived from a flower with spiral phyllotaxis, but *Berberidopsis* is unlikely to represent a transitional state, since it is linked with *Aextoxicon*, which is pentamerous (Ronse De Craene & Stuppy 2010). Many authors have suggested that the petals of Pentapetalae were derived from stamens, but this was questioned by Ronse De Craene (2007) on developmental grounds, and it seems unlikely from a phylogenetic perspective, since the more basal groups have so few stamens. However the transition occurred, it involved an increase in the total number of floral parts from that in basal eudicots, whether by addition of cycles, increase in the number of parts per cycle, or both.

Acknowledgments

I thank Judy Jernstedt, Stefan Little, and Yin-Long Qiu for useful discussions and help in obtaining literature.

References

Andrews, H.N. (1963) Early seed plants. *Science*, **142**, 925–931.
Arber, A. (1925) *Monocotyledons.* Cambridge University Press, Cambridge.
Arber, E.A.N. & Parkin, J. (1907) On the origin of angiosperms. *Journal of the Linnean Society, Botany*, **38**, 29–80.
Bailey, I.W. (1944) The development of vessels in angiosperms and its significance in morphological research. *American Journal of Botany*, **31**, 421–428.
Bailey, I.W. (1949) Origin of the angiosperms: need for a broadened outlook. *Journal of the Arnold Arboretum*, **30**, 64–70.
Bailey, I.W. & Swamy, B.G.L. (1951) The conduplicate carpel of dicotyledons and its initial trends of specialization. *American Journal of Botany*, **38**, 373–379.
Balasubramanian, S. & Schneitz, K. (2000) *NOZZLE* regulates proximal-distal pattern formation, cell proliferation and early sporogenesis during ovule development in *Arabidopsis thaliana*. *Development*, **127**, 4227–4238.
Banks, H.P. (1981) Peridermal activity (wound repair) in an Early Devonian (Emsian) trimerophyte from the Gaspé Peninsula, Canada. *The Palaeobotanist*, **28–29**, 20–25.
Banks, H.P. & Colthart, B.J. (1993) Plant-animal-fungal interactions in Early Devonian trimerophytes from Gaspé, Canada. *American Journal of Botany*, **80**, 992–1001.
Banks, H.P., Leclercq, S. & Hueber, F.M. (1975) Anatomy and morphology of *Psilophyton dawsonii*, sp. n. from the late Lower Devonian of Quebec (Gaspé) and Ontario, Canada. *Palaeontographica Americana*, **8**, 75–127.
Barkman, T.J., Chenery, G., McNeal, J.R., *et al.* (2000) Independent and combined analyses of sequences from all three genomic compartments converge on the root of flowering plant phylogeny. *Proceedings of the National Academy of Sciences of the United States of America*, **97**, 13166–13171.
Basinger, J.F., Rothwell, G.W. & Stewart, W.N. (1974) Cauline vasculature and leaf trace production in medullosan pteridosperms. *American Journal of Botany*, **61**, 1002–1015.
Bateman, R.M. & DiMichele, W.A. (1994) Heterospory: the most iterative key innovation in the evolutionary history of the plant kingdom. *Biological Reviews*, **69**, 345–417.

Bateman, R.M., DiMichele, W.A. & Willard, D.A. (1992) Experimental cladistic analyses of anatomically preserved arborescent lycopsids from the Carboniferous of Euramerica: an essay in paleobotanical phylogenetics. *Annals of the Missouri Botanical Garden*, **79**, 500–559.

Batygina, T.B., Shamrov, I.I. & Koleseva, G.E. (1982) Embriologiya nimfeynykh i lotosovykh. II. Razvitie zhenskikh embrional'nykh struktur. *Botanicheskii Zhurnal*, **67**, 1179–1195.

Beck, C.B. (1960) Connection between *Archaeopteris* and *Callixylon*. *Science*, **131**, 1524–1525.

Beck, C.B. (1966) On the origin of gymnosperms. *Taxon*, **15**, 337–339.

Beck, C.B. (1970) The appearance of gymnospermous structure. *Biological Reviews*, **45**, 379–400.

Beck, C.B. (1971) On the anatomy and morphology of lateral branch systems of *Archaeopteris*. *American Journal of Botany*, **58**, 758–784.

Becker, B. & Marin, B. (2009) Streptophyte algae and the origin of embryophytes. *Annals of Botany*, **103**, 999–1004.

Berry, C.M. & Stein, W.E. (2000) A new iridopteridalean from the Devonian of Venezuela. *International Journal of Plant Sciences*, **161**, 807–827.

Bharathan, G. (1996) Does the monocot mode of leaf development characterize all monocots? *Aliso*, **14**, 271–279.

Bierhorst, D.W. (1971) *Morphology of Vascular Plants*. Macmillan, New York.

Blackmore, S. & Crane, P.R. (1998) The evolution of apertures in the spores and pollen grains of embryophytes. In: *Reproductive Biology in Systematics, Conservation and Economic Botany* (eds S.J. Owens & P.R. Rudall). pp. 159–182. Royal Botanic Gardens, Kew.

Bonamo, P.M. & Banks, H.P. (1967) *Tetraxylopteris schmidtii*: its fertile parts and its relationships within the Aneurophytales. *American Journal of Botany*, **54**, 755–768.

Bowe, L.M., Coat, G. & dePamphilis, C.W. (2000) Phylogeny of seed plants based on all three genomic compartments: extant gymnosperms are monophyletic and Gnetales' closest relatives are conifers. *Proceedings of the National Academy of Sciences of the United States of America*, **97**, 4092–4097.

Bower, F.O. (1890) On antithetic as distinct from homologous alternation of generations in plants. *Annals of Botany*, **4**, 347–370.

Bower, F.O. (1908) *The Origin of a Land Flora*. Macmillan, London.

Bower, F.O. (1935) *Primitive Land Plants*. Macmillan, London.

Boyce, C.K. (2005) Patterns of segregation and convergence in the evolution of fern and seed plant leaf morphologies. *Paleobiology*, **31**, 117–140.

Boyce, C.K. & Knoll, A.H. (2002) Evolution of developmental potential and the multiple independent origins of leaves in Paleozoic vascular plants. *Paleobiology*, **28**, 70–100.

Bremer, K. (1985) Summary of green plant phylogeny and classification. *Cladistics*, **1**, 369–385.

Brenchley, W.E. (1913) On branching specimens of *Lyginodendron oldhamium*. *Journal of the Linnean Society, Botany*, **41**, 349–356.

Burleigh, J.G. & Mathews, S. (2004) Phylogenetic signal in nucleotide data from seed plants: implications for resolving the seed plant tree of life. *American Journal of Botany*, **91**, 1599–1613.

Campbell, D.H. (1924) A remarkable development of the sporophyte in *Anthoceros fusiformis*, Aust. *Annals of Botany*, **38**, 473–483.

Cantino, P.D., Doyle, J.A., Graham, S.W., *et al.* (2007) Towards a phylogenetic nomenclature of *Tracheophyta*. *Taxon*, **56**, 822–846.
Carlquist, S. (1996) Wood, bark, and stem anatomy of Gnetales: a summary. *International Journal of Plant Sciences*, **157** (Suppl. 6), S58–S76.
Carlquist, S. & Schneider, E.L. (2009) Do tracheid microstructure and the presence of minute crystals link Nymphaeaceae, Cabombaceae and Hydatellaceae? *Botanical Journal of the Linnean Society*, **159**, 572–582.
Chaloner, W.G. (1970) The evolution of miospore polarity. *Geoscience and Man*, **1**, 47–56.
Chaloner, W.G. & Sheerin, A. (1981) The evolution of reproductive strategies in early land plants. In: *Evolution Today* (eds G.G.E. Scudder & J.L. Reveal). pp. 93–100. Carnegie-Mellon University, Pittsburg, PA.
Chamberlain, C.J. (1935) *Gymnosperms: Structure and Evolution*. University of Chicago Press, Chicago, IL.
Chang, Y. & Graham, S.W. (2011) Inferring the higher-order phylogeny of mosses (Bryophyta) and relatives using a large, multigene plastid data set. *American Journal of Botany*, **98**, 839–849.
Chase, M.W., Soltis, D.E., Olmstead, R.G., *et al.* (1993) Phylogenetics of seed plants: an analysis of nucleotide sequences from the plastid gene *rbc*L. *Annals of the Missouri Botanical Garden*, **80**, 526–580.
Chaw, S.M., Parkinson, C.L., Cheng, Y., *et al.* (2000) Seed plant phylogeny inferred from all three plant genomes: monophyly of extant gymnosperms and origin of Gnetales from conifers. *Proceedings of the National Academy of Sciences of the United States of America*, **97**, 4086–4091.
Cichan, M.A. (1985) Vascular cambium and wood development in Carboniferous plants. II. *Sphenophyllum plurifoliatum* Williamson and Scott (Sphenophyllales). *Botanical Gazette*, **146**, 395–403.
Clement-Westerhof, J.A. (1988) Morphology and phylogeny of Paleozoic conifers. In: *Origin and Evolution of Gymnosperms* (ed. C.B. Beck). pp. 298–337. Columbia University Press, New York.
Cordi, J. & Stein, W.E. (2005) The anatomy of *Rotoxylon dawsonii* comb. nov. (*Cladoxylon dawsonii*) from the Upper Devonian of New York State. *International Journal of Plant Sciences*, **166**, 1029–1045.
Cornet, B., Phillips, T.L. & Andrews, H.N. (1976) The morphology and variation in *Rhacophyton ceratangium* from the Upper Devonian and its bearing on frond evolution. *Palaeontographica Abteilung B*, **158**, 105–129.
Crane, P.R. (1985) Phylogenetic analysis of seed plants and the origin of angiosperms. *Annals of the Missouri Botanical Garden*, **72**, 716–793.
Cronquist, A. (1968) *The Evolution and Classification of Flowering Plants*. Houghton Mifflin, Boston, MA.
Crum, H. (2001) *Structural Diversity of Bryophytes*. University of Michigan Herbarium, Ann Arbor, MI.
Dahlgren, R. & Bremer, K. (1985) Major clades of angiosperms. *Cladistics*, **1**, 349–368.
Dahlgren, R.M.T., Clifford, H.T. & Yeo, P.F. (1985) *The Families of the Monocotyledons*. Springer-Verlag, Berlin.
Degtjareva, G.V., Samigullin, T.H., Sokoloff, D.D., *et al.* (2004) Gene sampling versus taxon sampling: is *Amborella* (Amborellaceae) a sister group to all other extant angiosperms? *Botanicheskii Zhurnal*, **89**, 896–907.

Delwiche, C.F. & Palmer, J.D. (1997) The origin of plastids and their spread via secondary endosymbiosis. *Plant Systematics and Evolution, Supplement*, **11**, 53–86.

Dittrich, H.S., Matten, L.C. & Phillips, T.L. (1983) Anatomy of *Rhacophyton ceratangium* from the Upper Devonian (Famenian) of West Virginia. *Review of Palaeobotany and Palynology*, **40**, 127–147.

Donoghue, M.J. & Doyle, J.A. (1989) Phylogenetic analysis of angiosperms and the relationships of Hamamelidae. In: *Evolution, Systematics, and Fossil History of the Hamamelidae*, Vol. **1** (eds P.R. Crane & S. Blackmore). pp. 17–45. Clarendon Press, Oxford.

Donoghue, M.J., Doyle, J.A., Gauthier, J., *et al.* (1989) The importance of fossils in phylogeny reconstruction. *Annual Review of Ecology and Systematics*, **20**, 431–460.

Donoghue, M.J. & Kadereit, J.W. (1992) Walter Zimmermann and the growth of phylogenetic theory. *Systematic Biology*, **41**, 74–85.

Doran, J.B. (1980) A new species of *Psilophyton* from the Lower Devonian of northern New Brunswick, Canada. *Canadian Journal of Botany*, **58**, 2241–2262.

Doyle, J. (1945) Developmental lines in pollination mechanisms in the Coniferales. *Scientific Proceedings of the Royal Dublin Society*, **24** (5), 43–62.

Doyle, J.A. (1978) Origin of angiosperms. *Annual Review of Ecology and Systematics*, **9**, 365–392.

Doyle, J.A. (1994) Origin of the angiosperm flower: a phylogenetic perspective. *Plant Systematics and Evolution, Supplement*, **8**, 7–29.

Doyle, J.A. (1996) Seed plant phylogeny and the relationships of Gnetales. *International Journal of Plant Sciences*, **157** (Suppl. 6), S3–S39.

Doyle, J.A. (1998) Phylogeny of vascular plants. *Annual Review of Ecology and Systematics*, **29**, 567–599.

Doyle, J.A. (2005) Early evolution of angiosperm pollen as inferred from molecular and morphological phylogenetic analyses. *Grana*, **44**, 227–251.

Doyle, J.A. (2006) Seed ferns and the origin of angiosperms. *Journal of the Torrey Botanical Society*, **133**, 169–209.

Doyle, J.A. (2007) Systematic value and evolution of leaf architecture across the angiosperms in light of molecular phylogenetic analyses. *Courier Forschungsinstitut Senckenberg*, **258**, 21–37.

Doyle, J.A. (2008) Integrating molecular phylogenetic and paleobotanical evidence on origin of the flower. *International Journal of Plant Sciences*, **169**, 816–843.

Doyle, J.A. (2009) Evolutionary significance of granular exine structure in the light of phylogenetic analyses. *Review of Palaeobotany and Palynology*, **153**, 198–210.

Doyle, J.A. (2010) Function and evolution of saccate pollen. *New Phytologist*, **188**, 6–9.

Doyle, J.A. & Donoghue, M.J. (1986) Seed plant phylogeny and the origin of angiosperms: an experimental cladistic approach. *Botanical Review*, **52**, 321–431.

Doyle, J.A. & Donoghue, M.J. (1993) Phylogenies and angiosperm diversification. *Paleobiology*, **19**, 141–167.

Doyle, J.A. & Endress, P.K. (2000) Morphological phylogenetic analysis of basal angiosperms: comparison and combination with molecular data. *International Journal of Plant Sciences*, **161** (Suppl. 6), S121–S153.

Doyle, J.A. & Endress, P.K. (2010) Integrating Early Cretaceous fossils into the phylogeny of living angiosperms: Magnoliidae and eudicots. *Journal of Systematics and Evolution*, **48**, 1–35.

Doyle, J.A., Endress, P.K. & Upchurch, G.R., Jr (2008) Early Cretaceous monocots: a phylogenetic evaluation. *Acta Musei Nationalis Pragae, Series B, Historia Naturalis*, **64** (2–4), 59–87.

Doyle, J.A. & Hickey, L.J. (1976) Pollen and leaves from the mid-Cretaceous Potomac Group and their bearing on early angiosperm evolution. In: *Origin and Early Evolution of Angiosperms* (ed. C.B. Beck). pp. 139–206. Columbia University Press, New York.

Doyle, J.A. & Hotton, C.L. (1991) Diversification of early angiosperm pollen in a cladistic context. In: *Pollen and Spores: Patterns of Diversification* (eds S. Blackmore & S.H. Barnes). pp. 169–195. Clarendon Press, Oxford.

Doyle, J.A., Van Campo, M. & Lugardon, B. (1975) Observations on exine structure of *Eucommiidites* and Lower Cretaceous angiosperm pollen. *Pollen et Spores*, **17**, 429–486.

Drinnan, A.N., Crane, P.R. & Hoot, S.B. (1994) Patterns of floral evolution in the early diversification of non-magnoliid dicotyledons (eudicots). *Plant Systematics and Evolution, Supplement*, **8**, 93–122.

Eames, A.J. (1952) Relationships of the Ephedrales. *Phytomorphology*, **2**, 79–100.

Edwards, D.S. (1980) Evidence for the sporophytic status of the Lower Devonian plant *Rhynia gwynne-vaughanii* Kidston and Lang. *Review of Palaeobotany and Palynology*, **29**, 177–188.

Edwards, D.S. (1986) *Aglaophyton major*, a non-vascular land-plant from the Devonian Rhynie Chert. *Botanical Journal of the Linnean Society*, **93**, 173–204.

Eggert, D.A. & Gaunt, D.D. (1973) Phloem of *Sphenophyllum*. *American Journal of Botany*, **60**, 755–770.

Endress, P.K. (1994) *Diversity and Evolutionary Biology of Tropical Flowers.* Cambridge University Press, Cambridge.

Endress, P.K. & Doyle, J.A. (2009) Reconstructing the ancestral angiosperm flower and its initial specializations. *American Journal of Botany*, **96**, 22–66.

Endress, P.K. & Igersheim, A. (2000) Gynoecium structure and evolution in basal angiosperms. *International Journal of Plant Sciences*, **161** (Suppl. 6), S211–S223.

Feild, T.S., Brodribb, T. & Holbrook, N.M. (2002) Hardly a relict: freezing and the evolution of vesselless wood in Winteraceae. *Evolution*, **56**, 464–478.

Florin, R. (1951) Evolution in cordaites and conifers. *Acta Horti Bergiani*, **15**, 285–388.

Florin, R. (1954) The female reproductive organs of conifers and taxads. *Biological Reviews*, **29**, 367–389.

Floyd, S.K. & Bowman, J.L. (2007) The ancestral developmental tool kit of land plants. *International Journal of Plant Sciences*, **168**, 1–35.

Friedman, W.E. (1990) Double fertilization in *Ephedra*, a nonflowering seed plant: its bearing on the origin of angiosperms. *Science*, **247**, 951–954.

Friedman, W.E. (1993) The evolutionary history of the seed plant male gametophyte. *Trends in Ecology and Evolution*, **8**, 15–20.

Friedman, W.E. (1994) The evolution of embryogeny in seed plants and the developmental origin and early history of endosperm. *American Journal of Botany*, **81**, 1468–1486.

Friedman, W.E. (2006) Embryological evidence for developmental lability during early angiosperm evolution. *Nature*, **441**, 337–340.

Friedman, W.E. & Carmichael, J.S. (1996) Double fertilization in Gnetales: implications for understanding reproductive diversification among seed plants. *International Journal of Plant Sciences*, **157** (Suppl. 6), S77–S94.

Friedman, W.E. & Floyd, S.K. (2001) The origin of flowering plants and their reproductive biology – a tale of two phylogenies. *Evolution*, **55**, 217–231.
Friedman, W.E., Gallup, W.N. & Williams, J.H. (2003) Female gametophyte development in *Kadsura*: implications for Schisandraceae, Austrobaileyales, and the early evolution of flowering plants. *International Journal of Plant Sciences*, **164** (Suppl. 5), S293–S305.
Friedman, W.E., Moore, R.C. & Purugganan, M.D. (2004) The evolution of plant development. *American Journal of Botany*, **91**, 1726–1741.
Friedman, W.E. & Ryerson, K.C. (2009) Reconstructing the ancestral female gametophyte of angiosperms: insights from *Amborella* and other ancient lineages of flowering plants. *American Journal of Botany*, **96**, 129–143.
Friedman, W.E. & Williams, J.H. (2004) Developmental evolution of the sexual process in ancient flowering plant lineages. *The Plant Cell*, **16** (Suppl. 2004), S119–S132.
Frohlich, M.W. & Chase, M.W. (2007) After a dozen years of progress the origin of angiosperms is still a great mystery. *Nature*, **450**, 1184–1189.
Galtier, J. (1988) Morphology and phylogenetic relationships of early pteridosperms. In: *Origin and Evolution of Gymnosperms* (ed. C.B. Beck). pp. 135–176. Columbia University Press, New York.
Galtier, J. (2010) The origins and early evolution of the megaphyllous leaf. *International Journal of Plant Sciences*, **171**, 641–661.
Galtier, J. & Holmes, J.C. (1982) New observations on the branching of Carboniferous ferns and pteridosperms. *Annals of Botany*, **49**, 737–746.
Gaussen, H. (1946) *Les Gymnospermes, actuelles et fossiles. Travaux du Laboratoire Forestier de Toulouse*, Tome II, Vol. 1, Fasc. 3, Ch. 5. pp. 1–26. Faculté des Sciences, Toulouse.
Gauthier, J., Kluge, A.G. & Rowe, T. (1988) Amniote phylogeny and the importance of fossils. *Cladistics*, **4**, 105–209.
Gensel, P.G. (1984) A new Lower Devonian plant and the early evolution of leaves. *Nature*, **309**, 785–787.
Gensel, P.G., Kotyk, M.E. & Basinger, J.F. (2001) Morphology of above- and belowground structures in Early Devonian (Pragian–Emsian) plants. In: *Plants Invade the Land: Evolutionary and Environmental Perspectives* (eds P.G. Gensel & D. Edwards). pp. 83–102. Columbia University Press, New York.
Gifford, E.M. & Foster, A.S. (1989) *Morphology and Evolution of Vascular Plants*. W.H. Freeman, New York.
Goremykin, V.V., Hirsch-Ernst, K.I., Wölfl, S., *et al.* (2003) Analysis of the *Amborella trichopoda* chloroplast genome sequence suggests that *Amborella* is not a basal angiosperm. *Molecular Biology and Evolution*, **20**, 1499–1505.
Goremykin, V.V., Viola, R. & Hellwig, F.H. (2009) Removal of noisy characters from chloroplast genome-scale data suggests revision of phylogenetic placement of *Amborella* and *Ceratophyllum*. *Journal of Molecular Evolution*, **68**, 197–204.
Graham, L.E. (1993) *Origin of Land Plants*. John Wiley & Sons, New York.
Graham, L.E. & Wilcox, L.W. (1983) The occurrence and phylogenetic significance of putative placental transfer cells in the green alga *Coleochaete*. *American Journal of Botany*, **70**, 113–120.
Graham, L.E., Wilcox, L.W., Cook, M.E., *et al.* (2004) Resistant tissues of modern marchantioid liverworts resemble enigmatic Early Paleozoic microfossils. *Proceedings of the National Academy of Sciences of the United States of America*, **101**, 11025–11029.

Graham, S.W. & Olmstead, R.G. (2000) Utility of 17 chloroplast genes for inferring the phylogeny of the basal angiosperms. *American Journal of Botany*, **87**, 1712–1730.

Gray, J. (1985) The microfossil record of early land plants: advances in understanding of early terrestrialization, 1970-1984. *Philosophical Transactions of the Royal Society of London, Series B*, **309**, 167–195.

Hagemann, W. (1999) Towards an organismic concept of land plants: the marginal blastozone and the development of the vegetation body of selected frondose gametophytes of liverworts and ferns. *Plant Systematics and Evolution*, **216**, 81–133.

Haig, D. (2008) Homologous versus antithetic alternation of generations and the origin of sporophytes. *Botanical Review*, **74**, 395–418.

Hajibabaei, M., Xia, J. & Drouin, G. (2006) Seed plant phylogeny: gnetophytes are derived conifers and a sister group to Pinaceae. *Molecular Phylogenetics and Evolution*, **40**, 208–217.

Hamby, R.K. & Zimmer, E.A. (1992) Ribosomal RNA as a phylogenetic tool in plant systematics. In: *Molecular Systematics of Plants* (eds P.S. Soltis, D.E. Soltis & J.J. Doyle). pp. 50–91. Chapman and Hall, New York.

Harris, T.M. (1937) The fossil flora of Scoresby Sound East Greenland. Part 5: Stratigraphic relations of the plant beds. *Meddelelser om Grønland*, **112**(2), 1–114.

Harris, T.M. (1940) *Caytonia*. *Annals of Botany*, **4**, 713–734.

Harris, T.M. (1951) The relationships of the Caytoniales. *Phytomorphology*, **1**, 29–39.

Hébant, C. (1977) *The Conducting Tissues of Bryophytes*. J. Cramer, Vaduz.

Hennig, W. (1966) *Phylogenetic Systematics*. University of Illinois Press, Urbana, IL.

Hermsen, E.J. & Hendricks, J.R. (2008) W(h)ither fossils? Studying morphological character evolution in the age of molecular sequences. *Annals of the Missouri Botanical Garden*, **95**, 72–100.

Hernandez-Castillo, G.R., Rothwell, G.W. & Mapes, G. (2001) Thucydiaceae fam. nov., with a review and reevaluation of Paleozoic walchian conifers. *International Journal of Plant Sciences*, **162**, 1155–1185.

Hesse, M. (2001) Pollen characters of *Amborella trichopoda* (Amborellaceae): a reinvestigation. *International Journal of Plant Sciences*, **162**, 201–208.

Hilton, J. & Bateman, R.M. (2006) Pteridosperms are the backbone of seed plant evolution. *Journal of the Torrey Botanical Society*, **133**, 119–168.

Hofmeister, W. (1862) *On the Germination, Development and Fructification of the Higher Cryptogamia and on the Fructification of the Coniferae*. Ray Society, London.

Huynh, K.-L. (1976) L'arrangement du pollen du genre *Schisandra* (Schisandraceae) et sa signification phylogénique chez les Angiospermes. *Beiträge zur Biologie der Pflanzen*, **52**, 227–253.

Igersheim, A. & Endress, P.K. (1997) Gynoecium diversity and systematics of the Magnoliales and winteroids. *Botanical Journal of the Linnean Society*, **124**, 213–271.

Imaichi, R., Kato, M. & Okada, H. (1995) Morphology of the outer integument in three primitive angiosperm families. *Canadian Journal of Botany*, **73**, 1242–1249.

Jansen, R.K., Cai, Z., Raubeson, L.A., *et al.* (2007) Analysis of 81 genes from 64 plastid genomes resolves relationships in angiosperms and identifies genome-scale evolutionary patterns. *Proceedings of the National Academy of Sciences of the United States of America*, **104**, 19369–19374.

Judd, W.S., Campbell, C.S., Kellogg, E.A., *et al.* (2008) *Plant Systematics: a Phylogenetic Approach*, 3rd edn. Sinauer Associates, Sunderland, MA.

Kaplan, D.R. (1973) The problem of leaf morphology and evolution in the monocotyledons. *Quarterly Review of Biology*, **48**, 437–457.

Kaplan, D.R. (1977) Morphological status of the shoot systems of Psilotaceae. *Brittonia*, **29**, 30–53.
Kaplan, D.R. (2001) The science of plant morphology: definition, history, and role in modern biology. *American Journal of Botany*, **88**, 1711–1741.
Karol, K.G., McCourt, R.M., Cimino, M.T., *et al.* (2001) The closest living relatives of land plants. *Science*, **294**, 2351–2353.
Keeling, P.J. (2004) Diversity and evolutionary history of plastids and their hosts. *American Journal of Botany*, **91**, 1481–1493.
Kenrick, P. (1994) Alternation of generations in land plants: new phylogenetic and palaeobotanical evidence. *Biological Reviews*, **69**, 293–330.
Kenrick, P. (2002) The origin of roots. In: *Plant Roots: the Hidden Half* (eds Y. Waisel, A. Eshel & U. Kafkafi), 3rd edn. pp. 1–13. Marcel Decker, Basel.
Kenrick, P. & Crane, P.R. (1991) Water-conducting cells in early fossil land plants: implications for the early evolution of tracheophytes. *Botanical Gazette*, **152**, 335–356.
Kenrick, P. & Crane, P.R. (1997) *The Origin and Early Diversification of Land Plants: a Cladistic Study*. Smithsonian Institution, Washington, DC.
Kidston, R. & Lang, W.H. (1917) On Old Red Sandstone plants showing structure, from the Rhynie Chert Bed, Aberdeenshire. Part I. *Rhynia Gwynne-Vaughani* Kidston & Lang. *Transactions of the Royal Society of Edinburgh*, **51**, 761–784.
Kidston, R. & Lang, W.H. (1920) On Old Red Sandstone plants showing structure, from the Rhynie Chert Bed, Aberdeenshire. Part II. Additional notes on *Rhynia Gwynne-Vaughani*, Kidston & Lang; with descriptions of *Rhynia major*, n. sp. and *Hornea lignieri*, n. g., n. sp. *Transactions of the Royal Society of Edinburgh*, **52**, 603–627.
Leinfellner, W. (1969) Über die Karpelle verschiedener Magnoliales. VIII. Überblick über alle Familien der Ordnung. *Österreichische Botanische Zeitschrift*, **117**, 107–127.
Lemieux, C., Otis, C. & Turmel, M. (2007) A clade uniting the green algae *Mesostigma viride* and *Chlorokybus atmophyticus* represents the deepest branch of the Streptophyta in chloroplast genome-based phylogenies. *BMC Biology*, **5**, 2.
Leslie, A.B. (2010) Flotation preferentially selects saccate pollen during conifer pollination. *New Phytologist*, **188**, 273–279.
Lewis, L.A. & McCourt, R.M. (2004) Green algae and the origin of land plants. *American Journal of Botany*, **91**, 1535–1556.
Loconte, H. & Stevenson, D.W. (1991) Cladistics of the Magnoliidae. *Cladistics*, **7**, 267–296.
Long, A.G. (1966) Some Lower Carboniferous fructifications from Berwickshire, together with a theoretical account of the evolution of ovules, cupules and carpels. *Transactions of the Royal Society of Edinburgh*, **66**, 345–375.
Long, A.G. (1975) Further observations on some Lower Carboniferous seeds and cupules. *Transactions of the Royal Society of Edinburgh*, **69**, 267–293.
Lu, P. (1996) Meristem development and organ formation in *Selaginella martensii*. PhD thesis, University of California, Davis, CA.
Lucas, W.J., Ding, B. & Van der Schoot, C. (1993) Plasmodesmata and the supracellular nature of plants. *New Phytologist*, **125**, 435–476.
Magallón, S. & Sanderson, M.J. (2002) Relationships among seed plants inferred from highly conserved genes: sorting conflicting phylogenetic signals among ancient lineages. *American Journal of Botany*, **89**, 1991–2006.
Manhart, J.R. (1995) Chloroplast 16S rDNA sequences and phylogenetic relationships of fern allies and ferns. *American Fern Journal*, **85**, 182–192.

Manos, P.S., Soltis, P.S., Soltis, D.E., *et al.* (2007) Phylogeny of extant and fossil Juglandaceae inferred from the integration of molecular and morphological data sets. *Systematic Biology*, **56**, 412–430.

Mapes, G. & Rothwell, G.W. (1984) Permineralized ovulate cones of *Lebachia* from late Palaeozoic limestones of Kansas. *Palaeontology*, **27**, 69–94.

Mapes, G., Rothwell, G.W. & Haworth, M.T. (1989) Evolution of seed dormancy. *Nature*, **337**, 645–646.

Mathews, S. (2009) Phylogenetic relationships among seed plants: persistent questions and the limits of molecular data. *American Journal of Botany*, **96**, 228–236.

Mathews, S., Clements, M.D. & Beilstein, M.A. (2010) A duplicate gene rooting of seed plants and the phylogenetic position of flowering plants. *Philosophical Transactions of the Royal Society, Series B*, **365**, 383–395.

Mathews, S. & Donoghue, M.J. (1999) The root of angiosperm phylogeny inferred from duplicate phytochrome genes. *Science*, **286**, 947–950.

McAbee, J.M., Hill, T.A., Skinner, D.J., *et al.* (2006) *ABERRANT TESTA SHAPE* encodes a KANADI family member, linking polarity determination to separation and growth of Arabidopsis ovule integuments. *The Plant Journal*, **46**, 522–531.

Meister, R.J., Kotow, L.M. & Gasser, C.S. (2002) SUPERMAN attenuates positive *INNER NO OUTER* autoregulation to maintain polar development of *Arabidopsis* ovule outer integuments. *Development*, **129**, 4281–4289.

Melkonian, M. (1989) Flagellar apparatus ultrastructure in *Mesostigma viride* (*Prasinophyceae*). *Plant Systematics and Evolution*, **164**, 93–122.

Mishler, B.D. & Churchill, S.P. (1984) A cladistic approach to the phylogeny of the "bryophytes." *Brittonia*, **36**, 406–424.

Mishler, B.D. & Churchill, S.P. (1985) Transition to a land flora: phylogenetic relationships of the green algae and bryophytes. *Cladistics*, **1**, 305–328.

Moore, M.J., Bell, C.D., Soltis, P.S., *et al.* (2007) Using plastid genome-scale data to resolve enigmatic relationships among basal angiosperms. *Proceedings of the National Academy of Sciences of the United States of America*, **104**, 19363–19368.

Mundry, M., Stützel, T. (2004) Morphogenesis of the reproductive shoots of *Welwitschia mirabilis* and *Ephedra distachya* (Gnetales), and its evolutionary implications. *Organisms, Diversity & Evolution*, **4**, 91–108.

Nishida, H., Pigg, K.B., Kudo, K., *et al.* (2004) Zooidogamy in the Late Permian genus *Glossopteris*. *Journal of Plant Research*, **117**, 323–328.

Nixon, K.C., Crepet, W.L., Stevenson, D., *et al.* (1994) A reevaluation of seed plant phylogeny. *Annals of the Missouri Botanical Garden*, **81**, 484–533.

Oliver, F.W. & Scott, D.W. (1903) On *Lagenostoma Lomaxi*, the seed of *Lyginodendron*. *Proceedings of the Royal Society of London*, **71**, 477–481.

Parkinson, C.L., Adams, K.L. & Palmer, J.D. (1999) Multigene analyses identify the three earliest lineages of extant flowering plants. *Current Biology*, **9**, 1485–1488.

Patterson, C. (1981) Significance of fossils in determining evolutionary relationships. *Annual Review of Ecology and Systematics*, **12**, 195–223.

Pettitt, J.M. & Beck, C.B. (1968) *Archaeosperma arnoldii* – a cupulate seed from the Upper Devonian of New York. *Contributions from the Museum of Paleontology, University of Michigan*, **22**, 139–154.

Phillips, T.L. & Galtier, J. (2005) Evolutionary and ecological perspectives of Late Paleozoic ferns Part I. Zygopteridales. *Review of Palaeobotany and Palynology*, **135**, 165–203.

Pickett-Heaps, J.D. (1969) The evolution of the mitotic apparatus: an attempt at comparative ultrastructural cytology in dividing plant cells. *Cytobios*, **1**, 257–280.

Pickett-Heaps, J.D. (1972) Variation in mitosis and cytokinesis in plant cells: its significance in the phylogeny and evolution of ultrastructural systems. *Cytobios*, **5**, 59–77.

Pickett-Heaps, J.D. (1975) *Green Algae*. Sinauer Associates, Sunderland, MA.

Pickett-Heaps, J.D. (1979) Electron microscopy and the phylogeny of green algae and land plants. *American Zoologist*, **19**, 545–554.

Pirozynski, K.A. & Malloch, D.W. (1975) The origin of land plants: a matter of mycotrophism. *BioSystems*, **6**, 153–164.

Poort, R.J., Visscher, H. & Dilcher, D.L. (1996) Zoidogamy in fossil gymnosperms: the centenary of a concept, with special reference to prepollen of late Paleozoic conifers. *Proceedings of the National Academy of Sciences of the United States of America*, **93**, 11713–11717.

Pryer, K.M., Schneider, H., Smith, A.R., *et al.* (2001) Horsetails and ferns are a monophyletic group and the closest living relatives to seed plants. *Nature*, **409**, 618–622.

Pryer, K.M., Smith, A.R. & Skog, J.E. (1995) Phylogenetic relationships of extant ferns based on evidence from morphology and *rbcL* sequences. *American Fern Journal*, **85**, 205–282.

Qiu, Y.-L. (2008) Phylogeny and evolution of charophytic algae and land plants. *Journal of Systematics and Evolution*, **46**, 287–306.

Qiu, Y.-L., Cho, Y., Cox, J.C., *et al.* (1998) The gain of three mitochondrial introns identifies liverworts as the earliest land plants. *Nature*, **394**, 671–674.

Qiu, Y.-L., Lee, J., Bernasconi-Quadroni, F., *et al.* (1999) The earliest angiosperms: evidence from mitochondrial, plastid and nuclear genomes. *Nature*, **402**, 404–407.

Qiu, Y.-L., Lee, J., Whitlock, B.A., *et al.* (2001) Was the ANITA rooting of the angiosperm phylogeny affected by long-branch attraction? *Molecular Biology and Evolution*, **18**, 1745–1753.

Qiu, Y.-L., Li, L., Hendry, T.A., *et al.* (2006a) Reconstructing the basal angiosperm phylogeny: evaluating information content of mitochondrial genes. *Taxon*, **55**, 837–856.

Qiu, Y.-L., Li, L., Wang, B., *et al.* (2006b) The deepest divergences in land plants inferred from phylogenomic evidence. *Proceedings of the National Academy of Sciences of the United States of America*, **103**, 15511–15516.

Qiu, Y.-L., Li, L., Wang, B., *et al.* (2007) A nonflowering land plant phylogeny inferred from nucleotide sequences of seven chloroplast, mitochondrial, and nuclear genes. *International Journal of Plant Sciences*, **168**, 691–708.

Qiu, Y.-L., Li, L., Wang, B., *et al.* (2010) Angiosperm phylogeny inferred from sequences of four mitochondrial genes. *Journal of Systematics and Evolution*, **48**, 391–425.

Rai, H.S., Reeves, P.A., Peakall, R., *et al.* (2008) Inference of higher-order conifer relationships from a multi-locus plastid data set. *Botany*, **86**, 658–669.

Raubeson, L.A. & Jansen, R.K. (1992) Chloroplast DNA evidence on the ancient evolutionary split in vascular land plants. *Science*, **255**, 1697–1699.

Remy, W. (1982) Lower Devonian gametophytes: relation to the phylogeny of land plants. *Science*, **215**, 1625–1627.

Remy, W., Gensel, P.G. & Hass, H. (1993) The gametophyte generation in some Early Devonian land plants. *International Journal of Plant Sciences*, **154**, 35–58.

Remy, W. & Remy, R. (1980) Devonian gametophytes with anatomically preserved gametangia. *Science*, **208**, 295–296.

Retallack, G. & Dilcher, D.L. (1981) Arguments for a glossopterid ancestry of angiosperms. *Paleobiology*, **7**, 54–67.

Rogers, C.E., Domozych, D.S., Stewart, K.D., *et al.* (1981) The flagellar apparatus of *Mesostigma viride* (*Prasinophyceae*)—multilayered structures in a scaly green flagellate. *Plant Systematics and Evolution*, **138**, 247–258.

Ronse De Craene, L.P. (2004) Floral development of *Berberidopsis corallina*: a crucial link in the evolution of flowers in the core eudicots. *Annals of Botany*, **94**, 741–751.

Ronse De Craene, L.P. (2007) Are petals sterile stamens or bracts? The origin and evolution of petals in the core eudicots. *Annals of Botany*, **100**, 621–630.

Ronse De Craene, L.P. (2008) Homology and evolution of petals in the core eudicots. *Systematic Botany*, **33**, 301–325.

Ronse De Craene, L.P. & Stuppy, W. (2010) Floral development and anatomy of *Aextoxicon punctatum* (Aextoxicaceae-Berberidopsidales): an enigmatic tree at the base of core eudicots. *International Journal of Plant Sciences*, **171**, 244–257.

Rothwell, G.W. (1972) Evidence of pollen tubes in Paleozoic pteridosperms. *Science*, **175**, 772–774.

Rothwell, G.W. (1981) The Callistophytales (Pteridospermopsida): reproductively sophisticated Paleozoic gymnosperms. *Review of Palaeobotany and Palynology*, **32**, 103–121.

Rothwell, G.W. (1982) New interpretations of the earliest conifers. *Review of Palaeobotany and Palynology*, **37**, 7–28.

Rothwell, G.W. (1988) Cordaitales. In: *Origin and Evolution of Gymnosperms* (ed. C.B. Beck). pp. 273–297. Columbia University Press, New York.

Rothwell, G.W., Crepet, W.L., Stockey, R.A. (2009) Is the anthophyte hypothesis alive and well? New evidence from the reproductive structures of Bennettitales. *American Journal of Botany*, **96**, 296–322.

Rothwell, G.W. & Karrfalt, E.E. (2008) Growth, development, and systematics of ferns: does *Botrychium* s.l. (Ophioglossales) really produce secondary xylem? *American Journal of Botany*, **95**, 414–423.

Rothwell, G.W. & Nixon, K.C. (2006) How does the inclusion of fossil data change our conclusions about the phylogenetic history of euphyllophytes? *International Journal of Plant Sciences*, **167**, 737–749.

Rothwell, G.W. & Scheckler, S.E. (1988) Biology of ancestral gymnosperms. In: *Origin and Evolution of Gymnosperms* (ed. C.B. Beck). pp. 85–134. Columbia University Press, New York.

Rothwell, G.W. & Serbet, R. (1994) Lignophyte phylogeny and the evolution of spermatophytes: a numerical cladistic analysis. *Systematic Botany*, **19**, 443–482.

Rothwell, G.W. & Stockey, R.A. (1989) Fossil Ophioglossales in the Paleocene of western North America. *American Journal of Botany*, **76**, 637–644.

Rudall, P.J. & Bateman, R.M. (2010) Defining the limits of flowers: the challenge of distinguishing between the evolutionary products of simple versus compound strobili. *Philosophical Transactions of the Royal Society, Series B*, **365**, 397–409.

Rudall, P.J. & Buzgo, M. (2002) Evolutionary history of the monocot leaf. In: *Developmental Genetics and Plant Evolution* (eds Q.C.B. Cronk, R.M. Bateman & J.A. Hawkins). pp. 431–458. Taylor & Francis, London.

Rydin, C., Källersjö, M. & Friis, E.M. (2002) Seed plant relationships and the systematic position of Gnetales based on nuclear and chloroplast DNA: conflicting data, rooting problems, and the monophyly of conifers. *International Journal of Plant Sciences*, **163**, 197–214.

Saarela, J.M., Rai, H.S., Doyle, J.A., *et al.* (2007) Hydatellaceae identified as a new branch near the base of the angiosperm phylogenetic tree. *Nature*, **446**, 312–315.

Sampson, F.B. (1993) Pollen morphology of the Amborellaceae and Hortoniaceae (Hortonioideae: Monimiaceae). *Grana*, **32**, 154–162.

Sanderson, M.J., Wojciechowski, M.F., Hu, J.-M., *et al.* (2000) Error, bias, and long-branch attraction in data for two chloroplast photosystem genes in seed plants. *Molecular Biology and Evolution*, **17**, 782–797.

Sarojam, R., Sappl, P.G., Goldshmidt, A., *et al.* (2010) Differentiating Arabidopsis shoots from leaves by combined YABBY activities. *The Plant Cell*, **22**, 2113–2130.

Scheckler, S.E. (1974) Systematic characters in Devonian ferns. *Annals of the Missouri Botanical Garden*, **61**, 462–473.

Scheckler, S.E. & Banks, H.P. (1971a) Anatomy and relationships of some Devonian progymnosperms from New York. *American Journal of Botany*, **58**, 737–751.

Scheckler, S.E. & Banks, H.P. (1971b) *Proteokalon* a new genus of progymnosperms from the Devonian of New York state and its bearing on phylogenetic trends in the group. *American Journal of Botany*, **58**, 874–884.

Schneider, H., Pryer, K.M., Cranfill, R., *et al.* (2002) Evolution of vascular plant body plans: a phylogenetic perspective. In: *Developmental Genetics and Plant Evolution* (eds Q.C.B. Cronk, R.M. Bateman & J.A. Hawkins). pp. 330–364. Taylor & Francis, London.

Schweitzer, H.-J. & Kirchner, M. (1998) Die rhäto-jurassischen Floren des Iran und Afghanistans. 11. Pteridospermophyta und Cycadophyta I. Cycadales. *Palaeontographica Abteilung B*, **248**, 1–85.

Scotland, R.W., Olmstead, R.G. & Bennett, J.R. (2003) Phylogeny reconstruction: the role of morphology. *Systematic Biology*, **52**, 539–548.

Scott, D.H. (1895) Nathanael Pringsheim. *Nature*, **51**, 399–402.

Serbet, R. & Rothwell, G.W. (1992) Characterizing the most primitive seed ferns. I. A reconstruction of *Elkinsia polymorpha*. *International Journal of Plant Sciences*, **153**, 602–621.

Shaw, J., Renzaglia, K. (2004) Phylogeny and diversification of bryophytes. *American Journal of Botany*, **91**, 1557–1581.

Skinner, D.J., Hill, T.A. & Gasser, C.S. (2004) Regulation of ovule development. *The Plant Cell*, **16** (Suppl. 2004), S32–S45.

Smith, G.M. (1955) *Cryptogamic Botany: Bryophytes and Pteridophytes*, Vol. 2., 2nd edn. McGraw-Hill, New York.

Sokoloff, D.D., Remizowa, M.V., Macfarlane, T.D., *et al.* (2008) Seedling diversity in Hydatellaceae: implications for the evolution of angiosperm cotyledons. *Annals of Botany*, **101**, 153–164.

Soltis, D.E., Albert, V.A., Savolainen, V., *et al.* (2004) Genome-scale data, angiosperm relationships, and 'ending incongruence': a cautionary tale in phylogenetics. *Trends in Plant Science*, **9**, 477–483.

Soltis, D.E., Soltis, P.S., Chase, M.W., *et al.* (2000) Angiosperm phylogeny inferred from 18S rDNA, *rbcL*, and *atpB* sequences. *Botanical Journal of the Linnean Society*, **133**, 381–461.

Soltis, D.E., Soltis, P.S. & Zanis, M.J. (2002) Phylogeny of seed plants based on evidence from eight genes. *American Journal of Botany*, **89**, 1670–1681.

Soltis, P.S., Soltis, D.E. & Chase, M.W. (1999) Angiosperm phylogeny inferred from multiple genes as a tool for comparative biology. *Nature*, **402**, 402–404.

Soria, A. & Meyer-Berthaud, B. (2003) Occurrence of whorled organotaxis in the cladoxylopsid *Pietzschia polyupsilon* Read and Campbell (Lower Carboniferous, USA). *Review of Palaeobotany and Palynology*, **124**, 29–49.

Springer, M.S., Teeling, E.C., Madsen, O., *et al.* (2001) Integrated fossil and molecular data reconstruct bat echolocation. *Proceedings of the National Academy of Sciences of the United States of America*, **98**, 6241–6246.

Stebbins, G.L. (1974) *Flowering Plants: Evolution above the Species Level.* Harvard University Press, Cambridge, MA.

Steemans, P., Le Hérissé, A., Melvin, J., *et al.* (2009) Origin and radiation of the earliest vascular land plants. *Science*, **324**, 353.

Stefanovic, S., Jager, M., Deutsch, J., *et al.* (1998) Phylogenetic relationships of conifers inferred from partial 28S rRNA gene sequences. *American Journal of Botany*, **85**, 688–697.

Stein, W.E. (1981) Reinvestigation of *Arachnoxylon kopfii* from the Middle Devonian of New York State, U.S.A. *Palaeontographica Abteilung B*, **147**, 90–117.

Stein, W.E. & Boyer, J.S. (2006) Evolution of land plant architecture: beyond the telome theory. *Paleobiology*, **32**, 450–482.

Stein, W.E., Wight, D.C. & Beck, C.B. (1984) Possible alternatives for the origin of Sphenopsida. *Systematic Botany*, **9**, 102–118.

Stevenson, D.W. (1990) Morphology and systematics of the Cycadales. *Memoirs of the New York Botanical Garden*, **57**, 8–55.

Stewart, K.D. & Mattox, K.R. (1975) Comparative cytology, evolution and classification of the green algae with some consideration of the origin of other organisms with chlorophylls a and b. *Botanical Review*, **41**, 104–135.

Stewart, W.N. (1951) A new *Pachytesta* from the Berryville locality of southeastern Illinois. *American Midland Naturalist*, **46**, 717–742.

Stewart, W.N. & Rothwell, G.W. (1993) *Paleobotany and the Evolution of Plants*, 2nd edn. Cambridge University Press, Cambridge.

Stockey, R.A. & Rothwell, G.W. (2003) Anatomically preserved *Williamsonia* (Williamsoniaceae): evidence for bennettitalean reproduction in the Late Cretaceous of western North America. *International Journal of Plant Sciences*, **164**, 251–262.

Strasburger, E. (1894) The periodic reduction of the number of the chromosomes in the life-history of living organisms. *Annals of Botany*, **8**, 281–316.

Takhtajan, A.L. (1969) *Flowering Plants: Origin and Dispersal*. Smithsonian Institution, Washington, DC.

Taylor, T.N. (1988) Pollen and pollen organs of fossil gymnosperms: phylogeny and reproductive biology. In: *Origin and Evolution of Gymnosperms* (ed. C.B. Beck). pp. 177–217. Columbia University Press, New York.

Taylor, T.N., Taylor, E.L. & Krings, M. (2009) *Paleobotany: the Biology and Evolution of Fossil Plants*, 2nd edn. Academic Press, London.

Taylor, W.A. & Strother, P.K. (2008) Ultrastructure of some Cambrian palynomorphs from the Bright Angel Shale, Arizona, USA. *Review of Palaeobotany and Palynology*, **151**, 41–50.

Thomas, H.H. (1925) The Caytoniales, a new group of angiospermous plants from the Jurassic rocks of Yorkshire. *Philosophical Transactions of the Royal Society of London, Series B*, **213**, 299–363.

Tomescu, A.M.F. (2009) Megaphylls, microphylls and the evolution of leaf development. *Trends in Plant Science*, **14**, 5–12.

Turmel, M., Otis, C. & Lemieux, C. (2006) The chloroplast genome sequence of *Chara vulgaris* sheds new light into the closest green algal relatives of land plants. *Molecular Biology and Evolution*, **23**, 1324–1338.

Umeda, A., Imaichi, R. & Kato, M. (1994) Ovular development and morphology of the outer integument of *Magnolia grandiflora* (Magnoliaceae). *American Journal of Botany*, **81**, 361–367.

Van Campo, M. & Lugardon, B. (1973) Structure grenue infratectale de l'ectexine des pollens de quelques Gymnospermes et Angiospermes. *Pollen et Spores*, **15**, 171–187.

van Heel, W.A. (1981) A S.E.M.-investigation on the development of free carpels. *Blumea*, **27**, 499–522.

Walker, J.W. (1976) Evolutionary significance of the exine in the pollen of primitive angiosperms. In: *The Evolutionary Significance of the Exine* (eds I.K. Ferguson & J. Muller). pp. 251–308. Academic Press, London.

Walker, J.W. & Skvarla, J.J. (1975) Primitively columellaless pollen: a new concept in the evolutionary morphology of angiosperms. *Science*, **187**, 445–447.

Walton, J. (1931) Contributions to the knowledge of Lower Carboniferous plants III. On the fossil-flora of the Black Limestones in Teilia Quarry, Gwaenysgor, near Prestatyn, Flintshire, with special reference to *Diplopteridium teilianum* Kidston sp. (gen. nov.) and some other fern-like fronds. *Philosophical Transactions of the Royal Society, Series B*, **219**, 347–379.

Wang, B., Yeun, L.H., Xue, J.-Y., *et al.* (2010) Presence of three mycorrhizal genes in the common ancestor of land plants suggests a key role of mycorrhizas in the colonization of land by plants. *New Phytologist*, **186**, 514–525.

Wang, X. (2010) Axial nature of the cupule-bearing organ in Caytoniales. *Journal of Systematics and Evolution*, **48**, 207–214.

Wanntorp, L. & Ronse De Craene, L.P. (2005) The *Gunnera* flower: key to eudicot diversification or response to pollination mode? *International Journal of Plant Sciences*, **166**, 945–953.

Wellman, C.H., Osterloff, P.L. & Mohiuddin, U. (2003) Fragments of the earliest land plants. *Nature*, **425**, 282–284.

Wiens, J.J. (2004) The role of morphological data in phylogeny reconstruction. *Systematic Biology*, **53**, 653–661.

Williams, J.H. & Friedman, W.E. (2002) Identification of diploid endosperm in an early angiosperm lineage. *Nature*, **415**, 522–526.

Williams, J.H. & Friedman, W.E. (2004) The four-celled female gametophyte of *Illicium* (Illiciaceae; Austrobaileyales): implications for understanding the origin and early evolution of monocots, eumagnoliids, and eudicots. *American Journal of Botany*, **91**, 332–351.

Wodniok, S., Brinkmann, H., Glöckner, G., *et al.* (2011) Origin of land plants: do conjugating green algae hold the key? *BMC Evolutionary Biology*, **11**, 103.

Worsdell, W.C. (1900) The structure of the female 'flower' in Coniferae. An historical study. *Annals of Botany*, **14**, 39–812.

Yamada, T., Imaichi, R. & Kato, M. (2001a) Developmental morphology of ovules and seeds of Nymphaeales. *American Journal of Botany*, **88**, 963–974.

Yamada, T., Imaichi, R., Prakash, N., *et al.* (2003) Developmental morphology of ovules and seeds of Austrobaileyales. *Australian Journal of Botany*, **51**, 555–564.

Yamada, T., Tobe, H., Imaichi, R., *et al.* (2001b) Developmental morphology of the ovules of *Amborella trichopoda* (Amborellaceae) and *Chloranthus serratus* (Chloranthaceae). *Botanical Journal of the Linnean Society*, **137**, 277–290.

Yoshida, O. (1960) Embryologische Studien über *Schisandra sinensis* Bailey. *Journal of the College of Arts and Sciences, Chiba University, Natural Sciences*, **3**, 459–462.
Zanis, M.J., Soltis, D.E., Soltis, P.S., *et al.* (2002) The root of the angiosperms revisited. *Proceedings of the National Academy of Sciences of the United States of America*, **99**, 6848–6853.
Zhou, Z. & Zhang, B. (1992) *Baiera hallei* Sze and associated ovule-bearing organs from the Middle Jurassic of Henan, China. *Palaeontographica Abteilung B*, **224**, 151–169.
Zhou, Z.-Y. (2009) An overview of fossil Ginkgoales. *Palaeoworld*, **18**, 1–22.
Zimmermann, W. (1930) *Die Phylogenie der Pflanzen*. Fischer, Jena.
Zimmermann, W. (1931) Arbeitsweise der botanischen Phylogenetik und anderer Gruppierungswissenschaften. In: *Handbuch der Biologischen Arbeitsmethoden*, Abt. 3, Teil 9 (ed. E. Abderhalden). pp. 941–1053. Urban & Schwarzenberg, Berlin.
Zimmermann, W. (1952) Main results of the "Telome Theory." *The Palaeobotanist*, **1**, 456–470.

Annual Plant Reviews (2013) **45**, 51–90 http://onlinelibrary.wiley.com
doi: 10.1002/9781118305881.ch2

Chapter 2

THE EVOLUTION OF BODY FORM IN BRYOPHYTES

Bernard Goffinet[1] and William R. Buck[2]

[1] *Department of Ecology and Evolutionary Biology, University of Connecticut, Storrs, CT, USA*
[2] *The New York Botanical Garden, Bronx, NY, USA*

Abstract: Bryophytes compose a paraphyletic group of land plants that mark the successful transition to land and paved the way for the evolution of plants with highly differentiated vascular tissues and multiple sporangia. Like all embryophytes, bryophytes follow a life cycle of alternating haploid and diploid generations, and hence, rely on the same fundamental set of genetic tools to develop their body. The vegetative plant of bryophytes is always the sex-bearing gametophyte, and upon fertilization, the embryo develops into a short-lived, unbranched sporophyte that remains attached to the maternal plant. Over the last 400 million years, bryophytes have undergone continuous and sometimes rapid diversification, leading to nearly 20 000 species exhibiting a broad range of vegetative architecture based on a simple modular Bauplan and critical innovations enhancing spore dispersal through time and space, and thereby insuring their persistence and ecological success.

Keywords: bryophytes; gametophyte; sporophyte; sporangium; reproduction; branching; apical cell; modular growth; peristome.

All land plants share a life cycle composed of two multicellular phases, one bearing the sex organs (the haploid gametophyte) and the other producing the spores (the diploid sporophyte). The sporophyte develops from a zygote that is always retained within the female sex organ, and is never shed: it is physically and physiologically dependent on the maternal gametophyte if only during the initial stages of its development. Most of today's vegetation types are dominated by plants characterized by sporophytes that are independent at maturity, branched and producing numerous sporangia, spores that undergo internal divisions (i.e., endospory), and megaspores that are retained, protected, and nourished by the sporophyte. The gametophytes are

Annual Plant Reviews Volume 45: The Evolution of Plant Form, First Edition.
Edited by Barbara A. Ambrose and Michael Purugganan.

highly reduced, no longer free-living or autotrophic, and the female gametophyte has become entirely dependent on the diploid phase, being enclosed in the megaspore. This extreme simplification of the female and male gametophytes, now reduced to eight and three cells, respectively, has seemingly led to a dead end, as their architecture is highly conserved among flowering plants.

By contrast, the sporophyte, which adopted the vegetative and—indirectly—the reproductive functions, has undergone significant modifications and adaptations in the architecture of its photosynthetic body and its sporangia, now confined to elaborate flowers. Five hundred million years ago, however, well before the origin of seeds and flowers, and even vascular tissues, plants exhibited complex gametophytic bodies and comparatively simple sporophytes. The direct descendants of these early land plants are represented today by the bryophytes, a heterogeneous group comprising liverworts, mosses, and hornworts.

To the evolutionary plant biologist, the significance of bryophytes is twofold: (1) phylogenetically they mark the transition from the algal ancestor of land plants to the vascular plants, and thereby the seed and flowering plants; and (2) morphologically and developmentally they may represent a stage in the gradual transformation from a gametophyte-dominant to a gametophyte-subordinate life cycle, and hence are essential to our understanding of the evolution of a complex independent sporophyte (Goffinet & Shaw 2009). The sporophyte of bryophytes being *merely* a diploid gametophyte, its specialization, beginning with the lack of actual sex organs and the initiation of sporogenous tissue, may result from the differential expression of specific genes, similar to the functional specialization of cells and tissues within a haploid or a diploid body. Furthermore, with the transfer of vegetative functions from the gametophyte to the sporophyte, identical genes may be required for conserved functions and account for the development of similar tissues or organs, such as leaves, in either phase of the life cycle. This thus leads to the concept of a fundamental set of genes necessary to develop a complex body, regardless of its ploidy level (Floyd & Bowman 2007). Observations of apogamy (in which the gametophyte develops sporangia in the absence of sexual reproduction) and apospory (in which the sporophyte grows vegetative plants; Figure 2.22) in some bryophytes confirm that the morphological and functional differentiation of the two generations, which share the same genome but differ in their ploidy level, is fundamentally guided simply by shifts in gene expression (reviewed in Goffinet *et al.* 2009; also Okano *et al.* 2009).

Application of such *simple* evolutionary processes would suggest that the functional homologies linking leaves, stems, and rhizoids in some bryophytes and vascular plants are dictated by genetic homologies with similar or identical sets of genes controlling the expression of these structures. The three lineages of bryophytes differ, however, significantly in the form of their photosynthetic body, and their precise phylogenetic relationships have remained controversial, although recent inferences suggest that they compose a paraphyletic group subtending the remainder of land plant lineages (Kenrick & Crane 1997; Qiu *et al.* 2006). Moreover, bryophytes are evolutionary dynamic lineages that have undergone significant transformations that have resulted in about 20 000 morphologically distinct extant species. Finally, the accurate reconstruction of the ancestral plant body for each lineage suffers from the ambiguity of the branching toward the root of their phylogenetic tree (e.g., in mosses), and the highly contrasting morphologies of lineages converging toward the root, two patterns suggestive of a rapid, potentially "adaptive" radiation. Hence, inferring homology relationships between structures developed by monosporangiophyte gametophyte-dominant bryophytes and their polysporangiophyte, sporophyte-dominant descendants, is, to say the least, challenging.

We will summarize the fundamental Bauplan of bryophytes, highlight their major differences, proceed to reviewing their phylogenetic relationships before detailing their morphological diversity, and remark on the major evolutionary trends that occur or remain to be elucidated. Ultimately, the recent progress in elucidating the genetic networks underlying the development of the plant body will be discussed.

2.1 Fundamental Bauplan of bryophytes

2.1.1 The apical meristem is unicellular and growth is modular

All bryophytes share a life cycle wherein the haploid gametophyte assumes the vegetative—photosynthetic and water absorptive—functions and the sporophyte is a single terminal sporangium. The gametophyte develops antheridia and archegonia, the male and female multicellular sex organs, respectively. The egg is always retained in the archegonium, where fertilization takes place. The resulting zygote develops into an embryo and ultimately an unbranched sporophyte anchored into and acquiring nutrients from the maternal gametophyte.

Growth of either the gametophyte or sporophyte is typically insured by the meristematic activity of a single apical cell, except for the liverwort and hornwort sporophyte (Crandall-Stotler 1980). Subsequent development is fundamentally metameric: the apical cell has generally several cutting faces and derivatives undergo parallel pathways, forming metamers composed of epidermal and cortical cells, and in some cases of leaf and branch initials. Metamers are assembled into modules. Unlike the sporophyte, the gametophyte is typically composed of multiple modules, with additional modules arising from the development of branch initials or through the splitting of the apical cell in hornworts and some thalloid liverworts. The vegetative body of bryophytes consists of a photosynthetic ribbon-like thallus or a leafy stem, and in either case, hair-like rhizoids anchor the plants to the substrate. Lignified vascular tissues are lacking, and stomata, when present, occur only in the sporangial wall, and may not be associated with photosynthetic tissues.

2.1.2 The architecture of the gametophyte varies within bryophytes (Figures 2.1, 2.2, 2.3, 2.4, 2.5, 2.6, 2.7, and 2.8)

The gametophyte of liverworts is either thalloid (i.e., ribbon-like) or leafy (i.e., a foliated axis), with the leaves being lobed or simple and never bearing a midrib (or costa). The mature gametophyte of mosses almost always develops leaves along a terete axis, and the leaves may bear a costa and are, except in *Takakia*, entire (i.e., undivided). The earliest ontogenetic stages may, however, be partially thalloid (i.e., *Sphagnum*; Figure 2.23), or include flattened appendages (e.g., *Andreaea* or *Tetraphis*). By contrast, hornworts are always thalloid, although their thallus can be divided into leaf-like appendages. Germination of the spore results in a small globose mass of cells that becomes flattened (Renzaglia 1978); except for rhizoids, a strictly filamentous stage is lacking. In liverworts the protonematal phase is short-filamentous, globose or cylindrical. As in hornworts, a single apical cell is differentiated and hence only one gametophore is produced per spore. By contrast, numerous buds may develop on the filamentous protonema of mosses (Figure 2.21) and hence each spore may give rise to many, genetically identical leafy gametophytes. Whether this innovation characterizes all mosses is not clear, as sporeling development is not known for *Takakia* and in *Sphagnum* multiple gametophores may occasionally arise from a single protonema. Finally, the filamentous stage of mosses typically comprises rhizoids, chloronema, and caulonema, which are lacking in liverworts and hornworts. These three types of filaments serve distinct functions: the green chloronema is the primary site for photosynthesis, the brown caulonema typically gives rise to the buds and hence to the leafy stems, whereas rhizoids anchor the plants (Duckett *et al.* 1998).

2.1.3 Bryophytes differ consistently in their sporophytes (Figures 2.9, 2.11, 2.12)

Liverworts, mosses, and hornworts differ in the architecture and ontogeny of their sporophytes, beginning with the first division of the zygote: transverse in the former two and longitudinal in hornworts. The sporophyte is always composed of two tissues, with the amphithecium surrounding the axial endothecium. The function of these varies among the three lineages. The amphithecial cells always form the capsule wall, the assimilative layer when present, and the spore sac in hornworts and peatmosses (*Sphagnum*). In peristomate mosses, the teeth lining the capsule mouth (see Section 2.4.2) are also derived from amphithecial layers. The endothecium gives rise to the columella in hornworts and mosses (lacking in all liverworts, and from some hornworts and mosses), and to the sporogenous tissue in liverworts and most mosses.

The sporophyte is composed of a foot, a stalk (except in hornworts), and a sporangium, and never bears any appendages. The ontogeny of these parts and their architecture at maturity distinguishes the three lineages. In liverworts, the sporangium or spore-bearing capsule is typically elevated on a stalk or seta (Figure 2.9) that elongates after spores have reached maturation, elevating the capsule above the protective leaves or sheaths developed by the maternal plant. The capsule dehisces typically along longitudinal lines and spores are generally released all at once. The sporangium of mosses (Figure 2.11) differentiates after the growth of the seta and spores are, in most cases, dispersed over extended periods of time. Dehiscence of the sporangium occurs along either one or more longitudinal lines or, in the case of most mosses, following the loss of the apical lid. The apical cell of the moss sporophyte ceases activity early in development, soon after the differentiation of the presumptive sporogenous region. The seta supporting the capsule grows from cells added by a distal but intercalary meristem. Sporogenesis is synchronous in both the liverwort and moss sporangia, whereas in hornworts, spore-mother cells (i.e., sporocytes) are formed continuously from a proximal intercalary meristem. Thus, in hornworts, spore development is basipetal and sporangial dehiscence proceeds from the apex down (Figure 2.12), allowing for a rather continuous release of spores.

2.2 Phylogenetic relationships of bryophytes

Bryophytes in a broad sense share a life cycle composed of a multicellular diploid sporophyte alternating with a haploid autotrophic gametophyte, but its three main lineages are characterized by distinct fundamental architectures. Bryophytes are clearly members of a monophyletic clade of land plants, the embryophytes, but their relationships among themselves and to other land plants have remained controversial, as virtually every phylogenetic hypothesis has gained support from one or more sources of characters (Goffinet 2000).

Extant land plants and their green algal ancestor exhibit three fundamental life cycles. In Charophycean algae the diploid phase consists of a single cell, the zygote that undergoes meiotic divisions resulting in four spores, or more if mitotic divisions occur (Graham *et al.* 1991). In bryophytes, both generations are multicellular, with the gametophytes remaining dominant. All living vascular plants produce multiple sporangia on a branched sporophyte, and the life span of the gametophyte is shorter than that of the sporophyte, although a few species, such as *Vittaria appalachiana*, develop perennial gametophytes. In all extant land plants, the haploid and diploid generations are heteromorphic: their fundamental functional differences (gamete vs. spore production) are accompanied by distinct life history roles and associated architectural differences, such as leaves, branches, and root or root-like structures.

On the basis of their shared life cycle, bryophytes have historically been considered closely related lineages, and implicitly maybe even most

2.1
2.2
2.3
2.4
2.5
2.6
2.7
2.8
2.9
2.10
2.11
2.12

Figure 2.1 The gametophyte of a complex thalloid, *Marchantia berteroana*, showing the pores on the upper epidermis and gemmae cups (photograph by Riccardo Rozzi (University of North Texas)). (For a color version of this figure, see Plate 2.1.)

Figure 2.2 The gametophyte of a simple thalloid, *Pallavicinia lyellii*. (For a color version of this figure, see Plate 2.2.)

Figure 2.3 The leafy liverwort, *Roivainenia jacquinotii*. (For a color version of this figure, see Plate 2.3.)

Figure 2.4 The carpocephalate receptacle of a complex thalloid, *Marchantia polymorpha*, showing the hyaline involucres protecting the archegonia. (For a color version of this figure, see Plate 2.4.) Photograph by Bernard Goffinet.

Figure 2.5 The leaves of *Gackstroemia magellanica* are lobed and the ventral lobe is modified into a water-sac. (For a color version of this figure, see Plate 2.5.)

Figure 2.6 The thalloid gametophyte of the hornwort, *Leiosporoceros dussii* (photograph by Juan Carlos Villarreal (University of Connecticut)). (For a color version of this figure, see Plate 2.6.)

Figure 2.7 The orthotropic leafy gametophyte of the acrocarpous moss, *Ulota crispa*. Note the hairy calyptra covering the developing sporophytes. (For a color version of this figure, see Plate 2.7.)

Figure 2.8 The pleurocarpous moss *Acroporium pungens* develops a plagiotropic monopodially branching gametophyte; some of the branches are reduced and develop the sex organs; the sporophyte (far left) appears directly attached to the stem. (For a color version of this figure, see Plate 2.8.)

Figure 2.9 The sporophyte of the liverwort, *Plagiochila*, with a delicate hyaline seta elevating the capsule. Note that some capsules have dehisced along longitudinal lines. (For a color version of this figure, see Plate 2.9.)

Figure 2.10 The sporophyte of *Pallavicinia lyellii* (see Figure 2.1) surrounded by a caulocalyx derived from thalline tissue. Shoot-calyptra enclosing the sporophyte is present but cannot be distinguished here. (For a color version of this figure, see Plate 2.10.)

Figure 2.11 The sporophyte of the moss, *Tayloria dubyi*, showing the elongate sporangium with an apical mouth through which spores are shed. (For a color version of this figure, see Plate 2.11.)

Figure 2.12 The sporophyte of the hornwort, *Nothoceros endiviaefolius*. Note the sporangium dehiscing from the apex downward into two hygroscopic valves (photograph by Juan Carlos Villarreal (University of Connecticut)). (For a color version of this figure, see Plate 2.12.)

2.13
2.14
2.15
2.16
2.17
2.18
2.19
2.20
2.21
2.22
2.23

Figure 2.13 The sporophyte of *Takakia*, exhibiting the characteristic spiral line of dehiscence (photograph by Jessica Budke (University of Connecticut)). (For a color version of this figure, see Plate 2.13.)

Figure 2.14 The capsule of *Sphagnum* (peatmosses) is elevated by a gametophyte stalk, the pseudopodium (photograph by Marianna Ricca (Duke University)). (For a color version of this figure, see Plate 2.14.)

Figure 2.15 Valvate capsule of *Andreaea rupestris* (photograph by Jessica Budke (University of Connecticut)). (For a color version of this figure, see Plate 2.15.)

Figure 2.16 Peristome of *Atrichum* (Polytrichaceae), composed of whole cells (i.e., nematodontous peristome) and with teeth connected by a membrane (i.e., epiphragm) (photograph by Neil Bell (University of Helsinki)). (For a color version of this figure, see Plate 2.16.)

Figure 2.17 Peristome of *Orthotrichum speciosum*, composed of cell remnants, allowing the teeth to bend (i.e., arthrodontous peristome). The outer ring, or exostome, is here composed of 16 teeth, which are fused in pairs; the inner ring, or endostome, is reduced to 8 teeth alternating with the outer pairs. (For a color version of this figure, see Plate 2.17.)

Figure 2.18 Plant body of *Micromitrium tenerum*, with a mature capsule immersed among the gametangial leaves (photograph by Jessica Budke (University of Connecticut)). (For a color version of this figure, see Plate 2.18.)

Figure 2.19 The sporophyte of *Physcomitrium coorgense*, illustrating the seta subtending the capsule, which is covered by a calyptra. (For a color version of this figure, see Plate 2.19.)

Figure 2.20 The capsules of *Splachnum rubrum* (left) and *Splachnum luteum* (right) are highly modified to attract insects. (For a color version of this figure, see Plate 2.20.)

Figure 2.21 Leafy shoots in mosses develop from a filamentous protonema (photograph by Jessica Budke (University of Connecticut)). (For a color version of this figure, see Plate 2.21.)

Figure 2.22 Aposporous protonema and leafy shot develop by a seta of *Funaria hygrometrica* (photograph by Jessica Budke (University of Connecticut)). (For a color version of this figure, see Plate 2.22.)

Figure 2.23 The protonematal stage of *Sphagnum* combines filamentous and thalloid structures. On the right a leafy gametophore develops (photograph by Jessica Budke (University of Connecticut)). (For a color version of this figure, see Plate 2.23.)

closely related groups, sharing a unique common ancestor. As early as 1984, however, application of cladistic principles and inferences from morphological characters in a broad sense led Mishler and Churchill (1984) to propose that bryophytes compose an evolutionary grade leading to the vascular plants. A single unique shared ancestry for the bryophytes has been proposed based on data from spermatogenesis (Garbary *et al.* 1993) and chloroplast gene sequences (Nishiyama *et al.* 2004), but this hypothesis has been repeatedly challenged (e.g., Kenrick & Crane 1997; Qiu *et al.* 2006). When only extant taxa are considered, the tempting interpretation of such a phylogenetic hypothesis is that the immediate ancestor of vascular plants with their sporophyte-dominated life cycle arose from a bryophyte-like group, suggesting a rapid shift in the imbalance between the haploid and diploid generations (Qiu *et al.* 2006). However, some early extinct lineages exhibit isomorphic generations (Taylor *et al.* 2005), that is, a sporophyte and a gametophyte that were similar in growth form and differed structurally only by the presence of sporangia or gametangia. The significance of such taxa in the evolution of a life cycle with alternating generations remains ambiguous, as the hypothesis that one or more lineages of bryophytes actually arose from such an ancestor through reduction of the branched sporophyte remains viable (Bateman 1996), and was considered by Schuster (1966) and earlier workers. A hypothesis of reverse evolution through simplification or loss of branching remains virtually impossible to prove, and is currently only rejected by the strong support obtained for the alternative hypothesis. Indeed, the divergence of any of the bryophyte lineages from an ancestor with isomorphic generations can only be demonstrated within a phylogenetic hypothesis that includes extinct taxa. The sole character shared by all extinct and extant polysporangiophytes is the ability of the sporophyte to branch. In the absence of additional diagnostic features, the bryophytes must be excluded from the polysporangiophytes. The convergence of phylogenetic inferences toward a shared and unique ancestry of all polysporangiophytes should thus not lead to an unconditional rejection of their paraphyly.

This controversy rests on the assumption that the axis of the bryophyte sporophyte is in fact homologous to that of the polysporangiophyte: the branched shoot system of vascular plants arose via ramification of the simple seta. Kato and Akiyama (2005) have rejected this hypothesis and argued that the sporangium and seta of the moss or liverwort sporophyte are, on the basis of developmental similarities, in fact homologous to the sporangial axis rather than the complete sporophyte of a polysporangiophyte. Under this scenario, the *vegetative* sporophytic axis of polysporangiophytes is the new innovation without a strict homolog in mosses, and potentially other bryophytes. Whether the sporophyte axis arose through amplification of an existing structure, the bryophyte seta, or *de novo*, is not clear, and the ambiguity may not be resolved unless developmental processes can be inferred from the earliest, leafless and nonvascular polysporangiophytes (see Sanders *et al.* 2007 for a similar problem: the origin of the leaf).

2.3 Evolution of plant form in liverworts

The Marchantiophyta compose the oldest lineage among extant land plants, yet can hardly be seen as mere fossils of the past. Approximately 6500 species are currently recognized (von Konrat *et al.* 2008). The diversity in the fundamental architecture of the vegetative body is unmatched among land plants (Schuster 1984; Crandall-Stotler *et al.* 2009), to the extent that the identification of a unique macroscopic feature of the gametophyte unifying all liverworts is virtually impossible. Liverworts are best defined by a rather simple architecture of the sporophyte, and the presence in their cells of oil bodies, membrane-bound organelles assembling and accumulating terpenoids, among others chemicals.

2.3.1 The gametophyte

Liverworts are typically organized into three groups based on their body form: (1) complex thalloid, (2) simple thalloid, and (3) leafy (Figures 2.1, 2.2, and 2.3). Complex thalloids derive their common name from the complex architecture of the gametophyte. The body is mainly composed of a storage tissue of hyaline parenchyma cells lined on the dorsal (upper) surface by a chambered tissue with photosynthetic filaments. The roof of these chambers typically has pores, but unlike stomata the opening is static and permanently open. Simple thalloids lack such anatomical differentiation. Leafy liverworts are, of course, defined by foliar appendages. Although such classification is convenient, none of these concepts defines unique monophyletic lineages, therefore suggesting multiple origins of leaves and simple thalloid gametophytes, as will be discussed in the following text.

The vegetative body is an assemblage of modules built from the divisions of derivative cells produced by the apical cell. The geometry of the single meristematic cell varies between wedge-shaped, hemidiscoid, lenticular, and tetrahedral. The number and orientation of the cutting faces is correlated with the shape of the apical cell: four (two lateral, one ventral, and one dorsal), three (two lateral and one basal axial), two and three (each oblique) faces, respectively. The metamers are always ranked. The correlation between geometry of the apical cell and growth form of the gametophyte follows a trend but is not perfect: (1) tetrahedral apical cells lead to the development of terete stems in all leafy Jungermanniopsida and Haplomitriopsida but leafy gametophytes are also developed from a lenticular apical cell in *Pleurozia*; (2) all other taxa with lenticular apical cells develop simple thalli but simple thalli may also arise from tetrahedral, wedge-shaped, and hemidiscoid apical cells; and (3) the wedge-shaped apical cells are otherwise diagnostic of all complex thalloid hepatics (Crandall-Stotler *et al.* 2009).

2.3.1.1 Branching

Branching follows a diversity of ontogenetic pathways. Within leafy liverworts alone, Crandall (1969) and Crandall-Stotler (1972) recognized 11 morphogenetic patterns and Thiers (1984) described yet another mode of branch formation. Branch development varies primarily in the timing of its initiation (at the three-celled merophyte stage or later), and in the origin of the initial cell (epidermal or medullary). The epidermal cell used to initiate branch formation is, in some cases, one of the leaf initials and the development of a branch is then marked by the occurrence of a partial leaf in its axil. In other lineages, leaf ontogeny is not affected by branch initiation, and development of a branch begins with the division of epidermal cells below the leaf. None of these branching modes is truly dichotomous. Splitting of the apical cell yielding two new diverging meristematic cells occurs only in complex and rarely in simple thalloid gametophytes. In the remaining simple thalloids, branch initials differentiate from epidermal cells of young merophytes or from ventral epidermal cells of the midrib in mature plants; only in the Metzgeriaceae may branches also arise endogenously from medullary initials (Renzaglia 1982).

2.3.1.2 Modifications of stems and branches

Vegetative axes may undergo specialization to favor clonal growth (creeping rhizomes), insure anchoring (leafless geotropic stolons), develop asexual diaspores (nearly leafless gemmiferous shoots), serve as propagules, increase gamete dispersal (antheridiophore), or enhance spore dispersal (carpocephalum). The latter two modifications characterize some members of the complex thalloids or Marchantiopsida, where gametangia are grouped on receptacles (Figure 2.4). These are derived either from subapical innovations, and hence exhibit the typical architecture of complex thalloid lobes with air chambers and rhizoids, or from the dorsal innovations, in which case they lack these traits (Schuster 1984). In either case, the receptacle is typically composed of two or more lobes, resulting from one or more dichotomous branching events. In some taxa, gametangial lobes alternate with sterile lobes. The gametangia are developed dorsally in the chambers (antheridia) or on the surface (archegonia) in which case they can, following the enhanced elongation of the dorsal tissues of the lobe, be pushed to the lower surface (Figure 2.4). Male receptacles may serve as splash-cups designed to enhance the dispersal of sperm cells. A stalk subtends the receptacle and may in some taxa elevate the gametangiophore above the vegetative thallus. Raising the female receptacles following sexual reproduction promotes dispersal of the spores, which are shed from the sporophytes hanging downward from the receptacle.

2.3.1.3 Leaves

Leaves are not strictly diagnostic of the Jungermanniidae (true leafy liverworts) and also occur in early diverging lineages such as *Treubia* and *Haplomitrium* and in some simple thalloids, but these differ in their leaf ontogeny. In the Jungermanniidae, the early divisions of the merophytes yield two presumptive leaf initials, which may undergo additional divisions resulting in secondary initials. In the Haplomitriopsida, a single initial is formed.

Annual Plant Reviews Volume 45: The Evolution of Plant Form, First Edition.
Edited by Barbara A. Ambrose and Michael Purugganan.

Plate 2.1 The gametophyte of a complex thalloid, *Marchantia berteroana*, showing the pores on the upper epidermis and gemmae cups (photograph by Riccardo Rozzi (University of North Texas)).

Plate 2.2 The gametophyte of a simple thalloid, *Pallavicinia lyellii.*

Plate 2.3 The leafy liverwort, *Roivainenia jacquinotii.*

Plate 2.4 The carpocephalate receptacle of a complex thalloid, *Marchantia polymorpha*, showing the hyaline involucres protecting the archegonia. Photograph by Bernard Goffinet.

Plate 2.5 The leaves of *Gackstroemia magellanica* are lobed and the ventral lobe is modified into a water-sac.

Plate 2.6 The thalloid gametophyte of the hornwort, *Leiosporoceros dussii* (photograph by Juan Carlos Villarreal (University of Connecticut)).

Plate 2.7 The orthotropic leafy gametophyte of the acrocarpous moss, *Ulota crispa.* Note the hairy calyptra covering the developing sporophytes.

Plate 2.8 The pleurocarpous moss *Acroporium pungens* develops a plagiotropic monopodially branching gametophyte; some of the branches are reduced and develop the sex organs; the sporophyte (far left) appears directly attached to the stem.

Plate 2.9 The sporophyte of the liverwort, *Plagiochila*, with a delicate hyaline seta elevating the capsule. Note that some capsules have dehisced along longitudinal lines.

Plate 2.10 The sporophyte of *Pallavicinia lyellii* (see Figure 2.2) surrounded by a caulocalyx derived from thalline tissue. Shoot-calyptra enclosing the sporophyte is present but cannot be distinguished here.

Plate 2.11 The sporophyte of the moss, *Tayloria dubyi*, showing the elongate sporangium with an apical mouth through which spores are shed.

Plate 2.12 The sporophyte of the hornwort, *Nothoceros endiviaefolius*. Note the sporangium dehiscing from the apex downward into two hygroscopic valves (photograph by Juan Carlos Villarreal (University of Connecticut)).

2.13
2.14
2.15
2.16
2.17
2.18
2.19
2.20
2.21
2.22
2.23

Plate 2.13 The sporophyte of *Takakia*, exhibiting the characteristic spiral line of dehiscence (photograph by Jessica Budke (University of Connecticut)).

Plate 2.14 The capsule of *Sphagnum* (peatmosses) is elevated by a gametophyte stalk, the pseudopodium (photograph by Marianna Ricca (Duke University)).

Plate 2.15 Valvate capsule of *Andreaea rupestris* (photograph by Jessica Budke (University of Connecticut)).

Plate 2.16 Peristome of *Atrichum* (Polytrichaceae), composed of whole cells (i.e., nematodontous peristome) and with teeth connected by a membrane (i.e., epiphragm) (photograph by Neil Bell (University of Helsinki)).

Plate 2.17 Peristome of *Orthotrichum speciosum*, composed of cell remnants, allowing the teeth to bend (i.e., arthrodontous peristome). The outer ring, or exostome, is here composed of 16 teeth, which are fused in pairs; the inner ring, or endostome, is reduced to 8 teeth alternating with the outer pairs.

Plate 2.18 Plant body of *Micromitrium tenerum*, with a mature capsule immersed among the gametangial leaves (photograph by Jessica Budke (University of Connecticut)).

Plate 2.19 The sporophyte of *Physcomitrium coorgense*, illustrating the seta subtending the capsule, which is covered by a calyptra.

Plate 2.20 The capsules of *Splachnum rubrum* (left) and *Splachnum luteum* (right) are highly modified to attract insects.

Plate 2.21 Leafy shoots in mosses develop from a filamentous protonema (photograph by Jessica Budke (University of Connecticut)).

Plate 2.22 Aposporous protonema and leafy shot develop by a seta of *Funaria hygrometrica* (photograph by Jessica Budke (University of Connecticut)).

Plate 2.23 The protonematal stage of *Sphagnum* combines filamentous and thalloid structures. On the right a leafy gametophore develops (photograph by Jessica Budke (University of Connecticut)).

Plate 3.1 Lycophyte architecture. (a) Radial dichotomies are found in Lycopodiaceae (*Huperzia squarrosa*). (b) The overall gross morphology of *Isoetes engelmannii* is distinct from other lycophytes. Flabellate dichotomies are found in (c) *Selaginella apoda* and (d) *Selaginella moellendorffii*.

Plate 3.2 Sporophyte reproductive morphology. (a) Sporangia are formed in between sterile regions of the plant in *Huperzia squarrosa*. Lines indicate fertile regions. Arrows indicate the sporangium associated with the adaxial side of each sporophyll. (b) Sporangia are arranged in distinct strobili in *Selaginella moellendorffii*. Arrowheads indicate some of the strobili that have formed at the tips of branches. (c) A dissection of *Isoetes engelmannii* shows the relatively large sporangium (arrow) that is found on the adaxial region of the sporophyll. The corm (Co) is labeled for orientation.

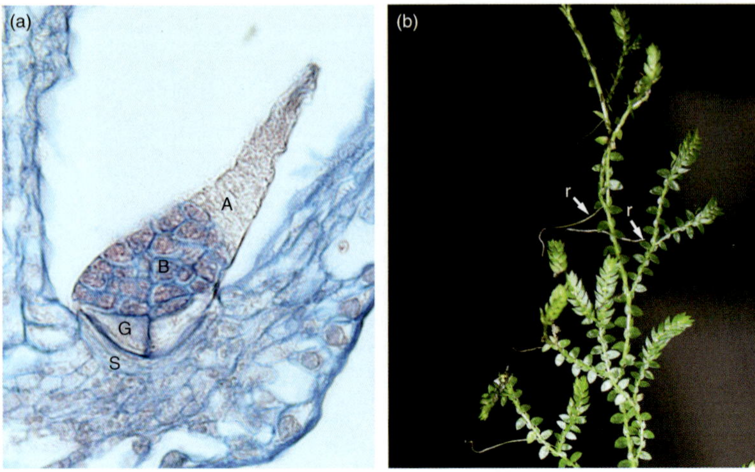

Plate 3.3 Enigmas in the lycophytes. (a) A stained histological section from *Selaginella moellendorffii* showing ligule morphology. The ligule is composed of sheath (S), glossopodium (G), main body, (B) and apex (A). (b) Some of the rhizophores (r) are indicated at the branching points in *Selaginella apoda*.

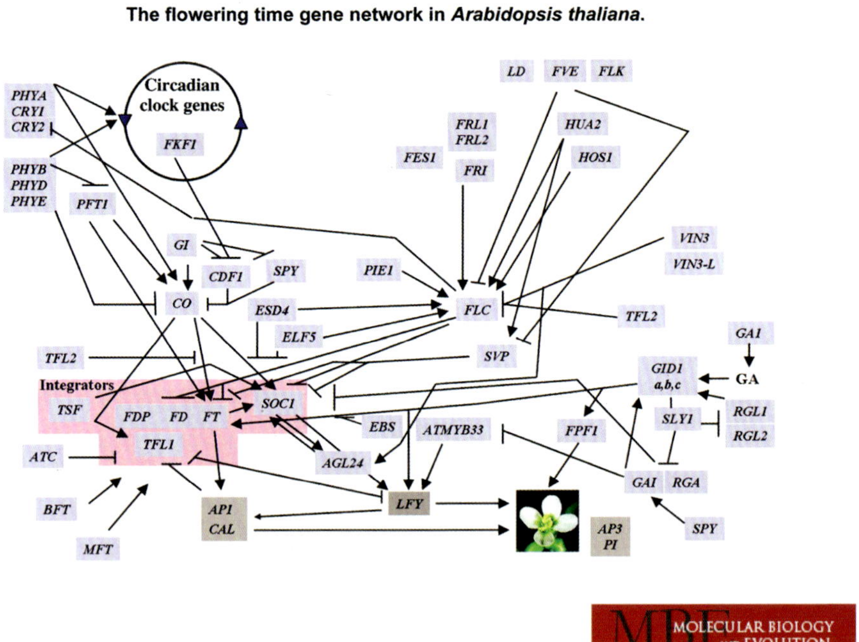

Plate 7.1 The flowering time genetic network in *Arabidopsis thaliana*. Reproduced from Flowers *et al.* (2009) with permission of Oxford University Press. This network highlights the relationships only among loci included in the studies by Ehrenreich *et al.* (2009) and Flowers *et al.* (2009). For a more comprehensive network, see Ehrenreich *et al.* (2009). Relationships among loci were inferred from the literature. The integrators (*FD, FDP, FT, SOC1*, *TFL1*, and *TSF*) integrate photoperiod, vernalization, and gibberellins signals to regulate activity of the inflorescence meristem. Genes that regulate floral meristem identity (*LFY, AP1*, *CAL*) or floral organ identity (*AP3, PI*) were not included in the flowering time association mapping or sequence diversity analyses work; see Shepard (2007) for more information about the evolutionary dynamics of these genes.

Plate 10.1 Phytochrome-regulated phenotypic expression. Phytochrome genes, arisen from gene duplication events, differ in their environment-dependent expression and environment-dependent gene products. In the simplest case, different environments, represented by different colors, enable the expression of particular phytochrome genes (e.g., dark green), and additional environmental conditions enable the activity of the gene product (e.g., light green). In this manner, different phytochromes can elicit a phenotype only under certain combinations or sequences of environments. The gene expression component actually consists of transcriptional regulation and posttranscriptional regulation. Various steps of gene expression can be environment-dependent and can depend on other gene products whose expression and activity might be environment-dependent. Regulation of the activity of the gene product also comprises many steps, which can be environment-dependent and can depend on other gene products whose expression and activity are environment-dependent. Colored dots beneath a process indicate that the process may be environment-dependent. In the hypothetical example of phytochromes, transcription occurs and mRNA is transcribed under "dark green" conditions. Hypothetical degrading microRNAs are formed under all but the "light-green" conditions, and alternate splice types that are not formed under light-green conditions are not formed, leaving only "light-green" stable mRNAs available for translation. That gene product combines with other gene products (the chromophore and perhaps other phytochrome apoprotiens for heterodimers). (*Continued*)

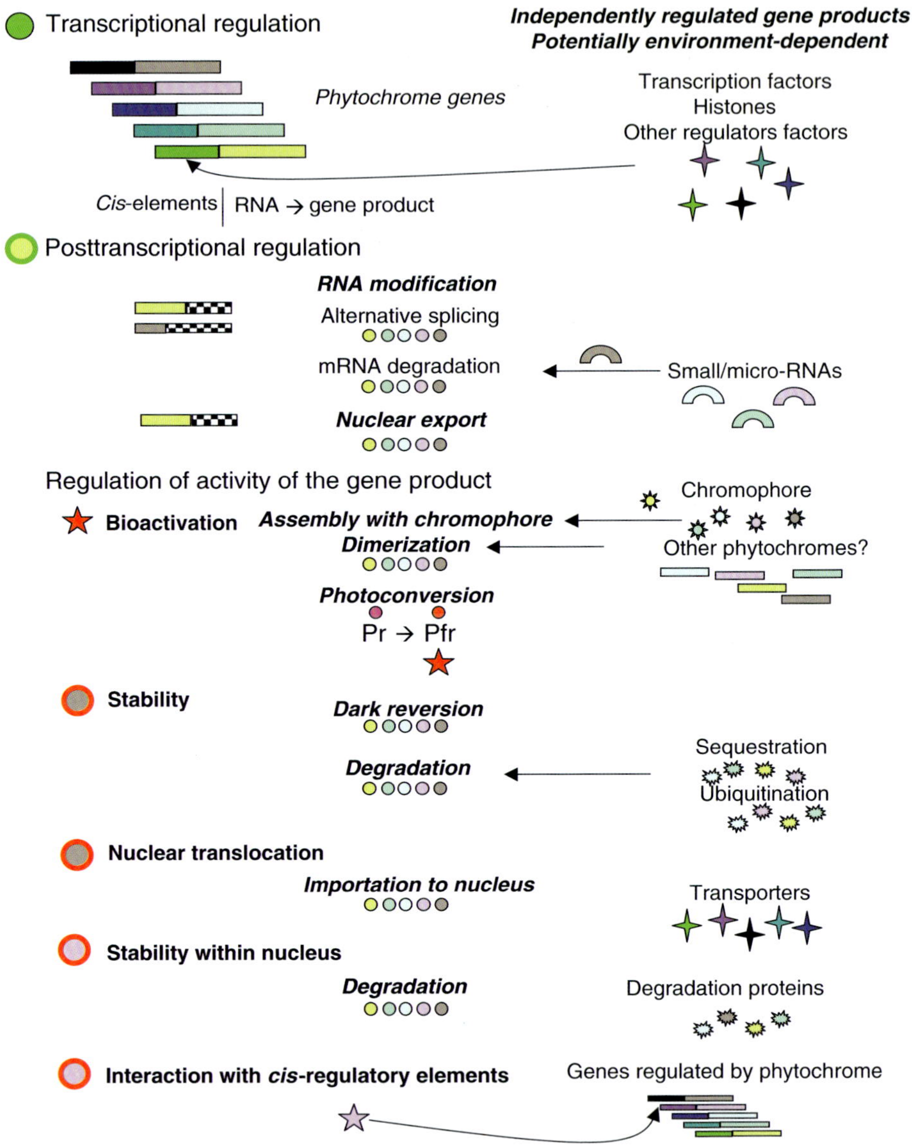

Plate 10.1 (*Continued*) Bioactivation of the assembled gene product occurs under "red" conditions, and that gene product is stable and translocated to the nucleus only under "gray" (plus "red") conditions. Within the nucleus, the bioactive gene product is stable only under "purple" conditions. It interacts with genes that are capable of being transcribed under "purple" conditions. These genes continue the signal transduction pathway. In this manner, biological processes occur only under particular combinations and sequences of environmental conditions. The sequence of the environments under which the process takes place is indicated by the colored symbols on the left.

In both cases the lamina is built from the activity of a basal meristem, but in the Jungermanniidae, apical cells derived from initials may contribute the distal portion of the leaf, in particular in species with lobed leaves. In species with unlobed leaves, apical cells may cease to divide early in the ontogeny, and the leaf grows exclusively from its base. Variations in the number of initials and in the meristematic activity of the various apical cells and the basal meristem account for much of the diversity in leaf shape exhibited by liverworts, from highly dissected leaves with filamentous lobes to entire laminae, from asymmetric leaves with unequal lobes to bilaterally symmetric bilobed leaves, from laminae that are longitudinally folded to those that are flat.

Leaves are always inserted along their entire base (i.e., lacking a petiole), but the angle of the insertion varies depending on the alignment of the median axis of the apical cell and of the shoot itself. A dorsal tilt leads to a *succubous* insertion, whereby the forward margin of the leaf points downward, whereas a ventral tilt leads to an *incubous* insertion with the margin pointing upward. When the alignment is perfect, the insertion is *transverse*. The mode of insertion is typically conserved within major lineages, and may be correlated to substrate (e.g., succubous in terricolous species and incubous in epiphytes). Leaves are always ranked along three straight parallel lines. Axes exhibit heteroblastic leaf series, with distinct proximal paedomorphic juvenile leaves (i.e., first leaves develop on a module and lack the typical traits of mature leaves higher up on the module). Leaf dimorphism is often more conspicuous around the axis. In such anisophyllous taxa, underleaves (amphigastria) are always transversely inserted, they are almost always much smaller than the lateral vegetative leaves, and may even be completely lacking. Leaves are typically unistratose, and always lack a costa, although in some taxa, cells with thickened walls compose a median band (vitta). Intralaminar cellular dimorphism is rarely pronounced, and often restricted to differentiated marginal cells, although some species differentiate ocelli, that is, cells lacking chloroplasts but with a single large oil body.

2.3.1.4 Modification of the leaf

Some species develop small gemmae along their margins or on the lamina or the entire leaf is deciduous and serves as a dispersal unit. More conspicuous are modifications of individual leaf lobes. In species with asymmetric complicate-bilobed leaves, the smaller ventral lobe may be modified into a water-sac (Figure 2.5), which often houses diverse invertebrates. The ontogeny of this modified lobule differs between the Lejeuneaceae and Porellales, involving growth from a submarginal or a median meristem, respectively, and at maturity the sac opens distally (upward) or proximally (downward).

2.3.1.5 Rhizoids

Rhizoids are always unicellular. They arise from epidermal cells or from basal laminal cells. Complex thalloids develop dimorphic rhizoids: smooth and pegged ones, the latter characterized by inward projections of the cell wall. Besides anchoring the plant, the rhizoids may act in water conduction (pegged) or serve as entry points for mycorrhizal fungi (smooth; Duckett *et al.* 2000).

2.3.1.6 Sex organs

Gametangia develop always from epidermal cells and above the surface, although in some derived complex thalloids, they may become secondarily embedded in chambers. Antheridial and archegonial development may consume the apical cell of the module (androgyny and acrogyny, respectively), and thereby terminate the growth of either the main stem or a branch, which then bear only a single cluster of gametangia. By contrast, when the formation of sex organs is initiated on metamers (anandrogyny and anacrogyny) multiple gametangia may occur on a single module. Most liverworts develop unisexual plants and even within monoicous species the sex organs are never mixed in a single cluster. Sex organs may be distributed along or clustered on the median dorsal surface in thalloid hepatics. In leafy taxa antheridia occur as singletons, or in pairs, and rarely in larger groups whereas archegonia are typically clustered. Various structures are developed to protect the sex organs, in particular the archegonium. In leafy species, gametangia are borne in the axils of unmodified or more generally modified leaves or underleaves (perichaetial bracts and bracteoles, respectively), which may fuse and form a perianth. In simple thalloids, dorsal scales form an involucre surrounding the female sex organs, whereas in complex thalloids, the thallus forms a tubular involucre around the archegonia. Following fertilization, protection of the developing sporophyte is enhanced by further growth of the outer bracts, or by thalline outgrowths (i.e., a caulocalyx; Figure 2.10) in the simple thalloid *Calycularia* or a pseudoperianth derived from the archegonial neck in the complex thalloid *Asterella*.

2.3.1.7 Asexual reproductive structures

Many species reproduce in the absence of fertilization by means of specialized vegetative diaspores, ranging from exogenous (or rarely endogenous) gemmae to entire caducous modules (i.e., branches). Diaspores form along the foliar or thalline margin or at the apex of modified shoots, and range from single-celled gemmae to small thalloid propagules. In *Marchantia* and a few other complex thalloids, bipolar disk-shaped gemmae are aggregated in cups (Figure 2.1), designed to favor the dispersal of the unit via splashing raindrops. Some diaspores insure local perenniality through an *in situ* dormant stage. These tubers resemble those of mosses but like in hornworts they are never rhizoidal in origin.

2.3.2 The sporophyte

The diploid generation of liverworts is composed of a foot, a seta, and a capsule. It completes its development inside a chamber built either from the original archegonium (i.e., a calyptra), from cauline or thalline tissues (i.e., a coelocaule) or from a combination of these (i.e., a shoot-calyptra). Successful fertilization thus initiates cell division not only in the archegonial wall but also in the maternal tissues below. In some cases, the perianth is replaced by a multistratose stem-perigynium growing from the meristematic activity below the archegonia. Some Jungermanniidae develop a marsupium, or pouch enclosing the entire sporophyte. In some complex thalloids, the sporophyte appears embedded in the thallus, as it is covered by a thick involucre that is a continuation of the thalline tissues. Typically a single archegonium is fertilized and only one sporophyte develops within a perichaetial cluster, but in *Marchantia* and related taxa, multiple sporophytes may emerge from a single involucre.

The shape of the sporangium varies from spherical to cylindrical, and its wall is devoid of stomata. It always lacks a central axis (or columella, as in other bryophytes), and is filled with a mass of spores mixed, most often, with diploid sterile filaments or elaters. The elaters are differentiated prior to meiosis, and distributed throughout the sporangium. Only in some simple thalloids are elaters derived from a specialized tissue at the base of the sporangial cavity. Elaters are elongate cells often with spiral wall thickenings, which through hygroscopic movement aid in spore dispersal. Meiosis is synchronous and its completion triggers the elongation of the seta, elevating the sporangium above the surrounding protective maternal tissues (Figure 2.9). The seta can grow to a length of a couple of centimeters literally overnight, and solely by cell elongation. The seta is composed of thin-walled cells and is therefore delicate and unable to withstand dehydration. The sporangium, raised above the maternal plant, quickly loses water and the tension in its epidermal cells mounts, resulting in the tearing of the capsule wall typically along two vertical suture lines extending all around the sporangium, resulting in four valves (Figure 2.9). The walls of the exothecial cells are unevenly thickened and, upon dehydration, the cells and hence the valves of the dehiscing capsule, bend backward rather rapidly, dispersing all the spores at once. In a few taxa, dehiscence is irregular or along a single subapical transverse line. Rarely are spores dispersed following the disintegration of the otherwise indehiscent capsule, such as in the Corsiniaceae, a lineage of cleistocarpous complex thalloids. In *Riccia*, also in the Marchantiopsida, the capsule is immersed within the maternal thallus; its wall disappears early during sporogenesis and spore dispersal follows the degradation of the gametophytic tissue.

2.3.3 Evolutionary trends

Historically, conspicuous macroscopic traits such as the organization of the gametophyte into complex and simple thalloids, and leafy plants provided the foundation for establishing major lineages. Phylogenetic inferences from molecular characters unambiguously reject the hypothesis that these traits define monophyletic lineages (Forrest *et al.* 2006; Figure 2.24). If species-poor lineages are ignored, the backbone phylogeny consists of complex thalloids

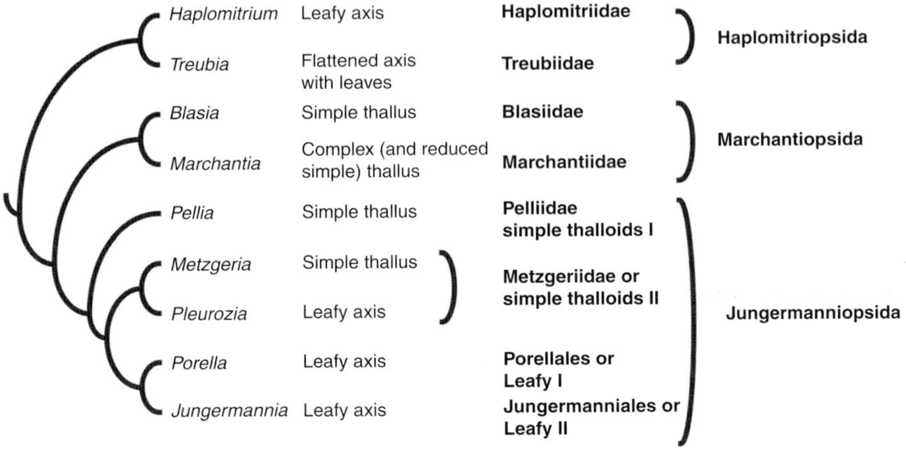

Figure 2.24 Summary phylogeny of liverworts modified from Vanderpoorten and Goffinet (2009), based on Forrest *et al.* (2006) with characterization of vegetative architecture based on Crandall-Stotler *et al.* (2009).

as the sister group to the remaining liverworts, with the simple thalloids forming a grade leading to the leafy ones, suggesting that a thalloid growth form is ancestral to a terete vegetative body. Inclusion of the species-poor lineages, however, results in the Blasiales and *Monoclea* with their simple thalloid architecture to be sister to or nested within the Marchantiidae, in the leafy liverwort *Pleurozia* to share a unique ancestor with a lineage of simple thalloids, and in the leafy *Haplomitrium* and "thalloid" *Treubia* composing the earliest diverging lineages of hepatics. Consequently, polarity of transformations in vegetative architecture remains ambiguous. However, observations of thalloid liverworts developing nearly cylindrical axes under low light and high humidity, may suggest that, fundamentally, as liverworts are ancestrally axial rather thalloid (Evans 1939).

Reverse evolution characterizes various lineages of liverworts and leads to the loss of one row of leaves in some Jungermanniidae, loss of one or more leaf lobes, the simplification of thallus complexity through loss of photosynthetic or assimilatory filaments in the chambers and reduction of the stalk subtending the female receptacle in some Marchantiidae, the thinning of the stem to a single ring of cells and a uniseriate central axis (e.g., *Cololejeunea*, Thiers 1988), or most conspicuously in the transfer of the photosynthetic function to a persistent and potentially modified protonema (e.g., *Metzgeriopsis*, Gradstein *et al.* 2006). The sporophyte, too, may undergo reduction, through the loss of seta elongation and even loss of the foot as in *Riccia*, a decrease in the number and even absence of elaters, or even the absence of regular dehiscence. Most of these shifts basically consist of paedomorphic and essentially neotenous developments of the plant: the premature termination of the differentiation of the vegetative body or seta in favor of accelerated maturation of the sex organs and dispersal of spores. A robust phylogeny of liverwort genera is currently under construction by A.J. Shaw (Duke University) and colleagues and will likely illustrate the polarity of morphological character transformations. Comparative developmental studies are, however, critically needed to dissect the diversity of heterochronic events underlying morphological reduction, and establish the homology of reduced traits.

2.4 Evolution of plant form in mosses

With approximately 12 000 species (Crosby *et al.* 2000), mosses rival ferns for the second most speciose lineage of land plants. Although much of the species diversity may be of fairly recent origin, the main lineages (superclasses sensu Goffinet *et al.* 2009) diverged perhaps as early as 400 million years ago (Newton *et al.* 2007). Not one macroscopic character is exhibited by and conserved among all mosses, indeed even the otherwise diagnostic multicellular rhizoids are lacking in *Takakia*, the likely sister lineage to all remaining mosses (Chang & Graham 2011). Mosses occur in a broad range of habitats, from leaves in the tropics to soil in deserts, from twigs in upper montane forests to submerged rocks in freshwater habitats. Although physiological adaptations may be critical to the colonization of extreme habitats, they may not be sufficient, and establishment may be optimized by critical modifications of the overall architecture of the plants through changes in branching and growth form.

2.4.1 The gametophyte

The vegetative body of mosses is almost always composed of a terete (but rarely 3–5-angled), cylindrical axis bearing leaves, and anchored to the substrate by rhizoids. The stem grows through the activity of a single tetrahedral apical cell with three cutting faces. The first periclinal division of the derivative cell isolates an inner and an outer cell, the former forming cortical and possibly conducting tissues and the latter developing into the epidermis with a branch and a leaf initial. All cell lines tracing back to a derivative cell compose a metamer and all metamers derived from a single apical cell constitute a module. Stems may be orthotropic, plagiotropic, or in the case of epiphytic taxa, pendent. In some mosses with persistent protonemata strictly vegetative stems are greatly reduced or lacking. In *Buxbaumia*, for example, the photosynthetic function is taken over by the chloronemata, while the small bracts (i.e., leaves) surrounding the sex organs lack chlorophyll. Among orthotropic mosses, *Dawsonia* develops the tallest stems, reaching nearly 80 cm in height (Burges 1949), but the longest single cauline module is likely to be found among pendent mosses, such as *Phyllogonium* and members of the Meteoriaceae, with stems hanging down one meter or more, or in aquatic mosses, such as *Fontinalis*.

2.4.1.1 Branches

Branches always develops from branch initials present on each metamer. This initial occurs at first one cell below or behind the leaf initial (i.e., the branch initial is always distal to the leaf initial), but intense divisions in this intermediate cell result in the initials being separated, with the leaf initial being pushed toward the branch initial of the metamer in front of it. As a consequence, a branch initial lies in the axil of the leaf of the next distal or older metamer (Berthier 1972). Except in taxa consistently lacking branches, the branch initial divides forming either a primordium composed of several cells or an actual bud, a tiny module with its own apical cell.

Branching is either sympodial or monopodial and never dichotomous, that is, the apical cell never splits into two new terminal meristematic cells. The consumption of the apical cell for the development of terminal sex organs (i.e., acrocarpy) terminates the growth of the module, and may in most acrocarpous mosses trigger the development of at least one subapical primordium or bud, a response suggestive of apical dominance inhibiting lateral bud development. The spatial distribution of branches is dictated by the pattern of dormancy and cell elongation between consecutive branch initials. Branches may be clustered in whorls near the apex of the otherwise unbranched stem (i.e., dendroid habit) or be regularly developed, giving the plant a feathery appearance (i.e., frondose habit). Only in *Sphagnum* are branches arranged into fascicles, as branches quickly form two to seven consecutive innovations.

Successive modules function as stems or primary modules, accounting for the growth of the plant. In many plagiotropic mosses, the stem typically grows continuously and gametangial formation is shifted laterally onto branches (i.e., clado- and pleurocarpy, see next paragraph). Lateral buds or primordia develop regularly along the stem, and may themselves produce innovations. Although each metamer bears a branch and a leaf initial, modules always develop fewer branches than leaves, and maintain dormancy of many primordia or buds. In plagiotropic plants, the stem may continue to serve a vegetative function and bear mature leaves, or function as a stolon or rhizome insuring clonal growth and developing branches. In addition to the specialization of stems and branches, consecutive branches along the stem may be functionally differentiated. The development of secondary modules is indeed either determinate (the apical cell ceases to divide) or terminated by the production of apical gametangia, which occurs either immediately or late in branch development. Consequently, branches may bear mature vegetative leaves, or only those modified leaves associated with the sex organs. Moss plants may thus be composed of a primary module only, or of successive modules that arise sympodially or monopodially or both (La Farge-England 1996; Newton 2007). Monopodial branching allows for the specialization of primary and secondary modules and for the functional differentiation of successive lateral branches.

The modular hierarchy often reflects a functional differentiation in monopodially branched plants. The primary module or stem insures the perenniality of the plant whereas secondary and subsequent modules fix carbon and energy needed to fuel growth and develop the sex organs. Gametangia always terminate growth of a module. In cladocarpous mosses, perichaetia develop at the end of vegetative branches, whereas in pleurocarpous mosses, they form on highly reduced branches that virtually lack any vegetative leaves. In pleurocarpous mosses, sexual branches are thus functionally specialized. In cladocarpous mosses, branches function first as primary modules, bearing mature vegetative leaves before gametangia are initiated at their apex.

In mosses with extreme sexual dimorphism, primary modules of female plants may form dense, extensive perennial mats or tufts with perichaetia (clusters of female sex organs) terminating stems or lateral branches, whereas male stems are dwarfed, lack branches, bear few and often reduced vegetative leaves, and develop perigonia (clusters of male sex organs). Such neotenous development of males, which are confined to growing on tomentose stems, leaves, or rhizoids of female plants, is genetically determined and irreversible or physiologically controlled with spores germinating into "normal" or dwarf males, depending on the substrate on which they land (Une 1985).

2.4.1.2 Modifications of the stem and branches

These are few and fulfill roles in the propagation of the plants. Leafless underground axes insure the longevity of plants with delicate aboveground shoots (e.g., Gigaspermaceae) or the clonal growth of plants with robust orthotropic branches (e.g., Polytrichaceae). In a few species, stems complete their growth by developing virtually leafless axes (i.e., pseudopodia) producing clusters of tiny terminal propagules (e.g., *Aulacomnium*). *Sphagnum* and *Andreaea* also develop pseudopodia, but here they elevate the sporophyte, and serve a function analogous to the setae (see Section 2.4.2). Specialized short caducous branches bearing highly reduced leaves serve as vegetative propagules in only a few species (e.g., *Orthodicranum flagellare* and *Wijkia extenuata*).

2.4.1.3 Leaves

Leaves of mosses share one characteristic also seen in leafy liverworts: the lack of a petiole, and thus are inserted along their entire base to the module. Ontogenetically leaves in these two lineages differ: they are built through the activity of two or more initials and a basal meristem in liverworts, versus a single initial in mosses. Phyllotaxy is first dictated by the shape of the apical cell. The tetrahedral apical cell produces derivatives at a 137° angle. Hence at best only every sixth metamer is aligned with the first one, and in most cases spirals of leaves reveal only inconspicuous axial ranks. Distichous arrangements result from the lenticular apical cell that has only two cutting faces (e.g., *Fissidens*) or from torsion of the stem. Complanate leaves may appear ranked but closer inspection often reveals that the phyllotaxy is spiral and only the blades are forced into a single plane (e.g., *Catagonium*). The distribution of leaves along the module is determined by the elongation of metamers and dormancy of leaf initials: leaves may be crowded, widely spaced, or reduced on much of the stem but forming distinct apical rosettes (e.g., *Rhodobryum*).

Leaves may vary in shape and architecture along or around the axis of the module, and between young and old, and juvenile and mature leaves (Mishler 1988). Young leaves found at the apex of a module often lack the characteristics, such as surface ornamentation or cellular dimorphism, of old mature leaves, which characterize typical shoots of a species. Juvenile leaves are developed on young modules and occur at the base of a mature module; when fully developed, juveniles may resemble the distal mature leaves, or be dimorphic, exhibiting traits of young leaves, in which case their development is paedomorphic. Heteroblastic series between paedomorphic juvenile leaves and mature leaves are present in various mosses. In some taxa, such juvenile leaves are the sole leaves developed on portions of the module, such as in stipitate plants with virtually naked stems and terminal rosettes of vegetative branches (e.g., *Hypnodendron*). Leaf dimorphism may also be pronounced in plagiotropic plants, with the leaf facing the substrate being highly reduced (i.e., amphigastria; e.g., *Epipterygium*) or with small leaves of creeping stems being paedomorphic and resembling juvenile leaves and only those on the erect branches developing into large leaves with the diagnostic traits (e.g., *Macromitrium*).

The leaf initial forms an apical cell that undergoes limited divisions. Each of these divides and forms a building block. The pattern and timing of cell divisions determines whether the cells are five- to six-sided or elongate, forming either a parenchymatous or prosenchymatous network, respectively (Frey 1970). In *Physcomitrella*, the number of derivatives produced by the initial seems to determine the size of the leaf and the width of the leaf is set by the number of longitudinal divisions in the derivatives (Harrison *et al.* 2009). At first the leaf is unistratose but at maturity a multistratose midrib is often present, and in some species, discrete or extensive regions of the blade may be composed of two or more layers of cells, especially marginally. The midrib (= costa) may be narrow or broad, and even span much of the leaf width and the demarcation of costa and lamina is ambiguous in the absence of developmental observations (e.g., Leucobryaceae). Costae may be composed of distinct cell types, including narrow, thick-walled cells (stereids) or narrow- or wide-lumened thin-walled cells, or be homogenous as in all pleurocarpous mosses. They hence serve in water and nutrient conduction, or function as a supporting rod for the blade. The differentiation of a costa can be experimentally inhibited when growing plants in submerged conditions (Zastrow 1934; Seppelt & Selkirk 1984).

Vegetative leaves are undivided, except in *Takakia* with deeply incised leaves, or in the Miyabeaceae (with *Homaliadelphus* having lateral leaves bearing a small, clasping lobe). Leaves exhibit broad variation in size, shape, architecture, and intralaminal cellular differentiation among species. Noteworthy deviations from the typical architecture, that is, a unistratose lamina bisected by a midrib, are the leaves of *Fissidens* and *Polytrichum*. In the former, the leaf is composed of dorsal and ventral laminae that clasp the stem and converge in the upper half of the leaf, an adaxial lamina extending beyond them, and an abaxial lamina extending from the apex to base or onto the stem. The vaginant laminae have been interpreted as corresponding to the true leaf whereas the dorsal and ventral laminae are outgrowths of the costa (Robinson 1970; Pursell 2007). Columns or blades of cells mounted typically on the adaxial side of the costa are not uncommon in mosses, but most conspicuous in the Polytrichaceae. The costa in *Polytrichum* is broad and bears numerous lamellae composed of small photosynthetic cells and lined with waxes preventing the intralamellar space from hydrating and thereby enhancing the diffusion of carbon dioxide into the leaf.

Cellular dimorphism within the leaf extends beyond the differentiation of costal cells. Cell shape and ornamentation may vary across the two main axes of the leaf, from elongate cells composing distinct borders (limbidia) to rectangular hyaline cells forming clasping bases. In *Leucobryum*, the large hyaline cells (i.e., leucocysts) compose two or more layers enclosing a single layer of small chlorophyllose cells, providing water and air reservoirs to sustain the photosynthetic activity in these inner cells. Similar functions are assumed by hyaline cells in *Sphagnum* but here, these cells alternate within a single layer with the narrow chlorophyllose cells. At least in *Sphagnum* the differentiation of hyaline cells can in some species be attenuated when growing the plants under high humidity and low light conditions. Differences between laminal cells may be inconspicuous and entirely microscopic, as changes in cell wall thickness or ultrastructure may account for the direction a leaf bends upon dehydration, particularly in plants with recurved margins, or incurved leaf tips.

Photosynthesis requires both water and carbon dioxide. In photosynthetic organs that are typically unistratose this dual need poses a dilemma: maintaining dry space for the diffusion of CO_2 while being hydrated. Like other bryophytes, mosses lack a thick cuticle and are ectohydric, that is, they acquire water through their whole body surface, and their leaves are then covered by a film of water preventing the rapid uptake of CO_2. Many mosses growing in dry habitats develop leaves designed to prolong hydration of the lamina: the cells bear one or more papillae that create extracellular capillary spaces between them, and whose tips may emerge above the water surface to allow for gas exchange. In some epiphytes, the leaves present a functional asymmetry of their surfaces: the abaxial surface is covered by a hydrophobic cuticle maintaining the surface dry, whereas the adaxial surface, appressed against the stem, is permeable (e.g., *Pilotrichella*; Proctor 2009). Microscopic rather than conspicuous architectural innovations allow species to survive in environments with only seasonal or diurnal water availability.

2.4.1.4 Modifications of the leaf

These reflect primarily strategies for the plant to reproduce clonally, by fragmentation of the leaf blade or by losing the apical portion of an excurrent costa. No case of fusion of leaves is known from mosses.

Other appendages are developed on modules, ranging from uniseriate hairs in the axils of leaves, or photosynthetic filaments coating the axis (i.e., paraphyllia) to tiny filamentous or foliose structures arching over the dormant buds (i.e., pseudoparaphyllia). Except for the shared epidermal origin of these appendages their homology remains ambiguous (Ignatov & Hedenäs 2007).

2.4.1.5 Rhizoids

Rhizoids are multicellular but uniseriate and branched filaments of cells separated by oblique cross walls. They are lacking only in *Takakia*, the sister lineage to all other mosses, and on the leafy plants, but not the thalloid protonemata, of *Sphagnum*. On the stem they are initiated by epidermal cells either only near the base of the stem or along the axis. In some species, rhizoids also emerge from specialized apical cells, nematogens, of the leaf. Their role is primarily to anchor the plant to the substrate but in orthotropic plants, where rhizoids may compose a spongy coat around the stem (i.e., tomentum), they may also contribute and even be essential to external water conduction. In some lineages rhizoids surround and thereby protect dormant branch initials. Although rhizoid development is influenced by environmental factors, the nature of these factors remains largely unknown (Duckett *et al.* 1998).

2.4.1.6 Sex organs

Gametangia are typically arranged in clusters surrounded by tight whorls of leaves that may be highly differentiated from the vegetative leaves; only in *Sphagnum* and in some unrelated lineages are antheridia or archegonia produced singly. Antheridia and archegonia are often segregated within (monoicy) or between (dioicy) plants, but in some species, both are present in a single cluster, within which they can be distributed randomly (synoicy) or in distinct whorls (paroicy). The use of monoicy or dioicy, rather than monoecy and dioecy, reflects the sexuality of a gametophyte rather than that of a sporophyte (Allen & Magill 1987). Perichaetia and perigonia may hold numerous archegonia and antheridia, respectively, along with sterile filaments called paraphyses. These hairs may protect the gametangia from dehydration, play a role in the dehiscence of the antheridia, or secrete substances to attract microarthropods to disperse sperm cells (Cronberg *et al.* 2006).

Gametangia are always produced at the apex of a module, and their development consumes the apical cell, and hence terminates the growth of the stem or branch. Mosses are always acrogynous and androgynous. In nearly half of mosses, perichaetia are located at the apex of the stem (acrocarpy, Figure 2.7). In pleurocarpous mosses gametangial production is shifted to short lateral branches (Figure 2.8), devoid of vegetative leaves. In some primarily acrocarpous lineages, a transformation from sympodial to monopodial growth is correlated with gametangia developing at the apex of secondary modules, a condition called cladocarpy.

2.4.1.7 Asexual reproductive structures

Clonal reproduction via asexual gemmae or vegetative propagules is widespread in mosses. Gemmae germinate following a pattern that recapitulates the ontogeny of the whole plant, whereas propagules already have a differentiated apical cell. Thus leaves and rhizoids may occur on propagules but not on gemmae. Many families of mosses develop some kind of mitotic diaspores (Correns 1899), and only in a few cases do these arise endogenously. Vegetative diaspores can occur on any of the organs of the leafy gametophore and as early as the protonematal phase (Imura & Iwatsuki 1990; Newton & Mishler 1994). Foliar gemmae may originate from any part of the leaf, but are often localized toward the margins and the end of the midrib. Although common in nature, rhizoidal tubers and protonematal gemmae may only be formed in culture in some groups (Arts 1987) and sometimes their development is triggered only by application of abscissic acid (Mallón *et al.* 2006). In some families abscission occurs at the tmema cell, a specialized cell that is unique among land plants (Duckett & Ligrone 1992; Ligrone *et al.* 1996).

2.4.2 The sporophyte

Following fertilization of the egg, a zygote is formed, and the embryo develops within a protective tissue derived from the archegonium and the stem. At maturity the sporophyte is composed of three parts: the foot, the seta, and the capsule. The seta, which may elevate the capsule above the protecting perichaetial leaves (Figure 2.19), lacks multicellular appendages (but may be variously ornamented, e.g., see *Calyptrochaeta cristata*), and is always unbranched, carrying a single terminal sporangium. Only in *Sphagnum* and *Andreaea* is the capsule elevated by a pseudopodium of gametophytic origin (Figure 2.14). Differential thickening of the epidermal and cortical cell walls results in torsion and hence movement of the seta upon hydration. The axial cells are often elongate, narrow, and thin-walled, composing a strand of putative water-conducting cells; such tissue may be differentiated in the sporophyte axis even if it is lacking in the gametophyte. The capsule is defined by an exothecium (outer tissue), which often bear stomata. Guard cells occur in pairs, even if the division between them is not perfect (e.g., *Funaria*). The capsule is unilocular, with a central axis, that is, the columella, penetrating and in most mosses extending to the operculum, that is, the lid of the capsule. In a few cases (e.g., *Schistidium*) the columella may be shed attached to the operculum (i.e., systylious). The single, continuous sporogenous tissue thus overarches (i.e., in *Sphagnum* and *Andreaea*) or surrounds the central columella. Only in *Archidium* is the sterile axis lacking in the sporangium (Snider 1975).

The shape of the capsule varies conspicuously among mosses, and even within families. Capsules may be spherical, cylindrical and radially symmetric, or elongate, curved and bilaterally symmetric. The capsule may be tiny such as *Micromitrium* (Figure 2.18) or massive, as in *Buxbaumia*. In the

Splachnaceae, the sporophyte exhibits the broadest variation in shape, color, and odor! Half of the species occur exclusively on dung or other animal remains and recruit insects to disperse their spores to this patchily distributed habitat. The spore dispersal syndrome includes brightly colored sporangia (Figure 2.20), sticky spores and the release of volatile compounds mimicking the odor of the substrate, sought by insects to feed on or upon which to lay their eggs (Marino *et al.* 2009).

2.4.2.1 Sporogenesis and dehiscence

Sporogenesis typically begins when the seta concludes its growth. Meiosis is nearly synchronous and all spores reach maturity simultaneously. The mode of dehiscence of the capsule varies among mosses. In *Takakia*, a single longitudinal line extends along a spiral between the poles of the capsule (Figure 2.13). In *Andreaea* (Figure 2.15) and *Andreaeobryum*, four (or more) straight lines define valves that often remain attached at their apex. In *Sphagnum* and most other mosses, the apical region of the capsule is differentiated into a lid (i.e., operculum), which falls off at maturity. The line of dehiscence is often defined by modified cells of the exothecium and adjacent inner cells. In operculate mosses, a ring of cells aiding in dehiscence, called the annulus, may vary from cells with thickened outer walls that remain attached to the exothecium (or operculum) to complex and multistratose structures that fall independently (i.e., revoluble). Although dehiscence is irreversible, the dispersal of spores is nearly instantaneous only in *Sphagnum* whose capsule explodes at maturity (Duckett *et al.* 2009). In most other mosses, spore dispersal is spread through time, either through the movement of the capsule wall closing or opening the space along the line of the dehiscence, or through the movement of tiny teeth lining the capsule mouth of operculate mosses (Pfaehler 1904). In the Polytrichopsida, a thin sheath bearing marginal perforations and spanning the sporangial opening (i.e., epiphragm) prevents the loss of the entire spore mass; the capsule is built like a salt-shaker (Figure 2.16).

2.4.2.2 Peristomes

Most mosses develop one or two rings of teeth around their capsule mouth, that is, the peristome (Figure 2.17). The teeth are built from entire cells in the Polytrichopsida and only from remnants of adjacent cell walls in the Bryopsida, the largest clade of mosses. At its most complex, three concentric layers of the amphithecium contribute to the formation of two rings, each of 16 teeth: the outer ring or exostome and the inner ring or endostome. The number of cells in each amphithecial layer varies and determines whether one, one and a half, or two columns of cells contribute to each tooth. The ontogeny of the teeth ends with the deposition of additional material on adjacent periclinal walls and the resorption of all anticlinal and the remaining periclinal walls (Goffinet *et al.* 1999). Following dehiscence of the operculum, one or two rings of triangular or linear teeth are exposed, forming a dome over the capsule mouth. The degree of thickening on the outer and inner surface determines the movement of the teeth as atmospheric moisture changes, with teeth bending toward the thicker side, upon dehydration. Peristomes that bend outward when moist and inward, closing the capsule mouth, when dry (i.e., hydrocastique) typically characterize epiphytes whereas peristomes that bend outward when dry and inward when moist (i.e., xerocastique) are most often encountered in terrestrial mosses (Mueller & Neumann 1988). Three main peristome types are distinguished at maturity, depending on the presence of an exostome, on the architecture of the teeth, and on the arrangement of inner and outer teeth. These peristome types define major evolutionary lineages within the Bryopsida and the ontogeny of the amphithecium is highly conserved within these (Goffinet *et al.* 1999), like fundamental floral architecture is within lineages of angiosperms.

2.4.2.3 The seta

The sporangium of mosses is typically elevated above the leaves surrounding the archegonia by a seta, which unlike in liverworts develops prior to rather than after sporangial differentiation. The initial mitotic division of the zygote is transverse. Only the upper or epibasal cell is subject to divisions contributing to the actual sporophyte, through the activity of a soon-differentiated apical cell with two cutting faces. This apical cell, however, soon ceases development, immediately after the differentiation of the presumptive tissues of the sporangium. The seta develops from the activity of a new meristem initiated below the presumptive sporangium, which is unique among land plants. The seta thus grows through the basipetal activity of this intercalary meristem (i.e., adding new cells downward) rather than through mere cell elongation as in liverworts. The intercalary meristem also yields some derivatives acropetally, contributing cells to the distal end of the seta and the base of the sporangium. In *Oedipodium*, representing the lineage sister to all peristomate mosses, the bipolar asymmetry of the meristematic activity results in a short seta and a long sterile axis, the pseudoseta, subtending the sporangium. The pseudoseta differs from the true seta by the presence of stomata and air-chambers, features otherwise only known from the capsule (Shimamura & Deguchi 2008).

2.4.2.4 The role of the maternal calyptra

Another unique feature of the moss sporophyte pertains to the role of the calyptra (Figure 2.7) in controlling its development. Upon fertilization the archegonium of mosses and the subtending cauline tissue resume growth forming a gametophytic protective tissue around the embryo. This epigonium is of determinate growth and will, as the sporophyte elongates, tear into a basal vaginula, forming a collar around the foot of the young sporophyte, and a hood or calyptra covering the apex of the developing sporophyte. At this time, the apical meristem has ceased activity and growth of the sporophyte is accounted for only by the activity of the intercalary meristem. Although the calyptra is detached from the maternal gametophyte, it remains a viable tissue. Removal of the calyptra prior to sporogenesis results in the abortion or malformation of the sporophyte (Zielinski 1909). Whether the calyptra exerts a physical pressure or a physiological control on the sporophyte was initially ambiguous (Bopp 1956). Bopp and Stehle (1957) noticed that water moved faster within sporophytes with the calyptra removed, suggesting that the calyptra may function to prevent water loss in the sporophyte. This hypothesis is supported by the presence of a multilayered cuticle on the calyptra at a time when the sporophyte is still poorly differentiated (Budke *et al.* 2011, 2012). The calyptra is truly a unique organ with an unparalleled role among land plants.

2.4.2.5 Branching

Mosses, like other bryophytes, are monosporangiophytes, with each sporophyte developing a single sporangium at the apex of the axis. Branched setae have, however, been reported as teratologies (e.g., Leitgeb 1876; Györffy 1929); seemingly in all cases the capsules are sessile on a single seta, suggesting that they grew following the division of the apical cell in the earliest stages of sporophyte development (i.e., prior to the differentiation of the intercalary meristem). Since such anomalies cannot be predicted, the ontogeny of the branches, in particular the origin of their apical meristem, remains unknown.

2.4.3 Evolutionary trends

Conspicuous transformations of the gametophyte body along the backbone phylogeny of mosses are rather ambiguous as revealed by the overall weakness of the lineages inferred from these characters (Newton *et al.* 2000). The macroevolutionary history of mosses continues to be resolved (Cox *et al.* 2010), but critical internal nodes lack strong character support (Figure 2.25), precluding meaningful reconstruction of ancestral states of critical characters and hence, inferences are limited and tentative.

Nearly half of all extant mosses arose following the transfer of perichaetia from stems or mature vegetative branches (acrocarpy) to short specialized branches (pleurocarpy; Figure 2.25). However, the significance of cladocarpy in this transition remains controversial (La Farge-England 1996; Newton 2007), and the exact innovation that may have triggered the rapid diversification of pleurocarpous mosses is ambiguous, although the evolution of the angiosperm forest has been suggested as the impetus for pleurocarp radiation (Kürschner & Parolly 1999).

The polarity in the transformation between unisexual and bisexual gametophytes is not evident at a macroevolutionary scale. However, sexuality is thought to be correlated to other characters such as size of the gametophyte and spores and chromosome number (Crawford *et al.* 2009). *Takakia*, potentially the sister taxon to all remaining mosses, has the lowest chromosome number (i.e., $n = 4$) and is dioicous, and may thus mark the plesiotypic condition but its gametophyte is tiny, a trait inconsistent with the prediction that unisexual mosses develop larger vegetative plants.

Simplification of the gametophyte architecture is rampant throughout the mosses, and ranges from the loss of a costa in the leaf, to the loss of papillae

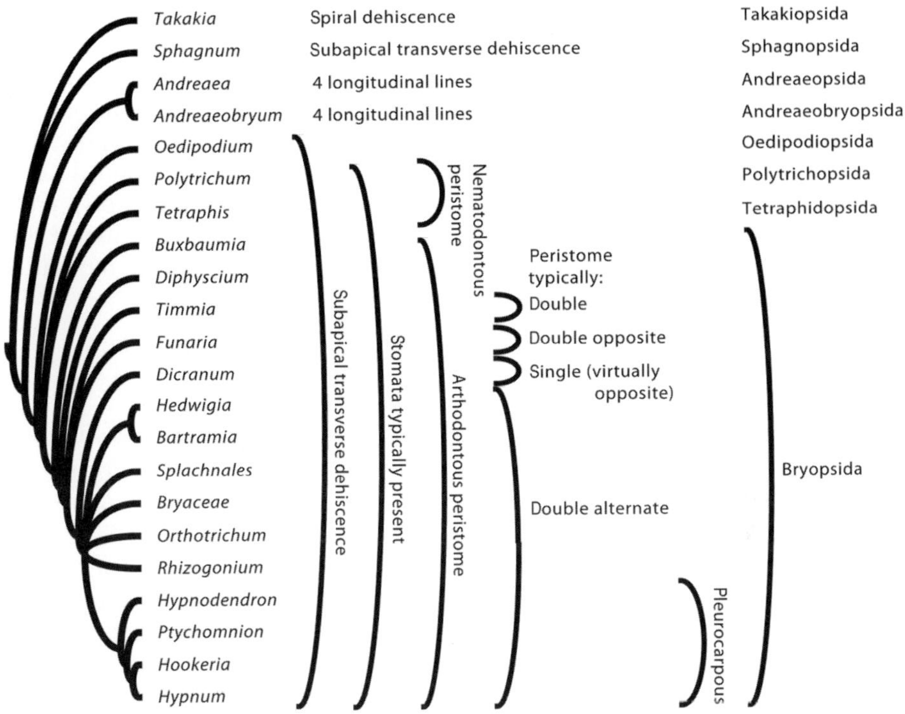

Figure 2.25 Summary phylogeny of mosses modified from Vanderpoorten and Goffinet (2009) and based on Cox *et al.* (2010) with characterization of morphological traits based on Goffinet *et al.* (2009).

on the laminal cells. Extreme reduction may result in the complete absence of foliate vegetative shoots, which is compensated for by the presence of a long-lived or perennial photosynthetic protonema, the juvenile stage in gametophyte development. Sex organs are developed on tiny buds as in *Ephemeropsis*, *Buxbaumia*, and in some species of *Pogonatum*, where the photosynthetic function is confined entirely to the protonemata, gametangial leaves/bracts and juvenile sporophytes.

Sporophyte evolution within mosses is characterized by increased complexity culminating in the elaboration of an arthrodontous peristome, and its subsequent modification into various types (Figure 2.25). Within orders and families of mosses, the architecture of the sporophyte, and in particular of its peristome may, however, not be universally conserved within subordinate lineages. Shortening of the seta resulting in the capsule immersed among the perichaetial leaves, and loss of precise sporangial dehiscence are common throughout mosses and are often linked to shifts in life strategy and habitat (e.g., Vitt 1984; Buck *et al.* 2000; Huttunen *et al.* 2004). Peristome architecture is often simplified through the loss of one or both rings, the reduction of the number of teeth, or the shortening of the teeth. In some cases, the simplification of the architecture has resulted from the paedomorphic development of the sporophyte (Shaw *et al.* 2000), by skipping the stages of peristome formation preceding sporangial differentiation. One pertinent example of a family of mosses exhibiting diverse sporophyte morphologies is the Funariaceae, with *Funaria* characterized by a long seta, a highly differentiated annulus releasing the operculum and a well-developed double peristome, and *Physcomitrella*, a highly derived taxon within the family, with an immersed capsule, dehiscing through the irregular disintegration of the capsule wall (i.e., cleistocarpy), and consequently lacking even the remnants of a peristome.

2.5 Evolution of plant form in hornworts

The Anthocerotophyta comprise approximately 200 species (Villarreal *et al.* 2010), and are the putative sister group, among extant land plants, to the polysporangiophytes, including the vascular plants. The species are annual or short-lived perennials, in some cases regenerating from tubers. They grow on soil or trees in mesic or hydric habitats, and seldom occur directly on rocks except in very wet environments. Hornworts may resemble thalloid liverworts but differ most conspicuously in the architecture of the sporophyte and microscopically by the immersed gametangia, lack of oil bodies in the cells, and in some species by the presence of single large chloroplasts with a pyrenoid, a region of the chloroplast characterized by the accumulation of proteins responsible for carbon fixation.

2.5.1 The gametophyte

The vegetative stage of the hornwort life cycle is always a dorsi-ventral plagiotropic thalloid individual (Figure 2.6), with little internal differentiation of tissues. The thallus of *Anthoceros* and its close relatives, as well as members of *Dendroceros* subg. *Apoceros*, is characterized by conspicuous schizogenous cavities filled with mucilage and ultimately harboring *Nostoc*, which either forms globose colonies or elongate axial strands (Villarreal & Renzaglia 2006). In other taxa, the thallus is solid at maturity because the small space below the ventral mucilage cleft cavities fails to expand unless penetrated by a cyanobacterium through the pore defined by two or more guard cells. The single apical cell is wedge-shaped with four cutting faces (two lateral, a ventral, and a dorsal), except in *Dendroceros* that has a hemidiscoid apical cell with three cutting faces (two lateral and an anterior axial one). The shape of the meristematic cell determines the growth form (rosette vs. ribbon) and typically also the architecture of the thallus (gradual thinning of the thallus toward the margin vs. presence of a prominent midrib). In *Nothoceros* the thallus wings may be partially incised but true leaves are always lacking.

2.5.1.1 Branches

Branching results from the division of the apical cell, and is thus truly dichotomous. Lateral branches are lacking and orbicular growth forms are achieved following multiple rapid terminal branching events.

2.5.1.2 Appendages

Appendages such as scales are lacking from the thallus surfaces, but several species develop numerous dorsal or marginal outgrowths, whose function is ambiguous, besides serving as potential diaspores.

2.5.1.3 Modification of thallus

The single conspicuous innovation through modification of the thallus is the development of a collar or involucre around the base of the sporophyte, derived from epidermal cells. The collar may protect the sporophyte from desiccation during the early ontogenetic stages, and possibly may prevent the mature sporophyte from breaking off the maternal thallus.

2.5.1.4 Rhizoids

Rhizoids are always unicellular but often branch at their apex. They arise through elongation or following the periclinal division of an epidermal cell. Confined to the ventral epidermis in wild populations, they can develop from dorsal cells in culture (Renzaglia 1978).

2.5.1.5 Sex organs

Gametangia develop on or below the dorsal epidermis. Archegonial initials derive from an anticlinal division of a surface cell. Archegonia develop surrounded by thalline cells, are sunken at maturity within the thallus, with the canal opening at the surface. Antheridial initials are borne from the periclinal division of an epidermal cell. The upper daughter cell gives rise to the roof of the chamber and forms the actual initial. These two cells are soon separated by a schizogenous cavity within which antheridia develop. Antheridial development has thus shifted from a strictly epidermal cell as in other embryophytes to a subepidermal cell in hornworts. The initial periclinal division of the epidermal cell may be followed by numerous additional divisions leading to the development of multiple antheridia per chamber in some taxa (Renzaglia *et al.* 2000). The mode of gametangial initiation precludes male and female sex organs forming mixed clusters, but many species are monoicous with archegonia and antheridia coexisting on single lobes.

2.5.1.6 Asexual reproductive structures

These are produced by various species but consist primarily of ventral or marginal tubers, marginal lobules, and dorsal gemmae. Specialized structures such as differentiated gemmiferous lobes or cups are never developed. Clonal propagation is, in some cases (such as in Southern Appalachian populations of *Nothoceros aenigmaticus*), the sole known mode of reproduction.

2.5.2 The sporophyte

The name "hornwort" refers to the horn-like sporophyte emerging from the maternal thallus (Figure 2.12). Unlike in typical liverworts and mosses, the sporangium of hornworts is not subtended by a sterile stalk, the seta. Growth of the sporophyte is not driven by the activity of an apical cell (as in liverworts); instead a meristematic tissue at the base of the sporangium adds cells acropetally (in the direction of the apex), unlike in mosses where the intercalary meristem adds cells to the seta below the sporangium. The capsule thus grows almost indeterminately, and shortly above the meristem four distinct tissues are differentiated: the wall, often bearing stomata, the assimilative layer, the sporogenous layer and the central columella. Like in liverworts, diploid sterile filaments, elaters (or also called pseudoelaters), are intermixed with the spores. Sporogenesis is asynchronous, and a complete basipetal ontogenetic gradient can be observed along the axis of the mature sporangium: from sporocyte differentiation near the base, to meiosis and ultimately spore maturation toward the apex. Not surprisingly, dehiscence proceeds from the distal end downward, against the gradient of spore ontogeny. Spore release is thus gradual, and spread through time. Sporangial dehiscence follows typically two longitudinal suture lines. The dispersal of the spores is facilitated by the hygroscopic movements of the elaters, the columella, and even the capsule valves.

2.5.3 Evolutionary trends

Thallus anatomy is fairly homogenous within hornworts, and conspicuous shifts in growth form or thallus architecture may be correlated with the shape of the apical cell. Recent progress in elucidating the relationships among hornworts (Figure 2.26) suggests that the ribbon-like thallus resulting from the activity of the hemidiscoid apical cell arose from the rosette growth form and the wedge-shaped meristematic cell. The number of antheridia per cavity decreases from 30 to 80 in *Leiosporoceros* to one in the Dendrocerotaceae, a trend preceded by a shift from tiered to nontiered organization of the cells of the antheridial jacket (Renzaglia *et al.* 2009).

Sporophyte reduction characterizes a single lineage of hornworts, the genus *Notothylas*. Its sporophytes are small and nearly completely enclosed in the gametophytic collar typically surrounding only the base of the hornwort sporophyte. These reduced sporophytes lack stomatal guard cells (Figure 2.26), a loss that also characterizes all members of the Dendrocerotaceae with tall sporophytes.

2.6 The ancestral developmental toolbox of land plants

All living plant cells perform universal metabolic functions such as protein synthesis, and the pathways likely rely on genes that are homologous across

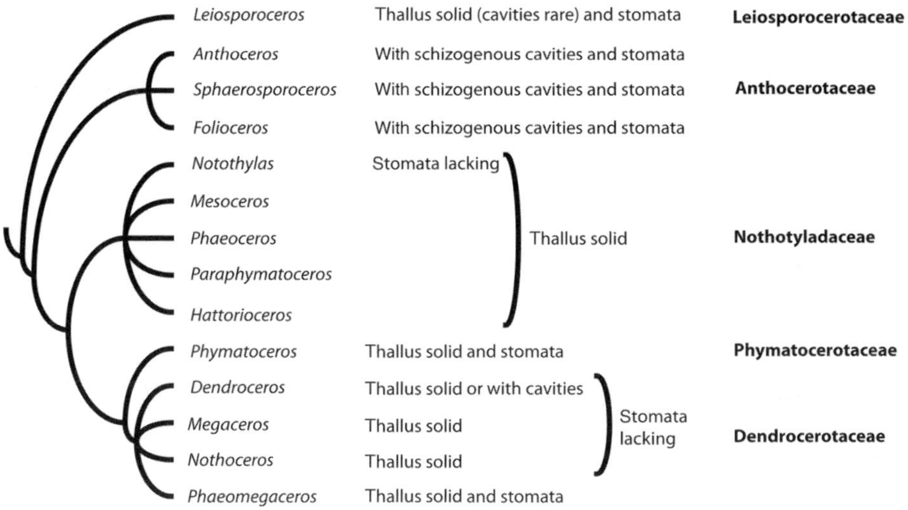

Figure 2.26 Summary phylogeny of hornworts modified from Vanderpoorten & Goffinet (2009) and based on Duff *et al.* (2007) with characterization of morphological traits based on Renzaglia *et al.* (2009).

land plants and expressed regardless of the ploidy level of the cell, as evidenced by the broad overlap in populations of expressed genes in a moss, a liverwort, and the flowering plant, *Arabidopsis* (Nishiyama *et al.* 2003; Chung *et al.* 2006). Whether this concept of a fundamental set of housekeeping genes can be extended to include genes involved in specific ontogenetic processes, such as the development of leaves in mosses and vascular plants, for example, is at the core of evolutionary developmental genetics or evo-devo in plant science: are specific ontogenetic processes controlled by similar genes in distinct lineages and how has gene evolution shaped morphological and developmental innovations in plants?

Cladogenic events, in particular at the macroevolutionary scale are marked by significant innovations in the ontogeny and architecture of the plant body. The fundamental concept in evolutionary biology is that of descent with modification, whereby traits are transformed overtime and hence innovations consist basically in the modification of preexisting or ancestral traits. In the context of developmental genetics, innovations arise from spatial or temporal shifts in gene expression, following or prior to gene duplication (Carroll 2000). Histological and thereby organ differentiation is triggered by homeotic genes that control the expression of genes involved in a developmental network (Nishiyama 2007; Langdale & Harrison 2008) or by small RNAs through posttranscriptional gene control (Axtell *et al.* 2007). Although various genes central to or shaping developmental networks in flowering plants have homologs in bryophytes (Axtell 2009; Rensing *et al.* 2009), their function remains to be elucidated.

The relative ease by which the genome of *Physcomitrella patens* can be transformed (Cove 2005; Cummings 2009) as well as the assembly of its nuclear genome (Rensing *et al.* 2008) has raised this species to the rank of a model organism in developmental biology (Cove *et al.* 2006). Similarly, the ability to introduce mutations into the genome of the complex thalloid liverwort *Marchantia polymorpha* (Ishizaki *et al.* 2008) and the upcoming assembly of its genome should allow for critical investigations in the nature and role of genes expressed during fundamental stages in the development of its body. Comparative genomic studies have highlighted the conservation of critical ontogenetic genes or gene families during the macroevolution of land plants and provided the foundation for the concept of an ancestral developmental tool kit in land plants (Floyd & Bowman 2007; Singer & Ashton 2007; Hedman *et al.* 2009), whereby the fundamental tools to develop distinct tissues or organs are inherited and modified in the course of evolution.

The best example illustrating this concept was provided by Menand *et al.* (2007) for the origin of root hairs in seed plants. These hairs are confined near the apex of the root, and consist of uniseriate filaments that resemble the rhizoids anchoring the vegetative body of bryophytes to the substrate. The development of these hairs is controlled in *Arabidopsis* by two transcription factors, whose coding genes are represented by homologs in the *Physcomitrella* genome, raising the possibility that these transcription factors may be involved in the formation of rhizoids in bryophytes. Reciprocal transformations of *Arabidopsis* and *Physcomitrella* revealed that the genes could be interchanged without affecting the phenotype of the plants, suggesting that the machinery controlling the development of gametophytic rhizoids in the bryophyte was co-opted for the development of root hairs on the sporophyte of *Arabidopsis*. Thus, the root hair *per se* is not strictly speaking an innovation, rather the shift in the expression of the transcription factors controlling uniseriate filaments from the haploid to the diploid phase is.

Whether such a scenario may also apply to the development of leaves is unlikely, considering that photosynthetic appendages have arisen multiple times among the vascular plants, that their developmental genetic pathways may not be homologous across lineages of polysporangiophytes (Tomescu 2008), and that the initiation of leaf ontogeny is determined by cells autonomously, suggesting that the genetic regulation of shoot development is fundamentally different in mosses and vascular plants (Harrison *et al.* 2009). Some of the genes critical to leaf development have homologs in *Physcomitrella* and may have provided at least some of the essential requirements for the evolution of the leaf, suggesting that the leaf itself may not be homologous across plants, but that parts of its ontogenetic pathways may be. However, presence of the gene alone is insufficient to infer functional homology as, for example, the KNOX genes that are implicated in leaf development in *Arabidopsis* are present in *Physcomitrella* but only expressed in the leafless sporophyte (Sakakibara *et al.* 2008).

The macroevolutionary history of land plants reflects not only mere shifts in gene and discrete trait expression but also significant neofunctionalization of genes: genes are conserved across embryophytes but their function changed through time. For example, the transcription factor that controls flower development by triggering the transition from a vegetative to a reproductive phase resulting in the differentiation of floral organs is also encoded in the *Physcomitrella* genome. In the moss, the gene coding for this factor (*LFY*) is duplicated and expressed in the haploid and diploid stages (Tanahashi *et al.* 2005). The role of *LFY* in gametophyte development is unclear; the genes are expressed in shoot apices and nascent archegonia, but their silencing has no phenotypic effect. By contrast, the disruption of *LFY* expression resulted in abnormalities of the sporophyte, and double mutants in fact failed to develop sporophytes despite successful fertilization: *LFY* expression is essential for the first division of the zygote. Homologs of other developmental genes have been identified in mosses, but their expression is sometimes broad throughout the plant and their precise role in the development of the moss body remains ambiguous, for example, as in the case of MIKC*-type genes, a set of class II MADS-Box genes (Hedman *et al.* 2009). These genes occur in all streptophytes (at least *Chara* and embryophytes) but their expression has shifted from the haploid phase only, to a broad expression in either phase in bryophytes, to a constrained, more discrete expression during the sporophyte ontogeny in vascular plants. In *Physcomitrella* their silencing alters gametangial formation, leaf shape, sporangial maturation, and capsule size.

Like other plants, bryophytes synthesize hormones that regulate growth of the sporophyte and the gametophyte (Cooke *et al.* 2003; Poli *et al.* 2003; Schwartzenberg 2009) through the activation of distinct ontogenetic pathways. Great progress has been made in identifying genes expressed during the development of the bryophyte, mostly the moss body, but these studies have essentially focused on the evolution of the function of these genes along the macroevolutionary tree of plant life. Bryophytes exhibit a broad spectrum of morphological innovations that distinguish liverworts, mosses, and hornworts, and also major lineages within these divisions. Elucidating the genetic pathways linked to the ontogeny of bryophyte-specific structures requires comparative studies within bryophytes. Primary targets may be the homology of the genetic pathways underlying the various modes of branching in leafy Jungermanniopsida, the transformation of the terete to the thalloid gametophyte in liverworts, the development of the moss peristome, and in particular of the distinct arthrodontous types, the differentiation of tissues within the sporangium and shifts in symmetry of the capsule, the adaxial/abaxial polarity in the costa of moss leaves, the translocation of sex organs to short lateral branches in pleurocarpous mosses, or the potential homology of gametophytic mucilage clefts and sporophytic stomatal guard cells in hornworts. The emergence of *Ceratodon purpureus*, a moss with an architecturally more complex vegetative and spore-producing body than that of *Physcomitrella*, as the next model system in evolutionary developmental genetics will hopefully encourage such comparative studies in the mosses.

Acknowledgments

The authors thank members of the Goffinet Lab (Jessica Budke, Laura Forrest, Lily Lewis, Yang Liu, and Juan Carlos Villarreal) for constructive criticism on an earlier version of this chapter. BG and WB acknowledge support from the National Science Foundation (DEB-0919284 & EF-0531557, and DEB-0948380, respectively). We thank Neil Bell, Jessica Budke, Marianna Ricca, Ricardo Rozzi, and Juan Carlos Villarreal for sharing pictures.

References

Allen, B.H. & Magill, R.E. (1987) In support of a distinct terminology for bryophyte sexuality. *Taxon*, **36**, 57–58.

Arts, T. (1987) Rhizoïdbroedkorrels of tubers bij acrocarpe mossen. *Muscillanea*, **6**, 5–17.

Axtell, M.J. (2009) The small RNAs of *Physcomitrella patens*: expression, function and evolution. In: *Annual Plant Reviews*, Vol. 36, *The Moss Physcomitrella patens*, (eds C. Knight, P.-F. Perroud & D. Cove). pp. 113–142. Blackwell Publishing Ltd., Oxford, UK.

Axtell, M.J., Snyder, J.A. & Bartel, D.P. (2007) Common functions for diverse small RNAs of land plants. *The Plant Cell*, **19**, 1750–1769.

Bateman, R.M. (1996) Nonfloral homoplasy and evolutionary scenarios in living and fossil land plants. In: *Homoplasy: the Recurrence of Similarity in Evolution* (eds M.J. Sanderson & L. Hufford). pp. 91–130. Academic Press, New York.

Berthier, J. (1972) Recherches sur la structure et le développement de l'apex du gamétophyte feuillé des mousses. *Revue Bryologique et Lichénologique*, **38**, 421–551.

Bopp, M. (1956) Die Bedeutung der Kalyptra für die Entwicklung der Laubmoossporogone. *Berichte der Deutschen Botanischen Gesellschaft*, **69**, 455–468.

Bopp, M. & Stehle, E. (1957) Zur Frage der Wasserleitung im Gametophyten und Sporophyten der Laubmoose. *Zeitschrift für Botanik*, **45**, 161–174.

Buck, W.R., Goffinet, B. & Shaw, A.J. (2000) Testing morphological concepts of orders of pleurocarpous mosses (Bryophyta) using phylogenetic reconstructions based on *trn*L-*trn*F and *rps*4 sequences. *Molecular Phylogenetics and Evolution*, **16**, 180–198.

Budke, J.M., Jones, C.S. & Goffinet, B. (2011) A hundred year old question: is the moss calyptra covered by a cuticle? A case study of *Funaria hygrometrica*. *Annals of Botany*, **107**, 1279–1286.

Budke, J.M., Goffinet, B. & Jones, C.S. (2012) The calyptra cuticle matures before the sporophyte cuticle in the moss *Funaria hygrometrica*. *American Journal of Botany*, **99**, 14–22.

Burges, A. (1949) The genus *Dawsonia*. *Proceedings of the Linnean Society of New South Wales*, **74**, 83–96.

Carroll, S.B. (2000) Endless forms: the evolution of gene regulation and morphological diversity. *Cell*, **101**, 577–580.

Chang, Y. & Graham, S.W. (2011) Inferring the higher-order phylogeny of mosses (Bryophyta) and relatives using a large, multigene plastid data set. *American Journal of Botany*, **98**, 839–849.

Chung, S.J., Takeshi, K., Sakai, A., *et al.* (2006) Hybridization of an *Arabidopsis* cDNA array with bryophyte cDNA. *Journal of the Hattori Botanical Laboratory*, **99**, 233–244.

Cooke, T.J., Poli, D. & Cohen, J.D. (2003) Did auxin play a crucial role in the evolution of novel body plans during the Late Silurian–Early Devonian radiation of land plants? In: *Evolution of Plant Physiology* (eds A.R. Hemsley & I. Poole). pp. 85–107. Academic Press, London.

Correns, C. (1899) *Untersuchungen über die Vermehrung der Laubmoose durch Brutorgane und Stecklinge*. Gustav Fischer, Jena.

Cove, D. (2005) The moss *Physcomitrella patens*. *Annual Review of Genetics*, **39**, 339–358.

Cove, D., Bezanilla, M., Harries, P., *et al.* (2006) Mosses as developmental systems for the study of metabolism and development. *Annual Review of Plant Biology*, **57**, 497–520.

Cox, C.J., Goffinet, B., Wickett, N.J., Boles, S.B. & Shaw, A.J. (2010) Moss diversity: a molecular phylogenetic analysis of genera. *Phytotaxa*, **9**, 175–195.

Crandall, B. (1969) Morphology and development of branches in the leafy Hepaticae. *Beihefte zur Nova Hedwigia*, **30**, 1–261.

Crandall-Stotler, B. (1972) Morphogenetic patterns of branch formation in the leafy Hepaticae—a résumé. *The Bryologist*, **75**, 381–403.

Crandall-Stotler, B. (1980) Morphogenetic designs and a theory of bryophyte origins and divergence. *BioScience*, **30**, 580–585.

Crandall-Stotler, B., Stotler, R. & Long, D.G. (2009) [2008] Morphology and classification of the Marchantiophyta. In: *Bryophyte Biology* (eds B. Goffinet & A.J. Shaw), 2nd edn. pp. 1–54. Cambridge University Press, Cambridge.

Crawford, M., Jesson, L.K. & Garnock-Jones, P.J. (2009) Correlated evolution of sexual system and life-history traits in mosses. *Evolution*, **63**, 1129–1142.

Cronberg, N., Natcheva, R. & Hedlund, K. (2006) Microarthropods mediate sperm transfer in mosses. *Science*, **313**, 1255.

Crosby, M.R., Magill, R.E., Allen, B., *et al.* (2000) *A Checklist of the Mosses*. Missouri Botanical Garden, St. Louis, MO.

Cummings, A.C. (2009) [2008] Mosses as model organisms for developmental, cellular, and molecular biology. In: *Bryophyte Biology* (eds B. Goffinet & A.J. Shaw), 2nd edn. pp. 199–236. Cambridge University Press, Cambridge.

Duckett, J.G. & Ligrone, R. (1992) A survey of diaspore liberation mechanisms and germination patterns in mosses. *Journal of Bryology*, **17**, 335–354.

Duckett, J.G., Ligrone, R., Andrews, N., *et al.* (2000) The enigma of pegged and swollen rhizoids in marchantialean hepatics: a functional explanation. *American Journal of Botany*, **87S**, 6–7.

Duckett, J.G., Pressel, S., P'ng, K.M., *et al.* (2009) Exploding a myth: the capsule dehiscence mechanism and the function of pseudostomata in *Sphagnum*. *New Phytologist*, **183**, 1053–1063.

Duckett, J.G., Schmid, A.M. & Ligrone, R.S. (1998) Protonemal morphogenesis. In: *Bryology for the Twenty-First Century* (eds J.W. Bates, N.W. Ashton & J.G. Duckett). pp. 223–246. Maney Publishing, Leeds.

Duff, R.J., Villarreal, J.C., Cargill, D.C. & Renzaglia, K.S. (2007) Progress and challenges toward developing a phylogeny and classification of the hornworts. *The Bryologist*, **110**, 214–243.

Evans, A.W. (1939) The classification of the Hepaticae. *Botanical Review*, **5**, 49–96.

Floyd, S.K. & Bowman, J.L. (2007) The ancestral developmental tool kit of land plants. *International Journal of Plant Sciences*, **168**, 1–35.

Forrest, L.L., Davis, E.C., Long, D.C., *et al.* (2006) Unravelling the evolutionary history of the liverworts (Marchantiophyta)—multiple taxa, genomes and analyses. *The Bryologist*, **109**, 303–334.

Frey, W. (1970) Blattentwickluung bei Laubmoosen. *Nova Hedwigia*, **20**, 464–556.

Garbary, D.J., Renzaglia, K.S. & Duckett, J.G. (1993) The phylogeny of land plants: a cladistic analysis based on male gametogenesis. *Plant Systematics and Evolution*, **188**, 237–269.

Goffinet, B. (2000) Origin and phylogenetic relationships of bryophytes. In: *Bryophyte Biology* (eds A.J. Shaw & B. Goffinet), 1st edn. pp. 124–149. Cambridge University Press, Cambridge.

Goffinet, B., Buck, W.R. & Shaw, A.J. (2009) [2008] Morphology, anatomy, and classification of the Bryophyta. In: *Bryophyte Biology* (eds B. Goffinet & A.J. Shaw), 2nd edn. pp. 55–138. Cambridge University Press, Cambridge.

Goffinet, B. & Shaw, A.J. (2009) [2008] *Bryophyte Biology*, 2nd edn. Cambridge University Press, Cambridge.

Goffinet, B., Shaw, A.J., Anderson, L.E., *et al.* (1999) Peristome development in mosses in relation to systematics and evolution. V. Orthotrichaceae. *The Bryologist*, **102**, 581–599.

Gradstein, S.R., Wilson, R., Ilkiu-Borges, A.L., *et al.* (2006) Phylogenetic relationships and neotenic evolution of *Metzgeriopsis* (Lejeuneaceae) based on chloroplast DNA sequences and morphology. *Botanical Journal of the Linnean Society*, **151**, 293–308.

Graham, L.E., Delwiche, C.F. & Mishler, B.D. (1991) Phylogenetic connections between the 'green algae' and the 'bryophytes.' *Advances in Bryology*, **4**, 213–244.

Györffy, I. (1929) Monstruoses Sporophyton von *Tetraplodon bryoides* aus Suomi. *Annales Botanici Societatis Zoologicæ-Botanicæ Fennicæ "Vanamo"*, **9**, 299–319.

Harrison, C.J., Roeder, A.H.K., Meyerowitz, E.M., *et al.* (2009) Local cues and asymmetric cell divisions underpin body plan transitions in the moss *Physcomitrella patens*. *Current Biology*, **19**, 461–471.

Hedman, H., Källman, T. & Lagercrantz, U. (2009) Early evolution of MFT-like gene family in plants. *Plant Molecular Biology*, **70**, 359–369.

Huttunen, S., Ignatov, M.S., Müller, K., *et al.* (2004) Phylogeny and evolution of epiphytism in the three moss families Meteoriaceae, Brachytheciaceae, and Lembophyllaceae. In: *Molecular Systematics of Bryophytes*, Vol. 98 (eds B. Goffinet, V. Hollowell & R. Magill). pp. 328–361. Missouri Botanical Garden, St. Louis, MO.

Ignatov, M.S. & Hedenäs, L. (2007) Homologies of stem structures in pleurocarpous mosses, especially of pseudoparaphyllia and similar structures. In: *Pleurocarpous Mosses: Systematics and Evolution*, Systematics Association Special Volume 71 (eds A.E. Newton & R.S. Tangney). pp. 269–286. CRC Press, Boca Raton, FL.

Imura, S. & Iwatsuki, Z. (1990) Classification of vegetative diaspores on Japanese mosses. *Hikobia*, **10**, 435–443.

Ishizaki, K., Chiyoda, S., Yamato, K.T., *et al.* (2008) *Agrobacterium*-mediated transformation of the haploid liverwort *Marchantia polymorpha* L., an emerging model for plant biology. *Plant & Cell Physiology*, **49**, 1084–1091.

Kato, M. & Akiyama, H. (2005) Origin of the sporophyte of land plants. *Taxon*, **54**, 443–450.

Kenrick, P. & Crane, P.R. (1997) *The Origin and early Diversification of Land Plants: a Cladistic Study*. Smithsonian Institution Press, Washington, DC.

von Konrat, M., Renner, M., Söderström, L., *et al.* (2008) Early land plants today: liverwort species diversity and the relationship with higher taxonomy and higher plants. *Fieldiana Botany*, **47**, 91–104.

Kürschner, H. & Parolly, G. (1999) Pantropical epiphytic rain forest bryophyte communities—coeno-syntaxonomy and floristic-historical implications. *Phytocoenologia*, **29**, 1–52.

La Farge-England, C. (1996) Growth form, branching pattern, and perichaetial position in mosses: cladocarpy and pleurocarpy redefined. *The Bryologist*, **99**, 170–186.

Langdale, J.A. & Harrison, C.J. (2008) Developmental transitions during the evolution of plant form. In: *Evolving pathways: Key Themes in Evolutionary Developmental Biology* (eds A. Minelli & G. Fusco). pp. 297–314. Cambridge University Press, Cambridge.

Leitgeb, H. (1876) Über verzweigte Moossporogonien. *Mitteilungen des Naturwissenschaftlichen Vereines für Steiermark*, **13**, 1–20.

Ligrone, R., Duckett, J.G. & Gambardella, R. (1996) Serial development of foliar gemmae in *Tortula* (Pottiales, Musci), an ultrastructural study. *Annals of Botany*, **78**, 305–315.

Mallón, R., Reinoso, J. & Rodríguez-Oubiña, J. (2006) *In vitro* development of vegetative propagules in *Splachnum ampullaceum*: brood cells and chloronematal bulbils. *The Bryologist*, **109**, 215–223.

Marino, P., Raguso, R. & Goffinet, B. (2009) The ecology and evolution of fly-dispersed dung mosses (Splachnaceae): manipulating insect behavior through odour and visual cues. *Symbiosis*, **47**, 61–76.

Menand, B., Yi, K., Jouannic, S., *et al.* (2007) An ancient function mechanisms controls the development of cells with rooting functions in land plants. *Science*, **316**, 1477–1480.

Mishler, B.D. (1988) Relationships between ontogeny and phylogeny, with reference to bryophytes. In: *Ontogeny and Systematics* (ed C.J. Humphries). pp. 117–136. Columbia University Press, New York.

Mishler, B.D. & Churchill, S.P. (1984) A cladistic approach to the phylogeny of the "bryophytes." *Brittonia*, **36**, 406–424.

Mueller, D.M.J & Neumann, A.J. (1988) Peristome structure and the regulation of spore release in arthrodontous mosses. *Advances in Bryology*, **3**, 135–158.

Newton, A.E. (2007) Branching architecture in pleurocarpous mosses. In: *Pleurocarpous Mosses: Systematics and Evolution*, Systematics Association Special Volume 71 (eds A.E. Newton & R.S. Tangney). pp. 287–307. CRC Press, Boca Raton, FL.

Newton, A.E., Cox, C.J., Duckett, J.G., *et al.* (2000) Evolution of the major moss lineages: phylogenetic analyses based on multiple gene sequences and morphology. *The Bryologist*, **103**, 187–211.

Newton, A.E. & Mishler, B.D. (1994) The evolutionary significance of asexual reproduction in mosses. *Journal of the Hattori Botanical Laboratory*, **76**, 127–145.

Newton, A. E., Wikström, N., Bell, N., *et al.* (2007) Dating the diversification of the pleurocarpous mosses In: *Pleurocarpous mosses: systematics and evolution*, Systematics Association Special Volume 71 (eds A.E. Newton & R.S. Tangney). pp. 337–366. CRC Press, Boca Raton, FL.

Nishiyama, T. (2007) Evolutionary developmental biology of nonflowering plants. *International Journal of Plant Sciences*, **168**, 37–47.

Nishiyama, T., Fujita, T., Shin-I, T., *et al.* (2003) Comparative genomics of *Physcomitrella patens* gametophytic transcriptome and *Arabidopsis thaliana*: Implication for land plant evolution. *Proceedings of the National Academy of Sciences of the United States of America*, **100**, 8007–8012.

Nishiyama, T., Wolf, P.G., Kugita, M., *et al.* (2004) Chloroplast phylogeny indicates that bryophytes are monophyletic. *Molecular Biology and Evolution*, **21**, 1813–1819.

Okano, Y., Aono, N., Hiwatashi, Y., *et al.* (2009) A polycomb repressive complex 2 gene regulates apogamy and gives evolutionary insights into land plant evolution. *Proceedings of the National Academy of Sciences of the United States of America*, **106**, 16321–16326.

Pfaehler, A. (1904) Étude biologique et morphologique sur la dissémination des spores chez les mousses. *Bulletin de la Société Vaudoise de Sciences Naturelles*, **40**, 41–132, pl. VI–XIV.

Poli, D.B., Jacobs, M. & Cooke, T.J. (2003) Auxin regulation of axial growth in bryophyte sporophytes: its potential significance for the evolution of early land plants. *American Journal of Botany*, **90**, 1405–1415.

Proctor, M.C.F. (2009) [2008] Physiological ecology. In: *Bryophyte Biology* (eds B. Goffinet & A.J. Shaw), 2nd edn. pp. 237–268. Cambridge University Press, Cambridge.

Pursell, R.A. (2007) Fissidentaceae. *Flora Neotropica Monograph*, **101**, 1–279.

Qiu, Y.-L., Li, L., Wang, B., *et al.* (2006) The deepest divergences in land plants inferred from phylogenomic evidence. *Proceedings of the National Academy of Sciences of the United States of America*, **103**, 15511–15516.

Rensing, S.A., Lang, D. & Zimmer, A.D. (2009) Comparative genomics. In: *Annual Plant Reviews*, Vol. 36, *The Moss Physcomitrella Patens* (eds C. Knight, P.-F. Perroud & D. Cove), pp. 42–75. Blackwell Publishing Ltd., Oxford, UK.

Rensing, S.A., Lang, D., Zimmer, A.D., *et al.* (2008) The *Physcomitrella* genome reveals evolutionary insights into the conquest of land by plants. *Science*, **319**, 64–69.

Renzaglia, K.S. (1978) A comparative morphology and developmental anatomy of the Anthocerotophyta. *Journal of the Hattori Botanical Laboratory*, **44**, 31–90.

Renzaglia, K.S. (1982) A comparative developmental investigation of the gametophyte generation in the Metzgeriales (Hepatophyta). *Bryophytorum Bibliotheca*, **24**, 1–253.

Renzaglia, K.S., Duff, R.J., Nickrent, D.L., *et al.* (2000) Vegetative and reproductive innovations of early land plants: implications for a unified phylogeny. *Philosophical Transactions of the Royal Society, Biological Sciences*, **355**, 769–793.

Renzaglia, K.S., Villarreal, J.C. & Duff, R.J. (2009) New insights into morphology, anatomy and systematics of hornworts. In: *Bryophyte Biology* (eds B. Goffinet & A. J. Shaw), 2nd edn. pp. 139–171. Cambridge University Press, Cambridge.

Robinson, H. (1970) Observations on the origin of the specialized leaves of *Fissidens* and *Schistostega*. *Revue Bryologique et Lichénologique*, **37**, 941–947.

Sakakibara, K., Nishiyama, T., Deguchi, H., *et al.* (2008) Class 1 KNOX genes are not involved in shoot developmental in the moss *Physcomitrella patens* but do function in sporophyte development. *Evolution and Development*, **10**, 555–566.

Sanders, H., Rothwell, G.W. & Wyatt, S. (2007) Paleontological context for the developmental mechanisms of evolution. *International Journal of Plant Sciences*, **168**, 719–728.

Schuster, R.M. (1966) *The Hepaticae and Anthocerotae of North America, East of the Hundredth Meridian*, Vol. 1. Columbia University Press, New York.

Schuster, R.M. (1984) Comparative anatomy and morphology of the Hepaticae. In: *New Manual of Bryology*, Vol. 2 (ed. R.M. Schuster). pp. 760–891. Hattori Botanical Laboratory, Nichinan.

Schwartzenberg, K. von (2009) Hormonal regulation of development by auxin and cytokinin in moss. In: *Annual Plant Reviews, The Moss Physcomitrella Patens* Vol. 36 (eds C. Knight, P.-F. Perroud & D. Cove). pp. 246–281. Blackwell Publishing Ltd., Oxford, UK.

Seppelt, R.D. & Selkirk, P.M. (1984) Effects of submersion of morphology and the implications of induced environmental modification on the taxonomic interpretation of selected Antarctic moss species. *Journal of the Hattori Botanical Laboratory*, **55**, 273–279.

Shaw, A.J., Anderson, L.E. & Mishler, B.D. (2000) Paedomorphic sporophyte development in *Bruchia flexuosa* (Bruchiaceae). *The Bryologist*, **103**, 147–155.

Shimamura, M. & Deguchi, H. (2008) Sporophyte anatomy of *Oedipodium griffithianum* (Oedipodiaceae). In: *Bryology in the New Millenium* (eds H. Mohamed, B.B. Baki, A. Nasrulhaq-Boyce *et al.*). pp. 319–325. University of Malaysia, Kuala Lumpur.

Singer, S.D. & Ashton, N.W. (2007) Revelation of ancestral roles of KNOX genes by functional analysis of *Physcomitrella* homologues. *Plant Cell Reports*, **26**, 2039–2054.

Snider, J.A. (1975) Sporophyte development in the genus *Archidium*. *Journal of the Hattori Botanical Laboratory*, **39**, 85–104.

Tanahashi, T., Sumikawa, N., Kato, M., *et al.* (2005) Diversification of gene function: homologs of the floral regulator *FLO/LFY* control the first zygotic cell division in the moss *Physcomitrella patens*. *Development*, **132**, 1727–1736.

Taylor, N.T., Kerp, H. & Hass, H. (2005) Life history biology of early land plants: Deciphering the gametophyte phase. *Proceedings of the National Academy of Sciences of the United States of America*, **102**, 5892–5897.

Thiers, B.M. (1984) Branch characters significant to subfamilial classification of Lejeuneaceae (Hepaticae). *Systematic Botany*, **9**, 33–41.

Thiers, B.M. (1988) The Australian species of *Cololejeunea*. *Beihefte zur Nova Hedwigia*, **90**, 113–146.

Tomescu, A.M.F. (2008) Megaphylls, microphylls and the evolution of leaf development. *Trends in Plant Sciences*, **14**, 5–12.

Une, K. (1985) Factors restricting the formation of normal male plants in the isosporous species of *Macromitrium* (Musci: Orthotrichaceae) in Japan. *Journal of the Hattori Botanical Laboratory*, **59**, 523–529.

Vanderpoorten, A. & Goffinet, B. (2009) *Introduction to Bryology*. Cambridge University Press, Cambridge.

Villarreal, A.J.C. & Renzaglia, K.S. (2006) Structure and development of *Nostoc* strands in *Leiosporoceros dussii* (Anthocerotophyta): a novel symbiosis in land plants. *American Journal of Botany*, **93**, 693–705.

Villarreal, J.C., Cargill, D.C., Söderström, L., *et al.* (2010) A synthesis of hornwort diversity: patterns, causes and future work. *Phytotaxa*, **9**, 150–166.

Vitt, D.H. (1984) Classification of the Bryopsida. In: *New Manual of Bryology*, Vol. 2 (ed. R.M. Schuster). pp. 696–759. Hattori Botanical Laboratory, Nichinan.

Zastrow, E. (1934) Experimentelle Studien über die Anpassung von Wasser und Sumpfmoosen. *Pflanzenforschung*, **17**, 1–70.

Zielinski, F. (1909) Beiträge zur Biologie des Archegoniums und der Haube der Laubmoose. *Flora*, **100**, 1–36.

Annual Plant Reviews (2013) **45**, 91–114
doi: 10.1002/9781118305881.ch3

Chapter 3

THE MORPHOLOGY AND DEVELOPMENT OF LYCOPHYTES

Barbara A. Ambrose

The New York Botanical Garden, Bronx, NY, USA

Abstract: During the Carboniferous Period around 350 million years ago, lycophytes dominated the landscape, comprising about 50% of the world's flora. Today, however, lycophytes are mostly inconspicuous plants, representing only about 0.5% of the flora. Studying the morphology of the lycophytes is integral to our understanding of land plant evolution and development because the lycophytes are sister to all other vascular plants. They represent an evolutionary lineage that has been distinct from other vascular plants, a lineage that has evolved growth forms and organs analogous to vascular plants and other organs that are homologous to those found in their sister group. As such, they provide a fascinating comparison for evolutionary developmental studies. The overall morphology of the lycophytes is relatively simple, with the sporophyte body composed of root, stem, microphyll, and sporangium. Three families comprise the present-day lycophytes: (1) Lycopodiaceae, (2) Isoetaceae, and (3) Selaginellaceae. Their unifying feature is the presence of a single sporangium per leaf, and this sporangium is located on the adaxial side of the leaf. Here, I briefly discuss the morphology and development of key lycophyte structures including meristems, sporangia, microphylls, rhizophores, and ligules.

Keywords: lycophyte; microphyll; Isoetaceae; Selaginellaceae; Lycopodiaceae; sporangia; heterospory; ligule; rhizophore.

3.1 Introduction

Lycophytes occupy a key phylogenetic position within the land plants as sister to all other vascular plants (Gifford & Foster 1989). The fossil record indicates that ancestral vascular plant lineages such as *Rhynia* had a

Annual Plant Reviews Volume 45: The Evolution of Plant Form, First Edition.
Edited by Barbara A. Ambrose and Michael Purugganan.

sporophyte-dominant life cycle, were naked and dichotomously branched, and had tracheids and terminal sporangia. In addition, fossils of early ancestors of lycophytes and their sister group both had naked branching axes that terminated in sporangia. Therefore, lycophytes comprise an evolutionary lineage that has evolved some organs and growth forms, such as leaves, tree habit, and cones, that are analogous to organs and forms that have evolved independently in other vascular plant lineages. While other growth forms and organs found in lycophytes, such as a sporophyte-dominant life cycle, branching, and sporangia, are homologous to those found in their sister group. Therefore, the lycophytes provide a fascinating comparison for evolutionary developmental studies. The extant lycophyte sporophyte body is relatively simple in terms of organ number and organ complexity with stem, roots, microphylls, and sporangia. In addition, lycophytes have several morphological features, including heterospory and endospory, that are considered important steps in the evolution and development of seeds. Finally, some lycophyte lineages contain two morphological novelties: the ligule and rhizophore.

Comparative studies will be important for understanding how land plants evolve and what role development plays in driving the evolution of plant form. Compared with the flowering plants, lycophytes have not only simple sporophyte bodies but also simple genomes. Many angiosperm genomes are characterized by polyploidization events with at least one whole genome duplication occurring at the base of the flowering plant lineage (Tang *et al.* 2008). Recent genome sequencing of the lycophyte *Selaginella moellendorffii* shows that there is no whole genome duplication characterizing this species (Banks *et al.* 2011). In addition, nearly 90% of the groups of genes underlying angiosperm development are present in the lycophyte genome (Banks *et al.* 2011). Together, these suggest that unraveling the molecular genetics underlying lycophyte development should be relatively straightforward. Elucidating the molecular genetic networks underlying the development of lycophytes will be integral to understanding the evolution and development of not only lycophytes but also all vascular plants.

More than 400 million years ago, the lycophytes, large and small, were a major feature of the terrestrial landscape. Although the morphology of some lineages has changed little for over 350 million years, all extant lycophyte species are small plants that are either terrestrial or epiphytic (Figure 3.1; Campbell 1895; Gifford & Foster 1989; Banks 2009). The vegetative sporophyte is composed of stem, leaves, and roots, with the stem or shoot branching dichotomously, either equally or unequally. Nearly all extant species have small sessile leaves with a single vascular strand and have traditionally been termed microphylls. The position of the sporangia on the adaxial region of the leaf is a defining feature of the lycophytes, either in the leaf axil or at the very base of the leaf on the adaxial surface. All sporangia are eusporangiate, arising from a group of superficial cells that divide periclinally, with the outer cells destined to form the sporangia wall and the inner cells the sporogenous tissue. Homospory and heterospory both exist in the lycophytes.

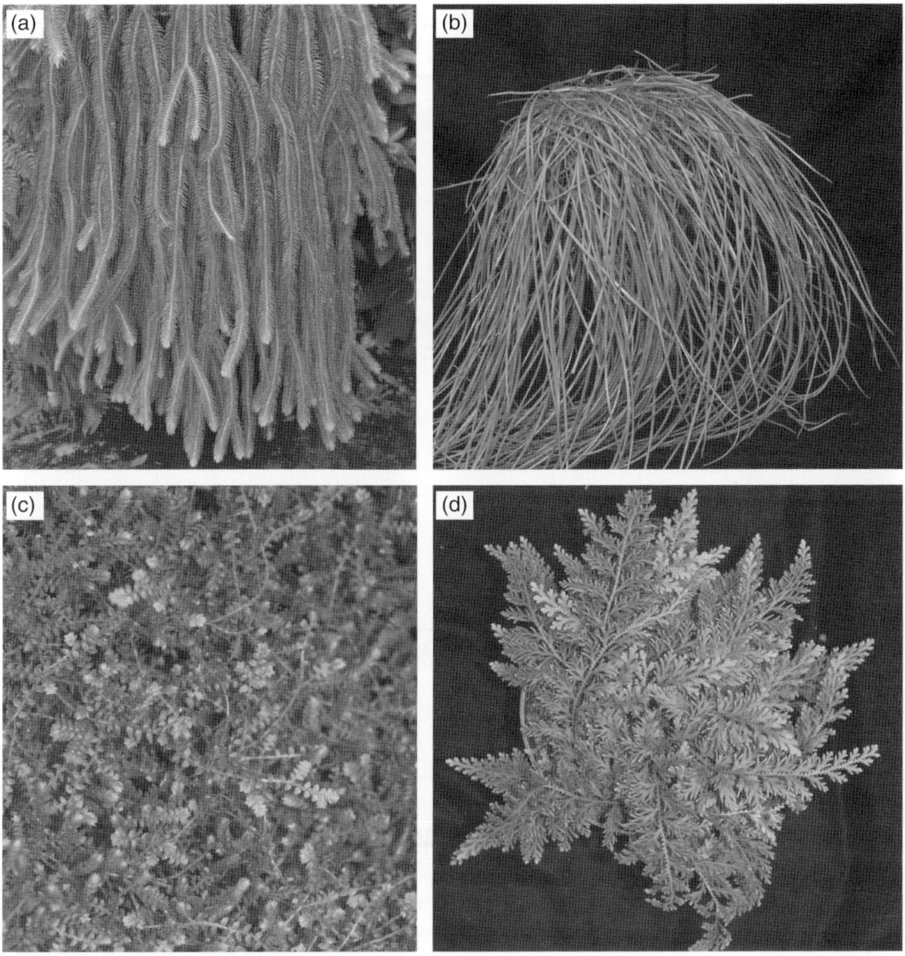

Figure 3.1 Lycophyte architecture. (a) Radial dichotomies are found in Lycopodiaceae (*Huperzia squarrosa*). (b) The overall gross morphology of *Isoetes engelmannii* is distinct from other lycophytes. Flabellate dichotomies are found in (c) *Selaginella apoda* and (d) *Selaginella moellendorffii*. (For a color version of this figure, see Plate 3.1.)

There are approximately 1300 species of extant lycophytes classified into three families: (1) Lycopodiaceae, (2) Selaginellaceae, and (3) Isoetaceae (Christenhusz *et al.* 2011), which can be traced back to the Paleozoic (Kenrick & Crane 1997). It is estimated that the Isoetaceae and Selaginellaceae split from a common ancestor in the Devonian some 377 million years ago. The morphology of Selaginellaceae has changed little since the Carboniferous, while much of the morphological diversification of Lycopodiaceae has occurred more recently (Bierhorst 1971; Kenrick & Crane 1997; Korall *et al.* 1999; Wikström 2001). Lycopodiaceae and Isoetaceae were families with diverse

Figure 3.2 Sporophyte reproductive morphology. (a) Sporangia are formed in between sterile regions of the plant in *Huperzia squarrosa*. Lines indicate fertile regions. Arrows indicate the sporangium associated with the adaxial side of each sporophyll. (b) Sporangia are arranged in distinct strobili in *Selaginella moellendorffii*. Arrowheads indicate some of the strobili that have formed at the tips of branches. (c) A dissection of *Isoetes engelmannii* shows the relatively large sporangium (arrow) that is found on the adaxial region of the sporophyll. The corm (Co) is labeled for orientation. (For a color version of this figure, see Plate 3.2.)

morphologies of which the present-day genera represent but a limited subset. Although all extant lycophytes are small plants, some fossil genera such as *Lepidodendron* included trees up to 50 m tall.

The taxonomy of extant lycophytes is relatively simple (Christenhusz *et al.* 2011). Lycopodiaceae are generally recognized as having four main clades recognized as genera: (1) *Lycopodium*, (2) *Huperzia*, (3) *Phylloglossum*, and (4) *Lycopodiella*. Many authors further divide *Lycopodium* into separate monophyletic genera, such as *Dendrolycopodium*, *Diphasiastrum*, and *Spinulum*. Isoetaceae has a single genus *Isoetes* (commonly known as quillwort), as does Selaginellaceae with a single genus *Selaginella* (commonly known as spike moss).

There are approximately 500 extant species of Lycopodiaceae, which mainly occupy tropical regions, though some are found in temperate and arctic regions. Lycopodiaceae may superficially resemble *Selaginella* species, but one obvious difference is that all spores in Lycopodiaceae species have the same size and function (homosporous). Other major differences found in Lycopodiaceae are: the phyllotaxy may be helical; well-defined strobili may or may not be formed; rhizophores and ligules are not found (Figures 3.1a and 3.2a).

There are approximately 800 extant species of Selaginellaceae that occupy diverse habitats in the tropics and temperate regions, including humid or dry locations (Korall *et al.* 1999). *Selaginella* species are dichotomously branched with small leaves (Figures 3.1c and 3.1d). Many species have leaves arranged in four rows—two rows of small leaves on the dorsal side and two rows of larger leaves on the ventral side. All the leaves have a tongue-like appendage, the ligule, present on the adaxial surface (Figure 3.3a). Most species have

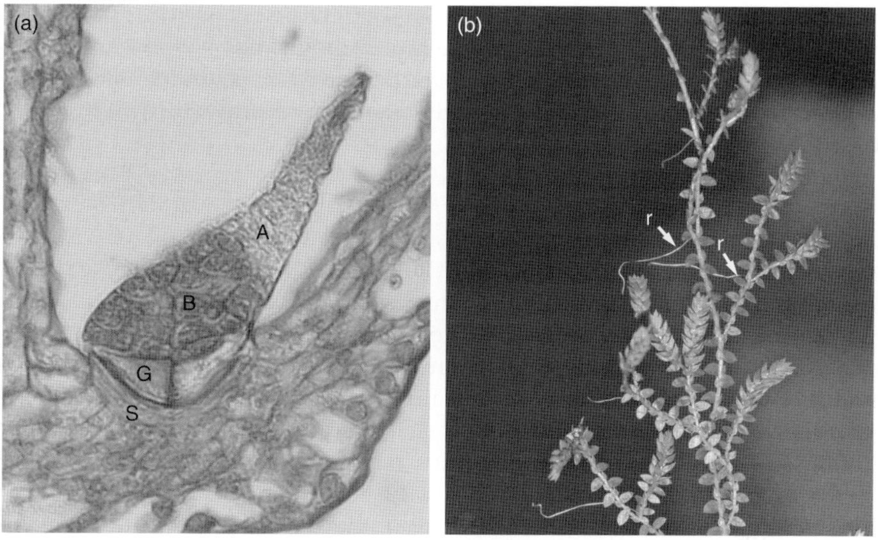

Figure 3.3 Enigmas in the lycophytes. (a) A stained histological section from *Selaginella moellendorffii* showing ligule morphology. The ligule is composed of sheath (S), glossopodium (G), main body, (B) and apex (A). (b) Some of the rhizophores (r) are indicated at the branching points in *Selaginella apoda*. (For a color version of this figure, see Plate 3.3.)

their sporangia arranged in a well-defined cone or strobilus (Figure 3.2b). All *Selaginella* species have two unequally sized male and female spores (heterosporous). Nearly all *Selaginella* species have rhizophores that arise at the branch forks (Figure 3.3b; see below for rhizophore morphology).

There are approximately 150 extant species of Isoetaceae found mainly in aquatic habitats in tropical and temperate regions. The gross morphology of Isoetaceae is quite distinct from Selaginellaceae and Lycopodiaceae (Smith 1938), and extant members of Isoetaceae superficially look more like aquatic monocots (Figures 3.1b and 3.2c). The leaves are broad at the base and then taper to a point and can be several centimeters long to nearly a meter. The leaves contain air canals as do many aquatic plants; however, *Isoetes* is the only plant that has four air canals. The Isoetaceae are sister to Selaginellaceae and shares with them the presence of ligules and heterospory. Morphological features uniting the Isoetaceae include Stigmarian rootlets, bipolar growth, and secondary growth. Fossil relatives include the tree *Lepidodendron*. Extant *Isoetes* have reduced cormose morphology with little stem elongation (Pigg 1992). The corm bears leaves upward and roots below, and secondary growth occurs from lateral and basal meristems.

Here, we will briefly discuss the vasculature, roots, plant architecture, meristems, microphylls, sporangia, as well as the unique structures—the rhizophore and ligule of the lycophytes. This overview is by no means comprehensive and is only meant to be an introduction to the interesting questions in lycophyte morphology and development that may now be addressed with molecular genetic tools. More extensive morphological overviews are available in several textbooks (Campbell 1895; Bower 1935; Schoute 1938; Smith 1938; Parihar 1967; Gifford & Foster 1989; Stewart & Rothwell 1993; Kenrick & Crane 1997).

3.2 Vasculature

The evolution of vasculature, which transports water and nutrients great distances around the plant and thereby releases major constraints on plant size, is one of the most important innovations in land plants. In addition, meristematic activity in the vascular tissue is important for secondary growth and an increase in the girth of plants. The vasculature of lycophytes is relatively simple and developmental genetic studies in lycophytes should provide great insight into vascular development in all land plants.

Interestingly, the arrangement and development of the vasculature as well as the sequence in which vascular tissues differentiate is the same in all vascular plants (Stewart & Rothwell 1993). The ancestral arrangement of the vasculature is protostelic with a central xylem core surrounded by phloem (Bower 1935; Smith 1938). Siphonostelic vasculature, central ground tissue surrounded by xylem and phloem, evolved later. Early in development, lycophyte vasculature is protostelic (Gifford & Foster 1989). In some lycophytes, later developing vasculature is siphonostelic. Although seed plants may have more complex arrangements of their vasculature, the early development of their vasculature is also protostelic, and vascular tissue differentiates in a particular direction (Beck *et al.* 1982). Exarch, where the direction of differentiation occurs toward the center, is the primitive condition for vascular plants and is found in lycophytes and all plant roots (Gifford & Foster 1989). In contrast, endarch vascular development, where differentiation occurs toward the periphery, is found in most seed plants. The tracheary cells of the xylem will differentiate into annular elements, spiral elements, or scalariform elements, with annular elements differentiating first and scalariform elements differentiating last—a sequence of development found in all vascular plants.

Secondary growth is only found in the Isoetaceae and the formation of trees is only found in fossil lycophytes (Schoute 1938; Stewart & Rothwell 1993). All trees are the result of secondary growth in the vascular cambium. However, unlike seed plants, fossil lycophyte trees had a unifacial cambium that produced secondary xylem and cortex but not secondary phloem.

3.3 Shoot apical meristems

The elaboration of the sporophyte body is first seen in the vascular plants. In the mosses, hornworts, and liverworts, the sporophyte consists of a single stalk bearing a terminal sporangium; it does not produce lateral organs and appears determinate. The shoot apical meristem (SAM) is responsible for generating all of the above ground organs of the sporophyte body throughout the life of the plant, and therefore, studies of the meristem will provide insight into how the sporophyte body is generated in the vascular plants. Meristems must maintain a balance between self-renewal and division and differentiation of organs on its flanks, and because of its importance, the anatomy of the SAM across the land plants has been well studied. SAM anatomy is diverse across the ferns and seed plants, and this same diversity is also found within the lycophytes (McAlpin & White 1974; Stevenson 1976a, 1976b; Steeves & Sussex 1989; Imaichi 2008). More recently, the molecular genetics of meristems has been an area of intense research in flowering plants (Dodsworth 2009). The extant lycophytes, as sister to other vascular plants, provide an ideal opportunity not only to examine the anatomy of the SAM but also to elucidate the molecular genetics underlying the diverse anatomies of the SAM in closely related groups.

In general, meristems are composed of initial(s) and highly dividing cells (McAlpin & White 1974; Stevenson 1976a, 1976b; Steeves & Sussex 1989; Imaichi 2008). The initial(s) of the meristem are the ultimate source of all cells and rarely divide, and they are surrounded by cytoplasmically dense, rapidly dividing cells. More specifically, researchers have categorized meristems into anywhere from two to seven different types based on anatomy (Popham 1951; Philipson 1990). Popham (1951) recognized seven different SAM anatomies based on the position and the number of initials and the division patterns in the SAM. The direction of the divisions of the initial(s) and their derivatives and how these divisions contribute to the body of the plant are also important to the classification of SAM types (Popham 1951). Many investigations on SAM anatomy include the number of cutting faces of the initial(s), which is important for understanding how the divisions contribute to the sporophyte body. Philipson (1990) recognized only two types of meristems: (1) the pteridophyte type, which has a prominent apical cell that does not divide periclinally so does not contribute directly to inner tissues, and (2) the seed plant type, which has initials that span more than one layer, and divisions of these do contribute to the inner tissues. Ferns and Selaginellaceae have the pteridophyte type of meristem, whereas seed plants, Lycopodiaceae and Isoetaceae, have the seed plant type (Philipson 1990).

The number and/or presence of prominent large apical initial(s) in lycophytes have been an area of intense research. The lycophyte SAM is either composed of a single apical cell, a group of apical cells, or without identifiable apical cells (Barclay 1931; Popham 1951; Stevenson 1976a; Imaichi 2008). The Selaginellaceae have either one or two initials; *Selaginella* species with a single apical cell can have an apical cell with two, three, or four cutting faces (Popham 1951). *Selaginella willdenovii* has been shown to have a single tetrahedral apical cell with three cutting faces, while the number of apical initials (one or two) in *Selaginella kraussiana* is currently being debated (Harrison *et al.* 2007; Jones & Drinnan 2009).

In *Lycopodium* no single apical initial has been recognized; instead, the SAM is composed of a group of apical initials (Campbell 1895; Stevenson 1976a). For example, *Huperzia lucidula* (Lycopodiaceae) was recently described as having four transient apical initials (Gola & Jernstedt 2011). The *Isoetes* SAM is broad and composed of a group of cells, although the presence of a conspicuous apical cell has been detected in some species (Karrfalt 1977).

In flowering plants, the molecular genetics of SAM maintenance has been well studied. Research has shown that a network of genes are arranged in a negative feedback loop and that communication between the cells is necessary for cell renewal, cell division, and cell differentiation (Dodsworth 2009). Cell communication is through plasmodesmatal networks, which are thought to be homologous across the land plants but can be formed during plant development in one of two ways (Imaichi & Hiratsuka 2007). Lineage-specific plasmodesmatal networks are formed in newly forming walls during cell division, while interface-specific plasmodesmatal networks are formed in existing cell walls. In ferns, SAMs with a single apical initial have lineage-specific plasmodesmatal networks, while seed plants with multiple initials across several layers have interface plasmodesmatal networks (Imaichi & Hiratsuka 2007). Selaginellaceae SAMs with single apical initials have plasmodesmatal networks similar to ferns, while Lycopodiaceae and Isoetaceae species with multiple initials have plasmodesmatal networks similar to seed plants (Imaichi & Hiratsuka 2007). This research on plasmodesmatal networks supports the meristem classification of Philipson (1990) that places Selaginellaceae into a fern type of meristem, and Lycopodiaceae and Isoetaceae into a seed plant type of meristem.

Although comparisons of SAM anatomy have been well studied across the land plants, little is known about the molecular genetics of SAM maintenance and signaling across the land plants. To better understand the evolution and development of meristems and the plant bodies they generate, comparative studies are necessary to know whether the SAM molecular genetic networks are the same or different in plants with a seed plant type SAM and those with a fern type SAM. The lycophytes can be a powerful resource for the study of meristem evolution and development. The extant lycophytes have diverse SAM anatomies yet generate similar sporophytes. For example, the Selaginellaceae and Lycopodiaceae have different types of SAMs, yet they have a similar gross morphology including the production of microphylls from the flanks of the SAM. It will be interesting to compare the molecular genetics of SAM maintenance not only between angiosperms and lycophytes but also between lycophytes with different SAM anatomy. In addition, it will be necessary to have anatomical data for bryophyte meristems and their plasmodesmatal networks to better understand the evolution of meristem organization and maintenance in land plants (Imaichi & Hiratsuka 2007). The

S. moellendorffii genome contains many of the transcription factor families necessary for SAM development in angiosperms (Banks *et al.* 2011), and molecular genetic studies of SAM initiation and maintenance in lycophytes will provide insight into the conservation, divergence, and plasticity of meristems in the land plants.

3.4 Sporophyte architecture

Branching generates the overall architecture of vascular plants, which can be dichotomous (splitting of shoot axis in two parts), axial (new shoot in lateral portion of shoot), or adventitious (new shoot formed from adult tissue) (Schoute 1938). Dichotomous branching is the predominant form of branching found in lycophytes, as opposed to axillary branching, which is found predominantly in seed plants (Philipson 1990). More specifically, branching and the direction of subsequent dichotomies of the shoot and whether the daughter branches grow equally (isotomous) or unequally (anisotomous) determine the overall architecture of lycophytes (Figure 3.1; Campbell 1895; Wigglesworth 1907). Once the shoot branches, the subsequent dichotomy can be at a right angle to the one below (cruciate dichotomy) or in the same plane (flabellate dichotomy or dorsiventral shoot) (Schoute 1938). Radial shoot architectures can also be generated by dichotomies occurring at any angle. In flabellate dichotomies, where the branches grow unequally—that is, where one branch is more dominant than the other—the resulting shoot system may resemble a fern leaf.

Members of Lycopodiaceae have cruciate and flabellate shoot architectures, whereas members of Selaginellaceae mainly have an anisotomous flabellate shoot architecture with one axis more dominant than the other (Parihar 1967). Morphological studies of branching in *Selaginella* have illustrated how the final shoot architecture is generated (Cusick 1953). These studies showed that the shoot architecture of *S. willdenovii* is generated by bifurcation of the main shoot axis that gives rise to a branch axis. The branch axis then bifurcates more frequently than the main axis to give the mature sporophyte architecture. The flabellate dichotomies found in *Lycopodium* and its segregates are in part defined by their particular branching patterns (e.g., *Diphasiastrum*, *Spinulum*, *Lycopodium s.s.*, and *Lycopodiella*).

The SAM is ultimately responsible for the elaboration of the sporophyte body plan. Depending on the species of lycophyte or fern examined, there are three ways that SAMs can be established during branching (Jernstedt *et al.* 1994). First, the apical initial can divide equally, generating two daughter apical initials. Second, the apical initial can persist in one of the new branches, while the apical initial of the other branch was a recent derivative of the parent SAM. Third, the apical initial can disappear and two new SAMs are formed in each new branch apex (Bierhorst 1977; Jernstedt *et al.* 1994). Although, there may be various ways to maintain or generate a new meristem in developing sporophytes, these are all found in sporophyte architectures generated by dichotomously branching axes. The successive divisions of the apical meristem without maintenance or increase in the rib zone can result in a dichotomously branched plant with little meristem left to continue plant growth. The rib zone is a group of rapidly dividing cells below the apical initials that contribute to stem development. For example, successive dichotomies in the fossil arborescent lycopod, *Lepidodendron*, result in a plant with thick branches at its base and thinner branches toward it apex. The SAM is diminished with each branching until eventually it is used up, a process called apoxogenesis. Another good and perhaps best-known example of apoxogenesis is found in *Psilotum*.

More studies are needed on the division of the SAM and the overall shoot architectures that they generate. Agnes Arber proposed that leaves are transformed shoots (Arber 1950). Recent molecular genetic studies in angiosperm leaf development have brought the partial shoot theory back into discussions of plant evolution and development (Barkoulas *et al.* 2008; Blein *et al.* 2008). In addition, it has been suggested that flabellate dichotomies with subsequent dominant branches were an important step in the evolution of axial branching and the evolution of compound leaves (Bower 1916; Schoute 1938). Lycophytes with their relatively simple yet diverse sporophyte architectures provide the perfect model group in which to elucidate the molecular genetic basis of shoot branching and development. These studies will be important for comparison with the molecular genetics of leaf development in land plants.

The overall architecture of *Isoetes* is very different from that of the Selaginellaceae and Lycopodiaceae. It lacks branched shoots (Bower 1935), and instead, the shoot is composed of a short corm with leaves arising apically and roots basally (Karrfalt 1977; Gifford & Foster 1989). Lateral and basal meristems generate several furrows in the corm. The SAM lies in the depression of the furrow, with leaf primordia arising around the SAM.

Studies of phyllotaxy have led to the proposal of the field theory to explain the arrangement of leaves in angiosperms (Steeves & Sussex 1989). The field theory proposes that leaf primordia arise in the first available space and are inhibited by leaves that have already formed. In lycophytes, phyllotaxy is mainly helical (Bhambie 1965), and studies in *S. willdenovii* have shown that the field theory does not sufficiently explain the phyllotactic pattern in this species (Cusick 1953).

In angiosperms, auxin maxima are necessary for primordium initiation, and mediators of auxin efflux carriers have been shown to be important for phyllotactic patterning (Reinhardt *et al.* 2000). Auxin also plays a role in apical dominance in angiosperms. Although it is not known if auxin plays a role in phyllotaxy in lycophytes, it is known that apical dominance operates in *Selaginella* and that this can be mediated by auxin (Jernstedt *et al.* 1994).

3.5 Microphylls

The evolution and development of leaves in vascular plants has been debated for more than a century and current molecular genetic evidence is providing data for a renewed debate. The leaves of bryophytes are generated during the gametophytic phase and are not considered homologous to the leaves found in the sporophytic phase of vascular plants. There are several hypotheses regarding the evolution of leaves in vascular plants and several different names have been put forth to try and accurately describe leaves in different groups. Alternative terms that have recently come into use for leaves, include "lycophylls" and "euphylls" (Pryer *et al.* 2004). However, until further evidence is provided on leaf homology in the vascular plants, we will use the more commonly used terms: microphylls and megaphylls (Gifford & Foster 1989).

Microphylls are found in lycophytes, and megaphylls are found in ferns and spermatophytes. Morphological and fossil evidence suggests that microphylls and megaphylls are not homologous because both lycophytes and euphyllophytes evolved separately from leafless ancestors in the Devonian (Doyle 1998). As the names suggest, microphylls and megaphylls are generally small and large, respectively, although there are exceptions in both cases. Microphylls are generally small, sessile, with an unbranched vasculature that does not reach the leaf apex, and no leaf gap (Smith 1938; Parihar 1967; Gifford & Foster 1989). However, leaves of fossil lycophytes were more morphologically diverse than extant lycophytes; some leaves were large and some were even divided (Grierson & Bonamo 1979; Wagner *et al.* 1982). There were even lycophyte leaves with complex venation in *Selaginella* (e.g., *Selaginella schaffneri*; Wagner *et al.* 1982).

There are several competing theories regarding the evolution and development of vascular plant leaves, with the most widely accepted being that proposed by Zimmermann in his telome concept (Wilson 1953; Zimmermann 1959; Gifford & Foster 1989). The telome theory of leaf evolution involves several processes that transform branches (telomes) into leaves (Wilson 1953; Parihar 1967; Gifford & Foster 1989). Zimmermann proposed the telome concept to explain the evolution of megaphylls and microphylls, with the number of telomes contributing to the development of megaphylls and microphylls as the main difference between the two. Microphylls have also been hypothesized to have evolved by the sterilization of sporangia (Gifford & Foster 1989; Kenrick & Crane 1997). However, the one problem with this hypothesis is that it does not explain why in extant lycophytes the sporangia are always associated with a leaf and with the upper surface of that leaf. Currently, the most widely accepted theory of microphyll evolution is the enation theory (Gifford & Foster 1989). The enation theory proposes that microphylls evolved by the superficial growth of the stem and then those outgrowths gradually became vascularized.

Selaginellaceae and ferns that have a prominent apical initial(s) produce leaves from the superficial layer, but in Lycopodiaceae, Isoetaceae, and seed plants, the leaves arise from multiple layers (Philipson 1990). In Selaginellaceae, the leaf primordia arise approximately ten cells below the SAM and appear as a horizontal ridge (Harvey-Gibson 1897). Divisions in the center of the primordia will give rise to the vein. Divisions occur at the apex of the leaf primordia but later growth occurs from divisions at the base of the leaf. The development of the microphyll in *Lycopodium planatum* also occurs by initial divisions in the microphyll apical cell but the final stage of development occurs by divisions at the base of the leaf (Wigglesworth 1907). In *Lycopodium*, the single vein of the microphyll differentiates basipetally and acropetally from the base of the leaf (Smith 1938). The *Isoetes* leaf primordium is initially round but later becomes flattened. *Isoetes* leaves also have an apical cell in the microphyll and a majority of its development and growth also occurs by divisions at the base (Campbell 1891).

All of the vegetative microphylls produced in Isoetaceae are morphologically similar to each other, whereas in *Lycopodium* and *Selaginella* two morphologically distinct microphylls (anisophylly) can form (Webster 1992). These two forms on the same plant are distinct from primordia inception (Dengler 1983a, 1983b). In anisophyllous species, the microphylls are arranged in four ranks (Smith 1938). In anisophyllous Selaginellaceae, the two different leaf forms occur in pairs (Dengler 1983a, 1983b; Webster 1992). Generally, the dorsal leaves are smaller and located on the other side of the stem, unlike the larger ventral leaf of the pair (Smith 1938). In Selaginellaceae, there are microphylls with similar adaxial and abaxial epidermis and species with dissimilar epidermis (Harvey-Gibson 1897). However, in anisophyllous species, stomata are only on the abaxial side of the microphyll (Smith 1938).

Recent comparative gene expression analyses between *S. kraussiana* and seed plants have resulted in diametrically opposed conclusions about the conservation of a common developmental mechanism underlying the development of microphylls and megaphylls (Harrison *et al.* 2005; Floyd and Bowman 2006). Harrison *et al.* (2005) found a conservation in the expression patterns of Class I KNOX and ARP orthologs in angiosperms and *S. kraussiana* and concluded that this genetic module was co-opted independently in each lineage for leaf development. The authors also suggested that the KNOX-ARP module may have had an ancestral role in shoot branching. However, Floyd and Bowman (2006) found differences in the expression patterns of Class III HD-Zip orthologs between seed plants and *S. kraussiana* and concluded that vascular plants do not share a common mechanism for leaf development, which supports the theory that microphylls and megaphylls have separate origins. These comparisons were performed between the Selaginellaceae that have a pteridophyte type of meristem (see Section 3.3) and whose leaves are specified from a superficial layer and angiosperms that have a seed plant type of meristem and whose leaves are specified from several layers. It will be interesting to compare the molecular genetics of meristem and leaf development between a lycophyte that has a seed plant type of meristem and whose leaves are specified from several layers, such as Lycopodiaceae species, and an angiosperm species with a similar type of meristem anatomy and leaf specification. Ultimately, a functional genomic system in a lycophyte will be necessary to help resolve this debate over the evolution and development of leaves.

3.6 Sporangia

The development of sporangia in the lycophytes, particularly the development of heterospory and endospory, are considered key steps in the subsequent evolution of the seed (Lyon 1901a, 1901b; Pettitt 1970; Bateman & DiMichele 1994). The defining feature of lycophytes is the presence of a single sporangium on the adaxial surface of a microphyll. Microphylls that bear sporangia are called sporophylls, and although there is a close association between sporangia and sporophylls, the development of one is not dependent upon the other. In *Isoetes* and *Selaginella*, the sporangium is generally located on the sporophyll between the ligule and stem axis (Bower 1935; Schoute 1938).

In *Selaginella* and some Lycopodiaceae, sporangia are organized into cones (strobili) and found distally on the branches (Figure 3.2b; Bower 1935; Gifford & Foster 1989). In other Lycopodiaceae, such as *Huperzia*, sporangia can alternate with sterile vegetative regions (Figure 3.2a). In Isoetaceae, every leaf has an associated sporangia and although some of these sporangia may abort early in development, the entire plant can be considered a strobilus (Figure 3.2c). In *Selaginella*, the sporophylls can be either identical to the microphylls or distinct. However, sporophylls are generally isophyllous even if microphylls are anisophyllous (Figure 3.2b).

The sporangia that develop in lycophytes and basal fern families (i.e., Psilotaceae, Ophioglossaceae, and Marattiaceae) are eusporangiate, whereas more derived ferns are leptosporangiate (Bower 1894, 1935; Campbell 1895; Gifford & Foster 1989). Sporangia develop from superficial initials in both eusporangiate and leptosporangiate species; there are, however, key differences in their development. Eusporangia generally develop from several superficial initials, which divide periclinally to give an inner and outer layer of cells. The outer layer of cells will give rise to the wall of the sporangia, while the inner layer of cells will give rise to the sporogenous cells. In *Isoetes*, sterilization of some of the sporogenous cells gives rise to plates (trabeculae) that may provide mechanical support to the large sporangia. In leptosporangiate species, a single superficial initial divides transversely to give inner and outer cells. The inner cells may not contribute to the development of the sporangia or may form part of the sporangial stalk, while the outer cell will continue to divide and differentiate into the sporangia wall and sporogenous cells. Selaginellaceae and Isoetaceae form two morphologically and functionally distinct sporangia and each contains either male or female spores (heterospory). In contrast, Lycopodiaceae form only one type of spore (homospory). Despite the differences, the initiation and early development of sporangia is very similar in Selaginellaceae and Lycopodiaceae.

The evolution of heterospory in lycophytes represents the first occurrence of organs in the sporophyte generation that will separate the sexes in the gametophyte generation (Pettitt 1970). Homosporous plants produce spores that are all alike, while heterosporous plants produce spores that are different in size and function (Lyon 1901a; Smith 1938). Homosporous Lycopodiaceae, for example, produce a single type of spore that will give rise to the male and female gametophytes, while the heterosporous lycophytes produce sporangia of different sexes: megasporangia that produce female gametophytes and microsporangia that produce male gametophytes (Bower 1894; Lyon 1901a, 1901b; Gifford & Foster 1989). The development of the gametophyte phase in homosporous lycophytes is always exosporic, while gametophyte development in heterosporous lycophytes is always endosporic (Gifford & Foster 1989).

In *Isoetes*, the megasporangia are produced early in the season and the microsporangia are produced later so that microsporangia are found on inner leaves and megasporangia on outer leaves (Campbell 1891; Smith 1900a, 1900b). In Selaginellaceae, the mega- and microsporangium occur in the same strobilus and there are four different arrangements depending on the species: (1) megasporangia basal and microsporangia apical; (2) two vertical rows of each; (3) two vertical rows of microsporangia and two vertical rows of mixed sporangia; and (4) strobili composed only of megasporangia (Mitchell 1910; Horner & Arnott 1963; Webster 1992).

In Selaginellaceae, the microsporangium produces thousands of small microspores, while the megasporangium typically produces four functional megaspores (Lyon 1901a, 1901b; Mitchell 1910). In early development of microsporangia and megasporangia, they are indistinguishable (Smith 1900a; Lyon 1901a, 1901b). However, before the sporocytes begin to differentiate, there is one more division of the sporogenous cells within the microsporangium compared with the megasporangium (Lyon 1901a, 1901b). The hormone ethylene may be involved in this process, as ethylene-treated strobili of *Selaginella* resulted in megasporangium developing in place of microsporangia (Brooks 1973). In the microsporangium, the sporocytes undergo meiosis and develop into microspores, and divisions and differentiation inside the spore (endospory) gives male gametophytes (Smith 1900a, 1900b; Lyon 1901a, 1901b; Pettitt 1970). In Selaginellaceae, the sperm are biflagellate, whereas in Isoetaceae the sperm are multiflagellate (Gifford & Foster 1989).

In the megasporangium, many of the sporocytes begin to disintegrate and the sporocytes that are left will enlarge and undergo meiosis to produce relatively large megaspores compared with the microspores (Smith 1900a,

1900b; Lyon 1901a, 1901b). In *Selaginella*, there is variation in the sizes of the megaspores in each tetrad (Lyon 1901a, 1901b; Mitchell 1910; Duerden 1929). In tetrads with small and large megaspores, only the large megaspore(s) are functional (Pettitt 1970). Typically there is a single tetrad formed with four megaspores. Divisions and differentiation in the megaspore (endospory) produce the female gametophyte (Lyon 1901a, 1901b; Gifford & Foster 1989). The female gametophyte is retained within the spore and the resulting embryo develops within the spore, which allows the embryo to develop with nutrition from the parent plant (Bower 1935).

The spores are the agents of dispersal in the lycophytes and there are a variety of sporangium morphologies that promote spore dispersal (Mitchell 1910; Bower 1935; Koller & Scheckler 1986; Giorgi *et al.* 1997). In some species, the sporangium wall has a band of thin-walled cells surrounded by thick-walled cells, and drying of the sporangium results in a forceful opening of the sporangium (Mitchell 1910; Bower 1935). In *Selaginella*, the spores are released forcibly from the sporangia with the megaspores traveling farther than the microspores (Mitchell 1910). This dispersal mechanism not only distributes the spores but also increases the chances of megaspores landing near microspores from another plant. Heterospory, endosporic development, and spore dispersal are all features that were important for the evolution and development of seeds in the spermatophytes (Lyon 1901a, 1901b; Bower 1935).

3.7 Roots

Fossil evidence indicates that roots are not homologous across the land plants (Doyle 1998). In ferns and lycophytes, the embryonic root arises laterally on the embryonic axis (homorhizic) as opposed to opposite the SAM on the embryonic axis (allorhizic) (Gifford & Foster 1989). In homorhizic plants, all postembryonic roots are adventitious, arising from the stem. Although the primary root of all lycophytes is ephemeral, later-forming roots appear to arise differently in each lineage (Parihar 1967; Gifford & Foster 1989). In many *Selaginella* species, the shoot-borne rhizophores are subsequently transformed into roots, while in Lycopodiaceae the roots arise from the stem. In *Huperzia*, for example, the root forms endogenously from the pericycle and will grow down through the cortex and emerge near the base of the plant. In *Isoetes*, the roots arise from the basal region of the corm where there is a root-producing meristem (Karrfalt 1984). This region of the corm in *Isoetes* has also been called a rhizophore or rhizomorph, and parallels have also been made between fossil Lepidodendrid rhizomorphs (Stigmarian root systems) and the *Isoetes* corm (Karrfalt 1984; Gifford & Foster 1989; Stewart & Rothwell 1993). Moreover, although Selaginellaceae and Lycopodiaceae roots are similar because both develop endogenously, similarities between the rhizophore of *Selaginella* and the Lepidodendrid rhizomorph have also been noted (Paolillo 1982).

Although lycophyte roots arise from seemingly different locations in various groups, they all arise from the stem and there are many similarities in their development (Paolillo 1982; Gifford & Foster 1989). In seed plants, the arrangement of the stem vasculature is more complex than that found in the roots. In contrast, the arrangement of the vasculature of the stem and root is very similar in lycophytes (Parihar 1967), and some investigators have hypothesized that there is little differentiation between the stem and root in lycophytes. Branching in lycophyte roots is dichotomous and subsequent branchings are usually at right angles, similar to what can found in the stem (Schoute 1938; Smith 1938; Parihar 1967). Yet interpretations of the root systems in the lycophytes remain controversial and intractable via classic approaches and these differing interpretations could possibly be solved with molecular genetic tools (Paolillo 1963; Parihar 1967; Karrfalt & Eggert 1977; Karrfalt 1980, 1981, 1984; Paolillo 1982).

3.8 Structural enigmas

Heterosporous lycophytes have two enigmatic structures: the ligule and (in *Selaginella*) the rhizophore (Figure 3.3). Both structures are considered evolutionary novelties; that is, these structures evolved *de novo*, not homologous with anything in outgroup taxa.

The ligule, which means "little tongue," is an appendage found on the adaxial side of the leaves (Figure 3.3a). It has been variously interpreted as an organ for storage or secreting of mucilage or water, or as a vestigial organ of unknown function (Horner *et al.* 1975; Webster 1992). Unraveling the molecular genetics of ligule development will help in determining the identity of the ligule. In *Selaginella*, the rhizophore arises from the angle meristem, the name given to the meristem at the fork of each branch (Figure 3.3b; Cusick 1954; Webster 1992; Jernstedt *et al.* 1994). The rhizophore is intriguing because it can develop into a shoot or a root depending on the environmental and developmental conditions, and the developmental plasticity of the rhizophore into either root or shoot is unique among plant meristems. Understanding the molecular genetic network underlying rhizophore development will illuminate mechanisms of stem cell maintenance and cell fate specification in plant meristems.

3.8.1 Ligules

The presence or absence of the ligule holds a prominent place in the classification of lycophytes, and Bower (1935) had recommended a division into two series: the *Ligulatae* comprised Selaginellaceae and Isoetaceae, and the *Eligulatae* comprised only the Lycopodiaceae. Ligules are always found in extant Selaginellales and Isoetales but not extant Lycopodiales, and it had been thought that the ligule was associated with heterosporous species only (Bower 1894; Campbell 1895; Harvey-Gibson 1896; Gifford & Foster 1989). A ligule, however, was found in a presumed homosporous fossil species, *Leclercqia complexa*, from the middle Devonian (Grierson & Bonamo 1979), and this suggests that the ligule evolved before heterospory.

Ligule morphology ranges from tongue-shaped to a stout body with a papillate margin to fan-shaped body with a lobed margin in Selaginellaceae (Figure 3.3a; Harvey-Gibson 1896). Ligules are found on the adaxial regions of microphylls (Campbell 1895; Harvey-Gibson 1896), and are also found on the adaxial side of the cotyledons in the embryo and on the adaxial side of sporophylls (Campbell 1895; Mitchell 1910; Paolillo 1963; Schulz *et al.* 2010).

Regardless of their overall shape, ligules are composed of four regions from bottom to top: (1) sheath, (2) glossopodium, (3) main body, and (4) the apex (Harvey-Gibson 1896; Smith 1900a). The sheath forms the base of the ligule and is composed of long tube-shaped cells (Smith 1938). In some species, a Casparian strip has been found in sheath cells. The glossopodium has large thin-walled cells that are largely vacuolated (Harvey-Gibson 1896; Smith 1938), while the main body of the ligule is composed of small cubical cells that are densely staining. The apex of the ligule is made up of a thin layer of cells that undergo cell death.

It has been reported that the ligules of Selaginellaceae have a multicellular origin, whereas those of Isoetaceae have a unicellular origin (Harvey-Gibson 1896). The development of the ligule has been well studied in Selaginellaceae (Harvey-Gibson 1896; Mitchell 1910), and Harvey-Gibson (1896) studied the morphology and development of ligules in more than 50 species. He found that ligules develop close to the stem but definitely develop on the surface of the microphyll. Ligules are specified early in development and originate from a long strip of epidermal cells, which undergo a periclinal division where the inner cells form the sheath cells and the outer cells form the remainder of the ligule. The sheath cells appear continuous with the microphyll epidermis. Subsequent divisions form the glossopodium cells, which quickly enlarge; the glossopodium may be one or two cells tall. There are then a rapid series of divisions that form the body of the ligule, which may be more than six cells tall, and then finally the apex is formed. Later in development, the sheath cells become cuticularized and the glossopodium cells may divide once more. The ligule develops precociously and matures before the rest of the microphyll. The development of the ligule on the sporophyll is similar to the development of the ligule on the microphyll except that the ligule that forms on the sporophyll tends to be larger (Mitchell 1910).

Ligules are conspicuous organs on the lycophyte body that have no readily apparent function and therefore have intrigued scientists and been variously interpreted for more than a century (Webster 1992). The granular contents of the ligule suggested a mucilage-secreting function to some researchers, while the continuity between the ligule and microphyll vasculature suggested a function in water absorption to others (Harvey-Gibson 1896; Smith 1938; Webster 1992). The large ligules found in some species suggested a protective function. Harvey-Gibson (1896) thought the ligules were a specialized scale based on their development, while his student thought the ligules on the sporophylls functioned to keep the sporangia moist during their development (Mitchell 1910). Finally, others consider the ligule *sui generis*, a structure of unique origin (Webster 1992). Hopefully, future molecular genetic studies can provide further evidence to development studies on the identity of the ligule.

3.8.2 Rhizophores

The rhizophore of *Selaginella* has been the subject of much research and yet its identity is still debated (Bower 1935; Schoute 1938; Cusick 1953, 1954; Bierhorst 1971; Gifford & Foster 1989; Imaichi & Kato 1989). The rhizophore has features of both shoot and root, and has been considered homologous with those organs, or as a novel organ of its own kind (*sui generis*). The shoot characteristics of the rhizophore include its exogenous origin from the branching point of the shoot, a SAM anatomy when it originates and dichotomous branching. The rhizophore also has root characteristics: leafless, positively geotropic, may lack chlorophyll, forms a root apical meristem endogenously later in its development, and later forms a root cap. The rhizophore will eventually form roots distally but will also form shoots under particular experimental or environmental conditions (Williams 1937).

Rhizophores develop from angle meristems that are formed at the shoot dichotomy (Figure 3.3b; Cusick 1953, 1954; Imaichi & Kato 1989; Jernstedt *et al.* 1992). In *Selaginella uncinata*, the rhizophore primordium is initiated at the second youngest dichotomy (Imaichi & Kato 1989). Two angle meristems form, a dorsal and a ventral, and these can have different fates depending on the species (Cusick 1953, 1954; Imaichi & Kato 1989; Jernstedt *et al.* 1992; Schulz *et al.* 2010). The ventral angle meristem, which usually develops first, typically produces a rhizophore, while the dorsal angle meristem can develop into a rhizophore or remain dormant. The angle meristems arise from a single apical cell and will be composed of one to several tetrahedral apical initial(s) surrounded by a group of cytoplasmically dense cells (Cusick 1954; Imaichi & Kato 1989; Jernstedt *et al.* 1992). Therefore, the angle meristem arises exogenously as does the SAM. However, the organization of the SAM and the rhizophore apical meristem are different. In *S. uncinata*, the SAM has a single lens-shaped apical initial with two cutting faces, while the rhizophore apical meristem has a tetrahedral apical initial with three cutting faces (Imaichi & Kato 1989). In addition, the vasculature of the rhizophore is similar to the roots that will be formed distally from the rhizophore (Harvey-Gibson 1902; Schoute 1938).

The rhizophore primordium gives rise to a cylindrical organ that has no lateral organs. Depending on the species, the rhizophore will grow anywhere from 0.1 cm to several centimeters or more before the rhizophore meristem disappears (Imaichi & Kato 1989, 1991; Lu & Jernstedt 1996). A new meristem forms several cell layers below the apex and will give rise to new cells apically and basally similar to a root apical meristem. A root cap will eventually form; however, Imaichi and Kato (1989, 1991) noted that initially the rhizophore apex was covered by a calyptra that eventually underwent cell death and was then replaced by a root cap. Depending on the species, the rhizophore will branch dichotomously and successive dichotomies will be at right angles (Imaichi & Kato 1989, 1991; Lu & Jernstedt 1996). This dichotomous branching is similar to what is found in the shoots of some lycophytes and *Psilotum*. The endogenous rhizophore meristem disappears and two new meristems are formed in each branch primordium. The rhizophore is positively geotropic and when it makes contact with a substrate the root cap and root hairs are formed (Lu & Jernstedt 1996).

Rhizophores will produce roots distally (especially after touching the soil) unless the growth of the plant is perturbed. For example, if the tips of the shoot are cut off then the rhizophores that normally would have produced roots instead produce leaf-producing shoots (Worsdell 1910; Williams 1937; Schoute 1938; Cusick 1954). Experiments in *Selaginella grandis* showed that if the shoots were placed horizontally on the substrate then the rhizophores developed as leaf-producing shoots (Schoute 1938). Experimental evidence indicates that auxin is important for the fate of the rhizophore (Williams 1937); if the main shoot is removed and replaced by auxin, then the rhizophore will develop into roots distally.

Thus, after more than 100 years of morphological and experimental studies, there is still disagreement about the identity of this organ, and new developmental methods might help resolve the disagreement. Recent molecular studies showed Class I KNOX expression in the rhizophore and SAM but not in the root suggesting that the meristem maintenance network involving KNOX proteins is shared between the rhizophore and SAM (Kawai *et al.* 2010). Different peptides have also been shown to present in the rhizophore and root (Jernstedt & Mansfield 1985). Further molecular genetic studies could help provide additional evidence as to the homology, or non-homology, of the rhizophore with roots or shoots. In addition, studies of the plasticity of the rhizophore meristem anatomy and identity could help illuminate the evolution and development of root and shoot meristems across the land plants.

3.9 Conclusions

The lycophytes are an intriguing lineage that demand more attention in the field of molecular genetics. The fossil record indicates that leaves, cones, roots, and secondary growth occurred independently in the lycophyte lineage, and comparative molecular genetic studies of these lycophyte organs and analogous organs in the seed plants will help to illuminate the role development plays in the evolution of form. It will be interesting to know whether similar molecular genetic modules have been co-opted independently to build analogous organs, if a core module is conserved and has been elaborated in seed plants, or whether analogous organs are built using completely different regulatory networks. Lycophytes also have organs and growth forms that are homologous to those that are found in the seed plants, particularly meristems, branching, and sporangia. Evolutionary developmental studies of these organs and forms in lycophytes will help us to better understand the development of these organs and forms in ferns and seed plants. Although a lycophyte genome has recently been sequenced, it will be necessary to develop a functional model system to unravel the evolution and development of lycophytes.

Acknowledgments

I am grateful to Robbin C. Moran and Dennis W. Stevenson for reading and commenting on this manuscript, and I am indebted to D.W. Stevenson, for sharing his knowledge and books and for his mentorship. I am thankful to Alejandra Vasco Gutierrez for contributing her beautiful photos.

References

Arber, A. (1950) *The Natural Philosophy of Plant Form*. Cambridge University Press, New York.

Banks, J.A. (2009) Selaginella and 400 million years of separation. *Annual Review of Plant Biology*, **60**, 223–238.

Banks, J.A., Nishiyama, T., Hasebe, M., *et al.* (2011) The Selaginella genome identifies genetic changes associated with the evolution of vascular plants. *Science*, **332**, 960–963.

Barclay, B.D. (1931) Origin and development of tissues in stem of Selaginella wildenovii. *Botanical Gazette*, **91**, 452–461.

Barkoulas, M., Hay, A., Kougioumoutzi, E., *et al.* (2008) A developmental framework for dissected leaf formation in the Arabidopsis relative Cardamine hirsuta. *Nature Genetics*, **40**, 1136–1141.

Bateman, R.M. & DiMichele, W.A. (1994) Heterospory: the most iterative key innovation in the evolutionary history of the plant kingdom. *Biological Reviews*, **69**, 345–417.

Beck, C.B., Schmid, R. & Rothwell, G.W. (1982) Stelar morphology and the primary vascular system of seed plants. *Botanical Review*, **48**, 691–815.

Bhambie, S. (1965) Studies in pteridophytes. *Proceedings: Plant Sciences*, **61**, 242–252.

Bierhorst, D.W. (1971) *Morphology of vascular plants*. Macmillan, New York.

Bierhorst, D.W. (1977) On the stem apex, leaf initiation and early leaf ontogeny in filicalean ferns. *American Journal of Botany*, **64**, 125–152.

Blein, T., Pulido, A., Vialette-Guiraud, A., *et al.* (2008) A conserved molecular framework for compound leaf development. *Science*, **322**, 1835–1839.

Bower, F.O. (1894) *Studies in the morphology of spore-producing members*. Published for the Royal Society. Paul, Trench, Trübner, London.
Bower, F.O. (1916) On leaf structure as illuminated by a study of the Pteridophyta. *Transactions of the Royal Society of Edinburgh*, **51**, 657–708.
Bower, F.O. (1935) *Primitive Land Plants, also known as the Archegoniatae*. Macmillan and Co., London.
Brooks, K.E. (1973) Reproductive biology of selaginella: I. determination of megasporangia by 2-chloroethylphosphonic acid, an ethylene-releasing compound. *Plant Physiology*, **51**, 718–722.
Campbell, D.H. (1891) Contributions to the life-history of isoetes. *Annals of Botany*, **os-5**, 231–258.
Campbell, D.H. (1895) *The Structure and Development of the Mosses and Ferns*. Macmillan and Co., London.
Christenhusz, M.J.M., Zhang, X.-C. & Schneider, H. (2011) A linear sequence of extant families and genera of lycophytes and ferns. *Phytotaxa*, **19**, 7–54.
Cusick, F. (1953) Experimental and analytical studies of Pteridophytes XXII. Morphogenesis in *Selaginella willdenovii* baker. *Annals of Botany*, **17**, 369–384.
Cusick, F. (1954) Experimental and analytical studies of pteridophyte XXV. Morphogenesis in *Selaginella willdenovii* Baker. *Annals of Botany*, **18**, 171–181.
Dengler, N.G. (1983a) The developmental basis of anisophylly in Selaginella martensii. I. Initiation and morphology of growth. *American Journal of Botany*, **70**, 181–192.
Dengler, N.G. (1983b) The developmental basis of anisophylly in Selaginella martensii. II. Histogenesis. *American Journal of Botany*, 193–206.
Dodsworth, S. (2009) A diverse and intricate signalling network regulates stem cell fate in the shoot apical meristem. *Developmental Biology*, **336**, 1–9.
Doyle, J.A. (1998) Phylogeny of vascular plants. *Annual Review of Ecology and Systematics*, **29**, 567–599.
Duerden, H. (1929) Variations in megaspore number in selaginella. *Annals of Botany*, **os-43**, 451–457.
Floyd, S.K. & Bowman, J.L. (2006) Distinct developmental mechanisms reflect the independent origins of leaves in vascular plants. *Current Biology*, **16**, 1911–1917.
Gifford, E.M. & Foster, A.S. (1989) *Morphology and Evolution of Vascular Plants*. W.H. Freeman and Co., New York.
Giorgi, A.F.-D., Holderegger, R. & Schneller, J.J. (1997) Aspects of spore dispersal in Selaginella. *American Fern Journal*, **87**, 93–103.
Gola, E.M. & Jernstedt, J.A. (2011) Impermanency of initial cells in Huperzia lucidula (Huperziaceae) shoot apices. *International Journal of Plant Sciences*, **172**, 847–855.
Grierson, J.D. & Bonamo, P.M. (1979) Leclercqia complexa: earliest ligulate lycopod (Middle Devonian). *American Journal of Botany*, **66**, 474–476.
Harrison, C.J., Corley, S.B., Moylan, E.C., *et al.* (2005) Independent recruitment of a conserved developmental mechanism during leaf evolution. *Nature*, **434**, 509–514.
Harrison, C.J., Rezvani, M. & Langdale, J.A. (2007) Growth from two transient apical initials in the meristem of Selaginella kraussiana. *Development*, **134**, 881–889.
Harvey-Gibson, R.J.H. (1896) Contributions towards a knowledge of the anatomy of the genus Selaginella, Spr. Part II. The Ligule. *Annals of Botany*, **os-10**, 77–88.
Harvey-Gibson, R.J.H. (1897) Contributions towards a knowledge of the anatomy of the Genus Selaginella, Spr. Part III. The leaf. *Annals of Botany*, **os-11**, 123–156.
Harvey-Gibson, R.J. (1902) Contributions towards a knowledge of the anatomy of the Genus Selaginella. Part IV. The root. *Annals of Botany*, **os-16**, 449–466.

Horner, H.T. & Arnott, H.J. (1963) Sporangial arrangement in North American species of *Selaginella*. *Botanical Gazette*, **124**, 371–383.
Horner, H.T., Beltz, C.K., Jagels, R., *et al.* (1975) Ligule development and fine structure in two heterophyllous species of Selaginella. *Canadian Journal of Botany*, **53**, 127–143.
Imaichi, R. (2008) Meristem organization and organ diversity. In: *Biology and Evolution of Ferns and Lycophytes* (eds TA Ranker & CH Haufler). pp. 75–103. Cambridge University Press, Cambridge.
Imaichi, R. & Hiratsuka, R. (2007) Evolution of shoot apical meristem structures in vascular plants with respect to plasmodesmatal network. *American Journal of Botany*, **94**, 1911–1921.
Imaichi, R. & Kato, M. (1989) Developmental anatomy of the shoot apical cell, rhizophore and root of *Selaginella uncinata*. *Botanical Magazine Tokyo*, **102**, 369–380.
Imaichi, R. & Kato, M. (1991) Developmental study of branched rhizophores in three Selaginella Species. *American Journal of Botany*, **78**, 1694–1703.
Jernstedt, J.A., Cutter, E.G., Gifford, E.M., *et al.* (1992) Angle Meristem origin and development in Selaginella martensii. *Annals of Botany*, **69**, 351–363.
Jernstedt, J.A., Cutter, E.G. & Lu, P. (1994) Independence of organogenesis and cell pattern in developing angle shoots of Selaginella martensii. *Annals of Botany*, **74**, 343–355.
Jernstedt, J.A. & Mansfield, M.A. (1985) Two-dimensional gel electrophoresis of polypeptides from stems, roots, leaves, and rhizophores of Selaginella kraussiana. *Botanical Gazette*, **146**, 460–465.
Jones, C.S. & Drinnan, A.N. (2009) The developmental pattern of shoot apices in Selaginella kraussiana (Kunze) A. Braun. *International Journal of Plant Sciences*, **170**, 1009–1018.
Karrfalt, E. (1984) The origin and early development of the root-producing meristem of Isoetes andicola L. D. Gomez. *Botanical Gazette*, **145**, 372–377.
Karrfalt, E.E. (1977) An apical cell in the shoot apex of Isoetes tuckermanii. *American Fern Journal*, **67**, 68–72.
Karrfalt, E.E. (1980) A further comparison of Isoetes roots and stigmarian appendages. *Canadian Journal of Botany*, **58**, 2318–2322.
Karrfalt, E.E. (1981) The comparative and developmental morphology of the root system of Selaginella selaginoides (L.) Link. *American Journal of Botany*, **68**, 244–253.
Karrfalt, E.E. & Eggert, D.A. (1977) The comparative morphology and development of isoetes L. II. Branching of the base of the corm in I. tuckermanii A. Br. and I. nuttallii A. Br. *Botanical Gazette*, **138**, 357–368.
Kawai, J., Tanabe, Y., Soma, S., *et al.* (2010) Class 1 KNOX gene expression supports the *Selaginella* Rhizophore concept. *Journal of Plant Biology*, **53**, 268–274.
Kenrick, P. & Crane, P.R. (1997) *The Origin and Early Diversification of Land Plants : a Cladistic Study*. Smithsonian Institution Press, Washington, DC.
Koller, A.L. & Scheckler, S.E. (1986) Variations in microsporangia and microspore dispersal in Selaginella. *American Journal of Botany*, **73**, 1274–1288.
Korall, P., Kenrick, P. & Therrien, J.P. (1999) Phylogeny of selaginellaceae: evaluation of generic/subgeneric relationships based on rbcL gene sequences. *International Journal of Plant Sciences*, **160**, 585–594.
Lu, P. & Jernstedt, J.A. (1996) Rhizophore and root development in Selaginella martensii: meristem transitions and identity. *International Journal of Plant Sciences*, **157**, 180–194.

Lyon, F.M. (1901a) A study of the sporangia and gametophytes of Selaginella apus and Selaginella rupestris. *Botanical Gazette*, **32**, 124–141.

Lyon, F.M. (1901b) A study of the sporangia and gametophytes of Selaginella apus and Selaginella rupestris (concluded). *Botanical Gazette*, **32**, 170–194.

McAlpin, B.W. & White, R.A. (1974) Shoot organization in the filicales: the promeristem. *American Journal of Botany*, **61**, 562–579.

Mitchell, G. (1910) Contributions towards a knowledge of the anatomy of the genus Selaginella, Spr Part V. The Strobilus. *Annals of Botany*, **os-24**, 19–33.

Paolillo, D.J., Jr. (1963) *The Developmental Anatomy of Isoetes*. University of Illinois Press, Urbana, IL.

Paolillo, D.J., Jr. (1982) Meristems and evolution: Developmental correspondence among the rhizomorphs of the lycopsids. *American Journal of Botany*, **69**, 1032–1042.

Parihar, N.S. (1967) *Pteridophytes*. Central Book Depot, Allahabad.

Pettitt, J. (1970) Heterospory and the origin of the seed habit. *Biological Reviews*, **45**, 401–415.

Philipson, W.R. (1990) The significance of apical meristems in the phylogeny of land plants. *Plant Systematics and Evolution*, **173**, 17–38.

Pigg, K.B. (1992) Evolution of isoetalean lycopsids. *Annals of the Missouri Botanical Garden*, **79**, 589–612.

Popham, R.A. (1951) Principal types of vegetative shoot apex organization in vascular plants. *The Ohio Journal of Science*, **51**, 249–270.

Pryer, K.M., Schuettpelz, E., Wolf, P.G., *et al.* (2004) Phylogeny and evolution of ferns (monilophytes) with a focus on the early leptosporangiate divergences. *American Journal of Botany*, **91**, 1582–1598.

Reinhardt, D., Mandel, T. & Kuhlemeier, C. (2000) Auxin regulates the initiation and radial position of plant lateral organs. *Plant Cell*, **12**, 507–518.

Schoute, J.C. (1938) *Manual of Pteridology*. Martinus Nijhoff, The Hague.

Schulz, C., Little, D.P., Stevenson, D.W., *et al.* (2010) An overview of the morphology, anatomy, and life cycle of a new model species: the lycophyte *Selaginella apoda* (L.) Spring. *International Journal of Plant Sciences*, **171**, 693–712.

Smith, G.M. (1938) *Bryophytes and Pteridophytes*. McGraw-Hill, New York.

Smith, R.W. (1900a) The structure and development of the sporophylls and sporangia of isoetes. *Botanical Gazette*, **29**, 225–258.

Smith, R.W. (1900b) The structure and development of the sporophylls and sporangia of isoetes (concluded). *Botanical Gazette*, **29**, 323–346.

Steeves, T.A. & Sussex, I.M. (1989) *Patterns in Plant Development*. Cambridge University Press, Cambridge; New York.

Stevenson, D.W. (1976a) Observations on phyllotaxis, stelar morphology, the shoot apex and gemmae of Lycopodium lucidulum Michaux (Lycopodiaceae). *Botanical Journal of the Linnean Society*, **72**, 81–100.

Stevenson, D.W. (1976b) The cytohistological and cytohistochemical zonation of the shoot apex of *Botrychium multifidum*. *American Journal of Botany*, **63**, 852–856.

Stewart, W.N. & Rothwell, G.W. (1993) *Paleobotany and the Evolution of Plants*. Cambridge University Press, New York.

Tang, H., Bowers, J.E., Wang, X., *et al.* (2008) Synteny and collinearity in plant genomes. *Science*, **320**, 486–488.

Wagner, W.H., Jr., Beitel, J.M. & Wagner, F.S. (1982) Complex venation patterns in the leaves of selaginella: megaphyll-like leaves in lycophytes. *Science*, **218**, 793–794.

Webster, T.J. (1992) Developmental problems in Selaginella (Selaginellaceae) in an evolutionary context. *Annals of the Missouri Botanical Garden*, **79**, 632–647.
Wigglesworth, G. (1907) The young sporophytes of Lycopodium complanatum and Lycopodium clavatum. *Annals of Botany*, **os-21**, 211–234.
Wikström, N. (2001) Diversification and relationships of extant homosporous lycopods. *American Fern Journal*, **91**, 150–165.
Williams, S. (1937) Correlation phenomena and hormones in *Selaginella*. *Nature*, **139**, 966–966.
Wilson, C.L. (1953) The telome theory. *Botanical Review*, **19**, 417–437.
Worsdell, W.C. (1910) The rhizophore of selaginella. *New Phytologist*, **9**, 242–253.
Zimmermann, W. (1959) *Die Phylogenie der Pflanzen*. Gustav Fischer Verlag, Stuttgart.

Annual Plant Reviews (2013) **45**, 115–140
doi: 10.1002/9781118305881.ch4
http://onlinelibrary.wiley.com

Chapter 4

EVOLUTIONARY MORPHOLOGY OF FERNS (MONILOPHYTES)

Harald Schneider

Department of Botany, Natural History Museum, London, UK

Abstract: Throughout its long history, concepts of plant morphology have been mainly developed by studying seed plants, in particular angiosperms. This chapter provides an overview to the morphology of ferns by exploring the evolutionary background of the early diversification of ferns, by discussing the main structures and organs of ferns, and finally by exploring our current knowledge of fern genomics and evolutionary developmental biology. Horsetails (Equisetopsida) and whisk ferns (Psilotales) are treated as part of the fern lineage. Throughout the chapter, I employ a process-oriented approach, which combines the process orientation of the Arberian Fuzzy Morphology with the process orientation of Darwinian evolution as reflected in current phylogenetics.

Keywords: Arberian Fuzzy Morphology; meristems; phylogenetics; evolutionary simplification; synapomorphy; root; telome theory.

4.1 Introduction

Ferns are one out of three extant lineages of vascular plants and form the sister clade to the seed plant lineage (Figure 4.1; Pryer *et al.* 2001). They comprise five extant lineages, namely, the leptosporangiate ferns, the marattioid ferns, the horsetails, the ophioglossoid ferns, and whisk ferns (Pryer *et al.* 2001; Smith *et al.* 2006). These five lineages show a remarkable range of morphological differentiation in the body plans of the sporophytes and to a lesser extent the gametophytes (Schneider *et al.* 2002, 2009; Pryer *et al.* 2004). It is, therefore, not surprising that the monophyly of this lineage was only recently widely accepted as a result of research efforts following the principle of reciprocal illumination by integrating fossil evidence (Stein *et al.* 1984; Stein 1993; Kenrick & Crane 1997), DNA sequence variation (Pryer *et al.* 2001, 2004;

Annual Plant Reviews Volume 45: The Evolution of Plant Form, First Edition.
Edited by Barbara A. Ambrose and Michael Purugganan.

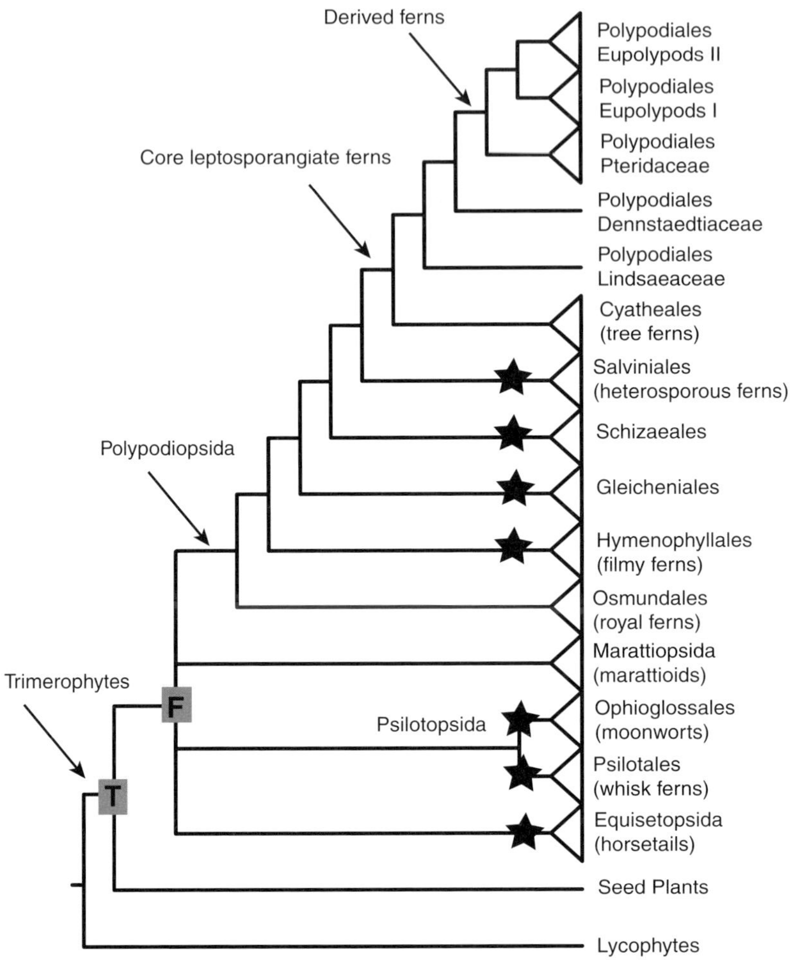

Figure 4.1 Phylogeny of ferns illustrating the current uncertainty about the early diversification of this lineage. Gray box T: indicating putative synapomorphies, for example, euphyll and RAM/SAM differentiation, shared by the offspring of the Trimerophyte lineage; gray box F: indicating putative synapomorphies, for example, apical growth of the euphyll, of the fern lineage. Stars indicate occurrences of deviations from a regular body plan differentiated into leaves, roots, and shoots. The body plan of horsetails is characterized by highly reduced euphylls that are arranged in whorls consisting of more than three leaves; the leaves of Gleicheniaceae and to some extent other families of Gleicheniales are characterized by an unusual pseudodichotomous branching and a habit resembling the shoot of some seed plants; filmy ferns (Hymenophyllaceae, Hymenophyllales) are not only characterized by the reduction of the leaf cuticula and lamina, but some clades show also a reduction of the roots or a loss of roots (e.g., *Crepidomanes*, *Didymoglossum*, *Gonocormus*) as well as a clear differentiation of leaves and shoots (*Gonocormus*); the winding leaf of the schizaealean fern *Lygodium* shows shoot-like features; moonworts (Ophioglossales) have a reduced root system and an unusual position of the sporangia; and whisk ferns (Psilotales) are characterized by the lack of roots, the reduction of leaves, unusual position of sporangia, and shoot-like organs best described as misfits.

Wikstroem & Pryer 2005; Qiu *et al.* 2006, 2007; Schuettpelz Korall & Pryer 2006; Karol *et al.* 2010; Rai & Graham 2010), and the morphological variation of extant taxa (Schneider *et al.* 2009). The lineage is sometimes called monilophytes based on the term Moniliformopsis introduced by Kenrick and Crane (1997) but I use here the term ferns instead following Pryer *et al.* (2004). Despite the differences of the body plans, ferns share several important features including a life cycle in which both generations, the usually haploid gametophyte and the diploid sporophyte, are free living and differ substantially in their body plans from each other (Schneider *et al.* 2002, 2009). The only exception is formed by the heterosporous waterferns, which are a rather species-poor lineage nested within the lineage of leptosporangiate ferns (Pryer *et al.* 2004). The gametophyte of ferns tends to be dimidiate and of a simple body plan, whereas the sporophyte develops a complex body plan that includes the differentiation into shoot, root, and leaves. In morphological complexity, the sporophyte of ferns is comparable to the sporophyte of seed plants, however, considerable differences exist such as the absence of secondary growth in extant ferns, the lack of a bipolar orientated embryo, and variation in shoot branching (Schneider *et al.* 2002, 2009).

In this review, I try to provide an overview on the current knowledge and to identify the major questions to be studied using phylogenetics and/or developmental genetics in the near future. I have approached this via two different perspectives: (1) a lineage and time-oriented perspective and (2) a morphological structure-based perspective. Finally, I review our limited knowledge on the developmental pathways in ferns and the need to sequence a whole fern genome.

Throughout the chapter, I refer to the vast body of literature on plant morphology. To avoid repetition, I want to point out some of the most comprehensive references concerning the morphology of ferns: Bierhorst (1971), Bower (1923, 1935), Campbell (1911), Goebel (1928/1930), Guttenberg (1965), Ogura (1972), Schoute (1938), Troll (1937), and Velenovsky (1905).

4.2 Context of evolutionary plant morphology

It is important to evaluate the context when interpreting fern morphology. Classical plant morphology has always experienced problems with ferns because the typological concept of roots and leaves had been developed on the model of the "Urpflanze," which is an essentialist representation of a eudicot angiosperm (Kaplan 2001a, 2001b). Ferns may be qualified as misfits in the context of typological/essentialistic morphology. I do not refer here only to extremes such as the sporocarp of waterferns (Nagalingum *et al.* 2006), the winding leaf of *Lygodium* (Mueller 1982), nonappendicular leaves found in some filmy ferns (Bierhorst 1973), and the "stolons" of *Nephrolepis* (Richards *et al.* 1983) but the fern body plan(s) in general. Some authors raised correctly the question if ferns are really cormophytes (Hagemann 1976, 1992, 1997).

I want to illustrate this with three examples. Firstly, the leaf of the majority of ferns share characteristics with shoots as a result of a dominant acropetal growth instead of a basipetal growth found in most angiosperm leaves (Reiser *et al.* 2000; Schneider *et al.* 2002, 2009). Secondly, ferns show a range of branching patterns in comparison to the highly conserved origin of secondary shoots in the leaf axils of seed plants (Bierhorst 1973; Hagemann 1976, 1989). Thirdly, the embryo is not organized in a bipolar fashion and, therefore, Goebel (1928/1930) introduced the term of primary "homorhiz" for ferns because its primary root is short-lived or perhaps not definable (Gutenberg 1965).

A more appropriate approach to interpret fern morphology is without a doubt offered by the concept of "continuum morphology" or "Aberian Fuzzy Morphology" (Sattler 1996; Rutishauser & Isler 2001; Kirchoff *et al.* 2008). It is remarkable that representatives of this school have not discussed ferns in greater detail yet. As an example, continuum morphology is well suited to provide an accurate description of the leaf of ferns because this concept allows a deviation from a single "Urform." Goethe's Urpflanze has three well-differentiated organs, the leaf, root, and shoot. These three organs are seen as the three distinct and invariant modules that build up higher plants (Figure 4.2). Recently, similarities have been stressed between classical morphology and classical invariant theory (Mavroodiev 2009). Continuum morphology employs a holistic view of the plant body in which the different organs, such as leaves, shoots, and roots, are linked by shared developmental processes. In comparison to angiosperms, the leaves of ferns share important similarities with shoots and thus the differentiation between the organs may

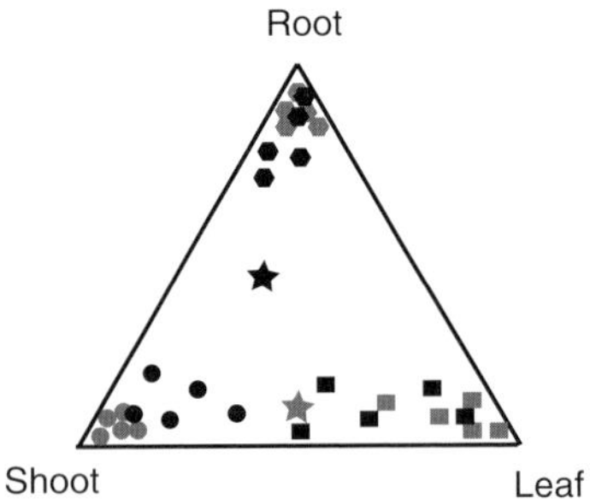

Figure 4.2 Continuum morphology concept of leaf (squares), root (diamonds), and shoot (circles) in seed plants (gray) and putative description of the differentiation of these organs in ferns (black). In general, the differentiation of shoot and leaf is lower in ferns than in seed plants despite some variation that exists especially in the context of leaves. However, leaves of ferns, for example, indeterminate growth frequently share features attributed to shoots in seed plants. Some fern leaves, for example, the climbing leaf of *Lygodium* may be best scored as a transitional form between leaf and shoot. The stars indicate the position of organs with ambiguous homology such as the stolon of the derived fern *Nephrolepis* (black star; Sanders *et al.* 2011) and the phylloclade of the monocot *Ruscus* (gray star; Hirayama *et al.* 2007).

be understood as less distinct (Figure 4.2). As an example, the leaves of many ferns show a relatively indeterminate growth that may be caused by a similar expression of transcription factors in the shoot apex and the apical meristems of the leaves/pinnae (Reiser *et al.* 2000; Harrison *et al.* 2005; Sanders *et al.* 2011). In addition, several unusual fern structures, for example, stolons of *Nephrolepis* (Richards *et al.* 1983; see also Sanders *et al.* 2011), can be best described as misfits that combine features of different organs.

It is important to keep the influence of classical morphology in mind because it is still very influential especially as a result of the remarkable efforts of Don Kaplan (2001a, 2001b) who provided access to this knowledge to a mainly English reading audience. However, classical morphology is not well aligned to the concept of Darwinian evolution. It is very important to keep in mind that the meaning of "primitive" in a typological context is not synonymous with "ancestral" in an evolutionary context. In the following, I address the evolution of ferns mainly from a perspective based on the insights of scholars, such as Willi Hennig (1965) and Walter Zimmermann (Donoghue & Kadereit 1992; Classen-Bockhoff 2001), focusing on the transformation of structures in time. This strictly phylogenetic approach is more likely to be compatible with the developmental, process-oriented perspective of the Arberian thinking than with the typological classification of classical plant morphology (Classen-Bockhoff 2001; Rutishauser & Isler 2001).

The importance of "transformation of forms" versus "fixed typology" can be illustrated with the discussion on the evolution of leaves. The origin of leaves is frequently discussed and the nonhomology of leaves of liverworts, mosses, lycophytes, ferns, and seed plants is widely accepted. However, these statements do not provide us with any understanding of the origin of leaves. Many proposed hypotheses require Goldschmidt's hopeful monsters as an explanation of the saltations proposed in the evolution of leaf-like structures as well as other structures in plants (Bateman & DiMichele 2002; Hintz *et al.* 2006; Theissen 2009). Several current scholars of plant evolution tend to underestimate the major progress achieved by Zimmermann's telome theory (Zimmermann 1959, 1965). The hypothesis proposes a sequence of transformations, for example, overtopping, plantation, and webbing/fusion, by which a dichotomously branching shoot system evolved into a megaphyll. This theory has serious limitations and some of the processes may look a bit simple in our improved understanding of the regulation of developmental processes as pointed out by many current authors (Floyd & Bowman 2006, 2007a, 2007b; Beerling & Fleming 2007; Tomescu 2008; Boyce 2010; Galtier 2010) but the "gradualistic" perspective breaks the evolution of complex organs into steps and thus it provides a concept that is actually testable, which cannot be achieved with the majority of typological-based hypotheses on the origin of leaves. Some of these processes may have involved fusion and/or substantial modifications of pathways and thus may have involved the establishment of hopeful monsters (Theissen 2009) especially if they transformed early ontogenetic stages (Rieppel 2001). I want to argue here for an approach that considers partial homology because leaves may not have originated by a single mutation creating the ultimate leaf but by a sequence of transformations. Of course, these hypothetical transformations need to be outlined under consideration of our current understanding of the regulation of plant developmental processes (Harrison *et al.* 2005; Floyd & Bowman 2006, 2007a, 2007b). Some of these transformations may have occurred before and others after the divergence of extant lineages of land plants but convergent evolution via co-opting the same ancestral gene pathways need to be considered (Harrison *et al.* 2005). A process- oriented approach, combining phylogenetic and developmental perspectives, instead of a typological-oriented approach is the most promising approach to identify the developmental pathways and their transformation in the evolution of land plants.

The phylogenetic relationships of the inferred model taxon have to be considered carefully to avoid positive misleading support or rejection of a hypothesis. Some interesting model systems, such as the highly modified body plans of some species of *Nephrolepis* (Sanders *et al.* 2011) and species belonging to the filmy fern genus *Gonocormus* (Schneider 2000) are derived members of particular fern lineages (Pryer *et al.* 2004; Smith *et al.* 2006). Thus, we need to consider the result as "secondary" modification similar to the situation of the evolution of misfits in angiosperms such as the phylloclade of the monocot *Ruscus* (Hirayama *et al.* 2007). Choosing the right model is always essential and thus the leaf development of more conservative taxa such as *Osmunda* may be more informative than highly derived systems. I point out *Osmunda* on purpose. The royal ferns, Osmundaceae, are not only the offspring of the Paleozoic radiation of leptosporangiate ferns but we also have evidence for a high conservation of the body plan of these ferns at least since the Triassic (Taylor *et al.* 2009).

4.2.1 Perspective 1: rapid radiation versus stasis in the evolution of fern body plans

The evolution of fern body plans is best summarized by employing the concept of two major phases as it was established in studies on the evolution of other major branches of life, for example, the evolution of angiosperm flowers (Endress 2001). The first phase is best visualized with the scenario of an explosive radiation of very divergent body plans whereas the second phase is characterized by conservation of a few of these body plans. The later process may be described as canalization. Ferns are especially suited to study these phases because in contrast to seed plants, the modern lineages still contain remnants of this explosive phase of morphological radiation whereas in seed plants older forms were consistently replaced by more derived forms (Figures 4.1 and 4.3).

The explosive phase started with the replacement of the trimerophyte by two daughter lineages, the seed plants and the ferns (Kenrick & Crane 1997;

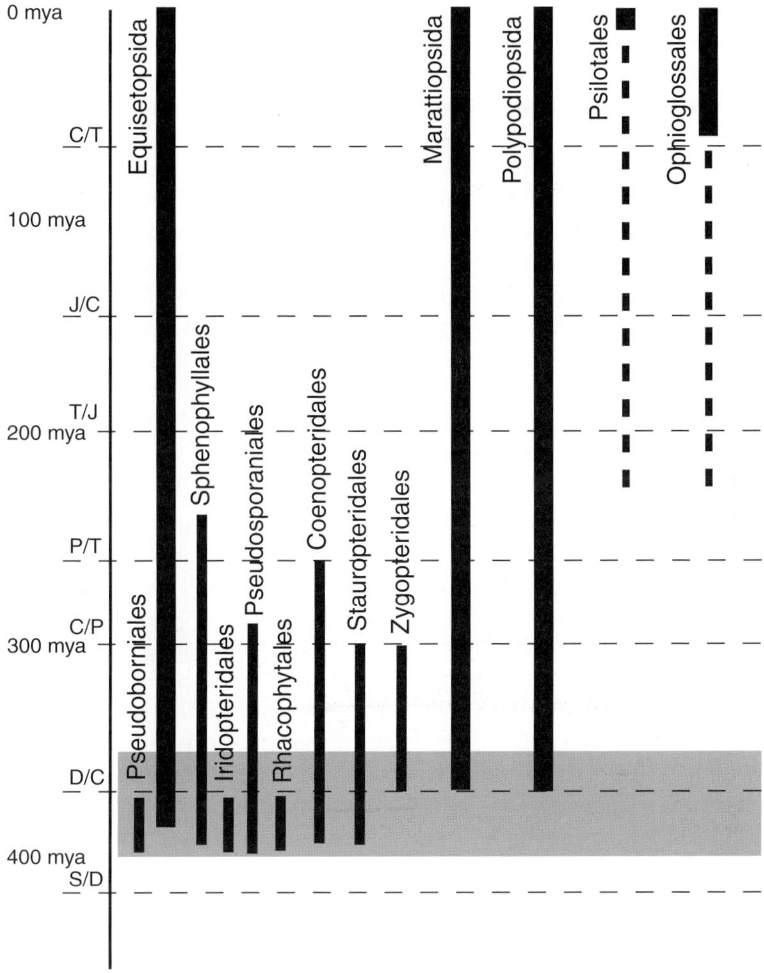

Figure 4.3 Overview about the time occurrence of lineages recognized within the monilophytes (following Taylor *et al.* 2009 with the exception of Ophioglossales and Psilotales). Gray box indicates the time of the early radiation of ferns. Dashed vertical lines indicate the borders between major geological periods: S/D, Silurian–Devonian Transition; D/C, Devonian–Carboniferous Transition; P/T, Permian–Triassic Transition; T/J, Triassic–Jurassic Transition; J/C, Jurassic–Cretaceous Transition; C/T, Cretaceous–Tertiary Transition (KT boundary).

Taylor *et al.* 2009). Both sister lineages underwent a rapid diversification in the Devonian in which a vast number of morphologically diverse lineages evolved (Figure 4.3). In the early Carboniferous, the body plans of the three lineages with a sufficient fossil record, namely, the leptosporangiate ferns, the marattioids, and the horsetails, had been established whereas many Devonian groups went extinct. The origin of the two other body plan groups, the ophioglossoids and the whisk ferns, is unclear because these sister lineages form a ghost lineage lacking a fossil record in the Mesozoic and Paleozoic (Schneider *et al.* 2002; Taylor *et al.* 2009). The identification of their Paleozoic ancestors would be one major breakthrough in our understanding of fern evolution.

The body plans of the five extant lineages share several major features such as the dominance of the sporophyte, the absence of secondary growth, relatively simple meristems, and shoot systems differentiated into a long-lived, mostly branched shoot axis and leaves (Schneider *et al.* 2002, 2009). However, they differ strongly in other aspects such as the morphology of the leaves (Section 4.4, see "Mature Sporophytes"), the location of the sporangia on the leaves, for example, dorsal or marginal position at the lamina, and the kind of indumenta. Ferns show a remarkable variation in the density and composition of the indumenta. Hairs and/or scales usually densely cover young fern leaves and the shoot apices as a protection. Hairs are unicellular to multicellular structures developed from a single epidermal basal cell, whereas scales are multicellular planar structures formed by a group of epidermal cells arranged in a line. Nearly all fern indumenta can be classified either as hairs or scales with the exception of some bristle like structures developed from more than one epidermal cell but without a dorsiventral organization as in scales found in a few fern groups. Most of the ferns also possess well-developed root systems (Schneider 1996) though they are absent in the whisk ferns, but this is arguably a secondary loss in the evolution of this lineage (see Section 4.2.2.4, page 128). The explosive phase shows an overwhelming morphological diversity when the fossil record of the group in the Devonian is taken into account. These groups show a remarkable range of variation in axis organization and more or less leaf-like appendages. A particularly well-known group, the zygopterids (Zygopteridales) had elaborate, three-dimensional fronds with pinnae arranged in four ranks. Thus, the rachis had a radially symmetric cross-section instead of the usual dorsiventral organized rachis of extant ferns (Taylor *et al.* 2009). The relationships of many of these groups are still poorly understood. In some cases, testable hypothetical relationships have been proposed based on the similarities of the shoot organization such as the putative sister relationship of the Sphenophyllales to the horsetails (Taylor *et al.* 2009) or Stein's hypothesis concerning the putative origin of horsetails of ferns from a common ancestor (Stein *et al.* 1984). I want to refer here to the exquisite review of our knowledge given in Taylor *et al.* (2009) for further reading. However, I also want to explicitly express the need for a cladistic study of ferns following the approach adopted by Kenrick and Crane (1997) to study the relationships of land plants. This kind of analysis will provide a hypothesis of character states transformation based on a rigorous distinction of apomorphic, homoplastic, and plesiomorphic character states as applied successfully since the introduction of cladistic methodology in evolutionary/ taxonomic sciences.

4.2.2 Perspective 2: key structures and organs of fern body plans

In this part, I address our knowledge on four key structures of fern morphology. The focus is on vegetative components and only limited consideration is given to reproductive structures and their evolution.

4.2.2.1 Meristems

Ferns show rather simple organized apical meristems (White & Turner 1995; Imaichi 2008). Usually, they contain a single apical cell. In a few taxa, such as Marattiaceae and Osmundaceae, the larger size of the apical meristem results in the formation of an apical cell group instead of a single one. At the sporophyte level, three kinds of apical meristems can be differentiated. The shoot apical meristem (SAM) has usually a prominent apical cell of a tetrahedral shape in most ferns. The SAM functions as the initial for the entire shoot. The organization of the SAM with a singe apical cell is correlated with the lineage specific plasmodesmatal networks, which contrasts ferns to seed plants with their multicellular SAMs with interface specific plasmodesmatal networks (Imaichi 2008). The leaves of ferns usually have meristems formed around a leaf apical cell with two cutting faces or very rarely three cutting faces. The leaf apical meristem (LAM) resembles fern SAMs in having high plasmodesmatal densities. The activity of the LAM results in the formation of a marginal blastozone, comparable to those found in seed plants (Hagemann & Gleissberg 1996; Hagemann 1997). However, this marginal meristem tends to be frequently fractionated resulting in the formation of pinnae and pinnules instead of simple blades. Future studies will need to discover the role of the regulation of the apical growth in this process. In ferns, entire leaf laminas are not uncommon but the majority of ferns tend to have leaves with highly divided laminas. The roots of ferns grow with an apical meristem (RAM, root apical meristem) formed by a single tetrahedral apical cell with four cutting faces (e.g., Guttenberg 1965; Ogura 1972; Gifford 1991; Schneider 1996). One face produces the initials for the calyptra (root cap) whereas the other faces form the initials for the root body. The division process is highly conserved and results in the formation of traceable cell lineages, which makes root development different from leaf and shoot development. Traceability of cell lineages in roots of euphyllophytes has been frequently taken as an advantage to study the control of root development in various organisms including *Arabidopsis* (e.g., Menant *et al.* 2007; Saint Savage *et al.* 2008; Sozzani *et al.* 2010). The gametophyte has again an apical meristem (gametophyte apical meristem (GAM)) with an apical cell with usually two cutting faces. The formation of a marginal blastozone is visible in some larger dorsiventrally organized gametophytes (Raghavan 1989; Hagemann 1997; Imaichi 2008).

4.2.2.2 The gametophytes

The morphological variation of fern gametophytes, called prothallia, is frequently ignored in developmental studies although a lot of documentation of this variation has been achieved (Nayar & Kaur 1971). The prothallia are usually small and ephemeral. They are simple plant bodies of either radially symmetric organization such as the gametophytes of ophioglossoid ferns and whisk ferns, filamentous such as the gametophytes of horsetails and some leptosporangiate ferns, or dorsiventrally organized such as the gametophytes of marattioids and most leptosporangiate ferns (Nayar & Kaur 1971). Most dorsiventral organized gametophytes show a symmetric heart shape but they can also be asymmetric or forming thalloid strap-shaped or filamentous structures (Nayar & Kaur 1971; Takahashi *et al.* 2009). Although, radially symmetric gametophytes are always multilayered, dorsiventrally organized gametophytes usually show only a single cell layer. However, a several cell layer thick central ridge can be found in the heart-shaped gametophytes of some basal leptosporangiate ferns (Osmundaceae) and the marattioids (Nayar & Kaur 1971). More recently, some researchers studied the role of the apical meristem and marginal growth in the alteration from cordate, symmetrical to strap-like asymmetrical gametophytes in the evolution of derived ferns such as *Leptochilus* (*Colysis*) *decurrens* (Takahashi *et al.* 2009).

The development of fern gametophytes has been studied for a long time, but we still do not understand the control of the phenotype development by the genotype, which comprises the information for both the gametophyte and sporophyte. This question has been targeted by many experimental studies and it is impossible to provide a full summary here (Raghavan 1989; Sheffield 2008). Some studies speculated about the impact of haploid versus diploid genomes but this has not been confirmed by observations in ferns that display apogamy (Raghavan 1989; Bell 1992; Park & Kato 2003). In these ferns, there is a lack of sexual reproduction and the gametophyte and sporophyte have the same chromosome number.

Besides the parenchymatic cells of the core body, fern gametophytes comprise a few other cell types. Firstly, they have rhizoids that are structurally similar to root hairs. Rhizoids are likely a plesiomorphic feature in vascular plants and root hairs may have evolved by a heterotopic mutation (Schneider *et al.* 2002). Rhizoid-like hairs are also found on the scales of some derived ferns (Bower 1923) and their origin may be a further case for heterotopic mutations in ferns. The origin of these structures has not been studied so far and hence this hypothesis was not fully developed previously to my knowledge. Fern gametophytes also may bear hairs, in particular glandular hairs (Nayar & Kaur 1971). The genes controlling the formation of these hairs may be the same controlling hair formation on the sporophyte shoot or leaf.

Other structures, such as archegonia and antheridia, are unique to gametophytes (Nayar & Karu 1971; Raghavan 1989). Gemmae and other forms of vegetative buds can be found in both generations (e.g., Bower 1923; Nayar & Kaur 1971; Raghavan 1989) but an exhaustive comparative study is still missing.

In summary, the gametophyte shares some cell types with the sporophyte, for example, undifferentiated parenchymatous cells forming the main body of the gametophyte, rhizoids, and hairs but the gametophyte usually lacks other cell types such as tracheids and phloem cells.

4.2.2.3 Embryos

Ferns show a remarkable variation of the structure of the embryo (Figure 4.4; Johnson & Renzaglia 2008, 2009). However, the embryo does not show a bipolar organization as found in seed plants (Guttenberg 1965; Bierhorst 1971; Schneider *et al.* 2002). Instead, fern embryos show a division into a usually large foot area and an embryo body that includes the formation of a somewhat bended vascular axis connecting the embryonic SAM and LAM with the RAM. The shoot meristem extends to the permanent shoot system whereas the embryonic root is short lived and gets replaced by other roots formed by tissue originating from the SAM. This situation was described as primary "homorhiz" (Goebel 1928/1930) because root systems of ferns lack a permanent primary root. In addition, the leaves gradually develop to more complex leaves starting from rather simple, often undivided leaves with a short petiole and a forked vein to more complex leaves, via step-by-step increase of the complexity in the sequence of leaf formation.

The variation among the embryos of vascular plants is remarkable although some structures appear highly conserved. All homosporous vascular plants appear to possess a foot, which plays a critical role to connect the young embryo and the gametophyte (Duckett & Ligrone 2003; Johnson & Renzaglia 2008, 2009). The foot is absent in seed plants, which is likely correlated with the evolution of seeds. Likewise, the conservation and variation of the suspensor is probably correlated with the evolution of seeds. It would be really interesting to compare the role of genes known to regulate embryo patterning in *Arabidopsis* to those in ferns (Izhaki & Bowman 2007). It is interesting to recognize the distinction of the embryos of horsetails (Equisetopsida) and whisk ferns (Psilotales) in comparison to their relatives (Figure 4.4). This may suggest an early ontogenetic deviation causing the highly distinct body plans of these lineages. The modification of embryo development may provide an explanation for the origin of body plans that appear less compatible with scenarios of gradualistic stepwise transformation. Thus, they may be "hopeful monsters" comparable to turtles (Rieppel 2001).

4.2.2.4 Mature sporophytes

The body plans of most ferns are composed of three organs: leaves, roots, and the shoot. The shoot is usually called a rhizome despite the range of variation in growth forms including long-creeping rhizomes, short-erect rhizomes, and tree-like rhizomes (Bierhorst 1954). This organ structure reminds one of seed

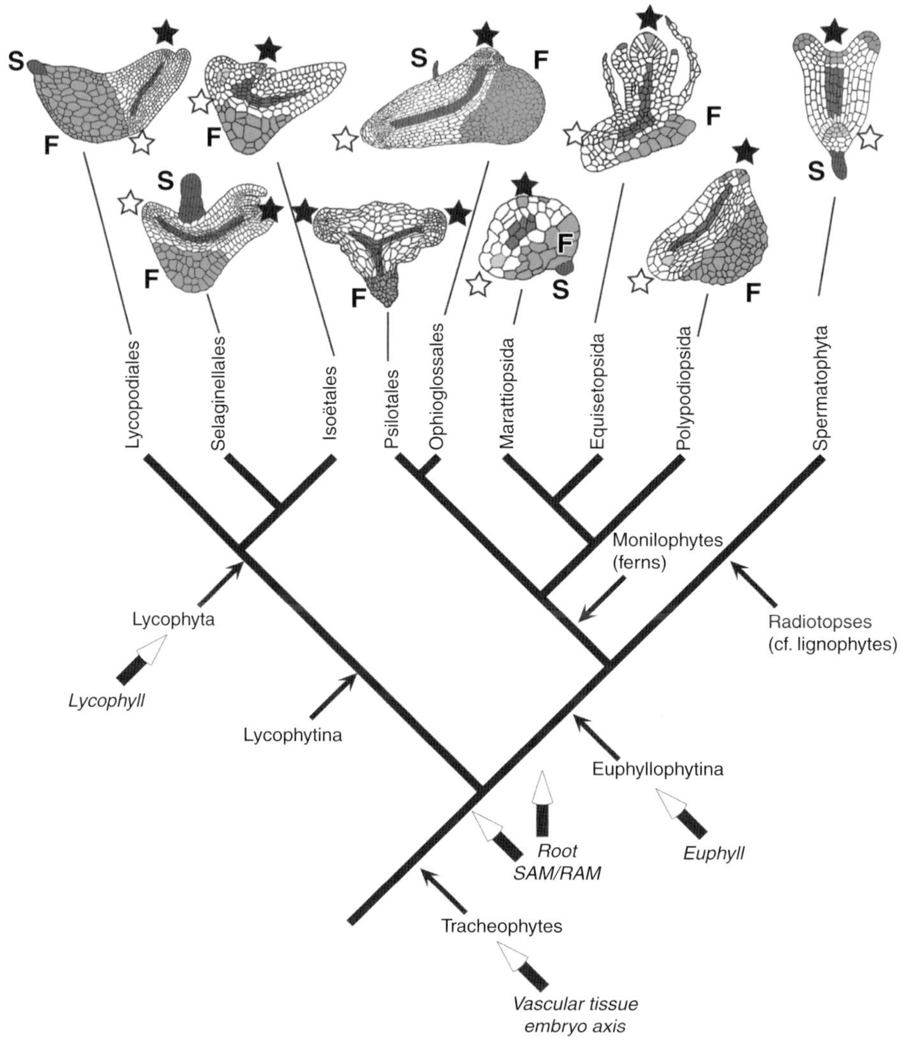

Figure 4.4 Evolution of embryos in vascular plants, plotted onto a phylogenetic hypothesis proposed by Pryer *et al.* (2001). The graphics have been redrawn from descriptions of fern embryos obtained in studies carried out in the nineteenth and twenty-first centuries (Guttenberg 1965; Bierhorst 1971). The examples were chosen to represent the variation among lineages of vascular plants. In lineages with considerable variation such as Ophioglossales, the chosen example represents the majority of embryos described so far. As an example, in the Ophioglossales embryos of closely related species may differ in the absence/presence of a suspensor, size of the foot, and the geometry of the embryo. In other lineages, for example, Polypodiopsida, the variation among described embryos is limited and embryo shown likely represents the ancestral state of the lineage. The bipolar embryo of seed plants is unique and characterized by the absence of a foot, which likely coincides with the evolution of seeds. Among seed plants, the embryo structure can be very different but the bipolar structure is one of the apomorphic characters of the seed plant lineage. *Note*: the embryos are not shown at the same stage of their development and are not shown proportionately.

plant shoots but it is still unclear if this differentiation happened before or after the separation of the monilophyte and seed plant lineages. It is important to recognize that leaves of ferns share the apical growth with the shoot and thus these two organs are less differentiated from each other than those found in extant seed plants. However, it is not clear yet if this difference is caused by independent origins as suggested by some authors or transformations that postdate a shared origin. Another major difference with seed plants is the variation in the processes of lateral shoot formation. In some ferns (e.g., *Gleichenia* and *Lygodium*), the rhizome branches dichotomously and thus leaf initiation is independent from lateral shoot initiation (Bierhorst 1973; Hagemann 1989). However, in the majority of ferns, the initiation of a lateral shoot is dependent on the formation of lateral leaves although they are usually not formed in the leaf axil. Most interesting are those ferns that seem to be able to form leaf–shoot complexes, such as in some Dennstaedtiaceae and filmy ferns, that are challenging the notion of two different developmental programs (Bierhorst 1973; Hagemann 1989). I need to stress again that these may be secondary modifications (character state is not ancestral for ferns), which could provide us with interesting insights into the evolution of developmental pathways. The formation of lateral branches is thus more variable than in seed plants, which may be best described as less constrained. The level of variation can be best illustrated by pointing out two unusual patterns in *Angiopteris* and ophioglossoids. Relatively little attention has, so far, been given to the formation of new shoots in the marattioid genus *Angiopteris*. Here, buds originate from serial dormant meristems along the stipules at the base of the leaves (Campbell 1911; Huang *et al.* 2011). This unique structure resembles similar arrangements in some angiosperms but it has very likely evolved independently. Similarly, the unusual body plan of ophioglossoids has not found as much attention as they require (Campbell 1911). The compact shoot produces usually only a single leaf per growing period despite the initials for the next leaves already being fully formed (Bierhorst 1977; Schneider *et al.* 2002). Ophioglossoid shoots rarely or never form lateral shoots but the leaves bear very unusual fertile structures. These also arise in the axil of the leaf and have been variously interpreted. Finally, it is important to

Figure 4.4 (*Continued*) It is important to note the high level of variation supporting the notion of limited conservation of embryology in the early diversification of ferns. It is interesting to note the distinctive embryos of the Equisetopsida and Psilotales, suggesting substantial changes in their early ontogeny that likely contributes to their distinct body plans. Putative shared synapomorphies are plotted on some deeper nodes. Roots originated either before the split of lycophytes and trimerophytes or evolved at least twice. Lycophylls and euphylls are two independent events in the evolution of vascular plants. SAM, black star—shoot apical meristem; RAM, open star—root apical meristem; F, foot; S, suspensor; dark central area of embryos—vascular tissue—embryo axis. Simple arrows indicate lineages, open arrows indicate putative origins of structures.

note that all extant monilophytes lack secondary growth despite the fact that some Paleozoic monilophytes may have had secondary growth (Taylor *et al.* 2009). In general, secondary growth was always rare in this lineage including monilophytes with tree-like habits, for example, tree ferns and the Paleozoic genus *Psaronius* (Taylor *et al.* 2009). Similar to palms, ferns evolved structural innovations that allowed them to form erect trunks of heights more than 20 m (Large & Braggins 2004). These innovations include the formation of root mantles, highly dissected steles, and organization of mechanical tissues (Ogura 1972; Large & Braggins 2004).

Nearly all monilophytes possess a root system. All roots can be classified as shoot born with the exception of the root of the embryo (Figure 4.4). However, the later is short lived and thus the root system of ferns is best described as rather uniform, for example, "homorhiz" (Goebel 1928/1930). Most fern roots branch via the lateral roots, which form endogenously from the endodermis or pericycle (Schneider 1996). This condition resembles the lateral root formation of seed plants. Dichotomously branching roots can sometimes be found in the ophioglossoid ferns but the roots of these ferns are usually unbranched (Campbell 1911; Bierhorst 1971; Schneider *et al.* 2002). All fern roots possess a root cap (calyptra) and nearly all form root hairs (Schneider 1996; Schneider *et al.* 2002, 2009). The later are lacking in the roots of ophioglossoids. Roots are completely lacking in the Psilotales, in the aquatic heterosporous fern genus *Salvinia* and some epiphytic/saxicolous species of filmy ferns (Schneider 2000; Schneider *et al.* 2002, 2009). The lack of roots in these cases is likely the result of evolutionary simplification of the body plan and not an ancestral condition. Many shared similarities indicate a shared origin of roots in the ancestor of ferns and seed plants but these similarities may be also explained by co-option of similar genes or developmental pathways. As an example, the pathway controlling the differentiation of the rhizodermis (epidermis of the root) is similar to the pathway controlling the differentiation of the epidermis (Schiefelbein *et al.* 2009; Pu & Brady 2010). In addition, the developmental pathway controlling the development of root hairs in angiosperms appears to have been co-opted from the pathway controlling the differentiation of rhizoids in bryophyte gametophytes (Menand *et al.* 2007; Jang *et al.* 2011).

Fern leaves grow mainly via an apical growth, which is most visible in the circinate development of the unfolding leaf (Schneider *et al.* 2002, 2009). The persistence of apical growth can result in the development of rather large leaves of several meters long as illustrated by the long twining leaf of the fern genus *Lygodium* (Mueller 1982). It is the only example in the plant kingdom where the plant climbs via a winding leaf instead of winding shoots. Again, this illustrates the fuzziness of the leaf concept defined for angiosperms when applied to ferns. However, the majority of fern leaves are well defined in their length and some are rather small. Extremely small leaves, such as the scale-like leaves of horsetails (Equisetopsida) and whisk ferns (Psilotales) have been the focus of previous controversies (e.g., Bierhorst 1977; Kaplan 2001b;

Schneider *et al.* 2002, 2009). They were classified as microphylls together with the leaves of lycophytes. However, Kaplan (2001b) showed in a comprehensive typological argument that the concept of microphylls is deeply flawed. This is consistent with the phylogenetic evidence of independent origins of so-called microphylls (Kenrick & Crane 1997). Here, I suggest refusing the division of the leaves of vascular plants into micro- and megaphylls. Both terms are problematic and contribute to confusions (e.g., Kaplan 2001b). Thus, I suggest that a new terminology, such as lycophylls and euphylls, would help to circumvent misunderstandings.

Lycophylls: Leaves of extant and extinct members of the lycophyte lineage that originated from ancestors such as *Drepanophycus*. Crane and Kenrick (1997) argued that these leaves might originate from sterile sporangia. This hypothesis still needs further testing especially in respect to the origin of genes controlling the differentiation of dorsiventral organs such as leaves (see Harrison *et al.* 2005). In angiosperms, the YABBY genes play a critical role but no copy of this family has been found in seed-free land plants (Efroni *et al.* 2010; Floyd & Bowman 2010; Sarojam *et al.* 2010). In this context, it is important to note that the sporangia of lycophytes show a dorsiventral organization in contrast to the sporangia of all other land plants, which show a radially symmetric organization (Kenrick & Crane 1997; Schneider *et al.* 2002, 2009). Lycophylls usually have a single unbranched vein but some exceptions exist within the genus *Selaginella* (Wagner *et al.* 1982).

Euphylls: Leaves of monilophytes and seed plants share features such as a more or less defined growth, the formation of leaf gaps in the stele of most taxa, and dorsiventral organization. The shared ancestral stages are described with the term pseudodichotomous branching or lateral branch systems (Floyd & Bowman 2010). The concept of euphylls differs substantially from the megaphyll concept because the concept aims to determine the shared ancestral character states (synapomorphy) and assumes subsequent transformations as independent such as the differentiation of lamina and petiole. This differentiation evolved at least twice during the diversification of the trimerophytes (ancestors of ferns and seed plants; Figure 4.1). It is important to note that this concept resembles the partial-shoot theory of the leaf and similar concepts of the Arberian Fuzzy Morphology (Rutishauer & Isler 2001; Kirchoff *et al.* 2008). The concept assumes explicitly that only early steps in the process of leaf evolution have been shared among ferns and seed plants among the different lineages of ferns. The joint origin is currently outlined as a shoot system with a pseudodichotomous branching (e.g., Beerling & Fleming 2007; Kidner 2007; Floyd & Bowman 2010).

The apical growth and morphogenesis of fern leaf lamina suggests a high similarity with the morphogenesis of shoots. Some authors discussed the limitation of ferns to form a blastozone (Zurakowski & Gifford 1988; Hagemann & Gleissberg 1996; Hagemann 1997) that would allow the development and evolution of entire laminas, which may explain the high frequency of fern leaves with highly dissected laminas. Ferns may evolve entire laminas mainly by reducing the space among secondary veins or by evolving complex anatomosed venations as found in many derived fern genera such as the majority of ferns belonging to the families Polypodiaceae and Tectariaceae (Zurakowski & Gifford 1988; Hagemann 1997; Schneider *et al.* 2009). Ferns can prove to be a powerful system to investigate the correlation between changes in developmental pathways and changes in leaf morphology. Sometimes, individuals with once pinnate or bipinnate laminas can be found in the same species (e.g., several species of *Asplenium*). In addition, hybrids can be generated between simple bladed species and bipinnate species such as the hybrid between the simple-bladed *Asplenium nidus* and the bipinnatifid *Asplenium prolongatum* (N. Murakami, personal communication) and *Asplenium* × *hybridum* that is the hybrid between the simple-bladed *Asplenium sagittatum* and the pinnatifid *Asplenium ceterach* (Pinter *et al.* 2002). In these hybrids, the lower part of the lamina resembles the leaf of the parent with the divided lamina, whereas the upper part of the lamina resembles the leaf of the parent with the undivided lamina.

The formation of complex induments on the leaf and shoot is a further character of great importance to ferns (Bower 1923). In general, two main types are recognized: hairs and scales. Hairs are classified as uni- to multicellular trichomes formed by cell divisions of a single epidermal cell. Similar hair structures can be also found on some gametophytes. Scales are classified as multicellular, dorsiventral structures that originate by cell divisions of several epidermal cells. Some authors have speculated that scales may represent reduced leaves because they sometimes show the phyllotaxis of leaves (Stuetzel & Geiling 1997) but some transitional forms between scales and hairs are also known (e.g., Bower 1923). These transitional forms include a multicellular hair-like structure formed by several epidermal cells but with a radial organization. Interestingly, the same type of indumentum can be found on the shoot and leaf supporting the view of a continuum of leaves and shoots in ferns (Ogura 1972). Hair-bearing scales are another notable morphological curiosity found in some ferns (Schneider *et al.* 2002, 2009; Wang *et al.* 2010). These structures can resemble rhizoids in their shape but neither the function nor the homology of these structures is currently understood. Their similarity to rhizoids of the gametophyte and root hairs suggest a heterotopic mutation that results in the expression of genes controlling the rhizoid and/or root hair development at the surface of shoot scales.

At least one putative *de novo* organ can be found in ferns. The origin of the sporocarps of heterosporous ferns is still mysterious (see Nagalingum *et al.* 2006 for references) despite the recent progress with the understanding of the phylogeny of water ferns and the recognition of the homology of all sporangia-bearing structures of these ferns (Nagalingum *et al.* 2006). Sporocarps are shoot-like structures bearing the sporangia either enclosed in walls resembling seeds (Marsileaceae) or along dichotomously branched axes (Salviniaceae). In ferns, sporangia are usually formed by leaves that are equally adapted to photosynthesis (trophosporophylls). However, in some ferns leaves or parts of the leaves are differentiated with the two functions segregated to either sporophylls (leaves forming sporangia) or trophophylls (leaves carrying out photosynthesis). Sporocarps are likely not homologous to sporophylls because many of their characteristics such as location are shoot-like (see Nagalingum *et al.* 2006). In the framework of Arberian Fuzzy Morphology, sporocarps need to be interpreted as misfits that are neither leaves nor shoots. In some cases (e.g., *Salvinia*), they also show positive geotropic growths resembling roots (Bierhorst 1971; Nagalingum *et al.* 2006).

Another mystery is the origin of the sporangia bearing structures of whisk ferns (Psilotales) and moonworts (Ophioglossales) (Figure 4.1). Both lineages differ from leptosporangiate ferns (Polypodiopsida) and marattioids (Marattiopsida) by not sharing a dorsal or marginal position of the sporangia (Bierhorst 1971; Schneider *et al.* 2009). Instead, the sporangia are located on the apical side of leaves (Bierhorst 1971; Schneider *et al.* 2009). Whisk ferns possess two (*Tmesipteris*) or three (*Psilotum*) partly fused sporangia located at the apical side of forked leaves (Bierhorst 1971, 1977; Schneider *et al.* 2009). Moonworts have a single spike, with the exception of *Cheiroglossa*, which can be branched (e.g., *Botrychium*, *Helminthostachys*) or unbranched (*Ophioglossum*) arising along the upper side of the petiole or sometimes lamina (*Ophioderma*) (Campbell 1911; Bierhorst 1971; Hauk *et al.* 2003). Given the sister relationship of whisk ferns and moonworts, the hypothesis of a homologous structure requires attention in future research (see Figure 4.5). However, the alternative hypothesis of homology between the aerial structure of Psilotales and the leaf-spike apparatus of Ophioglossales requires consideration as well (Bierhorst 1977). Bierhorst (1977) based on this hypothesis of homology on the aerial structures of *Psilotum* as compared to the leaves of the leptosporangiate fern *Stomatopteris*. Finally, we also need to consider the hypotheses regarding the nonaerial creeping rhizomes of *Psilotum* as de-novo organs or as

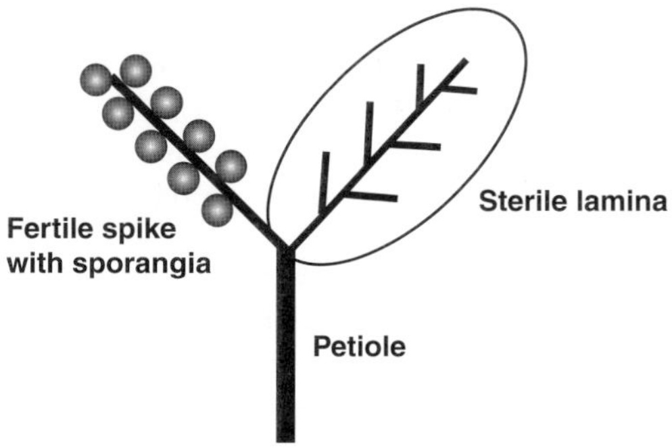

Figure 4.5 Model for the ancestral sporangiophyll of the sister lineage moonworts (Ophioglossales) and whisk ferns (Psilotales). The structure is composed of three main components: (1) a lamina, (2) a petiole, and (3) an axial structure bearing several sporangia. This morphology is found in most extant moonworts, whereas the whisk fern structure may have evolved by reduction of all three components.

homologous to rhizomes of other vascular plants (Bierhorst 1954; Takigochi *et al.* 1997).

In general, the body plan of Psilotales is very unlikely the result of the conservation of ancestral features but the result of transformations including simplifications and reductions (Schneider *et al.* 2002, 2009; Schneider 2007). The lack of roots is interpreted as a secondary reduction and not as a primary absence. This is consistent with the results of phylogenetic studies (Pryer *et al.* 2001, 2004; Schneider *et al.* 2009). In this context, it is interesting to consider the embryology of Psilotales that is likely also highly derived as indicated by very early branching (Guttenberg 1965) that is best described as pseudo-dichotomous (Bierhorst 1971; 1973, 1977). In this context, it is worth stressing that we still lack the description of embryos from Rhynie Chert fossils (H. Kerp, personal communication). It is evident that the lack of embryos of the otherwise completely recovered life cycle of *Aglaophyton* (Taylor *et al.* 2005) is a major draw back for the inference of ancestral embryo characters of vascular plants.

4.2.3 Perspective 3: genomics and evo-devo of ferns

With the publication of the whole genome of the spike moss *Selaginella moellendorffii*, whole genomes are now available for representatives of seed plant lineages and a lycophyte lineage (Banks *et al.* 2011). However, we are still lacking a whole genome for any fern and thus evo-devo studies on ferns rely either on candidate genes detected in the whole genome of angiosperms or the spike moss *Selaginella* or on EST or BAC libraries of selected species of ferns such as *Adiantum capillus-veneris* (Kawai-Toyooka *et al.* 2004; Yamauchi *et al.* 2005), *Ceratopteris richardii* (Rutherford *et al.* 2004), and *Pteridium aquilinum* (Der *et al.* 2011). Given the availability of whole genomes, it is not surprising that the majority of evo-devo studies are currently either focused on angiosperms such as *Arabidopsis thaliana* or the moss *Physcomitrella patens* (Rensing *et al.* 2008). In this context, it is worth noting the efforts to sequence the whole genome of the liverwort *Marchantia polymorpha*, which will have a critical impact on our ability to study the evolution of plant development and morphology (http://www.marchantia.org). The availability of whole genomes and other DNA sequencing technologies have revolutionized our understanding of plant development and its evolution such as the discovery of genes controlling the development of flowers and the discovery that the independence of the developmental pathways of gametophytes and sporophytes in *Physcomitrella* (Okano *et al.* 2009; Prigge & Bezanilla 2010). The publication of the genome of *Selaginella* will have a similar impact by providing access to the early evolution of plant development in vascular plants. However, many major questions concerning body plan evolution in euphyllophytes, the clade including angiosperms and ferns, will also require the study of ferns besides angiosperms and spike mosses (Figure 4.4). For example, comparative studies on the development of leaves in spike mosses and angiosperms will confirm the well-founded hypothesis of an independent origin of leaves in these two lineages despite the potential co-option of developmental genes in the convergent evolution of dorsiventral organized organs (Harrison *et al.* 2005, 2007; Langdale 2008). However, these insights will have a limited importance on the major challenge of leaf evolution, that is how often have leaves evolved in the euphyllophytes (Figure 4.4). This review will not focus on the evolution of plant organs in land plants because its focus is on ferns but addresses issues within ferns that can not be seen outside of the broader challenge, the evolution of plant body plans.

Several studies have assembled evidence for a shared genetic tool kit in land plants (Floyd & Bowman 2007a, 2007b) including various gene families controlling plant development such as MADS-box genes and other families of transcription factors (Hasebe *et al.* 1998; Muenster *et al.* 1997, 2008; Moyraud *et al.* 2010). Recent studies explored the genomes of the moss *Physcomitrella* and the liverwort *Selaginella* to explore the presence and copy number of genes in gene families known to be important in the regulation of plant development. In general, these studies recovered evidence for the origin of most gene families before the separation of seed plants from other land plants, with the YABBY genes as a notable exception, but also a general trend to seed plant specific gene duplications (Rensing *et al.* 2008; Banks *et al.* 2011). However, with the absence of a whole genome for any fern, we lack the ability to compare the genome of a representative of the sister lineage of seed plants that limits considerably the capability of these studies to identify seed plant specific gene duplications (Pryer *et al.* 2002).

Some studies have found evidence for independent recruitment of conserved developmental mechanisms in the evolution of plant morphology (Harrison *et al.* 2005, 2007; Langdale 2008) whereas others recovered evidence for highly conserved pathways (Schalau *et al.* 2008). Kaplan (2001a) argued for the importance of gene expression patterns in plant morphogenesis and their transformation in evolution. Ferns should be of particular interest to evolutionary developmental studies due to the mixture of highly conserved body plans as found in marattioids, highly derived transformations in other lineages such as the body plan of whisk ferns, as well as the morphological novelty, the sporocarp, found in heterosporous ferns (Nagalingum *et al.* 2006), or the climbing leaf of *Lygodium* (Mueller 1982). These studies will benefit from applying a robust phylogenetic framework as outlined in recent phylogenetic studies (e.g., Pryer *et al.* 2001, 2004; Schneider *et al.* 2009). However, the framework is still lacking the required resolution in the context of the position of horsetails, which may be sister to all other lineages of ferns or sister to one or two lineages of the fern clade (Schneider 2007; Schneider *et al.* 2009; Karol *et al.* 2010; Rai & Graham 2010). However, problems in identifying synapomorphies in the body plans of fern lineages (Schneider *et al.* 2002, 2009; Schneider 2007) may suggest a ubiquity of postlineage establishment transformation and co-option of conserved ancestral pathways in the evolution of key innovations in the evolution of fern lineages such as the innovation of sporocarps in the heterosporous water ferns (Nagalingum *et al.* 2006), rhizomes with cavities to be colonized by ants (Schneider *et al.* 2010), or the differentiation of litter collectors and trophosporophylls in the drynarioid ferns (*Drynaria*) and the genus *Platycerium* (Janssen & Schneider 2005; Schneider *et al.* 2010).

It is also worth noting that deviations from a generalized fern body plan are not restricted to horsetails and whisk ferns, but can also be found in some lineages of leptosporangiate ferns. In some cases, the existence of close relatives with regular body plans may allow the untangling of the evolution of pathways underlying these transformations such as the low level of differentiation of leaves and shoots in rootless filmy ferns of the genus *Gonocormus* (Schneider 2000). This pattern may indicate a lower level of conserved evolution of fern morphology if compared to seed plants, which will provide the opportunity to study the evolutionary processes underlying the transformation of plant morphogenesis on one hand, and will result in a challenge to differentiate between shared ancestral states, secondary simplifications, and transformation that co-opted genes of ancestral pathways to innovate new structures on the other hand. Of course, the latter question is at the center of the current discussion on the evolution of plant morphology, which resembles very much concepts of plant morphology ranging from typological concepts (Kaplan 2001a) to Aberian Fuzzy Morphology (Rutishauer & Isler 2001; Kirchoff *et al.* 2008). The integration of these concepts into comparative genomic analyses may provide us with the opportunity to address major questions of the evolution of body plans in all lineages of plants (Bowman *et al.* 2007).

Acknowledgments

Most of all, I want to thank the editors for the constant encouragement and their amazing patience with a very difficult author. I am deeply in dept to my teachers and many colleagues who discussed with me various aspects of plant morphology in particular Peter Endress (Zurich), Herbert Huber (Kaiserslautern), Don Kaplan (Berkeley), Mashiro Kato (Tsukuba), and Rolf Rutishauser (Zurich).

References

Banks, J.A., Nishiyama, T., Hasebe, M., *et al.* (2011) The *Selaginella* genome identifies genetic changes associated with the evolution of vascular plants. *Science*, **332**, 960–963.

Bateman, R.M. & DiMichele, W.A. (2002) Generating and filtering major phenotypic novelties: neoGoldschmidtian saltation revisited. In: *Developmental Genetics and*

Plant Evolution (eds Q.C.B. Cronk, R. Bateman and J.A. Hawkins). pp. 109–159. Taylor & Francis, London.

Beerling, D.J. & Fleming, A.J. (2007) Zimermann's telome theory of megaphyll leaf evolution: a molecular and cellular critique. *Current Opinion in Plant Biology*, **10**, 4–12.

Bell, P.R. (1992) Apospory and apogamy: implications for understanding the plant life cycle. *International Journal of Plant Sciences*, **153**, S123–S126.

Bierhorst, D.W. (1954) The subterranean sporophytic axes of *Psilotum nudum*. *American Journal of Botany*, **41**, 732–739.

Bierhorst, D.W. (1971) *Morphology of Vascular Plants*. MacMillan, New York.

Bierhorst, D.W. (1973) Non-appendicular fronds in the Filicales. *Botanical Journal of the Linnean Society*, **67** (Suppl. 1), 45–58.

Bierhorst, D.W. (1977) The systematic position of *Psilotum* and *Tmesipteris*. *Brittonia*, **290**, 3–13.

Bower, F.O. (1923) *The Ferns*. Vol. I. Cambridge University Press, Cambridge.

Bower, F.O. (1935) *Primitive Land Plants*. McMillan, London.

Bowman, J.L., Floyd, S.K. & Sakakibara, K. (2007) Green genes—comparative genomics of the green branch of life. *Cell*, **129**, 229–234.

Boyce, C.K. (2010) The evolution of plant development in a paleontological context. *Current Opinion in Plant Biology*, **13**, 102–107.

Campbell, D.H. (1911) *The Eusporangiate. The Comparative Morphology of the Ophioglossaceae and Marattiaceae*. Carnegie Institute, Washington, DC.

Classen-Bockhoff, R. (2001) Plant morphology: the historic concepts of Wilhelm Troll, Walter Zimmermann and Agnes Arber. *Annals of Botany*, **88**, 1153–1172.

Crane, P.R. & Kenrick, P. (1997) Diverted development of reproductive organs: a source of morphological innovation in land plants. *Plant Systematics and Evolution*, **206**, 161–174.

Der, J.P., Barker, M.S., Wickett, N.J., *et al.* (2011) *De novo* characterization of the gametophyte transcriptome in bracken fern, *Pteridium aquilinum*. *BMC Genomics*, **12**, 99.

Donoghue, M.J. & Kadereit, J.W. (1992) Walter Zimmernann and the growth of phylogenetic theory. *Systematic Biology*, **41**, 74–85.

Duckett, J.G. & Ligrone, R. (2003) The structure and development of haustorial placentas in leptosporangiate ferns provide a clear-cut distinction between euphyllophytes and lycophytes. *Annals of Botany*, **92**, 513–521.

Efroni, I., Eshed, Y. & Lifschitz, E. (2010) Morphogenesis of simple and compound leaves: a critical review. *Plant Cell*, **22**, 1019–1032.

Endress, P.K. (2001) Origins of flower morphology. *Journal of Experimental Zoology*, **291**, 105–115.

Floyd, S.K. & Bowman, J.L. (2006) Distinct developmental mechanisms reflect the independent origins of leaves in vascular plants. *Current Biology*, **16**, 1911–1917.

Floyd, S.K. & Bowman, J.L. (2007a) The ancestral developmental tool kit of land plants. *International Journal of Plant Sciences*, **168**, 1–35.

Floyd, S.K. & Bowman, J.L. (2007b) Gene expression patterns in seed plant shoot meristems and leaves. *Journal of Plant Research*, **123**, 43–55.

Floyd, S.K. & Bowman, J.L. (2010) Gene expression patterns in seed plant shoot meristems and leaves: homoplasy or homology? *Journal of Plant Research*, **123**, 43–55.

Galtier, J. (2010) The origins and early evolution of the megaphyllous leaf. *International Journal of Plant Sciences*, **171**, 641–661.

Gifford, E.M. (1991) The root apical meristem of *Asplenium bulbiferum*: structure and development. *American Journal of Botany*, **78**, 370–376.
Goebel, K. (1928/1930) *Organiographie der Pflanzen insbesondere der Archegoniaten und Bluetenpflanzen*, 3rd edn. G. Fischer Verlag, Jena.
Guttenberg, H. von (1965) *Histogenese der Pteridophyten. Handbuch der Pflanzenanatomie*, Vol. VII, Part 2. Borntraeger, Berlin.
Hagemann, W. (1976) Sind Farne Kormophyten? Eine Alternative zur Telomtheorie. *Plant Systematics and Evolution*, **124**, 251–276.
Hagemann, W. (1989) Acrogenous branching in pteridophytes. In: *Proceedings of the International Symposium on Systematic Pteridology* (eds K.H. Shing & K.U. Kramer). pp. 245–258. China Science & Technology Press, Beijing.
Hagemann, W. (1992) The relationships of anatomy to morphology in plants: a new theoretical perspective. *International Journal of Plant Sciences*, **153**, S38–S48.
Hagemann, W. (1997) Towards an organismic concept of land plants: the marginal blastozone and the development of the vegetation body of selected frondose gametophytes of liverworts and ferns. *Plant Systematics and Evolution*, **216**, 81–133.
Hagemann, W. & Gleissberg, S. (1996) Organogenetic capacity of leaves: the significance of marginal blastozones in angiosperms. *Plant Systematics and Evolution*, **199**, 121–152.
Harrison, C.J., Corley, S.B., Moylan, E.C., *et al.* (2005) Independent recruitment of a conserved developmental mechanism during leaf evolution. *Nature*, **434**, 509–514.
Harrison, C.J., Rezvani, M. & Langdale, J.A. (2007) Growth from two transient apical initials in the meristem of *Selaginella kraussiana*. *Development*, **134**, 881–889.
Hasebe, M., Wen, C.K., Kato, M., *et al.* (1998) Characterization of MADS homeotic genes in the fern *Ceratopteris richardii*. *Proceedings National Academy Sciences of the United States of America*, **95**, 6222–6227.
Hauk, W.D., Parks, C.R. & Chase, M.W. (2003) Phylogenetic studies of Ophioglossaceae: evidence from *rbcL* and *trnL-F* plastid DNA sequences and morphology. *Molecular Phylogenetics and Evolution*, **28**, 131–151.
Hennig, W. (1965) Phylogenetic Systematics. *Annual Review Entomology*, **10**, 97–116.
Hintz, M., Bartholmes, C., Nutt, P., *et al.* (2006) Catching a 'hopeful monster': shepherd's pursue (*Capsella bursa-pastoris*) as a model system to study the evolution of flower development. *Journal of Experimental Botany*, **57**, 3531–3542.
Hirayama, Y., Yamada, T., Oya, Y., *et al.* (2007) Expression patterns of class I KNOX and YABBY genes in *Ruscus aculeatus* (Aspergagaceae) with implications for phylloclad homogy. *Developmental Genes and Evolution*, **217**, 363–372.
Huang, Y.-M., Huang, M.H., Chen, C.-M., *et al.* (2011) Stipule propagation in five marattioid species native to Taiwan (Marattiaceae: Pteridophyta). In: *Working with Ferns: Issues and Applications* (eds H. Fernandez,A. Kumar & M.A. Revilla). pp. 127–134. Springer, New York.
Imaichi, R. (2008) Meristem organization and organ diversity. In: *Biology and Evolution of Ferns* (eds C.H. Haufler & T. Ranker). pp. 75–106. Cambridge University Press, Cambridge.
Izaki, A. & Bowman, J.L. (2007) KANADI and Class III HD-Zip gene families regulare embryo patterning and modulate auxin flow during embryogenesis in *Arabidopsis*. *The Plant Cell*, **19**, 395–508.
Jang, G., Yi, K., Pieres, N.D., *et al.* (2011) RSL genes are sufficient for rhizoid system development in early diverging land plants. *Development* **138**, 2273–2281.

Janssen, T. & Schneider, H. (2005) Exploring the evolution of himus collecting leaves in drynarioid ferns (Polypodiaceae, Polypodiidae). *Plant Systematics and Evolution*, **252**, 175–197.

Johnson, G. & Renzaglia, K. (2009) Evaluating the diversity of pteridophyte embryology in the light of recent phylogenetic analyses leads to new inferences on character evolution. *Plant Systematics and Evolution*, **283**, 149–164.

Johnson, G. & Renzaglia, K. (2008) Embryology of *Ceratopteris richardii* (Peridaceae, tribe Ceratopterideae), with emphasis on placental development. *Journal of Plant Research*, **121**, 581–592.

Kaplan, D.R. (2001a) Fundamental concepts of leaf morphology and morphogenesis: a contribution to the interpretation of molecular genetic mutants. *International Journal of Plant Sciences*, **162**, 465–474.

Kaplan, D.R. (2001b) The science of plant morphology: definition, history, and role in modern biology. *American Journal of Botany*, **88**, 1711–1741.

Karol, K.G., Arumuganathan, K., Boore, J.L., *et al.* (2010) Comoplete plastome sequence of *Equisetum arvense* and *Isoetes flaccida*: implications for phylogeny and plastid genome evolution of early land plant lineages. *BMC Evolutionary Biology*, **10**, 321. DOI: 10.1186/1471–2148.10–321.

Kawai-Toyooka, H., Kuramoto, C., Orui, K., *et al.* (2004) DNA interference: a simple and efficient gene-silencing system for high-throughput functional analysis in the fern *Adiantum*. *Plant & Cell Physiology*, **45**, 1648–1657.

Kenrick, P. & Crane, P.R. (1997) *The Origin and Early Diversification of Land Plants: A Cladistic Study*. Smithsonian Press, Washington, DC.

Kidner, C.A. (2007) Leaf evolution: working with what's on hand. *Evolution & Development*, **9**, 321–322.

Kirchoff, B.K., Pfeifer, E. & Rutishauser, R. (2008) Plant structure ontology: how should we label plant structures with doubtful or mixed identities? *Zootaxa*, **1950**, 103–122.

Langdale, J.A. (2008) Evolution of development mechanisms in plants. *Current Opinion in Genetics & Development*, **18**, 368–373.

Large, M.F. & Braggins, J.E. (2004) *Tree Ferns*. Timber Press, Portland.

Mavroodiev, E.V. (2009) Classical morphology of plants as an elementary instance of classical invariant theory. *PLoS ONE*, **4**, e6969. DOI: 10.1371.journal.pone.0006969.

Menand, B., Keke, Y., Jouannic, S., *et al.* (2007) An ancient mechanisms controls the development of cells with a root function in land plants. *Science*, **316**, 1477–1480.

Moyraud, E., Kusters, E., Monniaux, M., *et al.* (2010) LEAFY blossoms. *Trends in Plant Science*, **15**, 346–352.

Mueller, R.J. (1982) Shoot morphology of the climbing fern *Lygodium* (Schizaeaceae): general organography, leaf initiation, and branching. *Botanical Gazettte*, **143**, 319–330.

Muenster, T., Pahnke, J., DiRosa, A., *et al.* (1997) Floral homeotic genes were recruited from homologous MADS-box genes preexisting in the common ancestors of ferns and seed plants. *Proceedings National Academy Sciences of the United States of America*, **94**, 2415–2420.

Muenster, T., Faigl, W., Seadler, H., *et al.* (2008) Evolutionary aspects of MADS-box genes in the eusporangiate fern *Ophioglossum*. *Plant Biology*, **4**, 474–483.

Nagalingum, N.S., Schneider, H., Sakuj, M., *et al.* (2006) Comparative morphology of reproductive structures in heterosporous water ferns and a reevaluation of the sporocarp. *International Journal of Plant Sciences*, **167**, 805–815.

Nayar, B.K. & Kaur, S. (1971) Gametophytes of homosporous ferns. *Botanical Review*, **3**, 295–396.

Ogura, Y. (1972) *Comparative Anatomy of Vegetative Organs of the Pteridophytes. Handbuch der Pflanzenanatomie*, Vol. VII, Part 3, 2nd edn. Borntraeger, Berlin.

Okano, Y., Aono, N., Hiwatashi, Y., *et al.* (2009) A polycomb repressive complex 2 gene regulates apogamy and gives evolutionary insights into early land plant evolution. *Proceedings National Academy Sciences of the United States of America*, **106**, 16321–16326.

Park, C.-H. & Kato, M. (2003) Apomixis in the interspecific triploid bybrid fern *Cornopteris christenseniana* (Woodsiaceae). *Journal of Plant Research*, **116**, 93–103.

Pinter, I., Bakker, F., Barrett, J., *et al.* (2002) Phylogenetic and biosystematic relationships in four highly disjunct polyploid complexes in the subgenera *Ceterach* and *Phyllitis* in *Asplenium* (Aspleniaceae). *Organisms, Diversity & Evolution*, **2**, 299–311.

Prigge, M.L. & Bezanilla, M. (2010) Evolutionary crossroads in developmental biology: *Physcomitrella patens*. *Development*, **13**, 3535–3543.

Pryer, K.M., Schneider, H., Zimmer, E.A., *et al.* (2002) Deciding among green plants for whole genome studies. *Trends in Plant Science*, **7**, 550–554.

Pryer, K.M., Schneider, H., Smith, A.R., *et al.* (2001) Horsetails and ferns are a monophyletic group and the closest living relatives to seed plants. *Nature*, **409**, 618–622.

Pryer, K.M., Schuettpelz, E., Wolf, P.G., *et al.* (2004) Phylogeny and evolution of ferns (monilophytes) with a focus on the early leptosporangiate divergences. *American Journal of Botany*, **91**, 1582–1598.

Pu, L. & Brady, S. (2010) System biology update: cell type specific transcriptional regulatory networks. *Plant Physiology*, **152**, 411–419.

Qiu, Y.-L., Li, L., Wang, B., *et al.* (2007) A nonflowering land plant phylogeny inferred from nucleotides sequences of seven chloroplast, mitochondrial, and nuclear genes. *International Journal of Plant Sciences*, **168**, 691–708.

Qiu, Y.-L., Li, L., Wang, B., *et al.* (2006) The deepest divergences in land plants inferred from phylogenomic evidence. *Proceedings National Academy Sciences of the United States of America*, **103**, 15511–15516.

Rai, H.S. & Graham, S.W. (2010) Utility of a large, multigene plastid data set in inferring higher-order relationships in ferns and relatives (monilophytes). *American Journal Botany*, **97**, 1444–1456.

Raghavan, V. (1989) *Developmental Biology of Fern Gametophytes*. Cambridge University Press, Cambridge.

Reiser, L., Sanchez-Baracaldo, P. & Hake, S. (2000) Knots in the family tree: evolutionary relationships and functions of knox homeobox genes. *Plant Molecular Biology*, **42**, 151–166.

Rensing, S.A., Lang, D., Zimmer, A.D., *et al.* (2008) The *Physcomitrella* genome reveals evolutionary insights into the conquest of land by plants. *Nature*, **319**, 64–69.

Richards, J.H., Beck, J.Z. & Hirsch, A.M. (1983) Structural investigations of asexual reproduction in *Nephrolepis exaltata* and *Platycerium bifurcatum*. *American Journal of Botany*, **70**, 993–1001.

Rieppel, O. (2001) Turstles as hopeful monsters. *BioEsssays*, **23**, 987–991.

Rutherford, G., Tanurdzic, M., Hasebe, M., *et al.* (2004) A systematic gene silencing method suitable for high throughput, reverse genetic analyses of gene function in fern gametophytes. *BMC Plant Biology*, **4**, 6.

Rutishauser, R. & Isler, B. (2001) Developmental genetics and morphological evolution of flowering plants, especially bladderworts (*Urticularia*): Fuzzy Arberian morphology complements classical morphology. *Annals of Botany*, **88**, 1173–1202.

Saint Savage, N., Walker, T., Panciera, Y., *et al.* (2008) Mutual support of alternative cell fates through the directed movement of CAPRICE and GLABRA3 provides a mechanism for patterning the *Arabidopsis* root epidermis. *PLoS Biology*, **6**, e235.

Sanders, H.L., Darrah, P.R. & Langdale, J.A. (2011) Sector analysis and predictive modelling reveal iterative shoot-like development in fern fronds. *Development*, **138**, 2925–2934.

Sarojam, R., Sappi, P.G., Goldschmidt, A., *et al.* (2010) Differentiating *Arabidopsis* shoots from leaves by combined YABBY activities. *Plant Cell*, **22**, 2113–2130.

Sattler, R. (1996) Classical morphology and continuum morphology: opposition and continuum. *Annals of Botany*, **78**, 577–581.

Schalau, A., Kakhovskaya, I., Tewes, A., *et al.* (2008) Phylogenetic footprints in fern spore- and seed-specific promoters. *Plant Journal*, **53**, 414–424.

Schiefelbein, J., Kwak, S.-H., Wieckowski, Y., *et al.* (2009) The gene regulatory network for root epidermial cell-type pattern formation in *Arabidopsis*. *Journal of Experimental Botany*, **60**, 1515–1521.

Schneider, H. (1996) *Vergleichende Wurzelanatomie der Farne*. Shaker-Press, Aachen.

Schneider, H. (2000) Morphology and anatomy of roots in the filmy fern tribe Trichomaneae H. Schneider (Hymenophyllaceae, Filicatae) and the evolution of rootless taxa. *Botanical Journal of the Linnean Society*, **132**, 29–46.

Schneider, H. (2007) Plant morphology as the cornerstone to the integration of fossils and extant taxa in phylogenetic analyses. *Species, Phylogeny and Evolution*, **1**, 65–71.

Schneider, H., Kreier, H.-P., Janssen, T., *et al.* (2010) Key innovations versus key opportunities: identifying causes of rapid radiations in derived ferns. In: *Evolution in Action* (ed. M. Glaubrecht). pp. 61–75. Springer, Berlin.

Schneider, H., Pryer, K.M., Cranfill, R., *et al.* (2002) The evolution of vascular plant body plans – a phylogenetic perspective. In: *Developmental Genetics and Plant Evolution* (eds Q.C.B. Cronk, R. Bateman & J.A. Hawkins). pp. 330–364. Taylor & Francis, London.

Schneider, H., Smith, A.R. & Pryer, K.M. (2009) Is morphology really at odds with molecules in estimating fern phylogeny? *Systematic Botany*, **34**, 455–475.

Schoute, J.C. (1938) Morphology. In: *Manual of Pteridology* (ed. F. Verdoorn). pp. 1–64. Martius Nijhoff, Hague.

Schuettpelz, E., Korall, P. & Pryer, K.M. (2006) Plastid *atpA* data provide improved supports for deep relationships among ferns. *Taxon*, **55**, 897–906.

Sheffield, E. (2008) Alternation of generations. In: *Biology and Evolution of Ferns* (eds C.H. Haufler & T. Ranker). pp. 49–74. Cambridge University Press, Cambridge.

Smith, A.R., Pryer, K.M., Schuettpelz, E., *et al.* (2006) A classification for extant ferns. *Taxon*, **55**, 705–731.

Sozzani, R., Moreno-Risoueno, M.A., van Norman, J.M., *et al.* (2010) Spatiotemporal regulation of cell-cycle genes by SHORTROOT links patterning and growth. *Nature*, **466**, 128–132.

Stein, W.E., Jr., Wight, D.C. & Beck, C.B. (1984) Possible alternatives for the origin of Sphenopsida. *Systemaitc Botany*, **9**, 102–122.

Stein, W.E., Jr. (1993) Modelling the evolution of stellar architecture in vascular plants. *Paleobiology*, **28**, 70–100.

Stuetzel, T. & Gailing, O. (1997) Blaetter und Spreuschuppen bei Farnen: Unterschiede und Gemeinsamkeiuten unter morphogenetischen und phylogenetischen Gesichtspuntken. *Beitraete zur Biologie der Pflanzen*, **69**, 17–29.
Takahashi, N., Hashino, M., Kami, M., *et al.* (2009) Developmental morphology of strap-shaped gametophytes of *Colysis decurrens*: a new look at meristem development and function in fern gametophytes. *Annals of Botany*, **104**, 1353–1361.
Takigochi, Y., Imaichi, R. & Kato, M. (1997) Cell division patterns in the apices of subterranean axis and aerial shoot of *Psilotum nudum* (Psilotaceae): morphological and phylogenetic implications for the subterranean axis. *American Journal of Botany*, **84**, 588–596.
Taylor, T.N., Kerp, H. & Hass, H. (2005) Life history biology of early land plants, deciphering the gametophyte phase. *Proceedings National Academy of Sciences of the United States of America*, **102**, 5892–5897.
Taylor, T.N., Taylor, E.L. & Krings, M. (2009) *Paleobotany: The biology and Evolution of Fossil Plants*, 2nd edn. Academic Press, Amsterdam.
Theissen, G. (2009) Saltational evolution: hopeful monsters are here to stay. *Theory Bioscience*, **128**, 43–51.
Tomescu, A.M.F. (2008) Megaphylls, microphylls and the evolution of leaf development. *Trends in Plant Sciences*, **14**, 5–12.
Troll, W. (1937) *Vergleichende Morphologie der hoehren Pflanzen*. Vol. I, Part 1. Bortraeger, Berlin.
Velenovsky, J. (1905) *Vergleichende Morphologie der Pflanzen*. Vol. 1. Rivnac-Verlag, Praque.
Wagner, W.H., Beitel, J.M. & Wagner, F.S. (1982) Complex venation patterns in the leaves of *Selaginella*, mgephyll-like leaves in lycophytes. *Science*, **218**, 793–794.
Wang, L., Wu, Z.-Q., Xiang, Q.-P., *et al.* (2010) A molecular phylogeny and a revised classification of tribe Lepisoreae (Polypodiaceae) based on an analysis of four plastid DNA regions. *Botanical Journal of the Linnean Society*, **162**, 28–38.
White, R.A. & Turner, M.D. (1995) Anatomy and development of the fern sporophyte. *Botanical Review*, **61**, 281–305.
Wikstroem, N. & Pryer, K.M. (2005) Inconrgeunce between primary sequence data and the distribution of a mitochondrial atp1 group II intron among ferns and horsetails. *Molecular Phylogenetics Evolution*, **36**, 484–493.
Yamauchi, D., Sutoh, K., Kanegae, H., *et al.* (2005) Analysis of expressed sequence tags in prothallia of *Adiantum capillus-veneris*. *Journal of Plant Research*, **118**, 223–227.
Zimmermann, W. (1959) *Die Phylogenie der Pflanzen*, 2nd edn. G.Fischer, Stuttgart.
Zimmermann, W. (1965) *Die Telomtheorie. Fortschritte der Evolutionsforschung*, Vol. 1. G. Fischer Verlag, Stuttgart.
Zurakowski, K.A. & Gifford, E.M. (1988) Quantitative studies of pinnule development in the ferns *Adiantum raddianum* and *Cheilanthes viridis*. *American Journal of Botany*, **75**, 1559–1570.

Annual Plant Reviews (2013) **45**, 141–162
doi: 10.1002/9781118305881.ch5
http://onlinelibrary.wiley.com

Chapter 5
GYMNOSPERMS

Dennis Wm. Stevenson

The New York Botanical Garden, Bronx, NY, USA

Abstract: This chapter focuses on the oddities of gymnosperm anatomy, morphology, and life cycles. Of particular interest are those seemingly intractable aspects that now seem tractable by combining recent developments in plant molecular biology with plant anatomy, morphology, and development. In most examples, there is available data from plant morphology in the widest sense that can provide the framework for experimental and molecular approaches. These include, but are not limited to, leaf development, senescence, stem growth in diameter, phylloclade development, the structure of reproductive axes and their reciprocal transitions to vegetative growth, free nuclear gametophytes and embryos, and unique features associated with seeds.

Keywords: architecture; anatomy; conifers; cycads; embryology; Ginkgo; Gnetales; morphology; teratology.

5.1 Introduction

Gymnosperms are nonflowering seed plants. There is considerable controversy concerning whether the gymnosperms are a monophyletic group sister to the flowering plants or a paraphyletic grade (see Doyle this volume). There are four main groups that are generally recognized: (1) cycads (Cycadales), (2) *Ginkgo* (Ginkgoales), (3) conifers (Taxales plus Pinales), and (4) Gnetales (*Ephedra*, *Gnetum*, and *Welwitschia*). The relationships between these groups are also controversial and nearly every possible permutation of the six orders have been published (Burleigh & Mathews 2004; Lee *et al.* 2011). Regardless of the topology that is obtained or preferred, one can examine the comparative morphology of the five major groups. Therefore, the following treatment focuses on the comparative morphology and anatomy of the gymnosperms as flower-free seed plants with emphases on unusual structures and life cycle stages in an effort to elucidate those features that

Annual Plant Reviews Volume 45: The Evolution of Plant Form, First Edition.
Edited by Barbara A. Ambrose and Michael Purugganan.

have been considered intractable historically but now appear tractable using new approaches and methodologies.

5.2 Architecture

There are no known, extant or fossil, herbaceous gymnosperms and all are perennial woody plants even if some appear to be quite small and herbaceous as in some cycads (e.g., *Zamia pygmaea*). Gymnosperms often range from the typically pyramidal conifers to tall, unbranched pachycaulous cycads to vines, climbing in a majority of *Gnetum* species and scrambling in some species of *Ephedra*. Most conifers maintain a strong apical dominance in an orthotropic leader shoot with plagiotropic lateral shoots, which results in their pyramidal shape. This shoot dimorphism is so strong that plants of the Taxaceae that have been propagated from rooting plagiotropic branches continue to grow plagiotropically for over a half century without becoming or producing any orthotropic shoots. Some, such as *Araucaria*, spend the first years as tall pyramidal plants with very strong apical dominance and later lose that apical dominance and subsequently develop into a candelabra form.

Branching in the gymnosperms is primarily from axillary buds, the exception being the cycads, which lack axillary buds. Generally speaking, the lateral branches are morphologically similar to leader shoots. Some conifers and *Ginkgo* produce long and short shoots; the latter often termed spur shoots. *Ginkgo*, *Larix*, *Cedrus*, and *Pseudolarix* produce similar leaves on their spur shots and long shoots and it is the stems that differ, the short shoots being rosettes and thicker than the thinner long shoots with well-developed internodes. It is common for short shoots in *Ginkgo* to become long shoots and vice versa. In some conifers, such as *Pinus* and *Sciadopitys*, the long shoots produce only cataphylls and the axillary buds associated with those cataphylls produce the green leaved short shoots and thus the short shoots and long shoots are morphologically dissimilar. Unlike the majority of gymnosperms where leaves are abscised, the short shoots are abscised in *Pinus* and *Sciadopitys* and the lateral branchlets in the Cupressaceae, annually in the case of the deciduous genera, *Taxodium* and *Metasequoia*.

Ginkgo produces lignotubers (chi-chi) at the base of the trunk and also along the trunk and branches in some instances. These lignotubers can grow into the ground and form roots as well as aerial shoots. The first lignotubers are produced from buds in the axils of the cotyledons and later lignotubers are formed in the axils of the first few foliage leaves (Del Tredici 1997). Lignotubers are also formed in a similar fashion in *Sequoia* (Del Tredici 1997). Little is known concerning the processes that result in the formation of lignotubers beyond the fact that they seem to be a response to environmental stress and perhaps senescence of the tree.

In contrast to the axillary branching of *Ginkgo*, conifers, and the Gnetales, branching in cycads is by apical dichotomy and by adventitious buds

(Norstog & Nicholls 1997). Successive dichotomies are at right angles to each other and, thus, cruciate. Adventitious buds can form wherever there is meristematic activity. For example, in decapitated plants from the phellogen and/or vascular cambium where plantlets are often arranged in a circle and also from the pith, which expands by localized cell divisions and thus provides sites for plantlet formation. Another source of plantlet (bulbil) formation is the abscission layer (basically a localized phellogen) in the leaf bases (Stopes 1910).

Two of the most unusual conifers are *Parasitaxus* and *Phyllocladus* in the Podocarpaceae. *Parasitaxus* is a very rare monotypic genus from New Caledonia. Plants of *Parasitaxus usta* are woody and grow to about 2 m in height, have fleshy purple scale leaves, lack roots, and are always found attached to the roots of *Falcatifolium taxoides*, which itself is also a member of Podocarpaceae. Interestingly, even though parasitic, this species is clearly a perennial woody shrub, its parasitism functions via vesicular-arbuscular mycorrhizae and plants appear united with the host via a root graft (Woltz *et al.* 1994). The foliage and shoots do have chloroplasts, but lack any significant photosynthetic electron transport so that the transfer of carbon from the host plant is achieved mainly through the fungal intermediary. This is similar to that found in mycoheterotrophs such as the nonphotosynthetic angiosperm, Indian-pipe (*Monotropa*). Given its woody habit and association with vesicular-arbuscular mycorrhizae within the host union as well as its high stomatal conductance and low water potential, *Parasitaxus* functions in total unlike any known angiosperm mycoheterotroph and combines a unique suite of physiological features (Feild & Boddribb 2005).

Phyllocladus is unique within the gymnosperms because its photosynthetic surface is composed of flattened branch complexes (cladodes or phylloclades) that resemble simple to compound leaves. Seedlings have green needle foliage leaves comparable to those of other conifers. In contrast, in adult plants all leaves are ephemeral, nonphotosynthetic scales that usually do not subtend axillary buds. Shoot morphology of the adult phase is composed of succession of phylloclades each of which represents a branch complex of three branch orders (Tomlinson *et al.* 1989). Other than the descriptive nature of this change in development, nothing else is known about the process of leaf suppression and clade replacement, surely a fertile area for investigation in the relationship of leaves and axes and the switch between the two.

In *Sciadopitys*, long shoots comprise pseudowhorls of short shoots each of which is composed of two leaves fused along opposing edges (Bierhorst 1971). Usually, the apices of these leaves are not completely fused and one can observe the two tips. Rarely, an arrested bud may be observed between the apices. Thus, an issue arises. Is the area below such a bud the axis of the axillary shoot and, therefore, the whole structure is a cladode (Dörken & T. Stützel 2011) or is the bud simply displaced because of the intercalary growth of fused leaf bases?

Perhaps the most bizarre gymnosperm is *Welwitschia* plants, which comprise a single unbranched axis bearing two cotyledons and one pair of foliage leaves and one pair of scale leaves (Martens 1971). According to Martens and Waterkeyn (1963), the shoot apex of the *Welwitschia* seedling first initiates a pair of primordia, which later develops into the two permanent leaves of the sporophyte, and then later a second pair of appendages—the scaly bodies—at right angles to the foliage leaves. These have been interpreted as a second pair of foliar appendages excluding the cotyledons. Thus, the exposed portion of an old plant consists of a short, woody, unbranched stem, which is mostly two nodes, a cotyledonary node, a short hypocotyl, and a crown that bears only two huge strap-shaped leaves with basal intercalary meristems as the permanent photosynthetic organs. Little is known about the actual development of the scaly bodies and nothing about the basal intercalary meristem beyond its presence, which appears similar to the intercalary meristem found in monocots, particularly the grasses.

5.3 Shoots

The shoot apical meristem of cycads, *Ginkgo,* and the conifers is very similar and comprises an apical initial zone with periclinal divisions, a central mother cell zone, a peripheral zone, and a rib meristem. The shape of the shoot tip and volume of the various cytohistological zones in many conifers may change markedly during the annual growth sequence, but the basic zonation pattern remains the same. The shoot apex of cycads follows this pattern with massive apices up to 3.5 mm in diameter and often numerous periclinal divisions occurring in the surface cells. In contrast to conifers, *Ginkgo,* and cycads, the shoot apical meristem of the Gnetales has a true tunica where the surface layer has no periclinal divisions. The shoot apex of cycads is at the bottom of a bowl as the result of a primary thickening meristem (PTM) that produces derivatives centrifugally and extends distally from the rib meristem bordering the pith. This PTM elevates the leaf primordia above the shoot apex. This is similar to what is found in many monocots and tree ferns. However, in monocots and cacti, the PTM lies under the leaf bases and produces derivatives centripetally but also elevates the leaf primordium above the shoot apex. Tree ferns in contrast, have a PTM that appears to be diffuse, that is, it consist of several layers unlike the more restricted PTM found in cycads, monocots, and cacti. This sunken shoot apex appears to have developed independently in these plant groups and is related to the pachycaulous growth habit. There is a paucity of information about these meristems beyond anatomical and developmental descriptions.

The apices of long and short shoots, when present, have essentially the same structure. In both micro- and megasporangiate shoots, the organization of the apices are essentially the same as in vegetative shoots, although the dimensions may vary. The reproductive shoots vary in size during development and the apices of pollen cones lose their organization as the cones mature because the pollen cones are determinate axes. The same is true for the lateral ovule bearing axes of the ovulate cones. The main axis of ovulate cones are generally determinate, but in many genera of the Cupressaceae (e.g., *Cryptomeria, Cunninghamia,* and *Sequoia*), Pinaceae (*Pinus, Picea*), Sciadopityaceae, and Cephalotaxaceae, the main axis commonly reverts to a vegetative shoot so that one observes a vegetative shoot growing from the apex of the cone. This is occasionally seen in some cone bearing lycopods and also some flowers such as in roses.

Gymnosperms are heterosporous and microspores always give rise in the end to microgametophytes that produce sperm and megagametophytes that produce eggs, regardless of the actual size of the spores, gametophytes, or sporophytic structures that give rise to them. Further more, the sexual organs are not produced by the sporophytes, but rather by the gametophytes. Thus, it is erroneous to describe the sporophytic plants as male or female or bisexual plants and cones as male or female or bisexual cones. The more accurate and appropriate descriptors are microsporangiate and megasporangiate plants and cones or pollen and ovulate plants and cones. The cones of gymnosperms are typically monosporangiate, that is, either pollen cones or ovulate cones. Teratological bisporangiate cones are often found in Cupressaceae and Pinaceae, but without regularity either positionally or seasonally. In these conifers, the most common examples comprise pollen-bearing structures in the basal region and ovulate bearing structures in the apical region. In the transitional region between these, one can find cone scales with adaxial ovules and with abaxial microsporangia that appear to produce only abortive pollen. In *Larix,* there are cases where apical reversions to microsporophylls are known. In *Abies balsamea,* there are cases where there are basal ovule-bearing structures and apical microsporophylls (Rudall *et al.* 2011). Finally, in *Gnetum* and *Welwitschia* there are bisporangiate reproductive axes that are very dissimilar in organization and do not appear to be teratological but rather fixed features. In *Welwitschia,* pollen-producing plants produce two lateral bracteoles, two fused bracteoles forming an envelope, and six fertile microsporangiophores fused at their bases and that surround a sterile ovule. In contrast, seed-producing plants produce a fertile ovule surrounded by two sets of decussate bracteoles and no fertile or sterile microsporangiophores. In *Gnetum,* each cone comprises an axis bearing a fused bract pair (collar) in the basal region and perhaps a whorl of fused bracts in the upper region. In the axils of the collars of pollen cones, there are numerous fertile shoots each consisting of two fused bracteoles, enclosing a microsporophyll (or microsporangiophore) and the apical whorl consists of abortive ovules (Endress 1996). As in *Welwitschia,* there are no abortive pollen-bearing structures in ovulate cones in *Gnetum.* These unusual arrangements of fertile and sterile parts in the cones of Gnetales and the teratisms found in conifers are poorly understood in terms of function and development. In the case of the Gnetales, these can be studied as they are predictable. In conifers, this is more difficult because they are random. Finally, in *Ginkgo* and cycads, there are no records of bisporangiate axes or of bisporangiate plants.

The ovules of most living gymnosperms are interpreted as terminating axes with the exception of ovulate plants of cycads, where the ovules appear to be on megasporophylls. All ovules have a nucellus (megasporangium) enclosed by an integument. Some species have the ovules enclosed by other structures, in *Taxus* by an aril and in Gnetales by one (*Ephedra*) or two (*Gnetum* and *Welwitschia*) sets of bracteoles. For further discussion see Section "Seeds". The most problematic is the collar of *Ginkgo*. This structure, at the base of the ovule, has been variously interpreted as a reduced sporophyll, an aril, or a cupule (Douglas *et al.* 2007). Detailed developmental data have not elucidated this because the collar does not show a developmental pattern common to any of these interpretations. Thus, Douglas *et al.* (2007) have concluded that it is unique to *Ginkgo* and is a structure *sui generis*. Often there is, developmentally, an ephemeral organized flap of tissue between the collar and the ovule proper that is also considered to be a structure *sui generis* (Douglas *et al.* 2007). Clearly, *Ginkgo* ovules need to be studied using a molecular approach that may resolve these conundrums.

Stem anatomy is fairly uniform in conifers. All are basically woody and have secondary growth from a vascular cambium. The tracheary elements of the wood are composed only of tracheids and little parenchyma, and the wood is referred to as pycnoxylic. *Ginkgo* is fundamentally the same as the conifers as is *Ephedra*.

In contrast, the stems of cycads have copious parenchyma and are termed manoxylic. In fact, cycads are similar to cacti and have the same pachycaulous growth habit. The large pith of the cycads is the result of diffuse cell divisions in the developing pith and the broad cortex is the result of a PTM continuous with and distal to the apical meristem. One of the most unusual features of cycads is the girdling leaf trace. The two main leaf traces enter the stem and traverse the cortex in opposite directions and circumferentially to eventually become continuous with the stele 180 degrees from their entrance in the cortex. This may well be the result of the massive increase in stem diameter and no internodes. In this case, because the PTM is near the pith and producing most derivatives centrifugally, the procambium, which elongates in the direction of most cell divisions to maintain synchronous growth, will be elongating circumferentially rather than longitudinally.

Secondary, growth in some cycads is by successive cambia (multiple vascular cambia that form successively in the cortex). It is not known whether two or more of these cambia are active at the same time. This type of growth is also known for liana species of *Gnetum* and in *Welwitschia* (Martens 1971; Carlquist 1996). There is no discernable difference between successive cambia in cycads, Gnetales, and angiosperms such as the Caryophyllales beyond the eventual differentiation of tracheary elements themselves. This independent development should be examined from a morphogenesis approach and also in terms of water relations.

The Gnetales are unusual in the type of secondary xylem produced from the vascular cambium. The end walls have foraminate perforation plates that appear quite different from the scalariform perforation plates found in angiosperms. Moreover, the simple perforation plates found in *Gnetum* and *Welwitschia* appear to be derived from foraminate perforation plates (Carlquist 1996) and are most likely an independent origin of vessel elements. This concept of the independent origin of the gnetalean perforation plate is not without controversy (see, Carlquist 1996 and Muhammad & Sattler 1982). Perhaps, this could be elucidated by a comparative molecular approach to the development of the angiosperm vessel element and the gnetalean vessel element.

5.4 Leaves

The diversity of leaves in gymnosperms ranges from small scale-like leaves seen in many conifers and *Ephedra* to very large multipinnate leaves found in some cycads. As a rule, gymnosperms are evergreen, that is, leaves are retained for 2 or more years. *Ginkgo*, *Metasequoia*, and *Taxodium* are the exceptions. *Ginkgo* is deciduous by shedding its leaves each autumn, whereas *Metasequoia* and *Taxodium* shed branchlets with the current years leaves each autumn. Most gymnosperms produce two kinds of leaves; nongreen cataphylls, which are small and scale-like, and foliage leaves that are green and expanded at least to some extent. The exceptions to this are found in *Gnetum* and *Welwitschia*, which do not have cataphylls but do produce scale leaves in their strobili. Within cycads, *Stangeria* and *Bowenia* lack true cataphylls on vegetative axes and instead have cowl-like leaf bases that remain after abscission and function like cataphylls. These two genera do have scale leaves closely resembling cataphylls in association with the peduncles of their strobili. Small stipules are found with the leaf bases of some cycads. Glands are not found associated with the leaves of most gymnosperms. The exception to this is in the cycad genus, *Macrozamia*, where most species have a gland on the adaxial surface at the juncture of the pinnae with the rachis. These are bright yellow in some species (*Macrozamia fraseri*), white in some (*Macrozamia communis*), and red in others (*Macrozamia diplomera*). The structure and function of these intriguing variously colored glands are unknown. Likewise, this can be applied to *Zamia manicata* and *Zamia macrocheira*, which have petiolules with an adaxial encircling gland at the juncture of the blade and the petiolule. Nothing is known about the anatomy of these structures, but the one in *Z. macrocheira* appears as a flap that incurves to form a tunnel occupied by ants. In both species, the adaxial gland-like structure is absent in juvenile plants and in the first leaves of transplanted plants; thus, providing an opportunity for a study of a transformation series within a species.

There is an excellent body of literature on the general anatomy of gymnosperm leaves, both fossil and extant (Napp-Zinn 1966). General features are the presence of a thick cuticle, lignified epidermal and hypodermal cells, sunken stomata, presence of sclereids, and transfusion tissue. Differences mostly lie in the distribution of these tissues in the various gymnosperm families and orders and even from species to species in some genera. There is some controversy concerning the presence or absence of transfusion tissue in *Gnetum* (Martens 1971). Much is known about the anatomy of conifer leaves, both fossil and living (Napp-Zinn 1966). In contrast, the Gnetales and cycads are the least understood group of gymnosperms in terms of comparative leaf anatomy.

The foliage leaves of living conifers have been divided into four major categories: *type 1* includes needle-like leaves that are univeined and tetragonal in transection. This type is widely distributed among the Pinaceae, Araucariaceae, Podocarpaceae, and Cupressaceae; *type* 2 comprises univeined leaves, which are linear or lanceolate in contour and bifacially flattened. This type is the most common in living conifers as exemplified by Taxaceae, Cephalotaxaceae, many Podocarpaceae, many Pinaceae, and so on; *type 3* comprises all scale-like forms of leaves, for example, the adult foliage leaves of many Cupressaceae such as *Juniperus, Cupressus,* and *Callitropsis*; *type* 4 includes the broad, multiveined leaves of *Agathis*, and of some species of *Araucaria, Podocarpus*, and *Nageia*.

Most gymnosperm leaf venation patterns are based upon dichotomies. *Gnetum* with its apparent reticulate venation appears to be an exception to this but Rodin (1967) has demonstrated that this pattern is derived from anisotomous branching. The cotyledons of *Welwitschia* also appear to have reticulate venation (Martens 1971) but there have not been any developmental studies on these to demonstrate the origin of the pattern. The leaves of most conifer genera are vascularized by a single median bundle, for example, the leaves of *Abies, Picea, Pseudotsuga*, all members of the Cupressaceae, and certain species of *Pinus*. In contrast, the venation pattern that develop in the broad leaves of *Araucaria, Agathis*, and *Nageia* consist of a longitudinal series of dichotomously branched veins, not unlike those found in pinnae of most cycads. In the genus *Pinus*, there are the haploxylon or soft pines species (subgenus *Strobus*) with univeined leaves and the diploxylon or hard pines species (subgenus *Pinus*) with two-veined leaves, (Gifford & Foster 1989; Gernandt *et al.* 2005). Originally viewed as dichotomous venation, histogenetic studies have shown this not to be the case. Instead, in two-veined leaves, the young developing needle has a single median vein. As differentiation continues, a ray-like wedge of parenchyma is formed in the median region of the vein and eventually causes a division of the univein into two veins. In some instances, one or two additional rays are formed, dividing the original vein into three or four separated bundles Gifford and Foster (1989). Thus, ontogenetically the two-veined anatomy is not dichotomous but rather has a quite different development. This developmental pattern needs to be investigated in other conifers because the histogenesis of what appear as dichotomously branched veins in other species of conifers has not been studied.

While much is known about the anatomy of gymnosperm leaves particularly in conifers, both fossil and living (Napp-Zinn 1966), little is known about leaf development. There is only one paper on leaf development in cycads (Johnson 1943), two for Gnetales (Seelinger 1954; Rodin 1967), and only a few for conifers most of which focus on longitudinal sections from studying shoot apex organization (Napp-Zinn 1966; von Guttenberg 1961). Nevertheless, the scant literature does reveal some intriguing aspects in conifer leaf development as well as in *Ephedra* and *Ginkgo*.

In those conifers that have been studied (*Pinus*, *Taxodium*, *Cupressus*, *Cunninghamia*, and *Thujopsis*) and *Ephedra*, leaf primordia have a distinct set of apical initials and subapical initials as viewed in longitudinal sections. Transectional views of foliage leaves are available for only three of these genera (*Taxodium*, *Cunninghamia*, and *Ephedra*). In all cases there are both marginal and submarginal meristems and some adaxial cell divisions in *Taxodium* and *Cunninghamia* (von Guttenberg 1961, and literature cited therein). Cataphylls of *Pinus lambertiana* have very well-developed marginal meristems that produce a long series of biseriate derivatives (Sacher 1955). A very interesting comparison could be made between conifer foliage leaf development and that of lycophytes such as *Huperzia* where the anatomy is quite different but the morphological development appears descriptively similar.

Ginkgo produces broad fan-shaped leaves that are entire to shallowly bilobed to deeply notched and in some cases with secondary notches (lobes). Hara (1997 and literature cited therein) has shown that the divisions are derived from dichotomies in the adaxial–abaxial plane. The abaxial growth leads to the involute ptyxis most likely as a result of a confined space in the bud. Marginal expansion would appear from Hara's schematic drawings (Hara 1997) to be from at least submarginal meristems and one must assume then that there are marginal meristems in order for the protoderm to keep pace. A detailed study of the leaf development of all the forms of *Ginkgo* leaves including the rare cup-shaped leaves would be productive from both histogenetic and molecular development perspectives.

Most cycad species have pinnately compound leaves that can be very large, up to 7 m in *Encephalartos laurentianus*. There are at least three independent origins of multipinnate leaves in the cycads as these multipinnate leaves occur within three separate lineages. *Bowenia* (Zamioideae, Zamiaceae) has bipinnate leaves, as do some species of *Macrozamia* (Encephalartoideae, Zamiaceae) and *Cycas* (Cycadaceae). In *Bowenia*, the architecture of the bipinnate leaves appears to be monopodial, whereas, in *Macrozamia* and *Cycas* the divisions of the pinnae appear to be based upon dichotomies. In *Macrozamia*, there are species such as *Macrozamia polymorpha* where some leaves are pinnate and others have dichotomous leaflets and others, which are mixed within the same leaf. In *Macrozamia stenomera*, all leaves have dichotomous pinnae and often each of these dichotomies itself is dichotomously divided; thus, they are doubly dichotomous. In *Cycas*, those species with dichotomous pinnae, every leaflet exhibits dichotomies except for the first leaf of the seedlings. There is a clade in *Cycas* composed of *Cycas simplicipinna* as sister to a clade composed of *Cycas micholitzii*, and *Cycas multipinnata* (Hill 1999).

C. simplicipinna has undivided pinnae, *C. micholitzii* has pinnae that are once dichotomous, and *C. multipinnata* has pinnae that are two to five times dichotomous. Thus within the clade we have a transformation series from simple to dichotomous to multidichotomous. This unto itself is interesting and could be studied developmentally using a molecular approach. It becomes even more interesting when examining the development from seedling to adult of individual plants of *C. micholitzii* and *C. multipinnata*. In *C. micholitzii*, the first leaf has undivided pinnae, the second leaf has one dichotomous pinnae. In *C. multipinnata*, the first leaf has undivided pinnae, the second leaf has one dichotomous pinnae, the third through fifth leaves have a pinnae where one of the dichotomies is itself dichotomous resulting in a tripartite pinnae, and the sixth and subsequent leaves have pinnae with multiple dichotomies resulting in leaves that have a minimum of three dichotomies per pinna. Thus, we not only have a transformation series between species but also within a species. Combining the phylogeny with the known morphology provides an excellent opportunity to study the evolution of increasing leaf complexity in this system. Unfortunately, we have neither a phylogeny nor the morphological data to do a similar study in *Macrozamia*.

Gymnosperms, with the exception of *Ginkgo*, *Larix*, *Metasequoia*, *Pseudolarix*, *Taxodium*, and *Ephedra*, are evergreen with the leaves persisting for many years in some species. With the establishment of secondary growth via a vascular cambium in the stem, the primary xylem of the leaf trace is gradually ruptured. The conifers have developed a system of maintaining vascular continuity of the leaf traces across the cambium with the development of a fascicular cambium that produces mainly secondary xylem (Eames & MacDaniels 1947). Thus, one can observe buried leaf traces in the secondary wood of the stem. In some trees such as *Araucaria*, these can extend across several growth rings. The maintenance of leaf traces in cycads, Gnetales, and angiosperm trees is unknown, but must occur for the leaves to persist beyond the establishment of secondary growth.

5.5 Roots

The roots of gymnosperms are relatively uniform in their anatomy and development is comparable to flowering plants or ferns. The root apex is described as consisting of two groups of initials (Type V in Popham 1966). One group produces the stele and the second group the root cap and cortex. Gymnosperms are unusual because the dermatogen is not set off directly at the apex as in flowering plants but rather forms distal to the apex proper at the point where the base of the root cap is separated from the cortex (Eames & MacDaniels 1947; von Guttenberg 1961). As a rule, gymnosperm roots have diarch protoxylem poles.

Gymnosperms have a primary root that is long-lived and which produces secondary roots, which produce tertiary roots, and so on. *Ginkgo* and conifers have very uniform root morphology and anatomy. Root diversity is found in groups such as cycads and Gnetales. The primary root of the cycads is a fleshy taproot. In the seedling stage, the taproot is contractile and can contract up to 60% of its length (Rimbach 1929; Stevenson 1980; Norstog & Nicholls 1997). This root contraction then continues toward the root apex as the root grows in length. The root contraction is a result of the collapse of horizontal rows of parenchyma cells in the cortex and pith in the same manner as in some angiosperms such as *Oxalis* but different than that in monocots where inner cortical cells expand radially and then shorten longitudinally (Ruzin 1979; Jernstedt 1984a, 1984b). In contrast, root contraction is very different in *Botrychium* where lysigenous cavities form and then collapse. The genes involved in these developmental processes are unknown and the independent derivation of a similarly described developmental anatomy in cycads and species such as *Oxalis* that are clearly homoplasious would be quite interesting in the context of whether or not these represent using the same genes to achieve contraction as the only way to do it or if entirely different genes are involved in producing the same result.

Another unique feature of the roots of cycads is the production of ageotrophic coralloid roots that usually proliferate at or just below the soil surface and contain zone of cells containing nitrogen-fixing cyanobacteria, usually *Nostoc*. Located between the inner cortex and the loose outer cortex, the cyanobacterial zone consists of elongated cortical cells and large intercellular spaces. This portion of the coralloid occupied by cyanobacteria is known as the cyanobacterial zone. Initiated as a new meristematic layer in precoralloids long before cyanobacterial invasion, this specialized tissue is not found in normal lateral roots. This region is green when inhabited by cyanobacteria and visible to the naked eye in cross-section as a hollow cylinder extending from the base to just below the apex (Staff & Ahern 1993). There is some debate as to the origin of the cyanobacterial zone. Many authors have attributed the rhizoderm as the parent of the region of elongated cells occupied by the cyanobacteria, and a persistent root cap as the mantle that covers mature coralloids (Milindasuta 1975; Ahern & Staff 1994; Rai *et al.* 2000). However, in studies of coralloid developmental anatomy, Webb (1983)found evidence that the cyanobacterial zone is initiated in the cortex and that the mantle is originated not from a persistent root cap, but instead may represent a unique type of tissue. Other than a superficial study of the development of coralloid roots and their differentiation patterns, little is known in detail and nothing from a molecular perspective.

Whereas *Ginkgo*, conifers and *Ephedra* have secondary growth in their primary and successive orders of woody roots from a single vascular cambium as in most woody angiosperms, secondary growth in the roots of cycads and *Gnetum* and *Welwitschia* is by successive cambia (multiple vascular cambia that form successively in the cortex) as is also found in their woody stems. As in the case of the stem, it is not known whether any two or more of these cambia are active at the same time. Suffice it to say that as in stems there is histologically no discernable difference between successive cambia in cycads, Gnetales, and some angiosperms.

The cypress knee is a distinct structure of the root system of some members of the Cupressaceae (*Taxodium*, *Metasequoia*, and *Glyptostrobus*) and they generally occur on plants growing in swamps or other aquatic situations. The knees are vertical woody projections from the submerged roots. Knees are formed on the upper surface of these roots by the vascular cambium. One assumption of their function has been that they provided oxygen to the roots and thus act as pneumatophores; thus, this commonly used term for cypress knees. However, there is little actual evidence for this and in fact their function remains enigmatic (Briand 2001).

5.6 Seeds

The seeds of gymnosperms are of three types: (1) fleshy or baccate, (2) dry, and (3) winged. Fleshy seeds would appear to have developed at least three times because the fleshy part has different origins in different gymnosperm groups. In *Ginkgo* and the cycads, the fleshy tissue of the seed is from the outer part of the integument, the sarcotesta. In contrast, the fleshy part of the seed in *Taxus* is from an outgrowth of the ovule stalk that grows upward to eventually envelop the seed proper (Bierhorst 1971; Mundry 2000). Thus, it is an aril and quite distinct from a sarcotesta in origin. The fleshy part of the seed of other members of the Taxales, *Torreya* and *Cephalotaxus*, are more enigmatic because the fleshy part is neither free from the seed nor apparently postgenitally fused (Mundry 2000). Thus, it is interpreted as a congenitally fused aril with the seed proper. This of course cannot be demonstrated by conventional developmental approaches but may well be tractable from a molecular approach using the tools of evolution of development. Interestingly, *Phyllocladus* in the Podocarpaceae also has an aril-like structure that is fleshy when initiated and then becomes papery at maturity (Holloway 1937). The comparative development of the aril of the Taxaceae and *Phyllocladus* is an area that needs to be explored in a modern context.

The fleshy tissue associated with the seeds of *Juniperus* (Cupressaceae) and many Podocarpaceae (e.g., *Podocarpus* and *Nageia*) is derived from the whole cone being fleshy in the former and the lower receptacle portion of the cone plus the integument of the ovule forming a sarcotesta in the latter. Thus, we have two more methods of forming fleshy reproductive structures within the conifers.

Other examples of fleshy seeds are found in the Gnetales. Within *Ephedra*, there are species with fleshy cone bracts. That is, the ovule has an integument plus a vascularized seed envelope that is dry at maturity. The ovules are in the axils of cone bracts, one ovule per cone bract. It is the cone bracts that become fleshy and analogous to *Juniperus* in that the whole cone is fleshy. These fleshy cone bracts in *Ephedra* appear independently within the New

World *Ephedra* clade and again in the Old World *Ephedra* clade (Rydin *et al.* 2010), although anatomically the fleshy bracts are structurally the same in all *Ephedra* species where they are present.

Gnetum has fleshy seeds and in this case it is the seeds that are fleshy and not the whole cone as in *Ephedra*. *Gnetum* ovules have three enclosing structures: the innermost being the integument, which is surrounded by two successive seed envelopes to use the same terminology as for *Ephedra*. In *Gnetum*, the outermost seed envelope becomes fleshy and in some species both of the seed envelopes are fleshy.

Most conifers have dry seeds, many of which are winged. Winged seeds are common in the Cupressaceae and are independently developed within the Pinaceae and the Araucariaceae. The wings are developed from lateral growth of the integument in Pinaceae and Cupressaceae. In contrast, in *Welwitschia* the wings are developed from the seed envelope, which is interpreted as an inner pair of fused bracteoles or the neutral term, seed envelope that surrounds the integument. In Araucariaceae, it is unclear whether the wing is derived from the integument or from both the integument and the cone scale subtending the ovule.

5.7 Seedlings

The basic seedling has two cotyledons that do not appear to be produced by the embryonic shoot apex. In cycads and *Ginkgo*, germination is hypogeal with no hypocotyl and with haustorial cotyledons (cryptocotylar) that are often green although never exposed to light. In contrast, the conifers and Gnetales are epigeal with a well-developed hypocotyl and green expanded cotyledons (phanerocotylar). In *Gnetum* and *Welwitschia*, there is a lateral outgrowth of the hypocotyl that remains embedded in the megagametophyte. Whether or not the feeder is haustorial remains to demonstrated but it is interesting that these embryos appear to be able to have photosynthetic cotyledons and simultaneously have a haustorial feeder.

Variations in cotyledon number are common in the conifers and these variations run from 2 to 12 (Butts & Buchholz 1940; Bierhorst 1971), particularly in the Pinaceae. In cycads, there often appears to be only one cotyledon because the pair is so pressed together and *Ceratozamia*, in particular, is often described as having only one cotyledon. Close examination, however, reveals a longitudinal groove on both sides and when grown on a clinostat, there are always two very separate cotyledons (Dorety 1908).

Other than the variation described above, the embryos of the gymnosperms are very similar. Recently, Robbertse *et al.* (2011) have described the cycads as having a coleorhiza based upon development during seed germination. They found this coleorhiza is derived from the distal part of the suspensor and that the primary root grows through this tissue leaving a ring around the emergent root. Martens (1971) has reported the same thing for the three genera in the

Gnetales. Using these criteria, it would appear that all gymnosperms, for which we have data on embryo development, have a coleorhiza. Furthermore, this coleorhiza in the gymnosperms is identical in its development to that in monocotyledons such as the grasses where the term is commonly used. Using these criteria, it may well be that all plants with a suspensor and a primary root have a coleorhiza and that the derived condition is to have neither a suspensor nor a coleorhiza as in some of the angiosperms.

5.8 Embryology

It is not the intent here to provide a thorough review of the extensive literature on the embryology of the gymnosperms as that has been very well done by Singh (1978). Rather, I will focus on some of the more fascinating aspects of gymnosperm embryology that are in need of more detailed study. The gametophytes of gymnosperms are unisexual, dioicous, as opposed to the bisexual gametophytes, monoicous, of the homosporous ferns and lycophytes.

There are two types of microsporangium development in the gymnosperms. The first type, which occurs in most gymnosperms, is the usual eusporangium type where the sporangium is differentiated into an epidermis and a hypodermis from inception with the latter giving rise to the tapetum and sporogenous tissue. The second type, which occurs in a few genera of the Pinaceae (e.g., *Cedrus*, *Larix*, *Picea*, *Pinus*, and *Pseudotsuga*), actually has sporangial development more similar to a leptosporangium where surface initials give rise to the entire sporangium. Whether or not both conditions occur in the same species is an open question, however it certainly seems possible that the distinction between eusporangium and leptosporangium as pointed out by Bierhorst (1971) can be questioned in light of the intermediate nature of sporangial development in *Osmunda* and what appears in some of the conifers. Microsporogenesis is generally simultaneous, but in cycads both simultaneous and successive microsporogenesis have been reported even with the same species (Audran & Masure 1977). Pollen may be winged (many conifers), simple and sulcate (cycads, *Ginkgo, Welwitschia*), or leptomate (conifers and *Gnetum*). The microsporangia of the cycads are clustered into groups of two to five similar to the sorus of the ferns. Actually, there is a series from clusters with each sporangium having a vascular bundle as a branch of a common bundle as in *Cycas* to those where the cluster shows some basic degree of sporangial fusion that ranges from only a common stalk in the basal region in *Dioon* to more or less completely fused along their entire length in *Zamia* to completely fused stalks with some or all of the sporangia fused laterally in *Macrozamia.* Fusion of the sporangia and their stalks is correlated with simplification of vasculature. Thus, in *Macrozamia*, where the stalks are completely fused and some or all of the sporangia are fused, the common bundle enters the short common stalk and terminates without branching.

The wall of the mature microsporangium of all gymnosperms that have been examined consists of an epidermis with thickenings in the cell walls, an exothecium. In contrast, *Ginkgo* is the only exception and has an endothecium where one to three layers of subepidermal cells have thickening in their walls (Jeffery & Torrey 1916; Singh 1978) The former is also described as exokinetic dehiscence and the latter as endokinetic dehiscence of the microsporangium (Jeffery & Torrey 1916). Microsporangial dehiscence is longitudinal in cycads, *Ginkgo*, and conifers and poricidial in Gnetales. The lines of dehiscence appear to be where there are unthickened exothecial cells and perhaps unthickened endothecial cells in *Ginkgo*. More observations are needed for conifers, both in sporangial wall structure and in dehiscence.

With the exception of *Gnetum* and *Welwitschia*, microgametophyte development is quite similar among gymnosperms and the basic differences simply involve the number of cells produced before gamete production. Perhaps the most salient points are the development of the male gametes and pollen tube growth. Most gymnosperms produce two nonmotile male gametes per microgametophyte. Exceptions are found in two cycads, *Ceratozamia* with occasionally 4 and *Microcycas* with up to 16 (Norstog *et al.* 2004) and there are records of polyspermy in some species of *Cupressus* and *Juniperus* (Singh 1978; Schnarf 1933). Both *Ginkgo* and cycads have motile male gametes with numerous flagella. In cycads, the male gamete, which is mainly a nucleus, can be up to 0.5 mm in diameter with perhaps 40 000 flagella and is visible with the naked eye. In both *Ginkgo* and cycads these are easy to see and to collect. Developmentally, both are quite similar and only differ in details and the presence of a nucellar beak in *Ginkgo*, which is a distinctive outgrowth of the nucellar membrane of unknown function. Although much is known about their ultrastructure (Norstog *et al.* 2004), a comparison on a molecular basis, particularly small RNAs and with ferns would be most welcome.

Pollen tube emergence in the gymnosperms is via the sulcus in cycads, *Ginkgo*, and *Welwitschia* and via the leptoma in conifers, *Ephedra*, and *Gnetum*. The pollen tube with its flagellar sperm in *Ginkgo* and cycads is highly branched and haustorial and grows into the roof of the pollen chamber of the ovule. In contrast, the pollen tube of gymnosperms with nonmotile sperm is rarely branched, nonhaustorial, and grows toward the egg in the megagametophyte penetrating the archegonium when one is present. It appears that the pollen grains become oriented in such a way that the sulcus face the roof of the archegonium in those with motile sperm and in contrast face the megagametophyte in those with nonmotile sperm.

Gnetum and *Welwitschia* are considered unusual and more similar to angiosperms because only three cells are formed. These two genera form a clade and are considered to represent the end of a reduction series in the microgametophyte either as part of monophyletic gymnosperms or as a step to the angiosperms depending on the favorite topology for the seed plants.

Megasporogenesis generally results in a linear tetrad of megaspores, although isobilateral and tetrahedral are known to also occur in the *Sequoia*.

The gymnosperms have a monosporic megagametophyte except in *Gnetum* and *Welwitschia*, which will be discussed later. Generally, the megagametophyte development is correlated with pollination so that if pollination fails then the ovule aborts. In all gymnosperms, the germination of the megaspore results in a free nuclear stage, that is, the megagametophyte begins as a large coenocytic cell with nuclei from 256 to more than 3000. The variation in number of free nuclei seems to vary with the species and not so much within species, although this aspect has received little attention. The nuclear divisions appear synchronous and often appear in waves. At first the nuclei are randomly distributed in a thin layer of cytoplasm that surrounds a large central vacuole, but later this vacuole disappears as more and more nuclei are produced and becomes reduced in volume. At the end of the free nuclear period of divisions, aveolation begins except in *Gnetum* and *Welwitschia*. Aveolation is initiated when each peripheral nucleus becomes connected to six adjacent nuclei by microtubules. Anticlinal walls develop centripetally to form long, six-sided tubular cells. These cells are termed *alveoli* and each alveolus is hexagonal in transection with its nucleus positioned at the inner end, which is open to the cytoplasm next to the vacuole. The female gametophyte appears as a honeycomb with each hexagonal tube representing a single alveolus. The alveoli begin to extend centripetally toward the center of the gametophyte. After the alveoli have extended to the center of the gametophyte, each one becomes "closed" by the formation of an end wall. The nucleus of each alveolus then moves back toward the periphery of the gametophyte. Through repeated formation of periclinal cell walls (with reference to the gametophyte surface) each alveolus is converted into a file of cells. The process of aveolation can be visualized as a three-dimensional series of closing iris diaphragms. There are variations on this theme, but all gymnosperm megagametophytes go through a free nuclear phase and an aveolation phase (Singh 1978). Once the megagametophyte has become cellular, archegonia differentiate at the micropylar end with the exception again, as usual, of *Gnetum* and *Welwitschia*. The archegonia of *Ginkgo*, cycads, and most conifers are rather simple and at maturity are composed of an egg and an ephemeral ventral canal cell that have a common origin and a single tier of four neck cells. In contrast, the archegonia of conifers and *Ephedra* are much more elaborated and more like those found in lycophytes and ferns where there are two or more tiers of neck cells that, thus, form a ventral canal. Archegonia can occur singly with each enclosed in its own jacket layer as in cycads, *Ginkgo*, and most conifers or in archegonial complexes where several archegonia are enclosed in a single common jacket as in Cupressaceae sensu lato. Both conditions can occur in the same megagametophyte in *Ephedra*. The actual details of the formation of archegonial complexes are somewhat enigmatic. They can occur singly as one per megagametophyte or as several per megagametophyte. In some genera, the number of archegonial complexes seems to correlate with the number of pollen tubes growing in the megagametophyte (Singh 1978).

Often several sperm nuclei are found in the egg cytoplasm. This has been reported over the years for numerous genera including cycads and *Ginkgo* with motile sperm. Once fertilization has occurred as the result of the fusion of the sperm with the egg nucleus a zygote is formed and it is the first mitotic division of the zygote that is most peculiar. In all gymnosperms (except *Gnetum* and *Welwitschia*) where this division has been observed, there are two spindles that form within the single nucleoplasm, one corresponding to the sperm chromatin and the other to the egg chromatin. After the formation of these separate spindles, the two spindles merge. In contrast, the two sets of chromosomes are already merged before the first mitotic division of the zygote in *Gnetum* and *Welwitschia* so that only one spindle appears.

After fertilization, most gymnosperms have a free nuclear phase in the formation of the embryo with exceptions in *Sequoia*, *Gnetum*, and *Welwitschia*. The number of nuclei in this coenocytic phase of the embryo varies from 256 in Ginkgo to 512 and 1024 in cycads and from 4 to 64 in conifers and *Ephedra*. Cellularization follows.

Multiple embryos are formed in all gymnosperms by the fertilization of eggs in more than one archegonium or multiple eggs in a coenocytic megagametophyte (*Gnetum* and *Welwitschia*), simple polyembryony. There are also multiple embryos formed in the conifers and Gnetales by the cleavage of embryos, cleavage polyembryony. There are many forms of the latter in the conifers (Singh 1978). Within the Gnetales, *Ephedra* forms an embryo from each of the nuclei that become walled off from the free nuclear stage. *Gnetum* and *Welwitschia* form multiple cleavage embryos by a peculiar form of branching tubes with cell divisions at the tips of the tubes as described below.

As mentioned earlier, *Gnetum* and *Welwitschia* differ significantly in their megagametophyte development. All four megaspores produced survive so there is a coenomegaspore and the megagametophyte is tetrasporic in origin. In *Gnetum*, several coenomegaspores may be produced with only one megagametophyte surviving whereas in *Welwitschia* only one coenomegaspore is produced. Each of the four nuclei in the coenomegaspore undergoes a period of free nuclear divisions to produce up to 1000 nuclei. In *Gnetum*, wall formation begins at the chalazal end of the female gametophyte and proceeds toward the micropylar end. The resulting cells contain single nuclei or groups of two or more nuclei. In the chalazal end, the nuclei eventually fuse to produce a large single nucleus per cell. In the micropylar end, there are the egg nuclei, either singly or in coenocytic cells, which can be recognized by their large dense size and the cytoplasm surrounding them that becomes dense and striated. One or more pollen tubes enter the female gametophyte and come to lie next to the egg cells. One or both male gametes in a pollen tube may fertilize an egg (Friedman & Carmichael 1996). Multiple zygotes are formed per megagametophyte either from two sperm per pollen tube and/or by sperm from multiple pollen tubes. Following fertilization the micropylar end of the female gametophyte becomes cellular. Each zygote undergoes cell division, resulting in two repeatedly branching cells with each branch containing one nucleus. The cells elongate as they grow throughout the cellular female gametophyte. Eventually, an embryo is formed at each tip, but, as in other gymnosperms, only one embryo develops to maturity in the seed.

Welwitschia is similar to *Gnetum* but also has rather bizarre differences. After the free nuclear phase, wall formation follows with two to eight nuclei per cell in the micropylar end and six to twelve nuclei per cell in the chalazal end of the female gametophyte. In the chalazal end, the nuclei fuse to form large polyploid cells that undergo cell divisions as in *Gnetum*. The multinucleate cells at the micropylar end grow up into the nucellus and form female gametophytic tubes where the nuclei migrate. Pollen tubes growing downward in the nucellus meet and fuse with the upward growing female tubes with the dissolution of their common walls. Fertilization of one of the nuclei by a male gamete occurs in a common bulbous tip (a fertilization bulb). A zygote is then formed which, upon further growth, gives rise to a proembryo that grows toward the megagametophyte within the female gametophytic tube. As in other gymnosperms, many embryos are formed, but normally only one completes development.

Although seasonal, some aspects of embryology, especially in *Ginkgo*, cycads and conifers should be easy to study, in particular the free nuclear stages and cellularization of megagametophyte development and embryogenesis.

One of the interesting aspects of gymnosperm reproductive biology that is in need of further study is that of "double fertilization." Double fertilization has been observed in conifers and Gnetales. In the conifers, these events where one of the sperm nuclei of a pollen tube fuses with an egg in the archegonium and the other sperm nucleus in the same pollen tube apparently fuses with a ventral canal nucleus that has not degenerated. These instances appear to be teratological based upon the rarity of the observations. Double fertilization in *Ephedra* has been reported historically and has been well documented using modern advances in microscopy and microtechnique (Friedman & Carmichael 1996). In this case, it is also fertilization of an egg and of a ventral canal nucleus. Because the ventral canal nucleus often degenerates in *Ephedra*, one wonders whether this is a rare event like the conifers or whether it is common. It appears to be accepted as common, but given the rarity of earlier reports when researchers were looking for it, more studies need to be done to determine how ubiquitous it really is and if it is statistically significant. Double fertilization in *Gnetum* where both sperm from a pollen tube are involved in fertilization events in the same megagametophyte (Friedman & Carmichael 1996) is to be expected as the megagametophyte is tetrasporic and at lest four eggs are to be expected. What is interesting in both *Gnetum* and *Welwitschia* is the formation of fusion nuclei in the chalazal regions of the megagametophyte. This may well be an independent development of a fusion nucleus that is also found in angiosperms.

An area of neglected interest in the embryology of vascular plants is the comparative development of the free nuclear stages and their lack thereof.

The heterosporous lycophytes have a free nuclear phase of over 512 nuclei in megagametophyte development and the free nuclear divisions appear to be synchronous. Cellularization begins at the triradiate ridge of the megaspore and proceeds basipetally. The megagametophyte never becomes completely cellular even after fertilization. There is no free nuclear phase in embryogenesis in the lycophytes. In heterosporous ferns, there is no free nuclear phase in either megagametophyte development or embryogenesis. Comparisons of the stages of the free nuclear phases through cellularization between *Selaginella* and representative conifers, *Gnetum, Ginkgo*, and a cycad could be done fairly easily and should be quite interesting in terms of the genes involved.

It would seem that gymnosperms have synchronized pollination with megagametophyte development, which will be the nutrient and energy supply for the developing embryo and seedling. No pollination, no storage investment, and no waste (*Encephalartos* in the cycads may be an exception to this). However, often if the wrong pollen is deposited on a gymnosperm, the megagametophyte completes development for naught. In contrast, Angiosperms have developed a system of double fertilization whereby; the ovule aborts unless a zygote is present along with a triple-fusion nucleus, which will give rise to endosperm. Thus, there must actually be a zygote formed and no waste of storage tissue can presumably be formed. Perhaps, some of the peculiarities found in gymnosperm embryology are less than efficient answers to efficiency.

References

Ahern, C.H. & Staff, I. (1994) Symbiosis in cycads: the origin and development of coralloid roots in *Macrozamia communis* (Cycadaceae). *American Journal of Botany*, **81**, 1559–1570.

Audran, J. & Masure, E. (1977) Contribution a la connaissaoce de la composition des sporodermes chez les Cycadales (Prespennaphytes). Etude en microscopie electronique a transmission (M.E.T.) et abalayage (M.E.B.). *Palaeontographica* **162**, 115–158.

Bierhorst, D. (1971) *Morphology of Vascular Plants*. Macmillan, New York.

Briand, C.H. (2001) Cypress Knees: an enduring enigma. *Arnoldia*, **60**, 19–25.

Burleigh, J.G. & Mathews, S. (2004) Phylogenetic signal in nucleotide data from seed plants: implications for resolving the seed plant tree of life. *American Journal of Botany*, **91**, 1599–1613.

Butts, D. & Buchholz, J.T. (1940) Cotyledon numbers in conifers. *Transactions of the Illinois State Academy of Science*, **33**, 58–62.

Carlquist, S. (1996) Wood, bark, and stem anatomy of Gnetales: a summary. *International Journal of Plant Sciences*, **157**, S58–S76.

Del Tredici, P. (1997) Lignotuber formation in *Ginkgo biloba*. In: *Ginkgo biloba -'A Global Treasure*, (eds T. Hori, R.W. Ridge, W. Tulecke, P. Del Tredici, J. Tremouillaux-Guiller & H. Tobe), pp. 119–126, Springer-Verlag, Tokyo.

Dorety, H.A. (1908) The seedling of *Ceratozamia*. *Botanical Gazette*, **46**, 203–220.

Dörken, V.M. & T. Stützel. (2011) Morphology and anatomy of anomalous cladodes in *Sciadopitys verticillata* Siebold & Zucc. (Sciadopityaceae). *Trees*, **25**, 199–213.
Douglas, A., Stevenson, D. & Little, D. (2007) Ovule development in *Ginkgo biloba* L., with emphasis on the collar and nucellus. *International Journal of Plant Sciences*, **168**, 1207–1236.
Eames, A.J. & MacDaniels, L.H. (1947) *Plant Anatomy*. 2nd edn. McGraw-Hill, New York.
Endress, P.K. (1996) Structure and function of female and bisexual organ complexes in Gnetales. *International Journal of Plant Sciences*, **157**, S113–S125.
Feild, T.S. & Boddribb, T.J. (2005) A unique mode of parasitism in the conifer coral tree *Parasitaxus ustus* (Podocarpaceae). *Plant, Cell and Environment*, **28**, 1316–1325.
Friedman, W.E. & Carmichael, J.S. (1996) Double fertilization in Gnetales: implications for understanding reproductive diversification among seed plants. *International Journal of Plant Sciences*, **157**, S77–S94.
Gernandt, D.S., López, G.G., García, S.O., *et al.* (2005) Phylogeny and classification of *Pinus*. *Taxon*, **54**, 29–42.
Gifford, E.M. & Foster, A.S. (1989) *Morphology and Evolution of Vascular Plants*. W. Freeman, New York.
von Guttenberg, H. (1961) *Grundziige der Histogenese hoherer Pflanzen. II. Die Gymnospermen*. Handbuch der Pflanzenanatomie. Band VIII, Teil 4. Gebrüder Borntraeger, Berlin.
Hara, N. (1997) Morphology and anatomy of vegetative organs in *Ginkgo biloba*. In: *Ginkgo biloba -'A Global Treasure*, (eds T. Hori, R.W. Ridge, W. Tulecke, P. Del Tredici, J. Tremouillaux-Guiller & H. Tobe), pp. 119–126, Springer-Verlag, Tokyo.
Hill, K.D. (1999) *Cycas*—an evolutionary perspective. In: *Biology and Conservation of Cycads. Proceedings of the Fourth International Conference on Cycad Biology, 1–5 May 1996, China*, (ed. C.J. Chen), pp. 98–115, International Academic Publishers, Beijing.
Holloway, J.T. (1937) Ovule anatomy and development and embryogeny in *Phyllocladus alpinus* (Hook.) and in *P. glaucus* (Carr). *Transactions of the Royal Society of New Zealand*, **67**, 149–165.
Jeffery, E.C. & Torrey, R.E. (1916) Ginkgo and the microsporangial mechanisms of the seed plants. *Botanical Gazette*, **62**, 281–292.
Jernstedt, J.A. (1984a) Seedling growth and root contraction in the soap plant, *Chlorogalum pomeridianum* (Liliaceae). *American Journal of Botany*, **71**, 69–75.
Jernstedt, J.A. (1984b) Root contraction in hyacinth. I. Effects of IAA on differential cell expansion. *American Journal of Botany*, **71**, 1080–1089.
Johnson, M.A. (1943) Foliar development in *Zamia*. *American Journal of Botany*, **30**, 366–378.
Lee, E.K, Cibrian-Jaramillo, A., Kolokotronis, S.-O., *et al.* (2011) A functional phylogenomic view of the seed plants. *PLoS Genetics* **7** (12), e1002411. DOI: 10.1371/journal.pgen.1002411.
Martens, P. (1971) *Les Gnetophytes. Handbuch der Pflanzenanatomie, Band 12, Teil 2*. Gebrüder Borntraeger, Berlin.
Martens, P. & Waterkeyn, L. (1963) The shoot apical meristem of *Welwitschia mirabilis* Hooker. *Phytomorphology*, **13**, 359–363.
Milindasuta, B.-E. (1975) Developmental anatomy of coralloid roots in cycads. *American Journal of Botany*, **62**, 468–472.
Muhammad, A. & Sattler, R. (1982) Vessel structure of *Gnetum* and the origin of angiosperms. *American Journal of Botany*, **69**, 1004–1021.

Mundry, I. (2000). Morphologische und morphogenetische Untersuchungen zur Evolution der Gymnospermen. *Bibliotheca Botanica*, **152**, 1–90.
Napp-Zinn, K. (1966) *Anatomie des Blattes. II. Blattanatomie der Gymnospermen*. Handbuch der Pflanzenanatomie, Band 8, Teil 1. Gebrüder Borntraeger, Berlin.
Norstog, K.J., Gifford, E.M., & Stevenson, D.W. (2004) Comparative development of the spermatozoids of cycads and *Ginkgo biloba*. *Botanical Review*, **70**, 5–15.
Norstog, K.J. & Nicholls, T.J. (1997) *Biology of the Cycads*. Cornell University Press, Ithaca, New York.
Popham, R.A. (1966) *Laboratory Manual for Plant Anatomy*. C.V. Mosby, St. Louis.
Rai, A.N., Soderback, E., & Bergman, B. (2000) Cyanobacterium-plant symbiosis. *New Phytologist*, **147**, 449–481.
Rimbach, A. (1929) Die Verbreitung der Wurzelverkurzung im Pflanzenreich. *Berichte der Deutschen Botanischen Gesellschaft*, **47**, 22–31.
Robbertse, H.P.J., Grobbelaar, N. & duToit, E. (2011) Origin of the coleorhiza in cycad seedlings and its structural homology with that of the Poaceae. *Botanical Review*, **77**, 1–10.
Rodin, R.J. (1967) Ontogeny of foliage leaves in *Gnetum*. *Phytomorphology*, **17**, 118–128.
Rudall, P.J., Hilton, J., Vergara-Silva, F., *et al.* (2011) Recurrent abnormalities in conifer cones and the evolutionary origins of flower-like structures. *Trends in Plant Science*, **16**, 151–159.
Ruzin, S.E. (1979) Root contraction in *Freezia* (Iridaceae). *American Journal of Botany*, **66**, 522–531.
Rydin, C., Khodabandeh, A. & Endress, P.K. (2010) The female reproductive unit of *Ephedra* (Gnetales): comparative morphology and evolutionary perspectives. *Botanical Journal of the Linnean Society*, **163**, 387–430.
Sacher, J.A. (1955) Cataphyll ontogeny in *Pinus lambertiana*. *American Journal of Botany*, **42**, 82–91.
Schnarf, K. (1933) *Embryologie der Gymnospermen*. Borntraeger, Berlin.
Seelinger, I. (1954) Studien am Sprossvegetationskegel von *Ephedra fragilis var. campylopoda* (C. A. Mey.) Stapf. *Flora*, **141**, 114–162.
Singh, H. (1978) *Embryology of Gymnosperms*. Handbuch der Pflanzenanatomie, Band X, Teil 2. Gebrüder Borntraeger, Berlin.
Stevenson, D. (1980) Observations on stem and root contraction in cycads (Cycadales) with special reference to *Zamia pumila* L. *Botanical Journal of the Linnean Society*, **81**, 275–281.
Stopes, M. (1910) Adventitious budding and branching in *Cycas*. *New Phytologist*, **9**, 234–241.
Tomlinson, P.B., Takaso, T. & Rattenbury, J.A. (1989) Developmental shoot morphology in *Phyllocladus* (Podocarpaceae). *Botanical Journal of the Linnean Society*, **99**, 223–248.
Webb, D.T. (1983) Developmental anatomy of light-induced nodulation by *Zamia pumilia* L. seedlings in sterile culture. *American Journal of Botany*, **70**, 1109–1117.
Woltz, P., Stockey, R.A., Gondran, M. *et al.* (1994) Interspecific parasitism in the gymnosperms: unpublished data on two endemic New Caledonian Podocarpaceae using scanning electron microscopy. *Acta Botanica Gallica*, **141**, 731–746.

Annual Plant Reviews (2013) **45**, 163–188
doi: 10.1002/9781118305881.ch6 http://onlinelibrary.wiley.com

Chapter 6

IDENTIFYING KEY FEATURES IN THE ORIGIN AND EARLY DIVERSIFICATION OF ANGIOSPERMS

Paula J. Rudall

Royal Botanic Gardens, Kew, Richmond, Surrey, UK

Abstract: Understanding the origin and early diversification of the flowering plants (angiosperms) represent primary goals in biology. However, despite extensive knowledge of angiosperm biology and biodiversity, much remains to be determined about their evolutionary origin, partly because many of their characteristic (some supposedly diagnostic) features probably evolved iteratively among seed plants. This chapter focuses on the early evolutionary history of the "key" angiosperm reproductive structures: carpels, flowers, and inflorescences. It discusses some of the major morphogenetic markers that circumscribe deep nodes within the angiosperms, notably the base of the eudicot clade. Despite recent advances in comparative genomics and developmental genetics of an increasing range of model organisms, satisfactory resolution of longstanding and fundamental questions regarding angiosperm origins remain hampered by the problematic phylogenetic framework for seed plants, which is notable for strongly contrasting hypotheses based on molecular versus morphological data. Ongoing integrated comparative studies can potentially clarify floral evolution within angiosperms.

Keywords: angiosperm; canalization; carpel; development; evolution; flower; inflorescence.

6.1 Introduction: key features of flowering plants

Today, the flowering plants (angiosperms) represent by far the most species-rich group of land plants, dominating floras and habitats throughout most of the world. They are the primary food source for most land animals,

Annual Plant Reviews Volume 45: The Evolution of Plant Form, First Edition.
Edited by Barbara A. Ambrose and Michael Purugganan.

including humans. Several potentially key features are widely hypothesized to characterize the angiosperms, including flowers, carpels, columellate pollen, non-motile sperm, elongated pollen tubes with callose plugs, double fertilization leading to endosperm production, highly reduced megagametophytes and microgametophytes, an outer integument enclosing the ovule, paracytic stomata, and vessels in the secondary xylem (e.g., Doyle & Donoghue 1986; Williams & Friedman 2004; Rudall 2006; Rudall & Bateman 2007; Williams 2008; Doyle 2009). However, despite extensive knowledge of angiosperm biology and biodiversity, we remain unclear about precisely how to circumscribe the angiosperms (and hence even more perplexed about non-angiospermous seed plants, the gymnosperms), because many of these important angiospermous features are either difficult to interpret, absent from some early-divergent angiosperms, or probably evolved more than once among seed plants.

For example, pollen wall ultrastructure is widely used to distinguish angiosperm pollen from that of gymnosperms. Angiospermous exine typically possesses chambers between isolated islands of sporopollenin (columellae). When the pollen is released from the anther, the exine chambers are filled with proteins, glycoproteins, carbohydrates, and lipids that not only ensure pollen adhesion to the stigmatic surface but also allow recognition of compatible pollen by the receiving stigma (Knox 1984). Sporopollenin is highly resistant to degradation and the chambered exine layer is often well preserved in fossils, so pollen identification is critical in stratigraphic studies designed to date fossil sediments. Yet even this feature is not entirely reliable for angiosperm identification, because many columellate grains found in pre-Cretaceous (Triassic and Jurassic) deposits are probably not assignable to angiosperms. Furthermore, some early-divergent angiosperms (e.g., *Amborella* and some Nymphaeales) lack distinct columellae and possess a more granular tectum layer (Doyle 2005, 2009; Doyle & Endress 2010).

Two iconic angiosperm reproductive structures—flowers and carpels—represent the primary foci of this chapter. Both are widely perceived as key features of angiosperms, in contrast with gymnosperms, which are conventionally viewed as lacking flowers and bearing ovules/seeds that are incompletely enclosed. However, even these archetypal defining features are problematic because "flower" and "carpel" are interpretative terms (*sensu* Sattler 1975) rather than purely descriptive terms that would allow direct comparisons with other seed plants. Although they serve a useful purpose in helping us to describe species within the angiosperm clade, they encompass a wide range of structures, and hence effectively mask some important questions of homology between angiosperms and gymnosperms.

6.2 Patterning of flowers and inflorescences

Perhaps the most iconic feature of angiosperms is their typically colorful and ephemeral reproductive structure, the flower, which is intimately associated with insect-mediated microspore dispersal (pollination). The structure of the flower is governed by strong constraints that intuitively appear to follow geometric algorithms. In the majority of angiosperms, the flowers are organized into inflorescences, which themselves have a repetitive (modular) structure, involving iterative architectural branching patterns (Prenner *et al.* 2009; Endress 2010b). Two features that are fundamental to flowers are the restricted (determinate) growth of the primary floral axis and the spatial arrangement of their constituent organs in concentric zones, specifically in a sepal–petal–stamen–carpel sequence with the carpel zone in the center. Although some flowers lack one or more of the four floral organ types, the spatial arrangement of organs with respect to each other within a single flower is generally highly conserved. By contrast, the precise temporal sequence of organ initiation can be more variable; organ initiation is centripetal (acropetal) in the majority of angiosperms, but can be centrifugal (or basipetal) in a significant number of species, in which at least some organs develop in a distal-to-proximal sequence (Rudall 2010). Only two distantly related extant genera, the monocot *Lacandonia* and the early-divergent angiosperm *Trithuria*, are unique among extant angiosperms in possessing inside-out reproductive units in which carpels surround the stamens. These inside-out reproductive units represent putative flowers, though in both taxa they have been alternatively interpreted as inflorescences (e.g., Ambrose *et al.* 2006; Rudall & Bateman 2006; Rudall *et al.* 2009; Alvarez-Buylla *et al.* 2010).

Studies of laboratory-induced homeotic mutants in model angiosperm species—such as the eudicots *Arabidopsis* and *Antirrhinum*—have shown that at least three classes of MADS-box genes (termed A, B, and C, respectively) function in overlapping domains of the floral apex to determine floral organ identity (Coen & Meyerowitz 1991). These observations prompted the development of the influential ABC model (Figure 6.1) and subsequent variants such as the ABCE models and (A)BC models (Theissen *et al.* 2000;

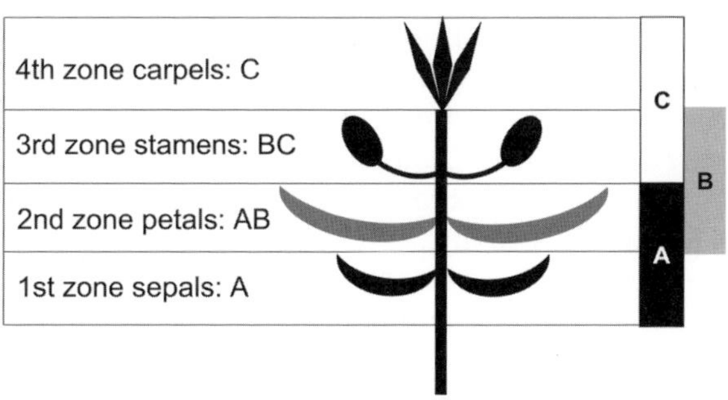

Figure 6.1 Diagram of ABC model of Coen and Meyerowitz (1991), in which three classes of MADS-box genes interact to determine the fates of organ primordia in *Arabidopsis*. A-function specifies sepal identity, a combination of A- and B-function specifies petal identity, a combination of B- and C-function specifies stamen identity, and C-function specifies carpel identity. The C-class gene *AGAMOUS* also ensures flower determinacy.

Figure 6.2 SEM of wild-type indeterminate inflorescence of *Arabidopsis thaliana*, the archetypal angiosperm model organism, showing two open flowers and several unopened flower buds at different stages of development.

Causier *et al.* 2010). In the archetypal angiosperm model organism *Arabidopsis* (Figure 6.2), the C-function MADS-box gene *AGAMOUS* (*AG*) specifies carpel identity, and a combination of C-function and B-function genes controls stamen identity. C-gene function also specifies determinacy of the apex. In extant gymnosperms, orthologs of angiosperm C-class genes are expressed in both male and female organs, and might therefore specify reproductive function, whereas B-function genes are expressed only in male structures (Theissen & Becker 2004; Zhang *et al.* 2004; Scutt *et al.* 2006). Another MADS-box gene that is implicated in floral patterning in *Arabidopsis* is the E-function gene *SEPALLATA3* (*SEP3*), which is required for the development of petals, stamens, and carpels (Castillejo *et al.* 2005). E-function is apparently highly conserved in angiosperms (e.g., Agrawal *et al.* 2005); the *SEP* gene family has undergone a series of duplications, of which the first occurred prior to the origin of extant angiosperms (Zahn *et al.* 2005).

The ABC model and its derivatives explain the genetic basis for floral organ identity and some aspects of floral architecture in model taxa, but the genetic bases for the precise radial arrangement of floral organs on the apex are less well understood. Most recent models favor a prepatterning interpretation, in which positional signals for all four organ zones are predetermined before actual organ formation (i.e., before activation of B-function and C-function genes on the floral apex). Radial floral patterning is controlled by a combination of *LEAFY* and its co-regulator *UNUSUAL FLORAL ORGANS*

(*UFO*) in a positive feedback loop (Parcy *et al.* 2002; Dong *et al.* 2005; Souer *et al.* 2008). Localized concentrations of auxin represent another significant factor in floral construction, influencing establishment of both organ patterning on the floral apex and the vasculature that will ultimately supply the organs (e.g., Cheng *et al.* 2006; Jönsson *et al.* 2006; Nuraliev *et al.* 2011). High auxin levels can promote carpel initiation and retard growth of stamen primordia (Steeves & Sussex 1989). In developing *Arabidopsis* flowers, stamens and petals are initiated at almost the same stage, but the petal primordia remain retarded relative to the stamens until Stage 9 of flower development, probably due to high auxin levels in the stamens (Aloni *et al.* 2006).

Inflorescences display a diverse range of architectural patterns across the angiosperms, though most observers recognize a fundamental distinction between determinate inflorescences, in which the primary axis is terminated in a flower, and indeterminate inflorescences, in which there is no terminal flower (Weberling 1992), as in *Arabidopsis* (Figure 6.2). Studies of the genetic control of inflorescence development have focused on *LEAFY*, *APETALA1* (*AP1*)—the primary promoters of floral meristem identity in *Arabidopsis*—and *TERMINAL FLOWER1* (*TFL1*), which acts to repress the floral meristem identity genes and specifies shoot identity (Parcy *et al.* 2002; Prusinkiewicz *et al.* 2007). The molecular genetics of inflorescence architecture has been studied most extensively in the grasses, particularly maize (e.g., Malcomber *et al.* 2006; McSteen 2006).

6.3 Eight extant lineages of flowering plants

Phylogeneticists presently assign all extant species of flowering plants to eight lineages (Figure 6.3): Amborellales, Austrobaileyales, Ceratophyllales, Chloranthales, eudicots, magnoliids, monocots, and Nymphaeales (APG III 2009). Of these lineages, two—eudicots (tricolpates) and monocots—contain the vast majority (>95%) of extant species. The precise relationships between these eight extant angiosperm lineages remain unresolved, particularly with respect to the "wildcard" taxa Ceratophyllaceae and Chloranthaceae, which are problematic to place in both morphological and molecular analyses (e.g., Doyle & Endress 2000; Moore *et al.* 2007). Cantino *et al.* (2007) assigned five lineages—Ceratophyllaceae, Chloranthaceae, eudicots, magnoliids, and monocots—to their "Mesangiospermae" (or core angiosperm) clade. The other three comparatively species-poor lineages (Amborellales, Nymphaeales, Austrobaileyales) are sometimes collectively termed the ANA-grade (or ANITA-grade). They form a stepwise grade that is placed immediately above the root node of the angiosperms in recent molecular analyses (e.g., Graham & Iles 2009). *Amborella trichopoda*, an endemic shrub from New Caledonia, is the sole extant species of Amborellales; this taxon is usually resolved as sister to all other extant angiosperms, either alone or combined with Nymphaeales. The water-lily clade, Nymphaeales, includes

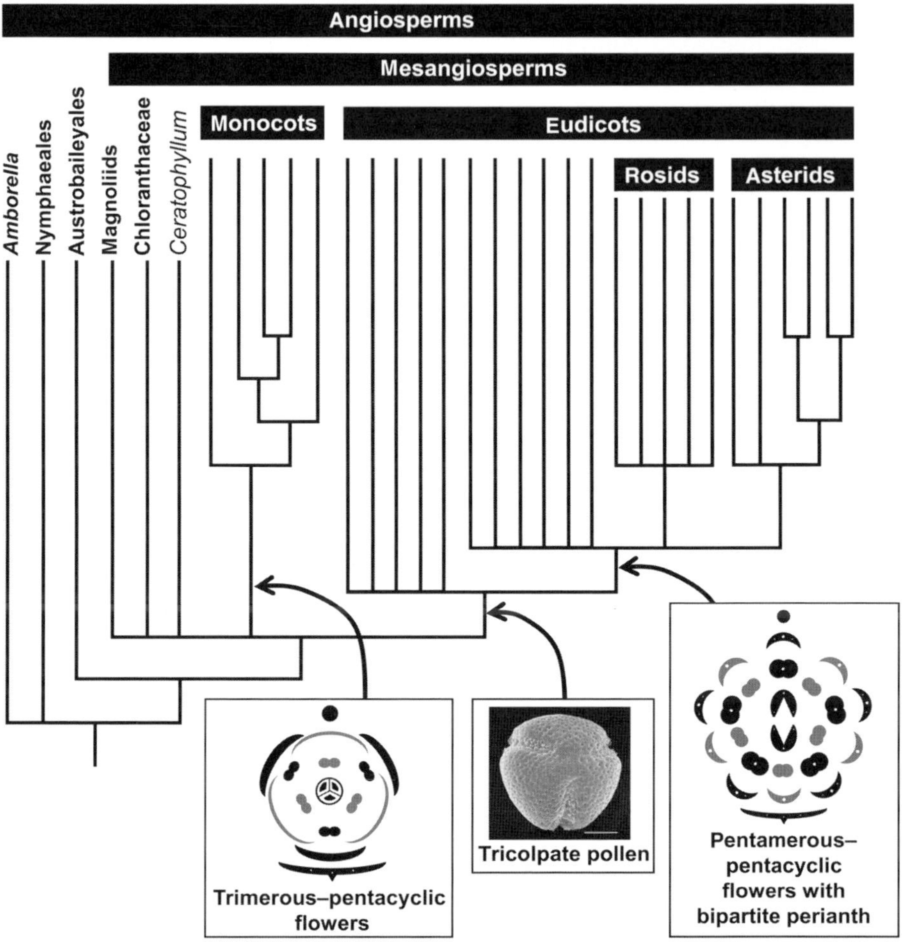

Figure 6.3 Diagram of phylogenetic relationships of the eight major angiosperm clades (Amborellales, Nymphaeales, Austrobaileyales, Chloranthales, Ceratophyllales, magnoliids, monocots, eudicots), based on Cantino *et al.* (2007) and APG III (2009). Superimposed are illustrations of deep-node transitions to trimerous–pentacyclic flowers in monocots, tricolpate pollen in eudicots, and trimerous–pentacyclic flowers in core eudicots.

three families of herbaceous aquatic plants: Cabombaceae, Hydatellaceae, and Nymphaeaceae, which together constitute approximately 90 species assigned to eight or nine genera. The final ANA-grade lineage, Austrobaileyales, consists of three families of woody plants: Austrobaileyaceae, Trimeniaceae, and Schisandraceae, which include approximately 70 species in five genera.

Molecular phylogenetic hypotheses of relationships within the eudicot clade have demonstrated a grade of early-divergent eudicots, including

Ranunculales and Gunnerales, and a relatively well-resolved crown group (Pentapetalae *sensu* Cantino *et al.* 2007) that encompasses two major lineages—the asterids and rosids—and some relatively isolated orders such as Caryophyllales (e.g., Soltis *et al.* 2003). Within the monocots, 11 or 12 orders are widely recognized. The monocot orders comprise a grade of two early-divergent lineages: Acorales (*Acorus*) and Alismatales (including Araceae), a grade of five "lilioid" orders: Petrosaviales, Dioscoreales, Pandanales, Liliales, and Asparagales (including Orchidaceae), and a clade of five commelinid orders: Arecales (palms), Commelinales, Dasypogonales, Zingiberales, and the largest order, Poales, which includes the grasses (e.g., Graham *et al.* 2006; reviewed by Remizowa *et al.* 2010).

6.4 Origin of the angiosperms: the phylogenetic framework

On July 22, 1879, Charles Darwin famously stated in a letter to Joseph Dalton Hooker that "the rapid development, so far as we can judge, of all the higher plants within recent geological times is an abominable mystery." Darwin was referring to the apparently abrupt evolutionary origin and subsequent rapid diversification of the angiosperms (Darwin & Seward 1903; Friedman 2009), topics that have remained among the most popular, yet still largely unresolved, discussion themes in evolutionary biology. Molecular analyses strongly support the monophyly of extant angiosperms (e.g., Qiu *et al.* 2000; Graham & Iles 2009), which probably had a single evolutionary origin from an unknown ancestral stem-group taxon (Hemiangiospermae *sensu* Arber & Parkin 1907) during the late Mesozoic era (Bell *et al.* 2010).

Regarding other major seed-plant lineages, an arguable total of 14 gymnosperm lineages are currently recognized in the fossil record (Bateman *et al.* 2006). Many are of more ancient origin than angiosperms and the majority of them are extinct, making assessment of their phylogenetic relationships far more problematic. Only four categories of gymnosperm are represented in extant vegetation: conifers, cycads, ginkgos, and gnetaleans. Of these, Gnetales ostensibly most closely resemble the angiosperms in several highly significant morphological characters, including double fertilization, reduced megagametophytes and microgametophytes, an outer integument on the ovule, vessels in the secondary xylem, and the possession of flower-like structures that are largely insect-pollinated (Doyle 1996; Labandeira 2010). Such similarities have ensured placement of Gnetales as a close relative of angiosperms in most cladistic analyses of seed plants based entirely on morphological data (Crane 1985; Doyle & Donoghue 1986; Nixon *et al.* 1994; Doyle 1996; Hilton & Bateman 2006). The topology in which angiosperms, Gnetales and some extinct groups (e.g., Bennettitales and Caytoniales) together form a clade is termed the anthophyte hypothesis, implying a close relationship between these disparate flower-bearing groups. In contrast, most molecular analyses of seed plants—which of necessity exclude all taxa known only from fossils—result in placement of Gnetales in a much more distant relationship with angiosperms, either as sister to (or embedded within) the conifer clade (respectively termed the Gnetifer and Gnepine hypotheses), sister to all other extant seed plants, or sister to all extant gymnosperms (e.g., Burleigh & Mathews 2004; Graham & Iles 2009; Mathews 2009).

In all of the contrasting hypotheses of their relationships, angiosperms occupy a long branch relative to other extant seed plants. Relaxed molecular clock analyses have obtained an estimated age of the angiosperms of 167–199 million years (Bell *et al.* 2010), suggesting that they originated sometime during the Jurassic period and diversified into the existing angiosperm orders during the subsequent Cretaceous period. Yet, if extant gymnosperms represent a single lineage that is sister to angiosperms (as indicated by most—though not all—molecular analyses), the two lineages must have diverged from each other much earlier, perhaps some 385 million years ago during the mid-Paleozoic (De Bodt *et al.* 2005; Zimmer *et al.* 2007). This vast disparity is sometimes attributed to mass extinction of numerous intervening lineages, or possibly to greatly elevated substitution rates along the branch leading to the angiosperms (Graham & Iles 2009). It also suggests that issues of seed-plant relationships and early angiosperm evolution cannot be satisfactorily resolved using molecular data alone (Bateman *et al.* 2006; Mathews 2009). Thus, it is highly likely that the closest relatives of angiosperms are groups that are known only from fossils. Analyses of morphological data consistently place certain extinct lineages along the branch leading to the angiosperms, notably Bennettitales, Caytoniales, Glossopteridales, and Pentoxylales (e.g., Crane 1985; Doyle & Donoghue 1986; Nixon *et al.* 1994; Doyle 1996; Hilton & Bateman 2006), even when the analysis is artificially constrained to make extant gymnosperms monophyletic (Doyle 2006). In particular, the extinct order Bennettitales, which was a dominant feature of Mesozoic vegetation, has been cited since the earliest days of paleobotanical research as potentially containing the angiosperm ancestor. This putative relationship was based partly on the highly flower-like reproductive structures of the Mesozoic bennettite *Williamsoniella* (for which Arber & Parkin (1907) coined the term "pro-anthostrobilus," compared with "anthostrobilus" for the modern angiosperm flower) and partly for the unusual paracytic stomata of bennettites, which resemble those of some extant angiosperms (Doyle & Donoghue 1986).

6.5 Resolving conflicting hypotheses of flower origin

Despite our considerable knowledge of comparative floral morphology and increasing understanding of floral developmental genetics, at least in model species, the evolutionary origin of flowers remains poorly understood. Flower-like structures are not restricted to angiosperms, and iterative

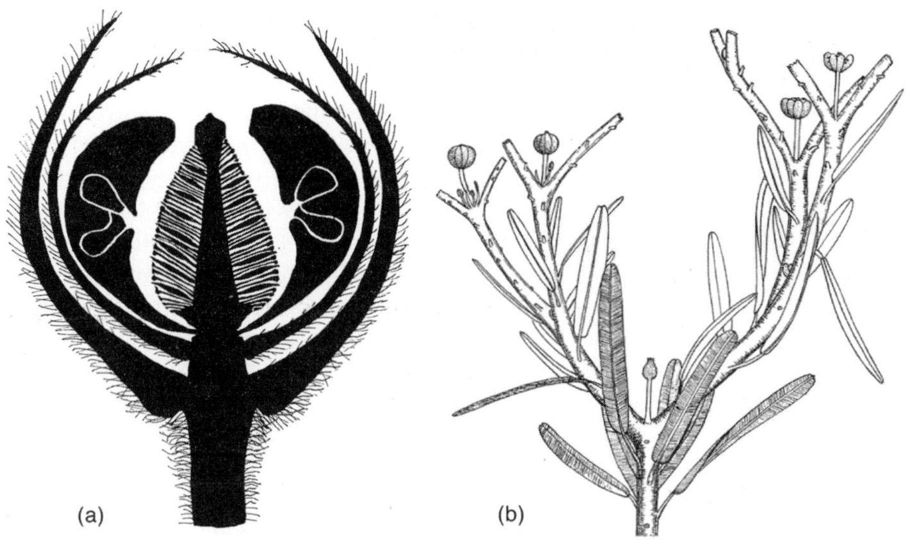

Figure 6.4 (a) Diagrammatic reconstruction (from Harris 1944) of longitudinal section of the flower of *Williamsoniella*, an extinct bennettitalean gymnosperm (see text regarding probable iterative evolution of flowers). (b) *Williamsoniella coronata*, reconstruction of determinate inflorescence with flowers probably arranged in cymose clusters; upper leaves removed (from Thomas 1916).

origins of flowers among seed plants appear highly likely. The extinct bennettite *Williamsoniella coronata* (Figure 6.4a) famously possessed flowers that were bisporangiate (bisexual): a perianth of helically arranged bracts surrounded a whorl of laterally flattened microsporophylls and a terminal cluster of megasporangia, each bearing numerous ovules (Thomas 1916; Harris 1944; Crane 1985). These flowers were probably arranged in inflorescences, though the reconstruction by Thomas (1916) appears equivocal as to whether the flowers were solitary or arranged in cymose clusters (Figure 6.4b). A morphological cladistic analysis found *Williamsoniella* to be highly derived within the bennettite clade (Crane 1985). Moreover, most other angiosperm stem-group candidates lack such readily identifiable flowers, including *Caytonia* and some of the putatively early-divergent bennettite lineages, suggesting that the flowers of *Williamsoniella* were derived independently of those of angiosperms (Bateman *et al.* 2006, 2011).

The longstanding debate as to whether the flower is derived from a simple or compound strobilus has been much discussed (reviewed by Doyle 1994; Bateman *et al.* 2006; Doyle 2008; Rudall *et al.* 2009). This debate is important in constructing models of floral evolution; despite the morphological convergence between flower-like structures derived from simple versus compound strobili, we might expect little commonality of patterning genes if they had evolved independently. The simple-strobilus model (Arber & Parkin 1907) interprets the flower as a single axis, typically bearing both megasporophylls (carpels) and microsporophylls (stamens). This model is termed "euanthial" or "uniaxial" to distinguish it from alternative "pseudanthial" or "polyaxial" models that perceive the flower as being derived from a condensed, multiaxial structure, and each organ as a condensed axis (e.g., Wettstein 1907; Melville 1960; Eames 1961; Meeuse 1972). The simple-strobilus model readily explains the majority of angiosperm flowers, and is widely adopted, mainly because most researchers consider all floral organs to be phyllome (leaf) homologs (but see Floyd & Bowman 2010, for a comparison of the developmental genetics of shoots and leaves in seed plants). On the other hand, the "flowers" of extant Gnetales are polyaxial, probably derived from reduced compound cones (e.g., Mundry & Stützel 2004). Even within angiosperms, polyaxial flowers could have evolved *de novo* in certain lineages. For example, based on analysis of comparative morphogenetic data in a relatively robust phylogenetic context, the supposed flowers of some early-divergent monocots (e.g., Cymodoceaceae and Zannichelliaceae) probably originated by secondary amalgamation of two or three *bona fide* flowers, resulting in a remarkably well-integrated structure that is not readily distinguishable from a true flower, even during development (Sokoloff *et al.* 2006).

The most noteworthy early Cretaceous macrofossils widely attributed to angiosperms encompass several fossil flowers, including the putative magnoliid *Archaeanthus* (Dilcher & Crane 1984) and some fossil water lilies (Friis *et al.* 2001). Chloranthaceae are also well represented in early Cretaceous sediments that contain fossil angiosperm flowers and fruits (Friis *et al.* 2001; Doyle & Endress 2010). One discovery that fuelled debate about the early evolution of flowers was the Chinese early Cretaceous fossil genus *Archaefructus* (Sun *et al.* 2002). *Archaefructus* was initially described as a close angiosperm relative (a stem-group angiosperm: Sun *et al.* 2002) but was subsequently placed among early-divergent angiosperms (Friis *et al.* 2003), most recently as sister to Hydatellaceae (Doyle 2008). These different phylogenetic placements of *Archaefructus* resulted partly from missing data (the genus is known only from compression fossils) and partly from contrasting interpretations of its reproductive unit as either uniaxial or polyaxial. Thus, the reproductive unit of *Archaefructus* has been interpreted either as a single attenuated perianthless bisexual flower, with carpels borne distal to paired or branched stamens on an elongated axis (Sun *et al.* 2002), or as a bractless racemose inflorescence consisting of perianthless male and female flowers, each highly reduced to only one or two organs (Friis *et al.* 2003). The relatively recent discovery of a "new" early-divergent angiosperm family, Hydatellaceae (Saarela *et al.* 2007), in which the inside-out reproductive units show features of both flowers or inflorescences, has raised the possibility of iterative hierarchical shifts between flowers and inflorescences during early angiosperm evolution (Rudall *et al.* 2007, 2008, 2009; Sokoloff *et al.* 2008).

Similarly, little is known about the evolutionary origin of the inflorescence; complex and iterative patterns of structural evolution within multiple angiosperm lineages have effectively masked this question (reviewed by Rudall & Bateman 2010). Assuming that the ancestral flower was a condensed bisexual simple strobilus, it is by definition a terminal structure, so the hypothesis of Parkin (1914) that the primitive angiosperm inflorescence was determinate remains widely supported. However, Endress and Doyle (2009) regarded the primitive angiosperm inflorescence type as equivocal (either determinate or indeterminate), based on phylogenetic optimizations of early-divergent extant lineages, because they interpreted the water-lily inflorescence in Nymphaeales as racemose (i.e., indeterminate) and the reproductive unit of *Archaefructus* as a racemose inflorescence bearing several highly reduced flowers.

Few authors have hitherto attempted to resolve the uniaxial versus polyaxial debate using evolutionary-developmental genetics, which of necessity requires use of living material. Even among extant angiosperms, distinguishing between polyaxial and uniaxial systems can be problematic, especially in species that lack the defining boundary of a perianth and/or other useful architectural markers such as bracts and prophylls. Reproductive structures of some early angiosperm lineages could have combined (or not yet adequately separated) features of flowers and inflorescences, allowing evolutionary experimentation that could have led to bidirectional homoplastic origins of flowers and inflorescences before "final" establishment of the flower (Rudall *et al.* 2009). Some comparative studies have targeted groups within angiosperms in which the inflorescence–flower boundary is obscure, as the eudicot family Euphorbiaceae (Prenner & Rudall 2007; Prenner *et al.* 2011) and the early-divergent angiosperm family Hydatellaceae (Rudall *et al.* 2009). Others have focused on reproductive units that show inside-out patterning with respect to a typical flower, as in Hydatellaceae (Rudall *et al.* 2009) and the monocot order Pandanales, most notably *Lacandonia* (Rudall & Bateman 2006; Alvarez-Buylla *et al.* 2010).

Another method of resolving the "uniaxial versus polyaxial" debate is to compare angiosperm flowers with cones of extant gymnosperms directly (e.g., Melzer *et al.* 2010). This approach is clearly problematic, given the vast phylogenetic gap between these putative lineages, the doubtful monophyly of extant gymnosperms, and the probable non-homology of at least some cones and flowers. On the other hand, many morphologists regard the conifer seed cone as polyaxial and the conifer pollen cone as uniaxial (the dimorphic cone hypothesis: Rudall *et al.* 2011), implying that the genetic controls of such condensed and highly modified structures could overlap. Theissen and Becker (2004) proposed that hermaphrodite flowers could have evolved either from a gymnosperm pollen cone, by reduction of B-gene expression in the distal region of the cone (the out-of-male hypothesis), or from a seed cone, by ectopic expression of B-genes in the proximal region (the out-of-female hypothesis). Based on comparative studies of frequent spontaneous mutant conifer cones, the out-of-female hypothesis appears more plausible, because most spontaneous bisexual cones appear to be modified seed cones rather than modified pollen cones (Bateman *et al.* 2011; Rudall *et al.* 2011).

6.6 Evolution of the perianth

The perianth represents the outermost zone of the flower, characterized by sterile, dorsiventrally flattened organs that surround the fertile floral organs. Typically, the perianth protects the immature fertile organs before anthesis, and after anthesis it helps to attract pollinators and direct them toward the reproductive structures. The ancestral perianth is putatively derived from sterile bracts surrounding the fertile organs. Among early-divergent angiosperms, some taxa (e.g., Chloranthaceae and Piperales) lack a perianth entirely. Even in cases where a perianth is present, precise delimitation between bracts and tepals is sometimes problematic (e.g., Buzgo *et al.* 2004; Kim *et al.* 2005). Furthermore, the perianth organs (tepals) of many early-divergent angiosperms and monocots show relatively little morphological distinction across the perianth zone, and in some water-lilies, the distinction between green and colorful perianth regions is partly environmentally controlled (Warner *et al.* 2009).

On the other hand, evidence from both morphology and developmental genetics indicates multiple origins of petals across the angiosperms, either from bracts (in which case they are termed bracteopetals) or from stamens (in which case they are termed andropetals). Expression of B-gene homologs is essential to petal identity in many angiosperms (e.g., Kramer *et al.*, 2007). In eudicots, the perianth is often differentiated into an outer whorl of green sepals (the calyx) and an inner whorl of colorful petals (the corolla). Baum and Hileman (2006) postulated that the corolla evolved by sterilization of the outer stamens, specifically by co-option of *WUSCHEL* as a coregulator of C-class genes, causing a shift in B-gene expression. Evolution of a bipartite perianth probably occurred at the base of the core eudicot clade, but was lost early in the evolution of the core eudicot clade Caryophyllales, in which secondary perianth differentiation into sepals and petals has occurred multiple times independently (Ronse de Craene 2007).

6.7 Carpels, gynoecia, and organ fusion

The name "angiosperm" is coined from the Ancient Greek words *angeíon* (receptacle, vessel) and *sperm* (seed), selected to indicate that the ovules are wholly enclosed within another organ (the carpel). This condition contrasts with that of gymnosperms, in which the ovules/seeds are not entirely enclosed within another structure and are usually described as "naked," though in truth the ovules of most gymnospermous groups are well protected. In contrast with the flower, the carpel is widely considered to be a structure that is unique to angiosperms. Thus, at least ostensibly, the carpel could potentially represent the most reliable defining feature of angiosperms, and is sometimes cited as a key innovation (Endress 2001a). However, this interpretation is problematic because the evolutionary origin of the carpel remains enigmatic; hitherto, it has proved impossible to conclusively determine its homology with any gymnosperm reproductive structure. The carpel has been variously considered either as directly homologous with a megasporophyll of other seed plants (with fused lateral margins) or as representing a more complex structure derived by fusion between an ovuliferous shoot and its subtending phyllome, as suggested by comparison with putative angiosperm relatives among seed ferns, such as *Caytonia* (Doyle 2008). Some models of floral evolution that seek to explain the evolutionary origin of bisexual flowers from unisexual cones have invoked homeosis for the origin of the carpel. For example, the Mostly Male Theory (Frohlich & Parker 2000) proposed that the carpel was derived by ectopic expression of ovules onto phyllomic structures derived from microsporophylls, which ultimately enclosed the ovules.

Regardless of whether it represents a megasporophyll or a more complex structure, the evolutionary origin of the carpel is linked with the process of carpel fusion, either within an individual organ (i.e., carpel closure) or between organs (e.g., ovuliferous shoot and subtending phyllome), or via an element of both. Endress and Igersheim (2000a) hypothesized that carpel margins were ancestrally "open" (i.e., their margins were unfused and the ovules were partly exposed, as in gymnosperms), noting that in many of the earliest extant lineages (including *Amborella*) the distal carpel margins are closed merely by mucilaginous secretory products rather than by actual fusion of tissues. In some species of Hydatellaceae, a short narrow distal canal remains open at anthesis (Rudall *et al.* 2007). In most other angiosperms with "free" (i.e., separate) carpels (termed apocarpous), the individual carpels are closed at anthesis, often by postgenital fusion of their margins; during earlier ontogenetic stages the developing ovules are exposed (though protected within the unopened bud). In syncarpous angiosperms, the carpels are fused together at anthesis to form a single complex structure (the syncarpous gynoecium), as in the majority of eudicots and monocots. In these cases, the individual carpels can either be initiated separately and later become fused postgenitally, or the entire gynoecium can arise congenitally as a single tubular unit, in which case the individual carpels are not themselves closed in all cases, as the ovules are enclosed within the entire gynoecium. Congenital carpel fusion represents by far the most common condition in eudicots, and postgenital fusion (or an element of both types) predominates in monocots (Remizowa *et al.* 2010). Endress (2011) emphasized the evolutionary importance of possessing a united pollen-tube-transmitting tract (compitum) that allows equal access of pollen tubes to all the carpels in a syncarpous gynoecium.

Some element of gynoecial diversity among extant angiosperms can be attributed to duplications in the ancestral C-function MADS-box lineage that is required for carpel development (Scutt *et al.* 2006; Zahn *et al.* 2006). In *Arabidopsis*, in addition to the C-function gene *AGAMOUS* (*AG*), other genes such as *CRABS CLAW* and *SPATULA* are also implicated in carpel development (Alvarez and Smyth 1999; Bowman & Smyth 1999). However, there exists considerable diversity of gynoecium morphology, some of which relates to organ fusion (Weberling 1992; Endress 2011). Carpels of both apocarpous and syncarpous taxa consist of a proximal ascidiate (tubular) region and a distal plicate (folded) region, and the degree of fusion between these respective regions—either congenital or postgenital—varies among different groups. The genetics of organ fusion incorporates a wide range of diverse factors. For example, the genes *LEUNIG* (*LUG*) and *FILAMENTOUS FLOWER* (*FIL*) are implicated in carpel fusion in *Arabidopsis*, influencing congenital and postgenital fusion respectively (Chen *et al.* 2001). Several other genes have also been implicated in fusion of various different organs in *Arabidopsis*, including *FIDDLEHEAD* (Pruitt *et al.* 2000, *HOTHEAD* (Krolikowski *et al.* 2003), and the *LATERAL ORGAN FUSION* genes *LOF1* and *LOF2* (Lee *et al.* 2009). The degrees of conservation of these genetic mechanisms across angiosperms remain to be determined.

Increased organ fusion has apparently promoted increased diversity in some angiosperm families. The highly species-rich monocot family Orchidaceae represents an outstanding example of complex organ fusion. Orchidaceae are characterized by congenital fusion between male and female floral structures, so that in both of the subfamilies Orchidoideae and Epidendroideae the single fertile anther and the sterile staminodes are borne on the same complex organ as the receptive stigma, a condition that is sometimes termed hyper-epigyny. Rudall and Bateman (2002) attributed at least the initial radiation in Orchidaceae to a combination of hyper-epigyny and floral zygomorphy, though subsequent radiations probably reflect more subtle innovations. On the other hand, Mondragón-Palomino and Theissen (2009) postulated that early duplications in *DEFICIENS*-like (DEF-like) MADS-box genes in Orchidaceae, followed by subfunctionalization and neofunctionalization, resulted in modularization of the orchid perianth that prompted subsequent diversification.

6.8 Origins of floral diversity: deep-node characters and genome duplications

Much of the striking diversity that exists among flowers of extant angiosperms relates to one or more of several dynamic factors governed by different gene complexes. In addition to differences in organ fusion (both postgenital and congenital), these factors encompass differences in floral ground plan, including increase or decrease in organ number, variation in patterns of floral symmetry, and incorporation of "novel" elements such as spurs and nectaries. A further potential factor in floral evolution is fasciation, which results either from incomplete division or partial fusion of neighboring axes and occurs when the shape and/or size of the apical meristem are radically altered, preferentially affecting certain families such as Araliaceae (e.g., Sokoloff *et al.* 2007).

Within the angiosperms, reproductive characters are of primary significance in taxonomic treatments, but relatively few of these characters precisely circumscribe deep nodes at the ordinal level or above. The majority of taxonomically useful characters are employed at the species and genus level, or at best the family level, because they are too homoplastic to be taxonomically applicable at deeper nodes. For example, species and genera are frequently delimited by more-or-less subtle differences in development that influence the degree of fusion between floral organs, both within and between floral whorls (especially petals and stamens). By contrast, a few exceptional characters define major clades within angiosperms. Examples of such "deep-node" characters include shifts in number of pollen apertures and number of ovule integuments, which occurred at the base of the eudicot clade and the asterid clades, respectively. A shift in pollen aperture position from polar to equatorial broadly defines the eudicot clade, coupled with an increase in aperture number from one to three, usually associated with simultaneous (as opposed to successive) microsporogenesis (Furness & Rudall 2004). Subsequent diversification in aperture patterns resulted in an increase in pollen aperture number in some groups, possibly reflecting adaptive evolution. On the female side, there was a major shift to unitegmy at the base of the asterid eudicot clade from the inferred ancestral bitegmic condition (e.g., Philipson 1974; Albach *et al.* 2001).

Whole-genome duplications played a key role in both the origin of the angiosperms and their early evolution during the Cretaceous period, when the major angiosperm lineages were established. These evolutionary events resulted in major changes in morphological complexity; they include doubling of regulatory factors, proliferation of transcription factor genes, and neofunctionalization of some of the "extra" genes (De Bodt *et al.* 2005; Bateman *et al.* 2006). Such duplication events—possibly associated with a single large-scale genomic duplication—took place before the divergence of extant angiosperms in at least three MADS-box gene lineages, representing B-, C-, and E-function, respectively (Zahn *et al.* 2005, 2006; Scutt *et al.* 2006). Similar major duplication events probably occurred at the base of the core eudicots, marked not only by characters such as triaperturate pollen and pentamery but also by possession of a bipartite perianth. The B-class *APETALA3* (*AP3*) lineage was duplicated to give rise to two distinct paralogs that have acquired subtly different functions in petal and stamen development (Kramer *et al.* 1998; Kramer & Hodges 2009). Thus, it appears likely that such major duplication events were responsible for (by definition) saltational evolution of "deep-node" characters in angiosperms.

6.9 Contrasting floral ground plans

The majority of eudicots and monocots possess pentacyclic flowers, characterized by four organ zones arranged in five distinct whorls. In monocots, the most widespread floral ground plan is trimerous–pentacyclic, consisting of five whorls, each composed of three floral organs: (1) six tepals in two alternating whorls (frequently all petaloid, rather than differentiated into petals and sepals), (2) six stamens in two alternating whorls, and (3) three carpels alternating with the inner stamen whorl (Remizowa *et al.* 2010; Ronse De Craene 2010). In contrast, the primary floral ground plan among core eudicots is pentamerous–pentacyclic, consisting of five whorls, each typically composed of five floral organs: ten perianth organs in two alternating whorls (usually differentiated into sepals and petals), ten stamens in two alternating whorls, and from two to five carpels in a single whorl (Soltis *et al.* 2003; Ronse De Craene 2010; Endress 2011). The flowers of the archetypal model angiosperm, *Arabidopsis thaliana* (a rosid eudicot), have a 4:4:2:4:2 arrangement, consisting of four sepals, four petals, six stamens in two whorls (two outer and four inner), and two carpels, the latter organs being fused together to form a central gynoecium that terminates the short floral axis (Hill & Lord 1989; Smyth *et al.* 1990; Endress 1992).

The apparently strong genetic constraints that govern the whorled bauplans of flowers of monocots and eudicots are not evident in the other six angiosperm lineages, which are less species-rich but exhibit considerable disparity in flower morphology, often lacking clear organ whorls and bearing increased or reduced organ numbers. In these lineages, the perianth is usually not differentiated into petals and sepals, or in some cases a perianth is entirely absent. All extant Chloranthaceae are perianthless except *Hedyosmum* (Chloranthaceae), which has three short, scale-like putative tepals (Endress 2001a; Li *et al.* 2005). The minute bisexual flowers of *Chloranthus* consist entirely of a single stamen and a single uniovulate carpel, contrasting strongly with (for example) the flowers of the water-lily genus *Nymphaea* (Nymphaeaceae), which possess numerous tepals, stamens, and multiovulate carpels. Flowers of the putative "basal" angiosperm *Amborella* are small and functionally unisexual, but female flowers possess staminodes surrounding the carpels (Endress & Igersheim 2000a, 2000b). Flowers of *Kadsura* (Schisandraceae) are unisexual and superficially cone-like, possessing a perianth that surrounds an elongated axis bearing numerous stamens or carpels. In the inside-out reproductive units (either flowers or inflorescences) of some species of *Trithuria* (Hydatellaceae), a perianth-like whorl surrounds uniovulate carpels and central stamens (Rudall *et al.* 2007). The aquatic genus *Ceratophyllum* bears minute unisexual flowers that are normally interpreted as lacking a perianth: pistillate (female) flowers consist of a single uniovulate carpel, whereas the staminate (male) reproductive structures contain 3–46 stamens that are initiated in a variable phyllotactic sequence (Iwamoto *et al.* 2003).

6.10 Iterative origins of floral symmetry patterns and floral novelties

Differences in floral symmetry patterns relate partly to differences in overall appearance, especially that of perianth whorls, and partly to the layout of individual organs in the flower (Endress 1999, 2001b; Rudall & Bateman 2004). Radial symmetry (actinomorphy, polysymmetry) is the most common floral condition, but flowers of many early-divergent angiosperms (e.g., *Amborella* and *Nymphaea*) are more correctly described as asymmetric (lacking any symmetry planes) because individual organs are spirally arranged. Flowers of many eudicots in which carpel number is reduced to two (including *Arabidopsis*) are strictly bilaterally symmetric (zygomorphic or monosymmetric), though their perianth—and hence their superficial appearance—is polysymmetric. Floral monosymmetry (zygomorphy) has evolved iteratively among both eudicots and monocots, occurring in several distantly related lineages. For example, strongly monosymmetric floral presentation characterizes two species-rich but distantly related lineages: the monocot family Orchidaceae and many members of the eudicot order Lamiales, including the model organism *Antirrhinum* (Rudall & Bateman 2004; Ronse De Craene 2010). Monosymmetric flowers frequently characterize insect-pollinated groups, but also occur in wind-pollinated groups such as grasses (Rudall & Bateman 2004).

When placed in a phylogenetic context, studies of homeotic mutants of *Antirrhinum* and other model taxa have demonstrated that such independent transitions to floral monosymmetry are each controlled by independent recruitment of homologs of the TCP-transcription factor *CYCLOIDEA* (*CYC*), at least among core eudicots (Preston & Hileman 2009; Citerne *et al.* 2010) and in the early-divergent eudicot order Ranunculales (Damerval & Nadot 2007; Damerval *et al.* 2007). Evidence for *CYC*-like control of floral monosymmetry is less clear in monocots. For example, a different genetic mechanism for floral zygomorphy has been postulated for orchids, involving differential expression of B-class (*DEF*-like) genes (Mondragón-Palomino & Theissen 2009). On the other hand, in rice, the *CYC* homolog *RETARDED PALEA1* (*REP1*) regulates palea structure during early flower development and is implicated in floral monosymmetry (Yuan *et al.* 2009). Thus, comparative studies suggest that *CYC* activity is probably highly conserved in angiosperms.

Structures that are supplementary to the usual complement of floral organs, such as nectaries, osmophores, and spurs, have evolved many times independently in angiosperms and display a wide range of morphologies (Endress & Matthews 2006; Bernardello 2007). They can be simple or complex, vascularized or nonvascularized, and in many cases they develop late in ontogeny. For example, the nectaries of wild-type *Arabidopsis* flowers are outgrowths located at the bases of the stamen filaments; they are largely nonvascularized (except for some phloem sieve elements at their bases) and are initiated relatively late in flower development (at Stage 9: Baum *et al.*

2001). *Arabidopsis* nectaries are established independent of ABC genes, and are partly regulated by the YABBY transcription factor *CRABS CLAW* (Baum *et al.* 2001), orthologs of which are now known to be associated with nectaries from diverse taxa across the core eudicots (Lee *et al.* 2005).

Floral spurs are tubular outgrowths of perianth organs (either petals or sepals) that frequently (though not invariably) secrete and/or collect nectar. Spurs have evolved *de novo* in several plant groups, notably not only in Orchidaceae and some Lamiales but also in other groups such as Berberidaceae (e.g., *Epimedium*) and Ranunculaceae (e.g., *Aquilegia* and *Delphinium*) (Endress & Matthews 2006). Although little is currently known about the genetics of spur development across the angiosperms, a possible clue comes from two independent spontaneous dominant mutations caused by transposon insertions near two KNOX transcription factors in *Antirrhinum majus*, a species that typically lacks spurs (Golz *et al.* 2002). The mutants possessed ectopic spur-like petal tubes similar to the floral spurs that occur naturally in other species of Antirrhineae, suggesting that KNOX gene expression could be responsible for the normal development of spurs.

The petaloid corona (sometimes termed a paracorolla) that occurs in some plant families (e.g., Amaryllidaceae, Gentianaceae, and Velloziaceae) probably functions as scent-producing organ, termed an osmophore. In most cases, the corona develops after the adjacent floral whorls, rather late in floral ontogeny. The homologies of the corona have been widely debated, especially in the daffodil family Amaryllidaceae, either as stamen-derived or petal-derived (e.g., Arber 1937); most current observers interpret it as a late elaboration of the zone between the petals and stamens (Endress & Matthews 2006; Sajo *et al.* 2010). The closely related family Alliaceae also possesses diverse petaloid appendages (putative osmophores); some are flattened and corona-like but others are filamentous (Rudall *et al.* 2002).

6.11 Constraints and canalization in floral evolution

Although many details remain to be determined regarding the evolutionary origin and subsequent diversification of angiosperms, it is clear that genomic duplication events must have had considerable implications for canalization of the structures that we now collectively term a flower (Rudall *et al.* 2009; Chanderbali *et al.* 2010). It is widely hypothesized that selection pressures leading to early angiosperm evolution were at least partly driven by climate change (Heimhofer *et al.* 2005; Feild *et al.* 2009) and insect diversification, perhaps reflecting experimentation with insect mutualisms (Labandeira 2010). Some observers attribute the considerable diversity that exists among extant angiosperms primarily to adaptive radiations. On the other hand, current evidence suggests that major seed-plant clades were marked by large-scale genome duplications. The existence of some major morphogenetic markers that circumscribe deep nodes (notably the base of the eudicot clade) not only represents a useful test of molecular-based angiosperm phylogenies but also potentially helps to identify candidates for key innovations—characters that were directly responsible for triggering major evolutionary diversification events. However, in practice, it is often difficult to determine whether such deep-node characters actually drive diversification or are (perhaps more likely) merely markers of radical genetic shifts resulting from large-scale saltational gene duplications.

Significant advances in the last decade have resulted from an integrated pluralistic approach to resolving these longstanding questions. When placed in a robust phylogenetic context within angiosperms, developmental-genetic studies using lab-induced mutants demonstrate that relatively simple genetic changes can both radically alter phenotype and potentially drive evolutionary radiations. Models that incorporate developmental morphology in a phylogenetic context will allow us to ask more accurate questions about the genetic bases of different phenotypes.

Acknowledgments

Thanks to Richard Bateman, Graça Sajo, Dmitry Sokoloff, and Margarita Remizowa for ongoing discussions on floral evolution. Richard Bateman provided helpful comments on the manuscript.

References

Agrawal, G.K., Kiyomi, A., Yamazaki, M., Miyao, A. & Hirochika, H. (2005) Conservation of the E-function for floral organ identity in rice revealed by the analysis of tissue culture-induced loss-of-function mutants of the *OsMADS1* gene. *Plant Molecular Biology*, **59**, 125–135.

Albach, D.C., Soltis, P.S. & Soltis, D.E. (2001) Patterns of embryological and biochemical evolution in the asterids. *Systematic Botany*, **26**, 242–262.

Aloni, R., Aloni, E., Langhans, M. & Ullrich, C. (2006) Role of auxin in regulating *Arabidopsis* flower development. *Planta*, **223**, 315–328.

Alvarez, J. & Smyth, D.R. (1999) *CRABS CLAW* and *SPATULA*, two Arabidopsis genes that control carpel development in parallel with *AGAMOUS*. *Development*, **126**, 2377–2386.

Alvarez-Buylla, E.R., Ambrose, B.A., Flores-Sandoval, E., *et al.* (2010) B-function expression in the flower center underlies the homeotic phenotype of *Lacandonia schismatica* (Triuridaceae). *Plant Cell*, **22**, 3543–3559.

Ambrose, B.A., Espinosa-Matias, S., Vazquez-Santana, S., *et al.* (2006) Comparative developmental series of the Mexican triurids support a euanthial interpretation for the unusual reproductive axes of *Lacandonia schismatica* (Triuridaceae). *American Journal of Botany*, **93**, 15–35.

APG III: Angiosperm Phylogeny Group III (2009) An update of the Angiosperm Phylogeny Group classification for the orders and families of flowering plants: APG III. *Botanical Journal of the Linnean Society*, **161**, 105–121.

Arber, A. (1937) Studies in flower structure. III. On the corona and androecium in certain Amaryllidaceae. *Annals of Botany*, **1**, 293–304.

Arber, E.A.N. & Parkin, J. (1907) On the origin of the angiosperms. *Botanical Journal of the Linnean Society*, **38**, 29–80.

Bateman, R.M., Hilton, J. & Rudall, P.J. (2006) Morphological and molecular phylogenetic context of the angiosperms: contrasting the "top-down" and "bottom-up" approaches to inferring the likely characteristics of the first flowers. *Journal of Experimental Botany*, **57**, 3471–3503.

Bateman, R.M., Hilton, J. & Rudall, P.J. (2011) Spatial separation and developmental divergence of male and female reproductive units in gymnosperms, and their relevance to the origin of the angiosperm flower. In: *Flowers on the Tree of Life* (eds L. Wanntorp, L.P. Ronse & De Craene), in press. Cambridge University Press, Cambridge.

Baum, D.A. & Hileman, L.C. (2006) A developmental genetic model for the origin of the flower. In: *Annual Plant Reviews*, Vol. 20, *Flowering and Its Manipulation* (ed. C. Ainsworth), pp. 3–27. Blackwell Publishing Ltd., Oxford, UK.

Baum, S.F., Eshed, Y. & Bowman, J.L. (2001) The *Arabidopsis* nectary is an ABC-independent floral structure. *Development*, **128**, 4657–4667.

Bell, C.D., Soltis, D.E. & Soltis, P.S. (2010) The age and diversification of the angiosperms re-revisited. *American Journal of Botany*, **97**, 1296–1303.

Bernardello, G. (2007) A systematic survey of floral nectaries. In: *Nectaries and Nectar* (eds. S.W. Nicolson, M. Nepi & E. Pacini), pp. 19–128. Springer, Dordrecht.

Bowman, J.L. & Smyth, D.R. (1999) *CRABS CLAW*, a gene that regulates carpel and nectary development in *Arabidopsis*, encodes a novel protein with zinc finger and helix-loop-helix domains. *Development*, **126**, 2387–2396.

Burleigh, J.G & Mathews, S. (2004) Phylogenetic signal in nucleotide data from seed plants: implications for resolving the seed plant tree of life. *American Journal of Botany*, **91**, 1599–1613.

Buzgo, M., Soltis, P.S. & Soltis, D.S. (2004) Floral developmental morphology of *Amborella trichopoda* (Amborellaceae). *International Journal of Plant Sciences*, **165**, 925–947.

Cantino, P.D., Doyle, J.A., Graham, S.W., *et al.* (2007) Towards a phylogenetic nomenclature of *Tracheophyta*. *Taxon*, **56**, 822–846.

Castillejo, C., Romera-Branchat, M. & Pelaz, S. (2005) A new role of the *Arabidopsis SEPALLATA3* gene revealed by its constitutive expression. *The Plant Journal*, **43**, 586–596.

Causier, B., Schwarz-Sommer, Z. & Davies, B. (2010) Floral organ identity: 20 years of ABCs. *Seminars in Cell and Developmental Biology*, **21**, 73–79.

Chanderbali, A.S., Yoo, M.J., Zahn, L.M., *et al.* (2010) Conservation and canalization of gene expression during angiosperm diversification accompany the origin and evolution of the flower. *Proceedings of the National Academy of Sciences of the United States of America*, **107**, 22570–22575.

Chen, C., Xu, Y., Zeng, M. & Huang, H. (2001) Genetic control by *Arabidopsis* genes *LEUNIG* and *FILAMENTOUS FLOWER* in gynoecium fusion. *Journal of Plant Research*, **114**, 465–469.

Cheng, Y., Dai, X. & Zhao, Y. (2006) Auxin biosynthesis by the *YUCCA* flavin monooxygenases controls the formation of floral organs and vascular tissues in *Arabidopsis*. *Genes and Development*, **20**, 1790–1799.

Citerne, H., Jabbour, F., Nadot, S. & Damerval, C. (2010) The evolution of floral symmetry. *Advances in Botanical Research*, **54**, 85–137.

Coen, E.S. & Meyerowitz, E.M. (1991) The war of the whorls: genetic interactions controlling flower development. *Nature*, **353**: 31–37.

Crane, P.R. (1985) Phylogenetic analysis of seed plants and the origin of angiosperms. *Annals of the Missouri Botanical Garden*, **72**, 716–793.

Damerval, C., Le Guilloux, M., Jager, M. & Charon, C. (2007) Diversity and evolution of *CYCLOIDEA*-like TCP genes in relation to flower development in Papaveraceae. *Plant Physiology*, **143**: 759–772.

Damerval, C. & Nadot, S. (2007) Evolution of perianth and stamen characteristics with respect to floral symmetry in Ranunculales. *Annals of Botany*, **100**, 631–640.

Darwin, F. & Seward, A.C. (eds) (1903) *More letters from Charles Darwin*, vol. **2**. John Murray, London, UK.

De Bodt, S., Maere, S. & Van de Peer, Y. (2005) Genome duplication and the origin of angiosperms. *Trends in Ecology and Evolution*, **20**, 591–597.

Dilcher, D.L. & Crane, P.R. (1984) *Archaeanthus*: an early angiosperm from the Cenomanian of the western interior of North America. *Annals of the Missouri Botanical Garden*, **71**, 351–383.

Dong, Z.C., Zhao, Z., Liu, C.W., *et al.* (2005) Floral patterning in *Lotus japonicus. Plant Physiology*, **137**, 1272–1282.

Doyle, J.A. (1994) Origin of the angiosperm flower: a phylogenetic perspective. *Plant Systematics and Evolution*, Suppl. 8, 193–208.

Doyle, J.A. (1996) Seed plant phylogeny and the relationships of Gnetales. *International Journal of Plant Sciences*, **157**, S3–S39.

Doyle, J.A. (2005) Early evolution of angiosperm pollen as inferred from molecular and morphological phylogenetic analyses. *Grana*, **44**, 227–251.

Doyle, J.A. (2006) Seed ferns and the origin of angiosperms. *Journal of the Torrey Botanical Society*, **133**, 169–209.

Doyle, J.A. (2008) Integrating molecular phylogenetic and paleobotanical evidence on origin of the flower. *International Journal of Plant Sciences*, **169**, 816–843.

Doyle, J.A. (2009) Evolutionary significance of granular exine structure in the light of phylogenetic analyses. *Review of Palaeobotany and Palynology*, **156**, 198–210.

Doyle, J.A. & Donoghue, M.J. (1986) Seed plant phylogeny and the origin of angiosperms: an experimental cladistic approach. *Botanical Review*, **52**, 321–431.

Doyle, J.A. & Endress. P.K. (2000) Morphological phylogenetic analysis of basal angiosperms: comparison and combination with molecular data. *International Journal of Plant Sciences*, **161**, S121–S153.

Doyle, J.A. & Endress. P.K. (2010) Integrating Early Cretaceous fossils into the phylogeny of living angiosperms: Magnoliidae and eudicots. *Journal of Systematics and Evolution*, **48**, 1–35.

Eames, A.J. (1961) *Morphology of the Angiosperms*. McGraw-Hill, New York.

Endress, P.K. (1992) Evolution and floral diversity: the phylogenetic surroundings of *Arabidopsis* and *Antirrhinum*. *International Journal of Plant Sciences*, **153**, S106–S122.

Endress, P.K. (1999) Symmetry in flowers: diversity and evolution. *International Journal of Plant Sciences*, **160**, S3–S23.

Endress, P.K. (2001a) The flowers in extant basal angiosperms and inferences on ancestral flowers. *International Journal of Plant Sciences*, **162**, 1111–1140.

Endress, P.K. (2001b) Evolution of floral symmetry. *Current Opinion in Plant Biology*, **4**, 86–91.

Endress, P.K. (2010a) Flower structure and trends of evolution in eudicots and their major subclades. *Annals of the Missouri Botanical Garden*, **97**, 541–583.

Endress, P.K. (2010b) Disentangling confusions in inflorescence morphology: patterns and diversity of reproductive shoot ramification in angiosperms. *Journal of Systematics and Evolution*, **48**, 225–239.
Endress, P.K. (2011) Evolutionary diversification of the flowers in angiosperms. *American Journal of Botany*, **98**, 370–396.
Endress, P.K. & Doyle, J.A. (2009) Reconstructing the ancestral angiosperm flower and its initial specializations. *American Journal of Botany*, **96**, 22–66.
Endress, P.K. & Igersheim, A. (2000a) Gynoecium structure and evolution in basal angiosperms. *International Journal of Plant Sciences*, **161**, S211–S213.
Endress, P.K. & Igersheim, A. (2000b) The reproductive structures of the basal angiosperm *Amborella trichopoda* (Amborellaceae). *International Journal of Plant Sciences*, **161**, S237–S248.
Endress, P.K. & Matthews, M.L. (2006) Elaborate petals and staminodes in eudicots: diversity, function, and evolution. *Organisms, Diversity and Evolution*, **6**, 257–293.
Feild, T.S., Chatelet, D.S. & Brodribb, T.J. (2009) Ancestral xerophobia: a hypothesis on the whole plant ecophysiology of early angiosperms. *Geobiology*, **7**, 237–264.
Floyd, S.K. & Bowman, J.L. (2010) Gene expression patterns in sed plant shoot meristems and leaves: homoplasy or homology? *Journal of Plant Research*, **123**, 43–55.
Friedman, W.E. (2009) The meaning of Darwin's "abominable mystery". *American Journal of Botany*, **96**, 5–21.
Friis, E.M., Doyle, J.A., Endress, P.K. & Leng, Q. (2003) *Archaefructus*—angiosperm precursor or specialized early angiosperm? *Trends in Plant Science*, **8**, 369–373.
Friis, E.M., Pedersen, K.R. & Crane, P.R. (2001) Fossil evidence of water lilies (Nymphaeales) in the Early Cretaceous. *Nature*, **410**, 357–360.
Frohlich, M.W. & Parker, D.S. (2000) The mostly male theory of flower evolutionary origins: from genes to fossils. *Systematic Botany*, **25**, 155–170.
Furness, C.A. & Rudall, P.J. (2004) Pollen aperture evolution – a crucial factor for eudicot success? *Trends in Plant Science*, **9**, 1360–1385.
Golz, J.F., Keck, E.J. & Hudson, A. (2002) Spontaneous mutations in *KNOX* genes give rise to a novel floral structure in *Antirrhinum*. *Current Biology*, **12**, 515–522.
Graham, S.W. & Iles, W.J.D. (2009) Different gymnosperm outgroups have (mostly) congruent signal regarding the root of flowering plant phylogeny. *American Journal of Botany*, **96**, 216–227.
Graham, S.W., Zgurski, J.M., McPherson, M.A., *et al.* (2006) Robust inference of monocot deep phylogeny using an expanded multigene plastid data set. *Aliso*, **22**, 3–21.
Harris, T.M. (1944) A revision of *Williamsoniella*. *Philosophical Transactions of the Royal Society, London, B*, **231**, 313–328.
Heimhofer, U., Hochuli, P.A., Burla, S., Dinis, J.M.L. & Weissert, H. (2005) Timing of early Cretaceous angiosperm diversification and possible links to major paleoenvironmental change. *Geology*, **33**, 141–144.
Hill, J.P. & Lord, E.M. (1989) Floral development in *Arabidopsis thaliana*: comparison of the wild type and the homeotic *pistillata* mutant. *Canadian Journal of Botany*, **67**, 2922–2936.
Hilton, J. & Bateman, R.M. (2006) Pteridosperms are the backbone of seed-plant evolution. *Journal of the Torrey Botanical Society*, **133**, 119–168.
Iwamoto, A., Shimuzu, A. & Ohba, H. (2003) Floral development and phyllotactic variation in *Ceratophyllum demersum* (Ceratophyllaceae). *American Journal of Botany*, **90**, 1124–1130.

Jönsson, H., Heisler, M.G., Shapiro, B.E., Meyerowitz, E.M. & Mjolsness, E. (2006) An auxin driven polarized transport model for phyllotaxis. *Proceedings of the National Academy of Sciences of the United States of America*, **103**, 1633–1638.

Kim, S., Koh, J., Ma, H., *et al.* (2005) Sequence and expression studies of A-, B-, and E-class MADS-box homologues in *Eupomatia* (Eupomatiaceae): support for the bracteate origin of the calyptra. *International Journal of Plant Sciences*, **166**, 185–198.

Knox, R.B. (1984) The pollen grain. In: *Embryology of Angiosperms* (ed. B.M. Johri), pp. 197–271. Springer-Verlag, Berlin.

Kramer, E.M., Dorit, R.L. & Irish, V. F. (1998) Molecular evolution of genes controlling petal and stamen development: duplication and divergence within the *APETALA3* and *PISTILLATA* MADS-box gene lineages. *Genetics*, **149**, 765–783.

Kramer, E.M. & Hodges, S.A. (2009) *Aquilegia* as a model system for the evolution and ecology of petals. *Philosophical Transactions of the Royal Society B*, **365**, 477–490.

Krolikowski, K.A., Victor, J.L., Wagler, T.N., Lolle, S.J. & Pruitt, R.J. (2003) Isolation and characterization of the Arabidopsis organ fusion gene HOTHEAD. *The Plant Journal*, **35**, 501–511.

Labandeira, C. (2010) The pollination of mid Mesozoic seed plants and the early history of long-proboscid insects. *Annals of the Missouri Botanical Garden*, **97**, 469–513.

Lee, D.K., Geisler, M. & Springer, P.S. (2009) *LATERAL ORGAN FUSION1* and *LATERAL ORGAN FUSION2* function in lateral organ separation and axillary meristem formation in *Arabidopsis*. *Development*, **136**, 2423–2432.

Lee, J.Y., Baum, S.F., Oh, S.H., Jiang, C.Z., Chen, J.C. & Bowman, J.L. (2005) Recruitment of *CRABS CLAW* to promote nectary development within the eudicot clade. *Development*, **132**, 5021–5032.

Li, G.S., Meng, Z, Kong, H.Z., Chen, Z.D., Theissen, G. & Lu, A.M. (2005) Characterization of candidate class A, B and E floral homeotic genes from the perianthless basal angiosperm *Chloranthus spicatus* (Chloranthaceae). *Development Genes and Evolution*, **215**(9), 437–449.

Malcomber, S.T., Preston, J.C., Reinheimer, R., Kossuth, J. & Kellogg, E.A. (2006) Developmental gene evolution and the origin of grass inflorescence diversity. *Advances in Botanical Research*, **44**, 425–481.

Mathews, S. (2009) Phylogenetic relationships among seed plants: persistent questions and the limits of molecular data. *American Journal of Botany*, **96**, 228–236.

McSteen, P. (2006) Branching out: the ramosa pathway and the evolution of grass inflorescence morphology. *Plant Cell*, **18**, 518–522.

Meeuse, A.D.J. (1972) Sixty-five years of theories of the multiaxial flower. *Acta Biotheoretica*, **21**, 167–202.

Melville, R. (1960) A new theory of the angiosperm flower. *Nature*, **118**, 14–18.

Melzer, R., Wang, Y.Q. & Theissen, G. (2010) The naked and the dead: the ABCs of gymnosperm reproduction and the origin of the angiosperm flower. *Seminars in Cell and Developmental Biology*, **21**, 118–128.

Mondragón-Palomino, M. & Theissen, G. (2009) Why are orchid flowers so diverse? Reduction of evolutionary constraints by paralogues of class B floral homeotic genes. *Annals of Botany*, **104**, 583–594.

Moore, M.J., Bell, C.D., Soltis, P.S. & Soltis, D.E. (2007) Using plastid genome-scale data to resolve enigmatic relationships among basal angiosperms. *Proceedings of the National Academy of Sciences of the United States of America*, **104**, 19363–19368.

Mundry, M. & Stützel, T. (2004) Morphogenesis of the reproductive shoots of *Welwitschia* mirabilis and *Ephedra distachya* (Gnetales), and its evolutionary implications. *Organisms Diversity and Evolution*, **4**, 91–108.

Nixon, K.C., Crepet, W.L., Stevenson, D.W. & Friis, E.M. (1994) A re-evaluation of seed plant phylogeny. *Annals of the Missouri Botanical Garden*, **81**, 484–533.

Nuraliev, M.S., Sokoloff, D.D & Oskolski, A.A. (2011) Floral anatomy of Asian *Schefflera* (Araliaceae, Apiales): comparing variation of flower groundplan and vascular patterns. *International Journal of Plant Sciences*, **172**, 735–762.

Parcy, F., Bomblies, K. & Weigel, D. (2002) Interaction of *LEAFY, AGAMOUS* and *TERMINAL FLOWER1* in maintaining floral meristem identity in *Arabidopsis. Development*, **129**, 2519–2527.

Parkin, J. (1914) The evolution of the inflorescence. *Linnean Journal of Botany*, **42**, 511–563.

Philipson, W.R. (1974) Ovular morphology and the major classification of the dicotyledons. *Botanical Journal of the Linnean Society*, **68**, 89–108.

Prenner, G., Cacho, I.N., Baum, D. & Rudall, P.J. (2011) Is *LEAFY* a useful marker gene for the flower–inflorescence boundary in the *Euphorbia* cyathium? *Journal of Experimental Botany*, **62**, 345–350.

Prenner, G. & Rudall. P.J. (2007) Comparative ontogeny of the cyathium in *Euphorbia* and its allies: exploring the organ–flower–inflorescence boundaries. *American Journal of Botany*, **94**, 1612–1629.

Prenner, G., Vergara-Silva, F. & Rudall, P.J. (2009) The key role of morphology in modelling inflorescence architecture. *Trends in Plant Science*, **14**, 302–309.

Preston, J.C. & Hileman, L.C. (2009) Developmental genetics of floral symmetry evolution. *Trends in Plant Science*, **14**, 147–154.

Pruitt, R.E., Vielle-Calzada, J.P., Ploense, S.E., Grossniklaus, U. & Lolle, S.J. (2000) FIDDLEHEAD, a gene required to suppress epidermal cell interactions in Arabidopsis, encodes a putative lipid biosynthetic enzyme. *Proceedings of the National Academy of Sciences of the United States of America*, **97**, 1311–1316.

Prusinkiewicz, P., Erasmus, Y., Lane, B., Harder, J.D. & Coen, E. (2007) Evolution and development of inflorescence architectures. *Science*, **316**, 1452–1456.

Qiu, Y.L., Lee, J., Bernasconi-Quadroni, F., *et al.* (2000) Phylogeny of basal angiosperms: analyses of five genes from three genomes. *International Journal of Plant Sciences*, **161** (Suppl. S6), S3–S27.

Remizowa, M.V., Sokoloff, D.D. & Rudall, P.J. (2010) Evolutionary history of the monocot flower. *Annals of the Missouri Botanical Garden*, **97**, 617–645.

Ronse De Craene, L.P. (2007) Are petals sterile stamens or bracts? The origin and evolution of petals in the core eudicots. *Annals of Botany*, **100**, 621–630.

Ronse De Craene, L.P. (2010) *Floral Diagrams*. Cambridge University Press, Cambridge.

Rudall, P.J. (2006) How many nuclei make an embryo sac in flowering plants? *Bioessays*, **28**, 1067–1071.

Rudall, P.J. (2010) All in a spin: centrifugal organ formation and floral patterning. *Current Opinion in Plant Biology*, **13**, 108–114.

Rudall, P.J. & Bateman, R.M. (2002) Roles of synorganisation, zygomorphy, and heterotopy in floral evolution: the gynostemium and labellum of orchids and other lilioid monocots. *Biological Reviews*, **77**, 403–441.

Rudall, P.J. & Bateman, R.M. (2004) Evolution of zygomorphy in monocot flowers, iterative patterns and developmental constraints. *New Phytologist*, **162**, 25–44.

Rudall, P.J. & Bateman, R.M. (2006) Morphological phylogenetic analysis of Pandanales: testing contrasting hypotheses of floral evolution. *Systematic Botany*, **31**, 223–238.

Rudall, P.J. & Bateman, R.M. (2007) Developmental bases for key innovations in the seed-plant microgametophyte. *Trends in Plant Science*, **12**, 317–326.

Rudall, P.J. & Bateman, R.M. (2010) Defining the limits of flowers: the challenge of distinguishing between the evolutionary products of simple versus compound strobili. *Philosophical Transactions of the Royal Society B*, **365**, 397–409.

Rudall, P.J., Bateman, R.M., Fay, M.F. & Eastman, A. (2002) Floral anatomy and systematics of Alliaceae with particular reference to *Gilliesia*, a presumed insect mimic with strongly zygomorphic flowers. *American Journal of Botany*, **89**, 1867–1883.

Rudall, P.J., Remizowa, M.V., Beer, A., *et al.* (2008) Comparative ovule and megagametophyte development in Hydatellaceae and water lilies reveal a mosaic of features among the earliest angiosperms. *Annals of Botany*, **101**, 941–956.

Rudall, P.J., Remizowa, M.V., Prenner, G., Prychid, C.J., Tuckett, R.E. & Sokoloff, D.D. (2009) Non-flowers near the base of extant angiosperms? Spatiotemporal arrangement of organs in reproductive units of Hydatellaceae, and its bearing on the origin of the flower. *American Journal of Botany*, **96**, 67–82.

Rudall, P.J., Sokoloff, D.D., Remizowa, M.V., *et al.* (2007) Morphology of Hydatellaceae, an anomalous aquatic family recently recognized as an early-divergent angiosperm lineage. *American Journal of Botany*, **94**, 1073–1092.

Rudall, P.J., Vergara-Silva, F., Hilton, J. & Bateman, R.M. (2011) Recurrent abnormalities in conifer cones and the evolutionary origins of flower-like structures. *Trends in Plant Science*, **16**, 151–159.

Saarela, J.M., Rai, H.S., Doyle, J.A., *et al.* (2007) Hydatellaceae identified as a new branch near the base of the angiosperm phylogenetic tree. *Nature*, **446**, 312–315.

Sajo, M.G., Mello-Silva, R. & Rudall, P.J. (2010) Homologies of floral structures in Velloziaceae, with particular reference to the corona. *International Journal of Plant Sciences*, **171**, 595–606.

Sattler, R. (1975) *Organogenesis of flowers – a photographic text-atlas*. University of Toronto Press, Toronto.

Scutt, C.P., Vinauger-Douard, M., Fourquin, C., Finet, C. & Dumas, C (2006) An evolutionary perspective on the regulation of carpel development. *Journal of Experimental Botany*, **57**, 2143–2152.

Smyth, D.R., Bowman, J.L. & Meyerowitz, E.M. (1990) Early flower development in *Arabídopsis*. *Plant Cell*, **2**, 755–767.

Sokoloff, D.D., Oskolski, A.A., Remizowa, M.V. & Nuraliev, M.S. (2007) Flower structure and development in *Tupidanthus calyptratus* (Araliaceae): an extreme case of polymery among asterids. *Plant Systematics and Evolution*, **268**, 209–234.

Sokoloff, D.D., Remizowa, M.V., Macfarlane, T.D., *et al.* (2008) Seedling diversity in Hydatellaceae: implications for the evolution of angiosperm cotyledons. *Annals of Botany*, **101**, 153–164.

Sokoloff, D.D., Rudall, P.J. & Remizowa, M.V. (2006) Flower-like terminal structures in racemose inflorescences: a tool in morphogenetic and evolutionary research. *Journal of Experimental Botany*, **57**, 3517–3530.

Soltis, D.E., Senters, A.E., Zanis, M., *et al.* (2003) *Gunnerales* are sister to other core eudicots: implications for the evolution of pentamery. *American Journal of Botany*, **90**, 461–470.

Souer, E., Rebocho, A.B., Bliek, M., Kusters, E., Bruin, R.A.M. & Koes, R. (2008) Patterning of inflorescences and flowers by the F-box protein *DOUBLE TOP* and the *LEAFY* homolog *ABERRANT LEAF AND FLOWER* in *Petunia*. *Plant Cell*, **20**, 2033–2048.

Steeves, T.A. & Sussex, I.M. (1989) *Patterns in plant development*. Cambridge University Press, Cambridge.

Sun, G., Ji, Q., Dilcher, D.L., Zheng, S., Nixon, K.C. & Wang, X. (2002) Archaefructaceae, a new basal angiosperm family. *Science*, **296**, 899–904.

Theissen, G. & Becker, A. (2004) Gymnosperm orthologues of class B floral homeotic genes and their impact on understanding flower origin. *Critical Reviews in Plant Sciences*, **23**, 129–148.

Theissen, G., Becker, A., Rosa, A.D., *et al.* (2000) A short history of MADS-box genes in plants. *Plant Molecular Biology*, **42**, 115–149.

Thomas, H.H. (1916) On *Williamsoniella*, a new type of bennettitalean flower. *Philosophical Transactions of the Royal Society, London, B*, **207**, 113–148.

Warner, K.A., Rudall, P.J. & Frohlich, M.W. (2009) Environmental control of sepalness and petalness in perianth organs of waterlilies – a new Mosaic Theory on the evolutionary origin of a differentiated perianth. *Journal of Experimental Botany*, **60**, 3559–3574.

Weberling, F. (1992) *Morphology of Flowers and Inflorescences.* Cambridge University Press, Cambridge.

Wettstein, R. (1907) *Handbuch der Systematischen Botanik, Band* 2. Deuticke, Vienna.

Williams, J.H. (2008) Novelties of the flowering plant pollen tube underlie diversification of a key life history stage. *Proceedings of the National Academy of Sciences of the United States of America*, **105**, 11259–11263.

Williams, J.H. & Friedman, W.E. (2004) The four-celled female gametophyte of *Illicium* (Illiciaceae; Austrobaileyales): implications for understanding the origin and early evolution of monocots, eumagnoliads, and eudicots. *American Journal of Botany*, **91**, 332–351.

Yuan, Z., Gao, S., Xue, D.W., *et al.* (2009) *RETARDED PALEA1* Controls palea development and floral zygomorphy in rice. *Plant Physiology*, **149**, 235–244.

Zahn, L.M., Kong, H., Leebens-Mack, J.H., *et al.* (2005) The evolution of the *SEPALLATA* subfamily of MADS-box genes: a preangiosperm origin with multiple duplications throughout angiosperm history. *Genetics*, **169**, 2209–2223.

Zahn, L.M., Leebens-Mack, J.H., Arrington, J.M., *et al.* (2006) Conservation and divergence in the *AGAMOUS* subfamily of MADS-box genes: evidence of independent sub- and neofunctionalization events. *Evolution and Development*, **8**, 30–45.

Zhang, P. Zhang, P., Tan, H.T., Pwee, K.H. & Kumar, P.P. (2004) Conservation of class C function of floral organ development during 300 million years of evolution from gymnosperms to angiosperms. *Plant Journal*, **37**, 566–577.

Zimmer, A., Lang, D., Richardt, S, Frank, W., Reski, R. & Rensing, S.A. (2007) Dating the early evolution of plants: detection and molecular clock analyses of orthologs. *Molecular Genetics and Genomics*, **278**, 393–402.

Annual Plant Reviews (2013) **45**, 189–226
doi: 10.1002/9781118305881.ch7 http://onlinelibrary.wiley.com

Chapter 7

GENOMICS, ADAPTATION, AND THE EVOLUTION OF PLANT FORM

Kristen Shepard

Department of Biological Sciences, Barnard College, New York, NY, USA

Abstract: A central question in the evolution of plant development is to understand how modifications of plant form contribute to adaptation within a species. Molecular population genomics and quantitative genetics offer complementary tools for answering this question. Quantitative trait locus mapping identifies genetic variation that underlies variation in plant form, while analyses of nucleotide diversity illuminate the evolutionary history of developmentally relevant genes. By combining these tools with developmental genetic characterization of natural sequence variants and with ecological experiments that test adaptive hypotheses, recent studies have begun to tease apart the genetic basis of adaptation. This chapter includes an overview of genome-wide mapping and population genomic methods, as well as case studies detailing the application of these techniques in root development, plant domestication, phytochrome-mediated phenotypic variation, and regulation of flowering time.

Keywords: plant development; molecular population genetics; QTL mapping; natural variation; evolution of development; adaptation; association mapping; phytochrome; flowering time; *BREVIS RADIX*.

7.1 Overview

A central goal of studies in the evolution of plant development is to understand how developmental mechanisms have been modified to generate the amazing diversity of plant form that we see around us. We know that diversity in the genome underlies the intraspecific and interspecific changes in plant development. Yet, even within a species, the genomes

Annual Plant Reviews Volume 45: The Evolution of Plant Form, First Edition.
Edited by Barbara A. Ambrose and Michael Purugganan.

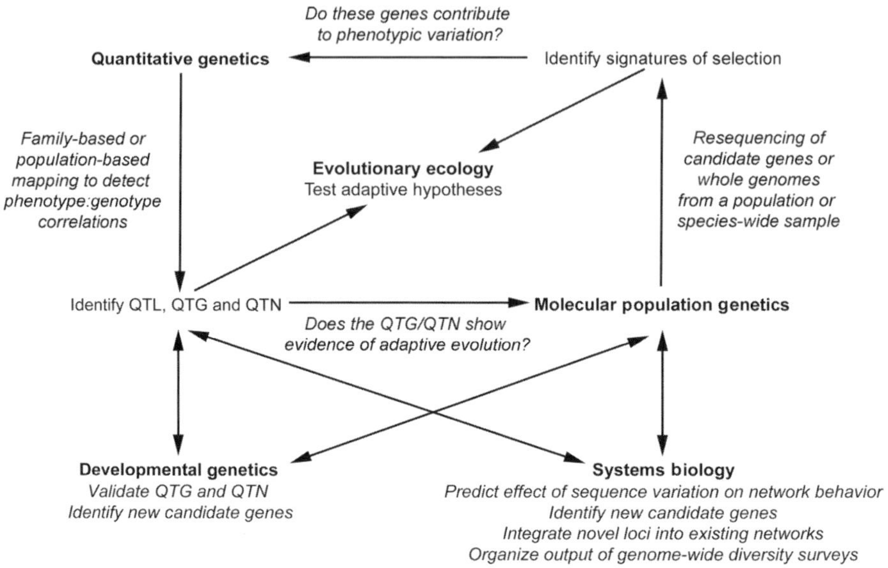

Figure 7.1 Overview of the roles of different biological disciplines in teasing apart the adaptive evolution of plant form. QTL, quantitative trait locus; QTG, quantitative trait gene; QTN, Quantitative trait nucleotide.

of any two individuals can differ in hundreds of thousands to millions of nucleotide sites as well as insertion/deletions (indels) and even gene copy number variants. Which of these genetic changes actually influences plant form? And, how do we determine whether this molecular variation is indeed adaptive?

Researchers interested in exploring these questions have a suite of powerful tools at their disposal (see Figure 7.1). The two general approaches are quantitative trait locus (QTL) mapping and molecular population genetic analysis, which are often designated as "top down" and "bottom up" approaches (Wright & Gaut 2005). QTL mapping begins with analyses of phenotypic variation and uses genetic mapping techniques to identify genomic regions that contribute to the phenotypic variation. In contrast, molecular population genetic studies begin with surveys of genetic variation within a species or a population. Analyses of the amount and type of nucleotide variation within and between species can detect whether genes bear a "signature of selection" consistent with adaptive evolution.

Neither QTL mapping nor molecular population genetics alone can conclusively demonstrate that a gene is involved in adaptive evolution. In combination, though, these techniques can build a strong circumstantial case. For example, if a QTL is identified in mapping studies, molecular population genetic analyses should be performed to assess whether the QTL falls in a genomic region that also has a signature of selection. Conversely, if a gene is deemed interesting based on the results of molecular population genetic tests, quantitative genetic studies of that gene's contribution to phenotypic variation should be performed. As these combined studies narrow the focus from a genomic region to a single gene and sometimes to particular nucleotide variants, functional analyses to confirm the predicted phenotypic effect of sequence variants are crucial. Finally—and this often presents the biggest challenge—adaptive hypotheses need to be explicitly tested in an ecological context (see Figure 7.1).

Both quantitative genetic and molecular population genetic approaches can be implemented to study particular genes of interest (candidate genes) or to evaluate the entire genome (e.g., through genome-wide diversity surveys and genome-wide association (GWA) mapping). To date, these approaches have proven most successful in model systems—such as *Arabidopsis thaliana*, rice, and maize—for which substantial genomic and germplasm resources are available. New mapping techniques and reduced costs for DNA sequencing and genotyping have greatly expanded the scale on which these studies can be implemented. In the near future, as sequencing costs further decline and phenotyping tools continue to improve, we should be able to tremendously expand the application of these tools to nonmodel plants and a broader range of evolutionary questions (Schneeberger & Weigel 2011).

7.2 The types of genetic variation present within species

Mutations create polymorphisms in the DNA sequences among individuals within a species. Some of these variants are single nucleotide polymorphisms (SNPs) that alter only one base pair. Nonsynonymous (replacement) SNPs occur in coding sequence and lead to a change in the encoded amino acid, while synonymous SNPs encode the same amino acid. SNPs in noncoding regions are typically referred to as silent substitutions because they do not affect the protein sequence; however, these noncoding SNPs can influence the regulation of gene expression or of intron splicing. Other mutations lead to changes in the number of nucleotides due to the insertion or deletion of base pairs. These indel polymorphisms may be as small as one base pair or may encompass large chromosomal segments, including the duplication or loss of entire genes (Clark *et al.* 2007). Developmentally relevant polymorphisms have been detected in both coding sequences and noncoding sites of genes (Alonso-Blanco *et al.* 2009).

Among 65 angiosperm genes for which the functional consequences of a natural polymorphism have been determined, more than half contain mutations in coding sequence (compiled by Alonso-Blanco *et al.* 2009). Nonsense mutations, frameshift indels, or splice site mutations produce a truncated protein in 31% of these genes, while replacement SNPs contribute to phenotypic variation in 28%. Copy number variation is also fairly common, with gene deletion polymorphisms occurring in 9%. The gene deletions and many of the truncated proteins are likely to act as natural null alleles. For example, in *A. thaliana*, deletion of the floral repressor *FLOWERING LOCUS M* (*FLM*) contributes to accelerated flowering time (Werner *et al.* 2005), and many independently derived loss-of-function mutations in *FRIGIDA* eliminate the vernalization requirement for flowering (reviewed by Shepard 2007). In most cases, the biochemical effects of replacement polymorphisms have not yet been ascertained; they may be null alleles, partial loss-of-function alleles, or even gain-of-function alleles.

The remaining well-characterized natural polymorphisms show variation in the regulation of gene expression, including altered level, location, or timing of transcription. These regulatory mutations include indels and SNPs in promoter sequences, indels in introns, transposable element insertions, and gene duplications (reviewed by Alonso-Blanco *et al.* 2009). In maize, two transposon insertions into an upstream enhancer region may account for the upregulation and altered spatial expression of *teosinte branched1* that accompanied selection for suppressed axillary branch growth during domestication (Clark *et al.* 2006; Zhou *et al.* 2011). In contrast, the increase in fruit size during tomato domestication depends to a large degree on variants that downregulate two genes: *fasciated* (*fas*) and *fw2.2*. A large insertion in the first intron decreases transcript levels of *fas*, a YABBY-like transcription factor, and correlates with increased locule number in large-fruited tomato cultivars (Cong *et al.* 2008). Decreases in the amount and duration of transcription of *fw2.2*, a negative regulator of cell division, also contribute to larger fruit size; however, the specific sequences associated with this transcription variation have not yet been identified (Cong *et al.* 2002).

Exciting recent studies suggest that epigenetic regulation may also exhibit intraspecific variation. Epigenetic regulation can be mediated by DNA methylation, chromatin modification, and/or small RNA activities (reviewed by Bossdorf *et al.* 2008). A dramatic example of the effect of differential DNA methylation is the conversion from bilateral to radial floral symmetry observed in wild populations of *Linaria vulgaris* (toadflax). In plants with radial flowers, the *Lcyc* gene is heritably hypermethylated and transcriptionally silenced (Cubas *et al.* 1999). Although natural examples of methylation variants remain rare, experimental approaches to studying epigenetic variation are beginning to bear fruit (Richards 2011). In *A. thaliana*, the extent of DNA methylation throughout the genome differs among individuals (Vaughn *et al.* 2007), and experimentally manipulated methylation states can be stably inherited for multiple generations and contribute to variation in plant height and flowering time in the laboratory (Johannes *et al.* 2009). Natural variation in chromatin modification has also been observed; for example, differences in the duration of cold exposure required to promote flowering in *A. thaliana* correlate with variation in histone modifications at *FLOWERING LOCUS C* (*FLC*) (reviewed in He 2009).

7.3 From phenotype to genotype: map-based approaches to identifying adaptive genes

7.3.1 The genetic architecture of quantitative traits

In studying the evolution of plant form, many of the traits of interest display phenotypes that continuously vary within a population or across a species. Examples of quantitatively varying traits include organ size, shape, or growth rate, as well as the timing of events such as germination or the onset of reproduction. This quantitative variation arises through the combined action of different alleles of many genes that contribute to the trait. A central goal of contemporary quantitative genetics is to understand the genetic architecture that underlies these complex traits: that is, the number of genes involved and the relative influence of different alleles on phenotype. Major effect alleles dramatically alter a plant's phenotype while minor effect alleles make more subtle contributions to phenotype. A further goal, particularly relevant to the study of adaptive evolution, is to better understand the importance of interactions among genes (epistasis) and of interactions between genes and environmental conditions (G × E) (Mackay 2001; Holland 2007).

QTL mapping searches for genomic regions that are strongly correlated with phenotypic variation in the trait of interest. In studies of the evolution of development, the focus is usually on plant structures; however, recent studies have also mapped expression QTL (eQTL) to identify natural variation in the level of transcription throughout the genome (Kliebenstein 2009). Mapping strategies fall into two categories: (1) family-based (also known as linkage mapping) and (2) population-based (also known as association mapping or linkage disequilibrium (LD) mapping; Myles *et al.* 2009). For many years, family-based mapping was the primary method used to detect QTL. More recently, population-based mapping has been implemented on a large scale. In both cases, after genomic regions have been shown to correlate with phenotypic variation, more fine-grained analyses can be used to identify the particular gene (quantitative trait gene, QTG) and even the precise DNA sequence variants (quantitative trait nucleotides, QTNs) responsible for trait variation (Holland 2007; Figure 7.1).

7.3.2 Family-based mapping

Family-based mapping begins by crossing two parent lines, often (but not always) using parents that differ in phenotype for the trait of interest. While it is possible to map traits in F2 or backcross progeny, it is increasingly common in short-lived annual species to use recombinant inbred lines (RILs). To construct these lines, two parents are crossed to generate F1 progeny that contain one copy of each parent's chromosomes. Repeated inbreeding of the F2s for four or more generations yields RILs, each of which is largely homozygous for chromosomes bearing different portions of the original parental chromosomes. In plants with short generation times (particularly self-pollinators), these RILs can be easily propagated and distributed to members of the research community (Jones *et al.* 2009). Each RIL is genotyped at a set of markers that are distributed throughout the genome, which detect polymorphisms in the nucleotide sequence of the two original parents, and thus identify the parental origins of different regions of the chromosomes in each RIL.

The phenotype of each RIL for the trait of interest is also quantified, and QTL are then identified using statistical methods that test for correlations between phenotypic variation and locations in the genetic map. These analyses reveal the relative contribution of each QTL to phenotypic variance as well as the effect of each parental allele of the QTL. For a more extensive discussion of the computational analyses involved in QTL mapping, see van Eeuwijk *et al.* (2010).

After a QTL has been identified, the real challenge begins—narrowing down to a particular gene or nucleotide that influences the trait. Because family-based mapping provides few opportunities for recombination between the parental genomes, a QTL can correspond to a large chromosomal region. A QTL mapping interval of approximately 10–30 cM long (which is a normal resolution in most QTL studies) may encompass megabases of DNA on a physical map and contain hundred of genes (Salvi & Tuberosa 2005). Considerable fine mapping may be required to isolate the QTG or QTN.

7.3.3 Advantages and disadvantages of family-based QTL mapping

Family-based QTL mapping provides many advantages in model and non-model systems. In many species, the crosses required to construct the mapping lines are relatively straightforward, and the density of genetic markers required is feasible even in species without extensive genomic resources. Within a species, family-based mapping is particularly useful if the goal is to understand the genetic basis of distinctly different phenotypes observed in two parental lines. In addition, if two different species can be mated, QTL mapping of hybrid lines can provide insight into the loci involved in phenotypic divergence (Frewen *et al.* 2000). These studies of interspecific hybrid lines can help to identify loci subjected to recent positive selection, which may display little intraspecific genetic variation at the selected locus.

Family-based mapping carries with it several practical disadvantages. First, extensive fine mapping is usually required to identify the actual gene underlying the developmental trait variation. Second, QTL studies are infeasible in species that propagate asexually or have a long generation time. Third, while family-based mapping has detected many QTL with strong effects on phenotype, this approach has been less effective in identifying QTL with smaller effects (Ross-Ibarra *et al.* 2007; Myles *et al.* 2009). Finally, when studying the evolution of plant form, a major conceptual disadvantage of family-based mapping is that it provides a very limited view into intraspecific variation. Only the genetic variation of the parents used to construct the mapping population can be explored, and loci that are invariant in the parents, but which make substantial contributions to phenotypic variation in other accessions, will go undetected. In addition, the alleles that are identified may not be particularly helpful in understanding the process of adaptation—they may simply be rare deleterious variants that are unlikely to make major contributions to adaptive evolution (Nordborg & Weigel 2008).

7.3.4 Population-based mapping

Population-based QTL mapping can be used either to study particular candidate genes or to conduct unbiased surveys through genome-wide association studies (GWAS). In population-based mapping, no crosses are required and instead a diverse group of natural accessions are genotyped (usually with SNPs) and phenotyped. Statistical analyses identify SNPs that correlate with phenotypic variation and are linked to a putative QTL for the trait of interest. SNPs that are located near a QTN will show nonrandom association (LD) with the QTN within the population and, thus, will appear to be associated with phenotypic variation. Because the population sample has a long history of recombination, SNPs that are further away will be in linkage equilibrium with the QTN and will not correlate with phenotypic variation.

The number of SNPs showing strong association with a QTN depends on a variety of factors including the species' breeding system, the populations sampled, and the rate of recombination in the chromosomal region carrying the QTN. In some cases, especially in outbreeding species, LD tends to be quite short, and the mapped QTL can correspond to a particular gene or even part of a gene. In other cases, particularly in inbreeders or in a region affected by positive selection, LD can be extensive, and the QTL may span more than 10–100 kb and include numerous possible candidate genes (Gaut & Long 2003; Myles *et al.* 2009).

GWAS typically require hundreds of thousands to millions of SNPs dispersed throughout the entire genome to achieve the resolution necessary to ensure good genomic coverage. The number of SNPs increases with genome size and in species with lower levels of LD (Myles *et al.* 2009). For a model system with a completely sequenced genome, obtaining these markers by resequencing the entire genome of each accession with next-generation sequencing techniques is fairly straightforward and increasingly cost-effective. Alternatively, SNPs can be initially detected by resequencing a smaller panel of individuals; genotyping these SNPs in the entire study population can then be accomplished using microarrays or mass spectrometry (Clark *et al.* 2007; Ehrenreich *et al.* 2009). For nonmodel systems with limited genomic resources, studies to date have usually focused on candidate genes identified through developmental genetic and systems biology work in model systems (see Ingvarsson *et al.* 2008).

7.3.5 Advantages and disadvantages of population-based QTL mapping

One of the greatest advantages of population-based mapping is the ability to sample a broad spectrum of genetic and phenotypic diversity. Because no crossing is required, population-based mapping is particularly useful in species that are slow-growing or difficult to propagate sexually (Ross-Ibarra *et al.* 2007). In addition, GWA mapping often locates QTL to smaller chromosomal regions than family-based maps, minimizing the need for extensive fine mapping. This higher resolution contributes to the potential of GWA mapping to identify more QTL, as well as QTL with smaller effects than family-based approaches (Myles *et al.* 2009).

Population-based mapping, however, is prone to both false positive and false negative detection of QTL. If the population includes relatively closely related individuals or displays population structure, then some SNPs that are *not* correlated with phenotypic variation will appear to be associated with the trait of interest. To help reduce false positives, structured association mapping includes estimates of kinship and structure in the statistical model used to search for associations between SNPs and phenotype (Yu *et al.* 2006). False negatives also arise, particularly if the QTL is associated with a rare allele. Although some members of the mapping population may bear an allele with a substantial effect on the trait of interest, that allele may go undetected because its phenotypic effect would be obscured by the many individuals bearing the alternative allele (Nordborg & Weigel 2008).

Another major challenge in population-based mapping is the execution of effective phenotyping experiments. The scale of these experiments must often be quite large in order to distinguish the signal of genotype–phenotype correlation from noise (Ross-Ibarra *et al.* 2007). In addition, obtaining phenotype data from diverse accessions in a "common garden" can be difficult if members of the mapping population show considerable local adaptation that hampers their ability to grow in a novel environment (Luquez *et al.* 2008).

7.3.6 Additional considerations in QTL mapping

QTL mapping holds great promise in identifying genetic bases of the evolution of plant form, but substantial challenges still remain. One important hurdle is improving ways to investigate the role of epistasis in phenotypic variation. Accumulating experimental evidence suggests that epistasis plays an important role in at least some species, including *A. thaliana* and rice; however, robust statistical methods for detecting epistasis, particularly in genome-wide studies, are still under development (Holland 2007). An additional difficulty in QTL analysis is evaluating the effect of genetic × environment interactions; because of G × E, replicated experiments may yield differing results, and experiments conducted in different environments may identify different sets of QTL (see *A. thaliana* flowering time studies by

Weinig *et al.* 2002). This G × E effect is particularly important to consider when designing laboratory experiments with the hope of extrapolating to natural conditions.

All map-based approaches require the collection of a great deal of phenotypic data, and collecting these data can be arduous and time consuming. As a result, most studies to date have focused on variables that are fairly easy to score and analyze, and this emphasis on one-dimensional measurements has severely limited the kinds of developmental traits studied by QTL mapping. Further progress in applying mapping approaches to the evolution of plant form will depend on improved high-throughput phenotyping techniques that facilitate rapid and precise quantification of complex phenotypes, particularly under field conditions.

7.3.7 Emerging approaches for detecting QTL

Perhaps the most important limitation of family- or population-based mapping is that neither approach seems effective in identifying a set QTL that adequately explains the phenotypic variation observed for a trait. In particular, QTL with small effects on phenotype have been difficult to detect. This lack of small effect QTL may be in part attributable to the emphasis on moderate-frequency markers in population mapping and to the low resolution of most family-based maps (Myles *et al.* 2009).

New approaches that combine family and population-based mapping may help to overcome the challenges of comprehensively detecting QTL contributing to a particular trait. One example of an experimental design that combines the best of both mapping approaches is the Nested Association Mapping (NAM) population created in maize. The NAM lines were created by crossing 25 genetically diverse accessions to a reference line. From each of these crosses, 200 RILs were generated (Yu *et al.* 2008). As a result, the lines capture a broad cross-section of maize diversity but avoid the complications of population structure that plague population-based approaches. Using the NAM lines, dozens of QTL with small effects on maize flowering time were detected, and additive models of the effects of these QTL successfully predicted the flowering time of each parental line (Buckler *et al.* 2009). It is worth noting, though, that this was a massive experiment, with nearly a million plants phenotyped across four field sites more than 2 years.

7.4 From genotype to phenotype: molecular population genetics and adaptive evolution

7.4.1 Overview of molecular population genetics

Molecular population genetic analyses directly assess genetic variation within a species or population. Although the conceptual underpinnings of molecular population genetics have a long history, empirical studies of DNA nucleotide sequence variation in plants have only been widespread in the last ~15 years. The general premise is that the amount and kind of nucleotide variation observed when comparing a gene in multiple individuals allows us to make inferences about the evolutionary history of that gene. Molecular population genetic analyses can be implemented either through large-scale multilocus surveys or by focusing on candidate genes identified through developmental genetics, systems biology, or QTL mapping.

7.4.2 Signatures of selection on DNA sequences

The null hypothesis for molecular population genetic analyses is based on the neutral theory of molecular evolution, which posits that most genetic variants observed in a population have no (or nearly no) effect on fitness (Kimura 1983). Neutral mutations are mainly affected by genetic drift, and a neutrally evolving gene should display a moderate level of nucleotide sequence diversity. Among the polymorphic sites in the gene, some variants will be rare (low frequency), more common (intermediate frequency), or very common (high frequency) within the population. In contrast, positive selection (directional selection) will cause a mutation that increases fitness to become more common. Eventually, this beneficial mutation will be fixed in all individuals within a population, or even throughout a species.

A region of the genome that has experienced positive selection will have low levels of nucleotide diversity both at the selected site and at flanking loci, the latter as a result of genetic hitchhiking of neutral sites with the positively selected gene. This local loss of variation adjacent to a positively selected site is often called a selective sweep, and the extent of the sweep along the chromosome depends on the strength of selection and on the recombination rate. Thus, common signatures of positive selection include low sequence diversity and an unusually large proportion of evolutionarily derived (as opposed to ancestral) SNPs that have accumulated to high frequencies. In addition, variants flanking the selected site will display extensive LD, and these variants will be inherited as a single haplotype consisting of many linked alleles along the chromosome (Przeworski 2002).

Another type of selection—balancing selection—results in a balanced polymorphism and occurs when more than one variant of a gene contributes to fitness so that no variant becomes fixed in the population or species. At the population level, alternative alleles may persist due to heterozygote advantage or frequency-dependent selection. In addition, some alleles may have differing fitness consequences depending on spatial or temporal variation in environmental conditions; in a variable environment, more than one allele may therefore be preserved. When two or more alternative alleles are maintained, neutral variants linked to the selected site will also be maintained by hitchhiking, but—unlike selective sweeps—the molecular signatures of a balanced polymorphism include elevated nucleotide diversity and an excess of intermediate frequency polymorphisms (Wright & Gaut 2005).

Although local adaptation is sometimes described as a type of balanced polymorphism (Nordborg & Innan 2003), it is useful to consider its signature of selection separately. In local adaptation, positive selection within a population leads to the accumulation of genetic variants that confer a phenotype better suited to the local environment. If gene flow into the population is sufficiently low, then the locally adapted population will show signs of a selective sweep at the beneficial locus, and the selected gene will display a low level of diversity among individuals within the population. In addition, because the selected population will now share fewer haplotypes with other populations, greater differentiation of sequence variation will be observed among populations (Schlötterer 2002).

7.4.3 Demographic factors can complicate inferences of selection

These theoretical predictions of the effects of drift and selection on nucleotide polymorphism frequencies assume that the population or species meets the criteria of an idealized Wright-Fisher model in which population size is constant and individuals mate randomly. These criteria are rarely met in real-world populations of plants; instead, demographic factors such as mating system and fluctuations in population size (bottlenecks, founder effects, and population expansions) must be taken into account. Deviations from the assumptions of the neutral model can profoundly influence genome-wide sequence diversity independently of a gene's selective history and can produce patterns of polymorphism that mimic the effects of selection. When only a single locus is sampled, it can be difficult to determine whether the observed pattern of variation results mainly from demographic processes, selection, or some combination of the two. However, demographic factors are expected to shape the genome as a whole, not just a particular gene. By comparing sequence diversity at putatively selected loci with a set of loci scattered throughout the genome, we can begin to tease apart the effects of selection and demography (Ross-Ibarra *et al.* 2007).

7.4.4 Gathering nucleotide sequence data

Obtaining the data for molecular population genetic studies is fairly straightforward if a fully sequenced genome is available or if a candidate gene approach is being undertaken. As sequencing technologies have become less expensive, the proportion of the genome included in resequencing studies has increased dramatically. Early molecular population genetic work in plants focused on fragments of coding sequence of candidate genes or anonymous loci; the assumption was that signatures of selection should be extensive enough that they would be evident even if the selected site was not included in the sequence sample. More recent studies have included the entire transcription unit plus ~1 kb of 5′ and 3′ flanking sequence; this expanded sampling in part reflects data suggesting that, for some genes, the pattern of polymorphism can be distinctly different in regulatory and coding sequences (Olsen *et al.* 2002).

Large chromosomal regions, and even entire genomes, have been resequenced in several model systems, including *A. thaliana*, rice, and maize. The first large-scale resequencing efforts used microarrays (Clark *et al.* 2007; McNally *et al.* 2009), but this technique is being superseded by next-generation sequencing technologies (Gore *et al.* 2009; for a review of sequencing technologies, see Deschamps & Campbell 2010). The 1001 Genomes Project for *A. thaliana*, an ambitious effort to resequence entire genomes both within local populations and across the species range, is currently underway (Weigel & Mott 2009).

For nonmodel systems, if large-scale resequencing remains impractical, amplified fragment length polymorphisms (AFLPs) or simple sequence repeats (SSRs, also known as microsatellites) can be used in genome-wide surveys of diversity. In particular, SSRs may be more useful for studies of closely related individuals where SNPs are expected to be relatively rare (Stinchcombe & Hoekstra 2008; Jones *et al.* 2009).

Early molecular population genetic studies in plants sampled ~12 individuals, but, as sequencing costs have declined, larger numbers of individuals (24–100) have been sampled. In addition, sequence from at least one individual of a closely related species is typically obtained because some tests to detect signatures of selection require outgroup information. Studies of domesticated crops and their wild relatives (either in the same species or a very closely related species) have demonstrated the utility of sampling multiple individuals in the outgroup (see Section 7.6.2). Data from close relatives are particularly useful in distinguishing genes with a signature of positive selection in one species from genes that simply accumulate little diversity in both species (Ross-Ibarra *et al.* 2007).

7.4.5 Interpreting the sequence data: summary statistics and tests of neutrality

To infer whether a gene bears a signature of selection, several tests have been developed that make use of the gene's overall level of polymorphism, the relative frequency of polymorphisms throughout the gene, and the extent of divergence from an outgroup species. Among the most commonly used are tests of neutrality, which evaluate whether the observed data are sufficient to reject the null hypothesis that a gene has evolved according to the predictions of the neutral theory. This section will focus on the summary statistics and tests of neutrality most commonly reported in the plant literature. For more extensive discussions of these tests and of additional molecular population genetic tests consult Oleksyk *et al.* (2010) and references therein.

One important consideration in choosing and interpreting tests of neutrality is that they detect the signatures of selection over differing time scales.

The tests that rely solely on intraspecific data can detect fairly recent selection, while those that incorporate interspecific data are more effective at detecting selection in the more distant past (Oleksyk *et al.* 2010). As a result, a single gene may show conflicting results when several different tests are implemented.

7.4.6 Nucleotide diversity and divergence

Nucleotide diversity summarizes how much a genomic region varies among individuals within a population or species. The two most common methods of estimating nucleotide diversity are Watterson's estimator (θ_W) and Tajima's estimator (θ_T, θ_π, or π) (Watterson 1975; Tajima 1983). For some analyses, diversity estimates are partitioned into categories based on the location of polymorphisms in protein coding sequences: synonymous (π_s), nonsynonymous (π_a), and silent (includes noncoding and synonymous, π_{silent}).

In contrast, nucleotide divergence describes how much a genomic region differs between species. Divergence is estimated as the number of nucleotide substitutions per site (K); several methods, differing in their assumptions about the rate of substitution for different classes of nucleotides, have been developed. Within coding regions, divergence is typically partitioned to focus separately on synonymous sites (Ks) and nonsynonymous sites (Ka) (for a review of methods, see Li 1997).

7.4.7 Analysis of the site frequency spectrum: Tajima's D and similar tests

One of the most commonly used tests of neutrality is Tajima's D, which contrasts the level of diversity estimated using θ_W and π (Tajima 1989). Whereas nucleotide diversity describes how much sequence variation exists in a population or species, Tajima's D summarizes a sequence's site frequency spectrum, which is the extent to which polymorphic sites consist of rare or common variants. The site frequency spectrum is important because positive selection is likely to have very different consequences on the frequency of variants compared to balancing selection. A negative value of Tajima's D indicates that many of the polymorphisms were low-frequency alleles observed in only a few individuals (which inflates θ_W). This excess of rare polymorphisms is consistent with fairly recent positive selection. A positive value of Tajima's D indicates that many of the polymorphisms were moderate-frequency alleles observed in many individuals (which inflates π). This excess of intermediate frequency polymorphisms is possible evidence for maintenance of a balanced polymorphism. Because Tajima's D has the potential to detect both positive and balancing selection candidates, this test is widely employed in molecular population genetic surveys. Several other tests of neutrality have been developed that "unfold" the site frequency spectrum by using outgroup information to identify derived alleles. These tests include Fay and Wu's H test to detect hitchhiking associated with positive selection (Fay & Wu 2000) and Fu and Li's D test (Fu & Li 1993).

However, one important caveat with Tajima's D (as well as tests that use the unfolded site frequency spectrum) is that demographic factors can also strongly affect the site frequency spectrum. Population expansion can yield a site frequency spectrum similar to positive selection, while the effects of a bottleneck can be difficult to distinguish from a balanced polymorphism. Analyses of Tajima's D in genome-wide polymorphism surveys can provide information about the effects of demographic processes on the site frequency spectrum. For example, in *A. thaliana*, the mean value Tajima's D is negative (–0.8 in Nordborg *et al.* 2005).

One of the challenges in employing Tajima's D is therefore to determine which values should be considered significant evidence of selection. Software used to calculate Tajima's D typically assesses significance based on a standard probability distribution or via coalescent simulations, both of which assume that there are no complicating demographic factors (Librado & Rozas 2009). In practice, most researchers in model systems now rely on empirical criteria to evaluate the significance of the Tajima's D test statistic.

7.4.8 Analyses of linkage disequilibrium: haplotype-based tests

As researchers began to sample molecular variation more densely along chromosomes, new statistical tests were developed to identify patterns of LD consistent with positive selection. As described above, positive selection will lead to hitchhiking of neutral variants that are linked to the selected site, and the extent of the shared haplotype will depend on the strength of selection, the timing of selection, and the amount of recombination. One complication in these tests is that closely related individuals are also expected to share extended haplotypes; thus, population structure can bias estimates of haplotype length (Oleksyk *et al.* 2010).

The pairwise haplotype sharing (PHS) test, which compares the length of haplotypes flanking the gene of interest to those observed throughout the genome, is commonly used in plant molecular population genetics. To minimize the effects of population structure, each allele's haplotype is standardized by the average haplotype length at all sampled loci. An empirical approach is used to assess significance; alleles that are in the >95th percentile of the PHS distribution may be candidates for further study (Toomajian *et al.* 2006).

7.4.9 Comparing diversity to divergence: McDonald-Kreitman and HKA tests

Another set of molecular population genetic tests seeks evidence of natural selection by comparing the amount of variation within a species to the level of divergence between species. Under the null hypothesis of neutrality, the neutral mutation rate should determine the accumulation both of intraspecific polymorphisms and of interspecific nucleotide substitutions. Thus, for a neutrally evolving gene, the level of intraspecific nucleotide diversity should be proportional to interspecific divergence. Deviations from this expectation may be due to the action of selection. The McDonald-Kreitman test (McDonald & Kreitman 1991) and the Hudson Kreitman Aguade (HKA) test (Hudson *et al.* 1987) are two commonly used methods to test this null hypothesis.

The HKA test asks whether two or more loci show the same ratio of intraspecific diversity to interspecific divergence, as would be expected under neutrality. If a candidate gene has a disproportionately low level of diversity relative to divergence, then the gene (or a closely linked site) may have been a target of positive selection. An elevated level of diversity relative to divergence may indicate maintenance of a balanced polymorphism. Although originally used to test for differences between two portions of the same gene (Hudson *et al.* 1987), the HKA test has been adapted to compare a candidate gene to one or more putatively neutral genes (multilocus HKA; Wang & Hey 1996). The multilocus HKA approach has been further expanded using a maximum likelihood method (MLHKA; Wright & Charlesworth 2004). MLHKA permits testing of alternative hypotheses regarding the action of selection within a multilocus data set (e.g., on particular genes within a pathway) in comparison to a set of neutral reference genes. In addition, the MLHKA provides an estimate of the selection parameter for each locus.

In contrast to the HKA test, the McDonald-Kreitman test focuses on polymorphism and divergence for a single coding sequence (McDonald & Kreitman 1991). If a coding sequence is evolving neutrally, the proportion of nonsynonymous to synonymous substitutions should be similar for intraspecific polymorphisms and for interspecific fixed differences. Often, the test is implemented with polymorphism data for only one species; significance of the test is evaluated using standard methods for a 2×2 contingency table.

In the McDonald-Kreitman test, a significant excess of fixed replacement substitutions may indicate positive selection, which contributes to divergence of the gene's sequence between the two species. An excess of polymorphic replacement substitutions may reflect maintenance of a balanced polymorphism. However, the presence of many polymorphic replacements can also be consistent with relaxed purifying selection. The McDonald-Kreitman approach has also been adapted to estimate the strength of selection on amino acid replacements using a Poisson random field model (MKPRF; Bustamante *et al.* 2002).

7.4.10 Detecting local adaptation: population differentiation and reduced variability

Detecting local adaptation requires sampling of sequence diversity both within and among populations; appropriate molecular data include SNPs,

AFLPs, or SSRs. Data must be obtained both from the locus of interest and from neutral markers. The most commonly used tests to detect local adaptation are based on estimates of F_{ST}, which summarizes the level of sequence differentiation between populations. A locus selected during local adaptation should display greater differentiation than neutrally evolving loci (Beaumont & Balding 2004). In addition, tests that summarize the level of diversity for SSR markers, such as repeat number variance (lnRV) and expected heterozygosity (lnRH), can be used to detect the regional decrease in sequence diversity expected to occur during local adaptation (Schlötterer & Dieringer 2005). The significance of these tests is typically assessed by comparing to an empirical distribution of the test statistic derived from putatively neutral loci (Stinchcombe & Hoekstra 2008). In addition to their application in studies of local adaptation, these tests have also been used to detect the effects of artificial selection during crop domestication (see Section 7.6.2).

7.5 Bringing it all together—the need for thorough testing of adaptive hypotheses

7.5.1 Techniques for testing the functional consequences of polymorphisms

Whether candidate genes are identified by QTL mapping or by finding loci that have signatures of selection (or ideally by a combination of both methods), one next needs to demonstrate that (1) the gene is polymorphic; (2) the gene contributes to the aspect of plant form being studied; (3) alleles are functionally divergent; and (4) the allelic differences in gene function contribute to phenotypic variation (Weigel & Nordborg 2005). An additional criterion, that the same QTL or gene is implicated in phenotypic variation in more than one species or population, has also been proposed (Weigel & Nordborg 2005); however, this criterion may be overly restrictive.

First and foremost, the candidate gene should contain intraspecific DNA polymorphisms that distinguish haplotypes. In the case of a species-wide selective sweep, in which there is little intraspecific diversity, divergence from a closely related outgroup may be considered instead. Molecular population genetic studies of adjacent loci can also provide insights into whether the candidate itself is a target of selection or whether its sequence variation reflects the correlated effects of selection at a nearby gene (Shepard & Purugganan 2003).

The next step is to show that that the candidate gene actually affects the trait of interest; this is particularly important if the candidate was identified based on its signature of selection. Evidence supporting the role of the candidate in the trait of interest includes loss-of-function analyses and gene expression studies. In model systems, many loss-of-function mutants are publicly available; alternatively, the candidate gene's expression can be knocked down stably using RNA interference (RNAi) approaches (Ossowski *et al.* 2008) or transiently using virus-induced gene silencing (VIGS) (Becker & Lange 2010). The VIGS approach is particularly helpful in nonmodel plants, as it does not require stable transformation and whole-plant regeneration. One caveat is that due to genetic redundancy, a single-gene knockdown may not generate an obvious phenotype. In this case, determining the candidate gene's spatial domain and temporal pattern of transcription via RNA *in situ* hybridization can at least demonstrate that the gene is expressed in the organ of interest.

Evidence that alternative alleles of the candidate gene differ functionally can be obtained using either bioinformatics or *in vitro* experiments. Bioinformatics approaches include predictions of the effect of amino acid replacements on protein secondary and tertiary structure. Comparisons to homologous proteins can also provide insights into the effect of replacements on protein motifs (e.g., PolyPhen, which was developed to assess the consequences of human sequence variants; Adzhubei *et al.* 2010). Finally, laboratory-induced point mutations available from forward genetic studies can provide invaluable information about structure–function relationships. If the likely function of the candidate gene is known, biochemical assays on purified proteins can test whether alternative alleles have different properties. Appropriate assays might include enzyme activity, DNA binding affinity, or protein stability. These sorts of *in vitro* assays, coupled with site-directed mutagenesis and protein expression in heterologous systems to test specific biochemical hypotheses, provide an excellent way to evaluate the possible effects of candidate QTN on protein function.

Similar approaches can be used to evaluate the functional consequences of regulatory polymorphisms. Differences in the level of transcription in plants carrying different haplotypes can be tested using quantitative reverse transcription PCR, while the spatial domain of expression can be studied using RNA *in situ* hybridization. Differences in possible cis-regulatory elements can be detected using promoter prediction software (Yamamoto & Obokata 2008); these *in silico* predictions can then be tested using reporter transgenes.

Ideally, *in planta* evidence of the functional consequences of polymorphisms would be obtained through targeted gene replacement in which alternative haplotypes of the candidate gene would be inserted into the plant genome at the gene's endogenous location. This replacement of one haplotype with another would avoid the confounding variable of position effect on gene expression. Unfortunately, homologous recombination approaches do not yet work reliably in higher plants. The use of zinc-finger nucleases to mediate gene replacement and site-directed mutagenesis *in planta* shows promise (Weinthal *et al.* 2010), but has not yet been widely implemented.

Quantitative complementation is a widely used approach to assess functional variation between polymorphic sequences *in vivo* (Mackay 2001). These experiments compare the phenotypic effect of alternative haplotypes when they are introduced into null mutants. Functionally divergent haplotypes should demonstrate different degrees of rescue of the null mutant phenotype. The alternative haplotypes can either be crossed into the null mutant or they can be introduced as transgene constructs. Advantages of crossing to the null mutant and observing the F1 are that the candidate gene is expressed from its endogenous location, and the experiment can be performed quickly. The transgenic approach usually takes longer and introduces potential variability in gene expression because of position effects. However, transgenes offer more flexibility in evaluating the effects of regulatory versus coding polymorphisms; many studies attempt to isolate the effects of coding sequence mutations by using the same promoter (e.g., CAMV 35S) to regulate expression of both haplotypes. If transgenic analysis does suggest divergence of haplotype function, site-directed mutagenesis of transgenes provides an outstanding opportunity to test the phenotypic effects of single QTN. To date, these single nucleotide tests have been performed only rarely; however, for an excellent example, see functional analyses of *ENHANCER OF TRY AND CPC 2*, a regulator of trichome number variation in *A. thaliana* (Hilscher *et al.* 2009). When null mutants are not available in a non-model system, some success has been achieved in cross-species expression studies in which alternative haplotypes are introduced as transgenes into *A. thaliana* mutants.

Introgression of alternative haplotypes into a common genetic background or reciprocal introgressions into the two alternative backgrounds can also be used to evaluate phenotypic effects. Near isogenic lines (NILs), which differ only in the genomic region containing the candidate gene, can be created by crossing lines bearing two different haplotypes followed by repeated backcrossing to each parental line; PCR-based markers can be used to track inheritance of the desired haplotype. A similar approach, heterogeneous inbred families (HIFs), is particularly useful when working with QTL mapping populations (Tuinstra *et al.* 1997). A limitation of NILs and HIFs is that it is very difficult to introgress only the candidate gene into each background; adjacent loci will also be carried along and can potentially modify the phenotype. In addition, these methods may not be feasible in species with long generation times.

A final method for evaluating the phenotypic consequences of polymorphisms is fine-scale association mapping. Using population-based mapping approaches, the correlation between particular SNP variants and phenotypic variation can be assessed. For example, a recent study in European aspen (*Populus tremula*) identified two SNPs in the *phytochrome B2* coding sequence that contribute to variation in the timing of bud set (Ingvarsson *et al.* 2008). This method is most useful for species in which LD decays over short distances; if LD is extensive, association mapping will not be able to distinguish among the effects of different SNPs within a locus.

7.5.2 Testing adaptive hypotheses

Perhaps the most challenging aspect of genomic studies of intraspecific adaptation in plant form is demonstrating that the variation observed is not only phenotypically relevant, but also ecologically and evolutionarily relevant. A general scheme for comparing the effect of alternative alleles has been proposed by Shindo and co-workers (2007): after the alternative alleles have been introgressed or transformed into a common genetic background, the plants can be grown in a common garden experiment to estimate fitness. Fitness can be estimated by both number of progeny and by the success of those progeny in surviving to reproduce. Competition experiments can be used to estimate the relative fitness of each genotype when grown in mixed populations. Ideally, these competition experiments would be performed in the field; however, practical considerations (including regulations on field trials involving transgenics) may make field trials difficult. If these competition experiments are conducted in the greenhouse, care should be taken to simulate realistic abiotic conditions. These sorts of experiments are useful in evaluating putative targets of either positive or balancing selection.

For traits that show evidence of positive selection contributing to local adaptation, two additional approaches can be used to test the adaptive hypothesis. First, if local adaptation has occurred, then correlations among phenotypic variation, genetic variation, and geographical variation will often be evident. For example, a latitudinal cline may be observed in which the phenotypes of individuals differ according to their latitude of origin. If the geographical distribution of genetic polymorphisms parallels the distribution of phenotypes, then the hypothesis adaptive evolution is supported. The gene underlying the trait variation would also be expected to show greater differentiation (i.e., higher F_{ST}) among populations than neutrally evolving loci. Second, the fitness of each accession (or NILs bearing alternative alleles) in reciprocal transplant experiments can be evaluated. In reciprocal transplants, the plants are grown in the field in the environments from which each accession was originally collected. If local adaptation has occurred, then the plants bearing the local allele should show greater relative fitness in their native environment (Mitchell-Olds & Schmitt 2006).

7.6 Case studies in molecular population genomic approaches to the evolution of plant form

The remainder of this chapter consists of four case studies, each applying different methods to study the evolution of plant form.

7.6.1 Case study 1: Identifying novel components of developmental regulatory networks—*BREVIS RADIX* in *Arabidopsis* roots

Although forward and reverse genetic approaches have provided us with a tremendous amount of information about the genes that regulate development in extensively studied model systems, there are clearly still gaps in our knowledge. Analysis of the genetic underpinning of natural phenotypic variation represents one tool available to close this gap. For example, mapping loci contributing to variation in primary root elongation in *A. thaliana* led to the discovery of *BREVIS RADIX* (*BRX*), the founding member of a class of plant-specific proteins that may act as transcriptional regulators. *BRX* influences both cell proliferation and elongation in roots, possibly by mediating cross talk between auxin and brassinosteroid signaling (Mouchel *et al.* 2004; Mouchel *et al.* 2006).

The identification of *BRX* began with a phenotypic survey of *A. thaliana* accessions originally collected from different parts of the world. Among these accessions, primary root length in young seedlings varied by nearly sixfold. Crossing the accession with the shortest roots (Uk-1 from Umkirch, Germany) to one with moderately long roots (Sav-0 from Slavice, Czech Republic) identified a single locus with a major effect on root elongation. Fine mapping followed by sequencing of the ten genes in the 45 kb mapping interval revealed that the *BRX* gene in Uk-1 ($BRX^{Uk\text{-}1}$) contains a premature stop codon and encodes a truncated protein. The effect of this QTN on root length has been verified both by introgression of the $BRX^{Uk\text{-}1}$ haplotype into long-rooted accessions to produce short-rooted plants and by complementation of short-rooted lines with a functional *BRX* transgene (Mouchel *et al.* 2004; Shindo *et al.* 2008).

Accumulating evidence suggests that $BRX^{Uk\text{-}1}$ is a rare, deleterious allele. The premature stop codon in Uk-1 has not been observed in any other accessions (Mouchel *et al.* 2004), and resampling in the Umkirch region failed to identify other individuals carrying this mutation. Competition experiments performed in the laboratory suggest that the $BRX^{Uk\text{-}1}$ loss-of-function allele may decrease whole-plant fitness. When near-isogenic lines bearing functional or nonfunctional *BRX* alleles were grown in mixed flats, plants with the loss-of-function allele bore substantially fewer and smaller fruits. Interestingly, the two NILs had very similar fruit production when grown in pure stands, suggesting that the fitness consequences of the $BRX^{Uk\text{-}1}$ allele could depend on the composition of the population in which a plant grows (Shindo *et al.* 2008).

Recent molecular population genetic analyses identified a more common natural variant of *BRX* that contributes to increased root elongation. Resequencing of 19 accessions revealed a haplotype ($BRX^{Lc\text{-}0}$), which bears two amino acid replacements as well as a 21 bp deletion that removes seven amino acids from the BRX protein. The $BRX^{Lc\text{-}0}$ haplotype was present in ~7% of the accessions studied. Both QTL mapping in RILs and transgenic complementation experiments suggest that the $BRX^{Lc\text{-}0}$ haplotype can increase root length by about 10% compared to haplotypes encoding the full-length protein (Beuchat *et al.* 2010).

The presence of natural *BRX* variants with contrasting effects on root elongation suggests that this gene could be a target of adaptive evolution in *Arabidopsis*. However, preliminary molecular population genetic analyses using a small number of individuals do not identify a clear signature of selection in the *BRX* coding sequence. The level of nucleotide diversity is fairly typical for an *Arabidopsis* gene ($\theta_W = 0.0107$, $\pi = 0.0060$), and Tajima's D is negative, but not unusually so ($D = -1.31$) (Beuchat *et al.* 2010). Additional molecular population genetic analyses with a larger species-wide sample may provide more insight into the evolutionary history of *BRX*. Population-level surveys of the frequency of hyperactive haplotypes in the areas from which these accessions were originally collected may also prove informative.

Although several studies in *A. thaliana* have used family-based mapping strategies to identify QTL contributing to root system architecture, they have used only six accessions, capturing only a fraction of the available phenotypic variation (Mouchel *et al.* 2004; Loudet *et al.* 2005; Fitz Gerald *et al.* 2006). Further mapping experiments, perhaps in more diverse mapping populations, should help uncover additional loci involved in root system architecture, plasticity, and growth rates. Molecular population genetic analyses of genes known to regulate root development will also identify loci that contribute to this broad phenotypic variation.

7.6.2 Case study 2: Identifying potential targets of positive selection via a genomic scan in a nonmodel species—signatures of selection in sunflower SSRs

The identification of recent targets of positive selection is one of the most common goals of genome-wide molecular population genetic surveys. These studies have been particularly successfully in model domesticated species, such as maize and rice, for which extensive genomic resources are available (see Wright *et al.* 2005). Application of these techniques in nonmodel systems has been hampered by difficulties in identifying appropriate polymorphisms for scanning throughout the genome. Recent work searching for signatures of positive selection during sunflower (*Helianthus annuus)* domestication and improvement provides insights into how genome-wide, anonymous scans can be implemented in a species for which less genomic information is available. The results of these scans can then be integrated with existing QTL maps and validated using SNP-based molecular population genetic analyses (Chapman *et al.* 2008).

The sunflower study identified putative targets of positive selection by comparing sequence diversity in 24 wild sunflower populations, 8 primitive landraces, and 16 improved cultivars. Rather than sequence entire genes, the study initially focused on SSRs. SSRs are tandem repeats of 1–5 bp (e.g., $(AT)_n$), for which the number of repeats (n) varies among alleles. Mutations leading to changes in repeat number occur frequently, making SSRs a rich source of polymorphisms within a species, and differences in the size of an SSR can be readily scored using PCR followed by high-resolution electrophoresis (Jones *et al.* 2009).

Although SSRs are common in intergenic DNA, they also occur within genic regions. To more efficiently identify signatures of selection in transcribed regions, the sunflower survey used bioinformatics tools to find SSRs in an expressed sequence tag (EST) database. Four plants of each accession were genotyped at 492 SSRs (Chapman *et al.* 2008).

An SSR in a genomic region subjected to positive selection during domestication should show a smaller range of variation in repeat number and fewer alleles in the selected population compared to the wild population. This signature of selection should be present even if the actual target of selection was a closely linked site and not the SSR itself. For each SSR, the wild, primitive, and improved lines were compared using three test statistics: repeat number variance (lnRV), expected heterozygosity (lnRH), and population differentiation (F_{ST}) (Beaumont & Balding 2004; Schlötterer & Dieringer2005). If a gene was positively selected during domestication or improvement, lnRH and lnRV should decrease while F_{ST} should increase. Only loci that were outliers in the genome-wide distribution for at least two test statistics were considered selection candidates; this dual criterion was intended to minimize false positives (Chapman *et al.* 2008).

Overall, SSRs in the primitive landraces and the improved cultivars were less diverse than in the wild populations, as would be expected after a bottleneck occurring during domestication. Among the 492 SSRs surveyed, ~7% met the dual criteria for signatures of positive selection during either domestication or improvement. These loci showed dramatically lower levels of diversity within groups and substantially more differentiation between groups compared to putatively neutral loci that lacked a significant signature of selection. Based on comparisons of the EST loci to sequences in GenBank, several of these candidate loci are likely to be involved in amino acid metabolism, flowering time, or seed development; in addition, several genes of unknown function were identified (Chapman *et al.* 2008).

Map-based and molecular population genetic approaches were used to further evaluate the SSR selection candidates. To test whether the candidate loci might contribute to phenotypic differences in wild versus improved sunflower, the ESTs were mapped in RILs that had been previously used in QTL mapping of domestication-related traits. Most of the candidate genes were located in chromosomal intervals containing QTL for important domestication traits such as flowering time and seed size; however, it is important to remember that these QTL mapping intervals are quite large and likely contain many genes, so colocalization cannot be considered strong evidence of a functional role. To conduct SNP-based tests for selection, 500–1000 bp fragments of six candidate loci and seven neutral loci were sequenced from ~15 individuals from the wild, domesticated, and improved lines. In the domesticated and improved lines, the low level of nucleotide diversity and the significant results of MLHKA tests for selection support the conclusion that the candidate loci may have been subjected to positive selection during domestication (Chapman *et al.* 2008).

The sunflower study convincingly demonstrates that SSR-based methods can begin the process of identifying genes that were positively selected during domestication. A particular strength of the approach is that it successfully identified signatures of selection in genes of unknown function. Because LD decays over fairly short distances (1–2 kb) in sunflower, it is likely that the gene containing a selected SSR was indeed the target of selection (Chapman *et al.* 2008). However, surveys of sequence variation in genes flanking the putatively selected candidate genes are required to verify that the signature of selection is not the result of selection on an adjacent gene. One important limitation of the EST-anchored SSR approach is that only ~9% of genes have SSRs in transcripts (Chapman *et al.* 2008). Despite this limitation, similar studies may prove fruitful in evaluating the process of domestication in non-model crops or in anonymously searching for signatures of positive selection or local adaptation in wild species for which EST libraries are available.

7.6.3 Case study 3: Microevolution of a small gene family—phytochromes in *Arabidopsis*

Plants have evolved sophisticated light sensing systems that allow them to monitor the amount, wavelength, and photoperiod of light in their environment. This monitoring provides a plant with information about the direction from which light is available, the extent to which neighboring plants are present, and the season of the year. The light environment then influences a variety of developmental processes, including germination, seedling de-etiolation, phototropism and shade avoidance, and the initiation of reproductive development (Franklin & Quail 2010). In perennial plants, light cues may also play an important role in bud flush (the resumption of shoot growth in the spring) and bud set (the cessation of growth in the fall) (Luquez *et al.* 2008).

Phytochromes are primarily responsible for sensing red and far-red light, and most angiosperm genomes contain at least three genes that encode phytochrome proteins: *PHYA*, *B*, and *C* (Franklin & Quail 2010). Intraspecific variation in phytochrome sequences has been detected in several plant species, including sorghum (White *et al.* 2004), *Pinus sylvestris* (Garcia-Gil *et al.* 2003), *Cardamine nipponica* (Ikeda *et al.* 2009), *P. tremula* (Ingvarsson *et al.* 2008), and *A. thaliana* (Maloof *et al.* 2001; Balasubramanian *et al.* 2006; Filiault *et al.* 2008). Species for which QTL that regulate flowering time and/or plant morphology have colocalized with phytochrome genes include tomato (Jimenez-Gomez *et al.* 2007), pearl millet (Saidou *et al.* 2009), and *A. thaliana* (Maloof *et al.* 2001; Balasubramanian *et al.* 2006; Filiault *et al.* 2008).

In *A. thaliana*, five phytochrome genes (*PHYA*, *B*, *C*, *D*, and *E*) encode apoproteins that have overlapping, but distinct, roles in the regulation of plant development and physiology. All of the phytochromes play roles in seedling de-etiolation and leaf architecture. PhyA, a light-labile protein, is primarily responsible for responses to very low levels of light and to bright far-red light during germination. PhyA also contributes to entrainment of the circadian clock and to photoperiod perception. PhyB is the predominant sensor of red light; it is both the most abundant of the light-stable phytochromes and the protein with the broadest array of physiological roles, including germination, suppression of shade avoidance, and repression of flowering. PhyE and PhyD appear to have arisen by duplication of PhyB, and their physiological roles overlap considerably with those of PhyB. PhyC is involved in photoperiod perception and in repression of flowering under short-day conditions. For an excellent, extensive review of the roles of each phytochrome in *A. thaliana*, see Franklin and Quail (2010).

Co-immunoprecipitation experiments have demonstrated that active phytochromes can consist of homodimers or heterodimers. In *A. thaliana*, PhyC and PhyE exist mainly in heterodimers: PhyC with PhyB, and PhyE with PhyD or PhyB. In addition to participating in these heterodimers, PhyB and PhyD are found in homodimers, and they can also dimerize with each other. PhyA seems to act exclusively as a homodimer (Clack *et al.* 2009). The number of genes and the fact that they act in different combinations suggest that signaling via phytochromes is complex and that epistasis is likely to be apparent in quantitative genetic studies involving phytochromes.

Much of the molecular population genomic and quantitative genetic work in *A. thaliana* has focused on variation in hypocotyl elongation (an aspect of seedling de-etiolation) and flowering time among accessions grown under varying light conditions. Intriguingly, plants originally from southern latitudes tend to have longer hypocotyls when grown under white light than those from northern latitudes (Maloof *et al.* 2001; Stenøien *et al.* 2002). Additional studies have suggested that an accession's flowering time also correlates with its latitude of origin (Lempe *et al.* 2005). The possibility that sensitivity to light cues follows a latitudinal cline prompted a series of studies of natural genetic variation in the *A. thaliana PHYA*, *PHYB*, and *PHYC* genes. *PHYD*, which has been studied less extensively, is discussed in Section 7.6.7.

7.6.4 Phytochrome A

The first natural variant in *PHYA* was identified in a survey of *A. thaliana* accessions germinated under far-red light. One accession (Lm-2) had unusually long hypocotyls, which indicates that it responds more weakly to far-red light than other accessions. Sequencing of $PHYA^{Lm\text{-}2}$ revealed a single amino acid replacement (Met548Thr; Maloof *et al.* 2001). This Met falls in the PHY domain of the protein and is otherwise highly conserved in all higher plant phytochromes (Rockwell *et al.* 2006), suggesting that it is functionally important. When the same substitution was engineered into the paralogous *PHYB* gene, a decrease in light sensitivity was also observed. In addition, biochemical studies showed that the $PhyA^{Lm\text{-}2}$ protein has different spectral properties, less autophosphorylation activity, and increased stability in the light compared to $PhyA^{Col}$ (Maloof *et al.* 2001).

Despite the identification of a natural variant with clear biochemical consequences, molecular population genetic analyses suggest that *PHYA* has not undergone adaptive evolution in *A. thaliana*. Neither the level of nucleotide diversity nor the site frequency spectrum are unusual for an *A. thaliana* gene, and no tests for positive selection showed significant results (Flowers *et al.* 2009).

7.6.5 Phytochrome B

Family-based mapping suggested that QTL for hypocotyl elongation in red and white light colocalized with *PHYB* (Borevitz *et al.* 2002). To identify possible QTN, *PHYB* was resequenced from 33 accessions; five replacement polymorphisms differed in the parents (L*er* and Cvi) of the original mapping population. Using association mapping, three of these amino acid substitutions were shown to significantly correlate with hypocotyl length in red light. To test the ability of the L*er* and Cvi haplotypes to confer differential hypocotyl elongation, quantitative complementation of a *phyB* mutant was performed with 35S: $PHYB^{Ler}$ and 35S: $PHYB^{Cvi}$ transgenes. Plants bearing the 35S: $PHYB^{Ler}$ construct had significantly shorter hypocotyls, suggesting that this haplotype encodes a PhyB protein with greater sensitivity to red light. In regression analyses, the amino acid replacement Leu143Ile (where Ile is derived and is present in PhyB^{Ler}) shows the strongest association with phenotypic variation (Filiault *et al.* 2008); however, no site-directed mutagenesis studies have explicitly tested this statistical correlation.

Molecular population genetic analyses of *PHYB* fail to provide compelling support for any particular selective scenario. The 143Ile allele associated with increased light sensitivity is fairly common (frequency = 18%) among the accessions in the hypocotyl-length association study (Filiault *et al.* 2008). However, there is strong evidence neither of an ongoing sweep (PHS test) nor of maintenance of a balanced polymorphism (moderate levels of diversity, negative value of Tajima's D) (Flowers *et al.* 2009). In two studies, results of McDonald-Kreitman tests were not significant (Filiault *et al.* 2008; Flowers *et al.* 2009); however, a third study did detect a significant excess of polymorphic replacements (Mathews & McBreen 2008). Although *PHYB* is known to affect flowering time in loss-of-function mutants (Blázquez & Weigel 1999), no significant correlations between *PHYB* SNPs and flowering time were observed in an association mapping experiment (Ehrenreich *et al.* 2009).

7.6.6 Phytochrome C

Phenotypic effects of natural variation in *PHYC* have also been observed. These studies began with the identification of an accession (Fr-2) that flowers fairly rapidly even under short-day conditions (Lempe *et al.* 2005), suggesting decreased sensitivity to day length. Analyses of an F2 mapping population followed by positional cloning showed that *PHYC* was the likely culprit: the

PHYC^{Fr2} haplotype includes a nonsense mutation that leads to a truncated protein (Balasubramanian *et al.* 2006).

A survey of diversity in 29 *A. thaliana* accessions showed that *PHYC* sequences cluster into two well-supported haplotype groups distinguished by eight amino acid replacement polymorphisms and an indel 500 bp upstream of the start of translation. One haplotype group was present in 72% of accessions, including L*er* and the *PHYC*$^{Fr\text{-}2}$ nonsense allele; the second haplotype group was present in 28% of accessions, including the Col accession commonly used in laboratory studies (Balasubramanian *et al.* 2006). Further molecular population genetic analyses of *PHYC* have not yet been published.

Both mapping and quantitative complementation experiments suggest that the two *PHYC* haplotypes have functionally diverged. QTL for flowering time and/or hypocotyl elongation colocalize with *PHYC* in three different RIL mapping populations that bear contrasting alleles at this locus (reviewed by Balasubramanian *et al.* 2006). Association mapping also supports diversification of function in *PHYC*; among 115 accessions, the *PHYC* genotype was significantly correlated with hypocotyl length variation in several different lighting conditions. In white light, the hypocotyls of plants bearing the *PHYC*^{Ler} haplotype were longer than those with the *PHYC*^{Col} haplotype, which suggests that PhyC^{Ler} is less sensitive to light than PhyC^{Col}. A quantitative complementation experiment that assayed the effect of alternative *PHYC* alleles on flowering time also supports the hypothesis that PhyC^{Ler} is less sensitive to light cues. When crossed into a *phyC* mutant in the Col background, *PHYC*^{Ler} plants flowered more rapidly than plants bearing the *PHYC*^{Col} haplotype (Balasubramanian *et al.* 2006).

The ecological significance of this variation in PhyC light sensitivity remains unclear. Initial reports provided some evidence for a latitudinal cline with the more sensitive *PHYC*^{Col} haplogroup being more common in accessions native to Northern latitudes (Balasubramanian *et al.* 2006). However, contradictory results were obtained in growth chamber experiments testing whether variation in *PHYC* contributes to clinal adaptation in different photoperiods. Under long-day conditions, no significant association was observed between flowering time and *PHYC* genotype (Samis *et al.* 2008), which is consistent with prior observations that *phyC* loss-of-function mutants flower normally in long days (Monte *et al.* 2003). In short-day chambers, plants bearing *PHYC*^{Ler} flowered more rapidly than *PHYC*^{Col} plants, but no significant association was observed between *PHYC* haplotype and latitude of origin. Instead of the expected latitudinal cline in *PHYC*, Samis *et al.* (2008) observed a somewhat puzzling longitudinal cline in which *PHYC*^{Ler} is more common among western accessions, but flowering tends to occur *later* among the western accessions (the reverse of what the *PHYC* genotype would predict). The later flowering time of western accessions does make some ecological sense, as it correlates with evidence that fall begins later in western Europe (Samis *et al.* 2008).

Intriguingly, although there was no significant latitudinal cline of *PHYC* genotypes, a significant interaction of *PHYC* × latitude was observed. This interaction effect suggests that flowering time variation may be mediated in part by epistasis between *PHYC* haplotypes and a second locus with variants distributed along a latitudinal cline. The interacting alleles of *PHYC* and this unidentified factor have opposing effects on phenotype, which may account for the counterintuitive longitudinal clines in which the early-flowering $PHYC^{Ler}$ haplotype is more common in the later-flowering western accessions (Samis *et al.* 2008).

In light of the independent detection of developmentally relevant variation in *PHYB* and *PHYC*, and the emerging evidence that *PHYC* forms an obligate heterodimer with *PHYB*, it is tempting to speculate that *PHYB* might be one of the factors that is epistatic to *PHYC*. As noted previously, the $PHYB^{Ler}$ haplotype confers greater light sensitivity while $PHYB^{Cvi}$ confers lower sensitivity when introduced into the Col genetic background (Filiault *et al.* 2008). Thus, the physiological activity of a PhyB-PhyC heterodimer might depend on which particular haplotypes are present for each gene. If this is the case, family-based mapping experiments using parents with contrasting *PHYB* and *PHYC* haplotypes should show evidence of both transgression and epistasis for *PHYC*-mediated phenotypic variation. Mapping populations with contrasting *PHYC* haplotypes already exist (see Balasubramanian *et al.* 2006), so conducting this experiment on a variety of phytochrome-regulated traits should be straightforward.

7.6.7 Case study 4: Combining association mapping and population genomics—the *Arabidopsis* flowering time network

Perhaps one of the most exciting opportunities arising from the integration of population genomics, systems biology, and developmental biology is the ability to study how microevolution shapes the genetic networks that regulate plant growth and development. Among the first of these explorations of network behavior are recent association mapping and molecular population genomic analyses of the flowering time regulatory network in *Arabidopsis* (Ehrenreich *et al.* 2009; Flowers *et al.* 2009).

The *Arabidopsis* flowering time network (see Figure 7.2) is one of the most intensively studied in plant developmental genetics, and its relevance to plant fitness is clear. As an annual, *A. thaliana* has only one chance to reproduce; the transition from vegetative to reproductive development must therefore be carefully synchronized with environmental conditions. The onset of reproduction, or flowering time, is regulated by a variety of endogenous and exogenous cues, including day length, temperature, plant age, and nutritional status (reviewed by Putterill *et al.* 2004; Engelmann & Purugganan 2006; Amasino 2010). Developmental genetic studies have identified >60 genes

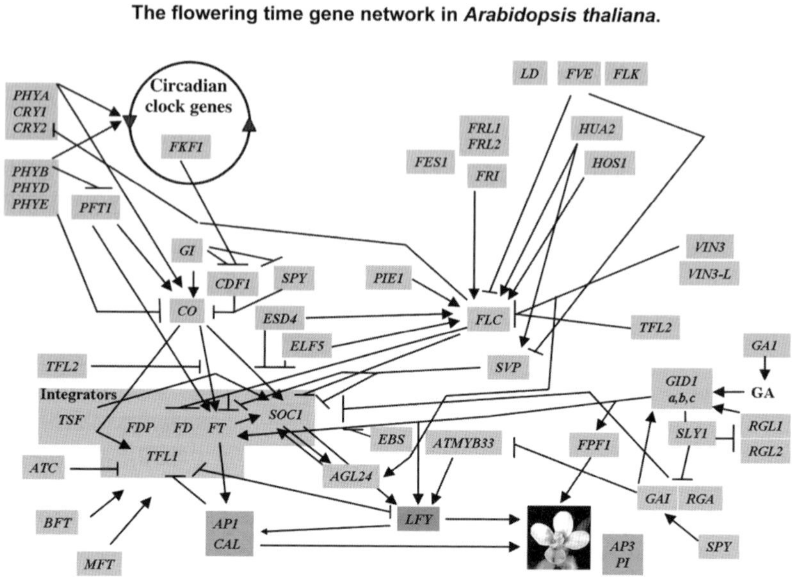

MOLECULAR BIOLOGY AND EVOLUTION

Figure 7.2 The flowering time genetic network in *Arabidopsis thaliana*. Reproduced from Flowers *et al.* (2009) with permission of Oxford University Press. This network highlights the relationships only among loci included in the studies by Ehrenreich *et al.* (2009) and Flowers *et al.* (2009). For a more comprehensive network, see Ehrenreich *et al.* (2009). Relationships among loci were inferred from the literature. The integrators (*FD, FDP, FT, SOC1, TFL1*, and *TSF*) integrate photoperiod, vernalization, and gibberellin signals to regulate activity of the inflorescence meristem. Genes that regulate floral meristem identity (*LFY, AP1, CAL*) or floral organ identity (*AP3, PI*) were not included in the flowering time association mapping or sequence diversity analyses work; see Shepard (2007) for more information about the evolutionary dynamics of these genes. (For a color version of this figure, see Plate 7.1.)

that are involved in this regulation, including genes that encode photoreceptors, transcriptional regulators, and novel proteins. The best characterized of these genes contribute to four pathways: (1) photoperiod sensing, which monitors day length; (2) vernalization, which permits flowering only after sustained exposure to cold; (3) gibberellic acid signaling, which serves as an endogenous regulator of the transition to flowering; and (4) floral integrators, which receive input from sensory pathways and act as effectors in the floral transition. To determine which members of the flowering time network bear signatures of selection or contribute to phenotypic variation, 52 genes were subjected to molecular population genetic analyses (Flowers *et al.* 2009) and evaluated in an association mapping study (Ehrenreich *et al.* 2009).

Overall, the flowering time genes as well as random background genes displayed a significant excess of replacement polymorphisms segregating within *A. thaliana* compared to replacement fixations between *A. thaliana* and *Arabidopsis lyrata*. To evaluate whether the replacement polymorphisms are likely to be neutral or deleterious, the frequency of the derived allele of each SNP was determined. Throughout the genome, derived replacement SNPs tended to be present in lower frequencies compared to the derived synonymous SNPs. The disproportionate number of low-frequency replacement SNPs may represent slightly deleterious substitutions that have not yet been removed by purifying selection. Noncoding DNA also contained an excess of low-frequency derived SNPs, suggesting that there is also an accumulation of potentially deleterious variants in intergenic regions (Flowers *et al.* 2009).

Population genetic statistics for the candidate genes in the flowering time network were compared to the background loci and to each other when partitioned according to pathway (photoperiod, vernalization, gibberellic acid, and floral integrator). When all candidates were compared to the background loci, no significant differences were observed in the distributions of nucleotide diversity (π), polymorphism frequency (Tajima's D), nonsynonymous divergence (Ka), the ratio of nonsynonymous : synonymous divergence (Ka/Ks), or the estimated strength of selection of amino acid mutations ($2N_es$, from the MKPRF test). Thus, the data fail to support the hypothesis that the flowering time network as a whole evolves differently from the rest of the genome. However, among the flowering time genes, genes involved in photoperiod sensing and gibberellin signaling displayed lower divergence (Ka/Ks) than those involved in vernalization or in the background dataset. One possible explanation for the lower divergence in the photoperiod and gibberellin pathways is that these genes are active in many developmental and physiological contexts; because mutations are more likely to have pleiotropic effects they may be subject to greater constraint (Flowers *et al.* 2009).

Each candidate gene was also analyzed individually using a battery of molecular population genetic tests, but strong evidence of positive selection or balanced polymorphisms was quite rare. None of the genes showed extraordinarily low or high nucleotide diversity (π), and only *FRL2*, which encodes a novel protein in the vernalization pathway, had a significantly negative value of Tajima's D. Some support for positive selection was observed in the PHS test; alleles of ten genes were located in unusually long haplotypes, but these long haplotypes might reflect strong selection at adjacent loci rather than on the candidate gene. Evidence of positive selection in either a McDonald-Kreitman test or MKPRK test was observed for three genes: *FLK*, which regulates RNA metabolism; *CO*, a transcription factor that integrates photoperiod signals; and *MYB33*, a transcription factor involved in gibberellin signaling. In addition, an excess of polymorphic replacements was observed at four loci: *PHYD*, a phytochrome; *PIE1*, a chromatin modification factor; *RGL1*, a negative regulator of gibberellin signaling; and *TFL1*, a regulator of inflorescence development. These abundant replacement polymorphisms may indicate relaxed purifying selection or a balanced polymorphism. Nonsense mutations, possibly leading to loss of function, were observed in *PHYD* and in three members of a gene family (*FRI*, *FRL1*, and *FRL2*) involved in vernalization (Flowers *et al.* 2009). While these results hint at some interesting evolutionary dynamics (particularly a possible signature of selection in *CO*), overall the study suggests that most genes in this network are evolving neutrally.

The companion association mapping study evaluated whether these candidate genes contribute to variation in flowering time under laboratory conditions (Ehrenreich *et al.* 2009). The mapping population consisted of 275 *A. thaliana* accessions for which the number of days until the onset of reproductive development (bolting) and the number of rosette leaves present at bolting were scored in growth chambers under long-day and short-day conditions. The markers used for mapping were haplotype-tagging SNPs (htSNPs) developed from the resequencing data for the candidate genes (Flowers *et al.* 2009) and 318 background loci (Nordborg *et al.* 2005). Rather than testing associations with every SNP, clusters of SNPs in perfect LD were identified, and then one of those SNPs was chosen to represent the haplotype. The htSNP approach minimizes the multiple testing concerns that arise when mapping many SNPs per gene and reduces the amount of genotyping required. In an effort to minimize false positives arising from relationships among the sampled accessions, the analysis model took into account estimates of population structure and kinship derived from htSNPs in the background loci (Ehrenreich *et al.* 2009).

More than half of the candidate genes had at least one significant htSNP association with flowering time or rosette leaf number in at least one environment; however, only ten of these genes had significant values using more stringent false discovery rate criteria. Each of the ten genes explained 2–9% of the phenotypic variance in days to flowering. In an attempt to validate the associations, the experiment was replicated in 360 MAGIC lines, which were generated by intercrossing 19 *A. thaliana* accessions and then creating RILs (Kover *et al.* 2009). Only 20% of the associations identified in the population sample were also observed in MAGIC lines, and 12 different genes showed significant associations with flowering time only in the MAGIC lines (Ehrenreich *et al.* 2009).

Based on the results of both mapping experiments, seven genes (14% of the candidates) showed reasonably strong and consistent associations with flowering time variation: *CO* and *PHYD* in the photoperiod pathway; *FRI*, *VIN3*, and *VIN3-L* in the vernalization pathway; *FLC*, a target of both the autonomous and vernalization pathways; and *GA1* in the gibberellin signaling pathway (Ehrenreich *et al.* 2009). The identification of *FRI* and *FLC* as contributors to flowering time variation is reassuring, as the roles of *FRI* loss-of-function alleles and *FLC* expression variants in modulating the vernalization response are well established (Johanson *et al.* 2000; Caicedo *et al.* 2004; Lempe *et al.* 2005). The evidence that the transcription factor *CO* might be involved in flowering time is exciting. *CO* plays a central role in integrating the inputs from photoreceptors in the photoperiodic control of flowering, but had not been identified as a candidate for adaptive evolution in previous mapping studies.

Integrating the results of these two studies illustrates of one the major challenges of the molecular population genomic approach: detecting consistent relationships between putative signatures of selection and phenotypic variation. Three candidates with significant associations (*FRI*, *CO*, and *GA1*) have alleles with unusually long haplotypes, which might indicate positive selection, and *CO* also has a selection coefficient consistent with positive selection. However, three other candidates associated with flowering time variation bear no significant signature of selection (*FLC*, *VIN3*, and *VIN3-L*). In *PHYD*, the presence of multiple nonsense alleles, an excess of replacement polymorphisms, and a significant association with flowering time variation could indicate that modulation of *PHYD* activity is adaptive. However, the molecular population genetic data are also consistent with *PHYD* being subject to less stringent selection, as might be expected in a fairly recent duplicate.

One possible explanation for the minimal detection of signatures of selection in the flowering time network is that much of the selection on flowering time acts at the population level. In a species-wide survey, the signatures of local adaptation may not be readily apparent. Investigations of flowering time variation and sequence diversity within and among *A. thaliana* populations will permit testing of this hypothesis, and the ongoing 1001 Genomes Project for *A. thaliana* (Weigel & Mott 2009) will soon give us the tools to undertake such an analysis.

7.7 Conclusion

In the last few years, we have made dramatic leaps in our ability to gather DNA sequence data rapidly and inexpensively. These technological advances leave us poised to make equally dramatic jumps in our understanding of how genotypic variation contributes to adaptive variation in plant form. Extensive resequencing, new QTL mapping strategies, and explorations of the epigenome, transcriptome, and proteome in model species such as *A. thaliana* and *P. tremula* will allow us to tackle fundamental questions about evolution in natural setting, while similar studies in domesticated crops will provide insights into the effects of strong artificial selection. More extensive population-based resequencing in *A. thaliana* will offer a new view into the genome-wide effects of local adaptation. The ever-lower costs of sequencing will facilitate the expansion of molecular population genomic approaches into many more nonmodel systems, each with its own set of fascinating developmental and evolutionary questions. In addition, improved methods for high-throughput phenotypic analysis will allow us to investigate aspects of plant form, such as leaf morphology, that have received comparatively little attention to date. Integrating the results of these studies into a systems biology framework will show how adaptation is abetted and constrained by the architecture of genetic networks.

References

Adzhubei, I.A., Schmidt, S., Peshkin, L., *et al.* (2010) A method and server for predicting damaging missense mutations. *Nature Methods*, **7**, 248–249.

Alonso-Blanco, C., Aarts, M.G.M., Bentsink, L., *et al.* (2009) What has natural variation taught us about plant development, physiology, and adaptation? *Plant Cell*, **21**, 1877–1896.

Amasino, R. (2010) Seasonal and developmental timing of flowering. *The Plant Journal*, **61**, 1001–1013.

Balasubramanian, S., Sureshkumar, S., Agrawal, M., *et al.* (2006) The *PHYTOCHROME C* photoreceptor gene mediates natural variation in flowering and growth responses of *Arabidopsis thaliana*. *Nature Genetics*, **38**, 711–715.

Beaumont, M.A. & Balding, D.J. (2004) Identifying adaptive genetic divergence among populations from genome scans. *Molecular Ecology*, **13**, 969–980.

Becker, A. & Lange, M. (2010) VIGS - genomics goes functional. *Trends in Plant Science*, **15**, 1–4.

Beuchat, J., Li, S.W., Ragni, L., *et al.* (2010) A hyperactive quantitative trait locus allele of *Arabidopsis BRX* contributes to natural variation in root growth vigor. *Proceedings of the National Academy of Sciences of the United States of America*, **107**, 8475–8480.

Blázquez, M.A. & Weigel, D. (1999) Independent regulation of flowering by Phytochrome B and gibberellins in *Arabidopsis*. *Plant Physiology*, **120**, 1025–1032.

Borevitz, J.O., Maloof, J.N., Lutes, J., *et al.* (2002) Quantitative trait loci controlling light and hormone response in two accessions of *Arabidopsis thaliana*. *Genetics*, **160**, 683–696.

Bossdorf, O., Richards, C.L. & Pigliucci, M. (2008) Epigenetics for ecologists. *Ecology Letters*, **11**, 106–115.

Buckler, E.S., Holland, J.B., Bradbury, P.J., *et al.* (2009) The genetic architecture of maize flowering time. *Science*, **325**, 714–718.

Bustamante, C.D., Nielsen, R., Sawyer, S.A., *et al.* (2002) The cost of inbreeding in *Arabidopsis*. *Nature*, **416**, 531–534.

Caicedo, A.L., Stinchcombe, J.R., Olsen, K.M., *et al.* (2004) Epistatic interaction between *Arabidopsis FRI* and *FLC* flowering time genes generates a latitudinal cline in a life history trait. *Proceedings of the National Academy of Sciences of the United States of America*, **101**, 15670–15675.

Chapman, M.A., Pashley, C.H., Wenzler, J., *et al.* (2008) A genomic scan for selection reveals candidates for genes involved in the evolution of cultivated sunflower (*Helianthus annuus*). *Plant Cell*, **20**, 2931–2945.

Clack, T., Shokry, A., Moffet, M., *et al.* (2009) Obligate heterodimerization of *Arabidopsis* Phytochromes C and E and interaction with the PIF3 basic helix-loop-helix transcription factor. *Plant Cell*, **21**, 786–799.

Clark, R.M., Wagler, T.N., Quijada, P., *et al.* (2006) A distant upstream enhancer at the maize domestication gene *tb1* has pleiotropic effects on plant and inflorescent architecture. *Nature Genetics*, **38**, 594–597.

Clark, R.M., Schweikert, G., Toomajian, C., *et al.* (2007) Common sequence polymorphisms shaping genetic diversity in *Arabidopsis thaliana*. *Science*, **317**, 338–342.

Cong, B., Barrero, L.S. & Tanksley, S.D. (2008) Regulatory change in YABBY-like transcription factor led to evolution of extreme fruit size during tomato domestication. *Nature Genetics*, **40**, 800–804.

Cong, B., Liu, J.P. & Tanksley, S.D. (2002) Natural alleles at a tomato fruit size quantitative trait locus differ by heterochronic regulatory mutations. *Proceedings of the National Academy of Sciences of the United States of America*, **99**, 13606–13611.

Cubas, P., Vincent, C. & Coen, E. (1999) An epigenetic mutation responsible for natural variation in floral symmetry. *Nature*, **401**, 157–161.

Deschamps, S. & Campbell, M.A. (2010) Utilization of next-generation sequencing platforms in plant genomics and genetic variant discovery. *Molecular Breeding*, **25**, 553–570.

Ehrenreich, I.M., Hanzawa, Y., Chou, L., *et al.* (2009) Candidate gene association mapping of *Arabidopsis* flowering time. *Genetics*, **183**, 325–335.

Engelmann, K. & Purugganan, M. (2006) The molecular evolutionary ecology of plant development: flowering time in *Arabidopsis thaliana*. *Advances in Botanical Research: Incorporating Advances in Plant Pathology*, **44**, 507–526.

Fay, J.C. & Wu, C.I. (2000) Hitchhiking under positive Darwinian selection. *Genetics*, **155**, 1405–1413.

Filiault, D.L., Wessinger, C.A., Dinneny, J.R., *et al.* (2008) Amino acid polymorphisms in *Arabidopsis* phytochrome B cause differential responses to light. *Proceedings of the National Academy of Sciences of the United States of America*, **105**, 3157–3162.

Fitz Gerald, J.N., Lehti-Shiu, M.D., Ingram, P.A., *et al.* (2006) Identification of quantitative trait loci that regulate *Arabidopsis* root system size and plasticity. *Genetics*, **172**, 485–498.

Flowers, J.M., Hanzawa, Y., Hall, M.C., *et al.* (2009) Population genomics of the *Arabidopsis thaliana* flowering time gene network. *Molecular Biology and Evolution*, **26**, 2475–2486.

Franklin, K.A. & Quail, P.H. (2010) Phytochrome functions in *Arabidopsis* development. *Journal of Experimental Botany*, **61**, 11–24.

Frewen, B.E., Chen, T.H.H., Howe, G.T., *et al.* (2000) Quantitative trait loci and candidate gene mapping of bud set and bud flush in *Populus*. *Genetics*, **154**, 837–845.

Fu, Y.X. & Li, W.H. (1993) Statistical tests of neutrality of mutations. *Genetics*, **133**, 693–709.

García-Gil, M.R., Mikkonen, M. & Savolainen, O. (2003) Nucleotide diversity at two phytochrome loci along a latitudinal cline in *Pinus sylvestris*. *Molecular Ecology*, **12**, 1195–1206.

Gaut, B.S. & Long, A.D. (2003) The lowdown on linkage disequilibrium. *Plant Cell*, **15**, 1502–1506.

Gore, M.A., Chia, J.-M., Elshire, R.J., *et al.* (2009) A first-generation haplotype map of maize. *Science*, **326**, 1115–1117.

He, Y. (2009) Control of the transition to flowering by chromatin modifications. *Molecular Plant*, **2**, 554–564.

Hilscher, J., Schlötterer, C. & Hauser, M.T. (2009) A single amino acid replacement in ETC2 shapes trichome patterning in natural *Arabidopsis* populations. *Current Biology*, **19**, 1747–1751.

Holland, J.B. (2007) Genetic architecture of complex traits in plants. *Current Opinion in Plant Biology*, **10**, 156–161.

Hudson, R.R., Kreitman, M. & Aguade, M. (1987) A test of neutral molecular evolution based on nucleotide data. *Genetics*, **116**, 153–159.

Ikeda, H., Fujii, N. & Setoguchi, H. (2009) Molecular evolution of phytochromes in *Cardamine nipponica* (Brassicaceae) suggests the involvement of *PHYE* in local adaptation. *Genetics*, **182**, 603–614.

Ingvarsson, P.K., Garcia, M.V., Luquez, V., *et al.* (2008) Nucleotide polymorphism and phenotypic associations within and around the *phytochrome B2* locus in European aspen (*Populus tremula*, Salicaceae). *Genetics*, **178**, 2217–2226.

Jiménez-Gómez, J.M., Alonso-Blanco, C., Borja, A., *et al.* (2007) Quantitative genetic analysis of flowering time in tomato. *Genome*, **50**, 303–315.

Johannes, F., Porcher, E., Teixeira, F.K., *et al.* (2009) Assessing the impact of transgenerational epigenetic variation on complex traits. *PLoS Genetics*, **5**, e1000530.

Johanson, U., West, J., Lister, C., *et al.* (2000) Molecular analysis of *FRIGIDA*, a major determinant of natural variation in *Arabidopsis* flowering time. *Science*, **290**, 344–347.

Jones, N., Ougham, H., Thomas, H., *et al.* (2009) Markers and mapping revisited: finding your gene. *New Phytologist*, **183**, 935–966.

Kimura, M. (1983) *The Neutral Theory of Molecular Evolution*, Cambridge University Press, Cambridge.

Kliebenstein, D. (2009) Quantitative genomics: Analyzing intraspecific variation using global gene expression polymorphisms or eQTLs. *Annual Review of Plant Biology*, **60**, 93–114.

Kover, P.X., Valdar, W., Trakalo, J., *et al.* (2009) A multiparent advanced generation inter-cross to fine-map quantitative traits in *Arabidopsis thaliana*. *PLoS Genetics* **5**, e1000551.

Lempe, J., Balasubramanian, S., Sureshkumar, S., *et al.* (2005) Diversity of flowering responses in wild *Arabidopsis thaliana* strains. *PLoS Genetics* **1**, e6.

Li, W.-H. (1997) *Molecular Evolution*, Sinauer Associates, Inc., Sunderland, MA.

Librado, P. & Rozas, J. (2009) DnaSP v5: a software for comprehensive analysis of DNA polymorphism data. *Bioinformatics*, **25**, 1451–1452.

Loudet, O., Gaudon, V., Trubuil, A., *et al.* (2005) Quantitative trait loci controlling root growth and architecture in *Arabidopsis thaliana* confirmed by heterogeneous inbred family. *Theoretical and Applied Genetics*, **110**, 742–753.

Luquez, V., Hall, D., Albrectsen, B.R., *et al.* (2008) Natural phenological variation in aspen (*Populus tremula*): the SwAsp collection. *Tree Genetics & Genomes*, **4**, 279–292.

Mackay, T.F.C. (2001) Quantitative trait loci in *Drosophila*. *Nature Reviews Genetics*, **2**, 11–20.

Maloof, J.N., Borevitz, J.O., Dabi, T., *et al.* (2001) Natural variation in light sensitivity of *Arabidopsis*. *Nature Genetics*, **29**, 441–446.

Mathews, S. & McBreen, K. (2008) Phylogenetic relationships of B-related phytochromes in the Brassicaceae: Redundancy and the persistence of phytochrome D. *Molecular Phylogenetics and Evolution*, **49**, 411–423.

McDonald, J.H. & Kreitman, M. (1991) Adaptive protein evolution at the *Adh* locus in *Drosophila*. *Nature*, **351**, 652–654.

McNally, K.L., Childs, K.L., Bohnert, R., *et al.* (2009) Genomewide SNP variation reveals relationships among landraces and modern varieties of rice. *Proceedings of the National Academy of Sciences of the United States of America*, **106**, 12273–12278.

Mitchell-Olds, T. & Schmitt, J. (2006) Genetic mechanisms and evolutionary significance of natural variation in *Arabidopsis*. *Nature*, **441**, 947–952.

Monte, E., Alonso, J.M., Ecker, J.R., *et al.* (2003) Isolation and characterization of *phyC* mutants in *Arabidopsis* reveals complex crosstalk between phytochrome signaling pathways. *Plant Cell*, **15**, 1962–1980.

Mouchel, C.F., Briggs, G.C. & Hardtke, C.S. (2004) Natural genetic variation in *Arabidopsis* identifies *BREVIS RADIX*, a novel regulator of cell proliferation and elongation in the root. *Genes & Development*, **18**, 700–714.

Mouchel, C.F., Osmont, K.S. & Hardtke, C.S. (2006) *BRX* mediates feedback between brassinosteroid levels and auxin signalling in root growth. *Nature*, **443**, 458–461.

Myles, S., Peiffer, J., Brown, P.J., *et al.* (2009) Association mapping: Critical considerations shift from genotyping to experimental design. *Plant Cell*, **21**, 2194–2202.

Nordborg, M., Hu, T.T., Ishino, Y., *et al.* (2005) The pattern of polymorphism in *Arabidopsis thaliana*. *PloS Biology*, **3**, e196.

Nordborg, M. & Innan, H. (2003) The genealogy of sequences containing multiple sites subject to strong selection in a subdivided population. *Genetics*, **163**, 1201–1213.

Nordborg, M. & Weigel, D. (2008) Next-generation genetics in plants. *Nature*, **456**, 720–723.

Oleksyk, T.K., Smith, M.W. & O'Brien, S.J. (2010) Genome-wide scans for footprints of natural selection. *Philosophical Transactions of the Royal Society B-Biological Sciences*, **365**, 185–205.

Olsen, K.M., Womack, A., Garrett, A.R., *et al.* (2002) Contrasting evolutionary forces in the *Arabidopsis thaliana* floral developmental pathway. *Genetics*, **160**, 1641–1650.

Ossowski, S., Schwab, R. & Weigel, D. (2008) Gene silencing in plants using artificial microRNAs and other small RNAs. *Plant Journal*, **53**, 674–690.

Przeworski, M. (2002) The signature of positive selection at randomly chosen loci. *Genetics*, **160**, 1179–1189.

Putterill, J., Laurie, R. & Macknight, R. (2004) It's time to flower: the genetic control of flowering time. *Bioessays*, **26**, 363–73.

Richards, E.J. (2011) Natural epigenetic variation in plant species: a view from the field. *Current Opinion in Plant Biology*, **14**, 204–209.

Rockwell, N.C., Su, Y.-S. & Lagarias, J.C. (2006) Phytochrome structure and signaling mechanisms. *Annual Review of Plant Biology*, **57**, 837–858.

Ross-Ibarra, J., Morrell, P.L. & Gaut, B.S. (2007) Plant domestication, a unique opportunity to identify the genetic basis of adaptation. *Proceedings of the National Academy of Sciences of the United States of America*, **104**, 8641–8648.

Saïdou, A.A., Mariac, C., Luong, V., *et al.* (2009) Association studies identify natural variation at *PHYC* linked to flowering time and morphological variation in pearl millet. *Genetics*, **182**, 899–910.

Salvi, S. & Tuberosa, R. (2005) To clone or not to clone plant QTLs: present and future challenges. *Trends in Plant Science*, **10**, 297–304.

Samis, K.E., Heath, K.D. & Stinchcombe, J.R. (2008) Discordant longitudinal clines in flowering time and *PHYTOCHROME C* in *Arabidopsis thaliana*. *Evolution*, **62**, 2971–2983.

Schlötterer, C. (2002) Towards a molecular characterization of adaptation in local populations. *Current Opinion in Genetics & Development*, **12**, 683–687.

Schlötterer, C. & Dieringer, D. (2005) A novel test statistic for the identification of local selective sweeps based on microsatellite gene diversity. In: *Selective Sweep* (ed. D. Niurminsky). pp. 55–64. Kluwer Academic Publishers, Boston.

Schneeberger, K. & Weigel, D. (2011) Fast-forward genetics enabled by new sequencing technologies. *Trends in Plant Science*, **16**, 282–288.

Shepard, K.A. (2007) The molecular population genetics of shoot development in *Arabidopsis thaliana*. *Genetica*, **129**, 19–36.

Shepard, K.A. & Purugganan, M.D. (2003) Molecular population genetics of the *Arabidopsis CLAVATA2* region: The genomic scale of variation and selection in a selfing species. *Genetics*, **163**, 1083–1095.

Shindo, C., Bernasconi, G. & Hardtke, C.S. (2007) Natural genetic variation in *Arabidopsis*: tools, traits and prospects for evolutionary ecology. *Annals of Botany*, **99**, 1043–1054.

Shindo, C., Bernasconi, G. & Hardtke, C.S. (2008) Intraspecific competition reveals conditional fitness effects of single gene polymorphism at the *Arabidopsis* root growth regulator *BRX*. *New Phytologist*, **180**, 71–80.

Stenøien, H.K., Fenster, C.B., Kuittinen, H., *et al.* (2002) Quantifying latitudinal clines to light responses in natural populations of *Arabidopsis thaliana* (Brassicaceae). *American Journal of Botany*, **89**, 1604–1608.

Stinchcombe, J.R. & Hoekstra, H.E. (2008) Combining population genomics and quantitative genetics: finding the genes underlying ecologically important traits. *Heredity*, **100**, 158–170.

Tajima, F. (1983) Evolutionary relationship of DNA sequences in finite populations. *Genetics*, **105**, 437–460.

Tajima, F. (1989) Statistical method for testing the neutral mutation hypothesis by DNA polymorphism. *Genetics*, **123**, 585–595.

Toomajian, C., Hu, T.T., Aranzana, M.J., *et al.* (2006) A nonparametric test reveals selection for rapid flowering in the *Arabidopsis* genome. *PLoS Biology* **4**, e137.

Tuinstra, M.R., Ejeta, G. & Goldsbrough, P.B. (1997) Heterogeneous inbred family (HIF) analysis: a method for developing near-isogenic lines that differ at quantitative trait loci. *Theoretical and Applied Genetics*, **95**, 1005–1011.

Van Eeuwijk, F.A., Bink, M.C.A.M., Chenu, K., *et al.* (2010) Detection and use of QTL for complex traits in multiple environments. *Current Opinion in Plant Biology*, **13**, 193–205.

Vaughn, M.W., Tanurdžíc, M., Lippman, Z., *et al.* (2007) Epigenetic natural variation in *Arabidopsis thaliana*. *PLoS Biology*, **5**, e174.

Wang, R.L. & Hey, J. (1996) The speciation history of *Drosophila pseudoobscura* and close relatives: Inferences from DNA sequence variation at the *Period* locus. *Genetics*, **144**, 1113–1126.

Watterson, G.A. (1975) Number of segregating sites in genetical models without recombination. *Theoretical Population Biology*, **7**, 256–276.

Weigel, D. & Mott, R. (2009) The 1001 Genomes Project for *Arabidopsis thaliana*. *Genome Biology*, **10**, 107.

Weigel, D. & Nordborg, M. (2005) Natural variation in *Arabidopsis*. How do we find the causal genes? *Plant Physiology*, **138**, 567–568.

Weinig, C., Ungerer, M.C., Dorn, L.A., *et al.* (2002) Novel loci control variation in reproductive timing in *Arabidopsis thaliana* in natural environments. *Genetics*, **162**, 1875–1884.

Weinthal, D., Tovkach, A., Zeevi, V., *et al.* (2010) Genome editing in plant cells by zinc finger nucleases. *Trends in Plant Science*, **15**, 308–321.

Werner, J.D., Borevitz, J.O., Warthmann, N., *et al.* (2005) Quantitative trait locus mapping and DNA array hybridization identify an *FLM* deletion as a cause for natural flowering-time variation. *Proceedings of the National Academy of Sciences of the United States of America*, **102**, 2460–2465.

White, G.M., Hamblin, M.T. & Kresovich, S. (2004) Molecular evolution of the phytochrome gene family in sorghum: Changing rates of synonymous and replacement evolution. *Molecular Biology and Evolution*, **21**, 716–723.
Wright, S.I., Bi, I.V., Schroeder, S.G., *et al.* (2005) The effects of artificial selection on the maize genome. *Science*, **308**, 1310–1314.
Wright, S.I. & Charlesworth, B. (2004) The HKA test revisited: a maximum-likelihood-ratio test of the standard neutral model. *Genetics*, **168**, 1071–1076.
Wright, S.I. & Gaut, B.S. (2005) Molecular population genetics and the search for adaptive evolution in plants. *Molecular Biology and Evolution*, **22**, 506–519.
Yamamoto, Y.Y. & Obokata, J. (2008) PPDB: a plant promoter database. *Nucleic Acids Research*, **36**, D977–D981.
Yu, J., Holland, J.B., McMullen, M.D., *et al.* (2008) Genetic design and statistical power of nested association mapping in maize. *Genetics*, **178**, 539–551.
Yu, J., Pressoir, G., Briggs, W.H., *et al.* (2006) A unified mixed-model method for association mapping that accounts for multiple levels of relatedness. *Nature Genetics*, **38**, 203–208.
Zhou, L., Zhang, J., Yan, J., *et al.* (2011) Two transposable element insertions are causative mutations for the major domestication gene *teosinte branched 1* in modern maize. *Cell Research*, **21**, 1267–1270.

Annual Plant Reviews (2013) **45**, 227–276 http://onlinelibrary.wiley.com
doi: 10.1002/9781118305881.ch8

Chapter 8

COMPARATIVE EVOLUTIONARY GENOMICS OF LAND PLANTS

Amy Litt

The New York Botanical Garden, Bronx, NY, USA

Abstract: The evolution of land plants is characterized by increasing adaptation to the terrestrial environment and increasing structural complexity. Flowering plants have more elaborate structure and life cycles than do mosses, and have diversified and colonized a wider variety of habitats. Nonetheless, analyses suggest that on average flowering plants have neither larger genomes nor more genes than mosses. Genome size varies dramatically among embryophytes but the most significant factor appears to be amplification of transposable elements, which not only mediate genome size but dynamically affect genome structure and function as well. Along with whole genome duplications and lineage-specific expansion of individual gene families, they are responsible for shaping land plant genomes, generating substantial differences in genome size, gene number, and gene order even among closely related cultivars. Comparative analyses of fully sequenced genomes have shed substantial light on the dynamics of genome and gene family evolution; they are just beginning, however, to help us understand the molecular basis of the key adaptations that have allowed plants to thrive on land.

Keywords: complexity; domestication; gene family expansion; genome size; transposable elements; whole genome duplication.

Green plants began to colonize land approximately 450 million years ago. During the course of land plant (embryophyte) evolution, morphology became more complex as plants developed adaptations that enhanced survival and diversification in a terrestrial environment, including conducting and support tissue, desiccation tolerance, true roots and leaves, seeds, and flowers among many others. Embryophytes colonized an extraordinarily broad range of environmental conditions, adapting to seasonal and extreme variation in temperature, sunlight, and rainfall and to different substrates and biotic communities. Plants expanded their repertoire of defensive compounds

Annual Plant Reviews Volume 45: The Evolution of Plant Form, First Edition.
Edited by Barbara A. Ambrose and Michael Purugganan.

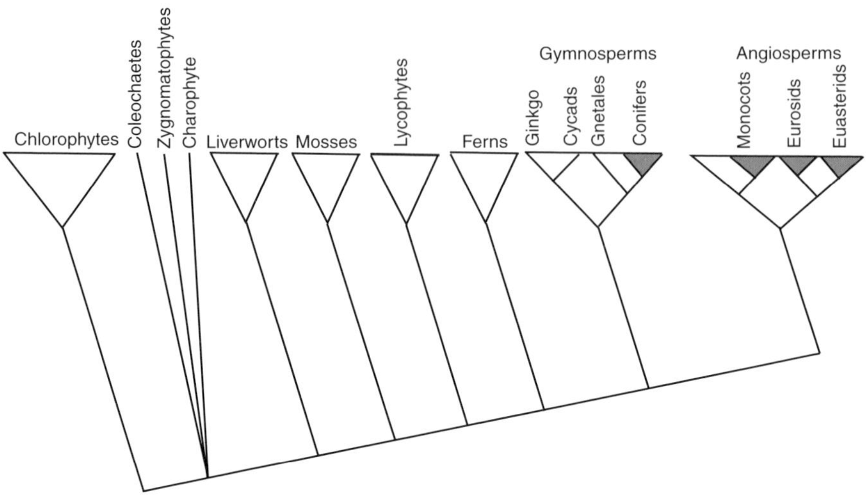

Figure 8.1 Simplified phylogeny of Viridiplantae. The sister group to land plants is unresolved. Gray triangles represent select clades within angiosperms and gymnosperms that are discussed in the text.

to enhance protection against herbivory, damage from the sun, microbial pathogens, and other environmental factors from which they are unable to escape physically. In addition, embryophytes also underwent a profound life history shift, from a dominant haploid gametophyte generation with a nonphotosynthetic sporophyte to a dominant diploid sporophyte and much reduced nonphotosynthetic gametophytes. Thus, the history of land plants is one of enhanced ability to survive in varied terrestrial environments as well as increasing complexity in vegetative and reproductive morphology. Reflecting this increase in complexity and diversity, the nuclear genomes of land plants probably are also more complex and include more sequence diversity than those of their ancestors. A comparative analysis of green plant genome sequences therefore can help us identify the unique features of the genomes of mosses, lycophytes, ferns, gymnosperms, and angiosperms that underlie the structural, physiological, and phytochemical features of each of these groups, including such innovations as vasculature, a sporophyte-dominant life cycle, and ovules and carpels, which provided increased adaptation to land and allowed increased diversification of land plant clades.

Land plants (embryophytes) evolved from green algae, although the sister group is unclear (Figure 8.1). Recent molecular phylogenetic studies based on mitochondrial, nuclear, or chloroplast sequence data have variously resolved the closest relatives of land plants as the Characeae, Coleochaetaceae, or Zygnemetaceae. These algal families are all members of the charophyte (or streptophyte) lineage (*sensu* Hall & Delwiche 2007), which includes several other algal families as well as the land plants (Hall & Delwiche 2007 and references therein; Qiu 2008; Becker & Marin 2009). Thus, an understanding of the structure and function of the genomes of charophyte algal species would provide a starting point for understanding the changes that occurred as embryophytes became more diverse and more variously adapted to the wide variety of terrestrial habitats. Unfortunately, as of this date, no charophyte algal nuclear genomes have been sequenced, so we must look somewhat further afield to the chlorophyte algae (Table 8.1; Figure 8.1) to help us understand the evolution of the land plant genome.

8.1 Evolution of nuclear genome size

An examination of seven sequenced chlorophyte nuclear genomes shows that most are considerably smaller than most embryophyte genomes and contain 30–50% fewer genes (Tirichine & Bowler 2011). *Ostreococcus tauri*, the smallest known single-celled eukaryotic organism, has a genome of 12.5 Mb with on the order of 8000 genes[1]; at the other end of the spectrum of sequenced green algae are *Chlamydomonas reinhardtii* and *Volvox carteri*, with genomes an order of magnitude larger and containing twice as many genes (120 Mb, 15 000 genes and 140 Mb, 14 500 genes, respectively) (Table 8.1; Derelle *et al.* 2006; Merchant *et al.* 2007; Prochnik *et al.* 2010). Embryophyte genomes sequenced to date range in size from the upper end of these chlorophyte genomes (the lycophyte *Selaginella moellendorffii*, Selaginellaceae, 110 Mb and 22 500 genes; referred to in this chapter as Selaginella; Banks *et al.* 2011) to the genome of maize (*Zea mays*, Poaceae, 2300 Mb, 32 500 genes; Schnable *et al.* 2009). These numbers based on sequenced genomes suggest a larger average genome size in embryophytes compared to chlorophytes, perhaps reflecting the greater complexity of the land plants. However, only a minute fraction of plant genomes have been sequenced; although sample size is still small, a better idea of the range of genome sizes can be obtained from cytological estimates.

Flow cytometry (Arumuganathan & Earle 1991) and Feulgen staining of nuclei (e.g., Bennett & Smith 1991) indicate that many fern and gymnosperm genomes are considerably larger than those of already sequenced algae and embryophytes, which goes a long way to explaining why no fern or gymnosperm genome has yet been fully sequenced. Fern genome size estimates range from 750 Mb (*Azolla microphylla*, Azollaceae) to 64 000 Mb (*Ophioglossum petiolatum*, Ophioglossaceae), with an average genome size of 11 000 Mb, two orders of magnitude larger than the chlorophyte genomes and one order of magnitude larger than the maize genome (e.g., Obermayer *et al.* 2002; Murray *et al.* 2010). Estimated gymnosperm genome sizes range from 2000–4000 Mb for species of Gnetum to 35 000 Mb for pine species (Pinaceae). The mean size of pine genomes is 23 000 Mb; the genomes of other gymnosperms are smaller on average but still relatively large (cycads, 14 500 Mb; Gnetales, 12 000 Mb; ginkgo, 10 000 Mb) (e.g., Ohri & Khoshoo 1986; Wakamiya *et al.* 1993; Davies *et al.* 1997; Leitch *et al.* 2001; Murray *et al.* 2010). In fact, nuclear

[1] Throughout this chapter, numbers have been rounded off to facilitate comparison.

Table 8.1 Nuclear genome size in fully sequenced genomes of algae and land plants

		Genome size (Mb)	# Chroms (*N*)	# Protein-coding genes	# WGD	%TE
Chlamydomonas reinhardtii	Chlorophyte (alga)	120	17	15 150		
Chlorella variabilis	Chlorophyte (alga)	45	12	9 800		
Ostreococcus tauri	Chlorophyte (alga)	12	20	8 000		
Volvox carteri	Chlorophyte (alga)	140	14	14 500		
Physcomitrella patens (Physcomitrella)	Bryophyte (moss)	480	27	32 300	1	50
Selaginella moellendorffii (Selaginella)	Lycophyte (club moss)	110	10	22 300	0	40
Arabidopsis thaliana (Arabidopsis)	Angiosperm (eudicot)	135	5	27 400	Hex +2	15
Carica papaya (papaya)	Angiosperm (eudicot)	370	9	24 700	Hex	50
Cucumis sativus (cucumber)	Angiosperm (eudicot)	370	7	26 700	Hex	25
Glycine max[a] (soybean)	Angiosperm (eudicot)	975	20	66 000	Hex +2	60
Malus X domestica (apple)	Angiosperm (eudicot)	740		57 400		40
Populus trichocarpa (cottonwood)	Angiosperm (eudicot)	400	19	40 700	Hex +1	40
Ricinus communis (castor bean)	Angiosperm (eudicot)	400		31 200	Hex	50
Vitis vinifera (grape)	Angiosperm (eudicot)	490	19	33 500	Hex	40
Brachypodium distachyon (Brachypodium)	Angiosperm (monocot)	270	5	25 500	2	20
Oryza sativa (rice)	Angiosperm (monocot)	390	12	40 600	2	35
Sorghum bicolor (sorghum)	Angiosperm (monocot)	730	10	34 500	2	55
Zea mays (maize)	Angiosperm (monocot)	2300	10	32 500	3	85

Source: Data from Arabidopsis Genome Initiative (2000); Blanc & Wolfe (2004); Paterson *et al.* (2004); International Rice Genome Sequencing Project (2005); Derelle *et al.* (2006); Tuskan *et al.* (2006); Jaillon *et al.* (2007); Merchant *et al.* (2007); Ouyang *et al.* (2007); Velasco *et al.* (2007); Lang *et al.* (2008); Ming *et al.* (2008); Rensing *et al.* (2008); Tanaka *et al.* (2008); Huang *et al.* (2009); Paterson *et al.* (2009); Schnable *et al.* (2009); Blanc *et al.* (2010); Chan *et al.* (2010); Murray *et al.* (2010); Vogel *et al.* (2010); Prochnik *et al.* (2010); Schmutz *et al.* (2010); Tang *et al.* (2010); Banks *et al.* (2011); Zhu and Buell (2007); http://rice.plantbiology.msu.edu/ (release 6.1); http://www.phytozome.net/sorghum; http://www.arabidopsis.org/.

Mb, Megabases; # Chroms, number of chromosomes; WGD, whole genome duplication; %TE, percentage of genome comprising transposable elements; Hex, paleohexaploid event (see the text).

All numbers except chromosome numbers are rounded off.

[a]Tetraploid.

genome sizes of sequenced species (Table 8.1) may not represent the range of genome sizes in a particular plant lineage, and may tend to be on the small size for more efficient and effective sequencing.

Cytological estimates of the genome size of other lycophyte genera show that genome size in this group varies by two orders of magnitude, from the 110 Mb genomes of *Selaginella* species to the 11 500 Mb genome of *Isoetes lacustris* (Isoetaceae) (Wang *et al.* 2005; Murray *et al.* 2010). Likewise, the smallest and largest flowering plant genomes known are not among the genomes that have been sequenced. Species of *Genlisea* and *Utricularia*, two parasitic genera in the Lentibulariaceae, have diminutive genomes of 65 Mb and 90 Mb, respectively (Greilhuber *et al.* 2006); at the other end of the spectrum is *Paris japonica* (Melianthaceae), in the monocot order Liliales, with an octoploid genome estimated at a staggering 150 000 Mb (Pellicer *et al.* 2010). The largest known diploid angiosperm genome is that of *Fritillaria platyptera* (Liliaceae), still a hefty 84 000 Mb. Although these species far outstrip estimates of even fern and gymnosperm genomes, the average angiosperm genome size is about 6000 Mb, suggesting these enormous genomes are atypical; indeed well over half the angiosperm genomes estimated to date are smaller than 5000 Mb and about one-third are under 1000 Mb (Murray *et al.* 2010). Nearly all sequenced angiosperm genomes are smaller than 1000 Mb (Table 8.1), and this biased representation should be borne in mind in discussions of genome evolution in land plants.

Although angiosperm genome size varies over 1–2000-fold (e.g., Bennett & Smith 1991; Schmidt 2002; Gaut & Ross-Ibarra 2008; Tang *et al.* 2008a), the skew toward smaller genomes has led to the hypothesis that large genomes are nonadaptive. This is manifest at multiple biological levels: species with large genomes may have lower rates of photosynthesis and growth as well as a lower speciation rate and reduced fitness in extreme environments (Knight *et al.* 2005; Gaut & Ross-Ibarra 2008). Thus, it would appear that smaller genomes have a fitness advantage. However, Smarda *et al.* (2010; see also Beaulieu 2010), in an experimental system, showed stabilizing selection on the genome of the grass *Festuca pallens* (Poaceae), which unlike the genomes of most species shows intraspecific variability. Selection was weak, but resulted in the elimination of both large and small genomes. They suggest that selection is primarily on genome functionality, particularly gene interactions and regulatory processes early in development, which will be less optimal in genomes at either of the size extremes. It is not clear whether these results can be extrapolated to other species, most of which do not show significant variation in genome size, or to differences in genome size among species or large plant groups.

Nonetheless, angiosperms, like mosses, lycophytes, ferns, and gymnosperms, appear to have expanded genomes relative to chlorophyte algae, suggesting adaptation to land and associated increased complexity favored increased genome size. However, genome size estimates for charophyte algae, the group that includes the sister group of angiosperms (Figure 8.1), range up to almost 20 000 Mb for *Chara contraria* (Characeae). In fact five of the six largest estimated genome sizes for green algae belong to species of *Chara* (Murray *et al.* 2010). It should be noted that we do not have information on gene content in *Chara* and it is suspected many of these species may be polyploid (Kapraun 2007), nonetheless the large genome sizes call into question the conclusion that the invasion of land led to a dramatic increase in genome size. Until we have resolved the sister-group of the embryophytes, and until we have sequenced genomes from the charophyte algae, we will not be able to evaluate changes in size, structure, and gene content of the green plant genome that occurred with the conquest of land.

Regardless of genomic changes that might have occurred with the colonization of land, a comparison of genome size and gene content among extant land plants shows that in spite of the apparent increase in organismal size and complexity that accompanied increased adaptation to terrestrial environments, there is no consistent trend among embryophytes toward larger genomes or greater gene number (Table 8.1). Gymnosperms have larger genomes than angiosperms, on average, and fern genomes are even larger. The genome of *Physcomitrella patens* (Funariaceae; referred to in this chapter as Physcomitrella), a moss that therefore belongs to one of the earliest diverging land plant lineages, is similar in size to that of the angiosperm black cottonwood tree (*Populus trichocarpa*, Salicaceae) (Table 8.1). There is also no correlation between genome size and chromosome number, although ferns set the upper limit on both. Ferns hold the record among living organisms with up to $N = 700$ (*Ophioglossum* species) (Khandelwal 1990; Barker & Wolf 2010), but pines, with relatively massive genomes, have $N = 12$, a number that is in line with chromosome counts characteristic of angiosperms and lycophytes with much smaller genome sizes and that is considerably smaller than chromosome counts in many mosses and green algae (Ahuja 2005; Morse *et al.* 2009; Murray *et al.* 2010 and references therein).

8.1.1 Gene number

A comparison of protein-coding gene number in sequenced genomes shows roughly similar numbers in the moss, lycophyte, and angiosperms; although the lycophyte genome has the smallest number of genes of the sequenced embryophytes, it is only slightly lower than estimates for angiosperm species such as papaya (*Carica papaya*, Caricaceae) or the grass *Brachypodium distachyon* (Poaceae; referred to in this chapter as Brachypodium), both of which are estimated to have fewer genes than the moss Physcomitrella (Table 8.1). Because whole genome sequences have not been generated for any fern or gymnosperm to date, estimates of gene complement have been based on indirect methods. Using methylation filtration to produce a gene-enriched fraction for sequencing, Rabinowicz *et al.* (2005) estimated the number of genes in the genome of the fern *Ceratopteris richardii* (Parkeriaceae) (for a fern, a modest 9000 Mb) at about 42 500, toward the high end of angiosperm numbers but within that range. In contrast, they estimated the gene complement of loblolly pine (*Pinus taeda*, Pinaceae, with a genome size of 21 500 Mb) at about 225 000, an order of magnitude greater than the number of genes in any genome sequenced to date (Rabinowicz *et al.* 2005). Their results suggest either that pines have strikingly high gene numbers, which would suggest significant differences in genome dynamics and function, or that this technique is ineffective in isolating the coding fragment in pine genomes. A sequenced pine genome is needed to solve this intriguing puzzle.

A comparison of the sequenced genomes of Physcomitrella, Selaginella, and 15 angiosperms suggests the last common ancestor of land plants had on the order of 6800 genes (Banks *et al.* 2011); this is probably an underestimate but compared to the numbers of genes in sequenced embryophyte genomes it still suggests that dramatic increases in gene number occurred with land plant diversification. The last common ancestor of flowering plants is thought to have had a gene complement about 50–100% larger than that of the land plant ancestor, at 10 000–14 000 (e.g., Maere *et al.* 2005; Vandepoele & Van de Peer 2005; Tuskan *et al.* 2006; Tang *et al.* 2008b; Van de Peer *et al.* 2009a; Paterson *et al.* 2010). Most sequenced land plant genomes have gene counts in the range of about 25 000–40 000 (Table 8.1); although there is some tendency for larger genomes to have a higher estimated gene count, there is no clear correlation among land plants between gene number and genome size (Table 8.1). Thus smaller land plant genomes in general have a higher density of genes rather than fewer genes.

Why are the genomes of ferns and gymnosperms so much larger than those of other land plant groups? And why do some species, such as maize or soybean (*Glycine max*, Fabaceae) among sequenced angiosperm genomes, have such large genomes compared to taxa such as sorghum (*Sorghum bicolor*, Poaceae) and rice (*Oryza sativa*, Poaceae), black cottonwood, cucumber (*Cucumis sativus*, Cucurbitaceae), or castor bean (*Ricinus communis*, Euphorbiaceae) (Table 8.1)? A number of factors have been implicated in differences in genome size including segmental duplication (e.g., Bancroft 2001; Blanc *et al.* 2000; Vision *et al.* 2000), variation in the length of introns (Deutsch & Long 1999) or tandem repeats (Ellegren 2002; Morgante *et al.* 2002), and the ongoing transfer of genes from organelles to the nuclear genome (Adams & Palmer 2003; Shahmuradov *et al.* 2003). However, the impact of these factors is relatively minor compared to the two main factors that are responsible for the relative increases in genome size in most groups studied thus far: whole genome duplications (WGDs) and expansion of transposable element (TE) families. In addition to influencing the size of genomes, these two factors can have profound effects on the structure and function of plant genomes.

8.2 Whole genome duplications

Plant genomes are highly dynamic, showing dramatic differences in size and organization even among relatively closely related species. WGDs appear to have occurred in nearly all land plant lineages, often multiple times (e.g., Leitch & Bennett 1997; Soltis & Soltis 1999; Wendel 2000; Cui *et al.* 2006; Van de Peer *et al.* 2009a; Tang *et al.* 2010). Such duplications, as well as smaller-scale duplications of chromosomes or chromosomal segments or even individual genes or gene clusters, are thought to supply the raw material for genetic diversification. With two identical copies of a gene in the genome, for instance, one copy can continue to perform the original function whereas the other may take on new functions, thus allowing for increased complexity and diversification of plant structure and function (Ohno 1970; Lynch & Conery 2000; Wendel 2000; Van de Peer *et al.* 2009b). Even though it is thought that many duplicates may be eliminated, others may remain in the genome, accumulating mutations and deteriorating into pseudogenes or acquiring novel functions in the process of diploidization of the genome (e.g., Lynch & Conery 2000; Hufton & Panopoulou 2009).

Genome duplication therefore provides a mechanism for increase in both genome size and gene number. Maere *et al.* (2005) determined that more than 90% of the increase in the number of developmental genes, transcription factors (TFs), and signal transducers in the genome of *Arabidopsis thaliana* (Brassicaceae; referred to in this chapter as Arabidopsis) was the result of three WGDs spread throughout the history of the species. In fact, studies have shown that duplicates in these gene families may be preferentially retained after WGDs and other large-scale duplications, but not after smaller duplication events (De Bodt *et al.* 2005; Maere *et al.* 2005; Gaut & Ross-Ibarra 2008; Van de Peer *et al.* 2009b; Paterson *et al.* 2010), suggesting that WGDs may be of specific importance for increasing gene groups that play a role in organismal complexity.

Large-scale duplications and WGDs are detected mainly using two different methods. In the first, pairs of paralogs in the genome are identified, and sequence divergence (*Ks*, synonymous substitutions) between the two members is estimated. *Ks* is used as a proxy for time; a greater number of substitutions is an indication of a more ancient duplication event, thus paralog pairs with similar *Ks* values may have arisen via the same duplication event. When the distribution of *Ks* values is plotted (Figure 8.2a), the resulting curve is expected to be L-shaped, reflecting a large number of recent small-scale duplications (Lynch & Conery 2000). A secondary peak in such a curve indicates a large number of ancient paralogs of similar age, suggesting a WGD at some timepoint in the past (e.g., Blanc & Wolfe 2004; Maere *et al.* 2005; Cui *et al.* 2006; Rensing *et al.* 2008; Van de Peer *et al.* 2009a; Tang *et al.* 2010). Various methods may also be applied to translate *Ks* values into dates of divergence in millions of years, thereby allowing estimates of when in embryophyte history the genome duplication occurred. However, Tang *et al.* (2008b) show that median *Ks* values vary among angiosperm lineages, perhaps related to generation time, and that much higher substitution rates in some lineages can lead to erroneous conclusions about when, in terms of evolution and phylogeny, duplications occurred.

In the second method, genomic synteny or colinearity is used to identify homologous segments within or among genomes (e.g., Lagercrantz & Lydiate 1996; Grant *et al.* 2000; Paterson *et al.* 2004; Jaillon *et al.* 2007; Tang *et al.* 2008b; Tang *et al.* 2010). Using this technique, genomic regions from one species can be mapped to homologous regions in the genome of another species to identify shared and independent duplication events. If numerous loci in species A are each found to map to two regions in species B (Figure 8.2b), that is evidence that the lineage of species B experienced a genome duplication after it separated from the lineage of species A (e.g., Jaillon *et al.* 2007; Ming *et al.* 2008; Tang *et al.* 2008b; Chan *et al.* 2010). This technique is particularly useful for identifying the evolutionary timing of WGDs, as the duplications can be mapped onto a phylogenetic tree to identify lineages that share a particular

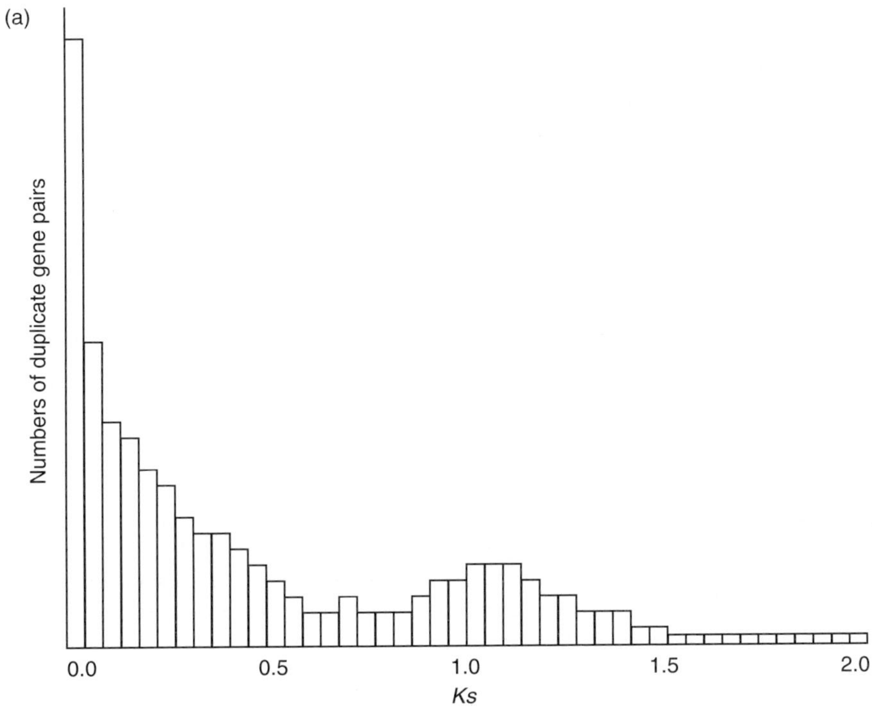

Figure 8.2 (a) Hypothetical *Ks* plot. Duplicate genes for the most part rapidly are eliminated, diverge, or deteriorate thus the steep drop-off at low *Ks* values, which correspond to recent duplications. The bulge around $Ks = 1.0$ corresponds to a potential whole genome duplication (WGD). (b) Hypothetical and simplified chromosome mapping to identify WGD. Species A has five chromosomes (black bars), species B has ten (gray bars). Each locus in the genome of species A corresponds to two loci in the genome of species B (only one locus shown per chromosome). This highly simplified example shows exact doubling of loci and chromosomes as would be seen in a contemporary tetraploid; however, most examples are far more complex (e.g., Jaillon *et al.* 2007).

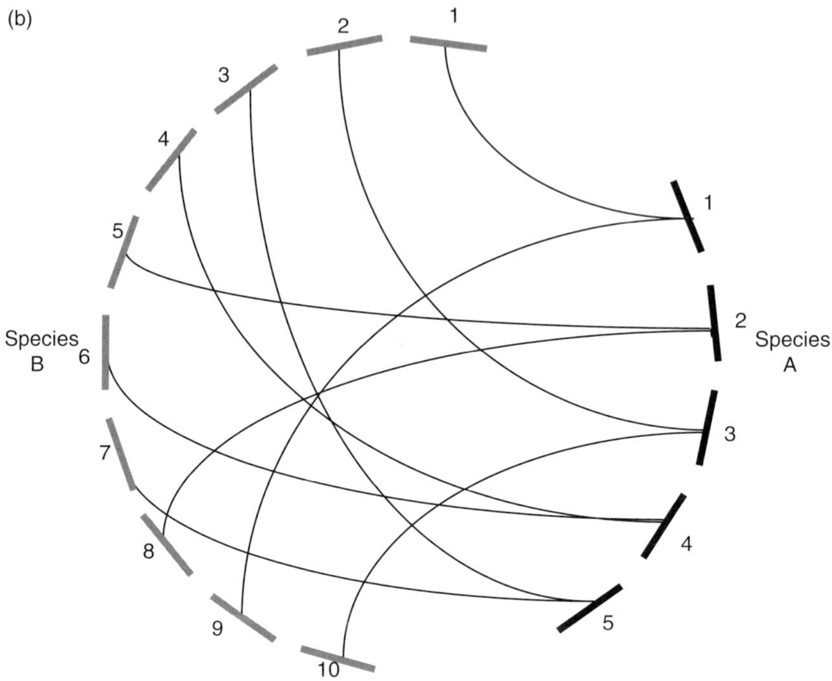

Figure 8.2 (*Continued*)

duplication event (Kellogg 2003). However, genomic rearrangements, gene loss, small-scale duplications, and many other processes of dynamic genomes may obliterate colinearity, making WGDs difficult to identify, particularly in long-diverged lineages (e.g., Tang *et al.* 2008a).

8.2.1 Whole genome duplications in non-flowering plants

Nearly all sequenced land plant genomes show evidence of at least one WGD, although these have not been documented from algal genomes. Rensing *et al.* (2008), based on the distribution of *Ks* values, identified one probable WGD in the history of the moss Physcomitrella. Although this finding confirms that WGD is a common phenomenon among land plants, the significance for moss evolution will not be clear until other moss genomes have been sequenced. In contrast, Banks *et al.* (2011) did not detect any genome-wide duplications in the genome of the lycophyte Selaginella; to date this is the only sequenced plant genome that appears not to have undergone any polyploidization in its history. Most other species of Selaginella that have been evaluated have similar sized genomes (Little *et al.* 2007; Murray *et al.* 2010); however, other lycophytes have nuclear genomes that are 20, 40, or more times larger (Wang *et al.* 2005; Murray *et al.* 2010); this raises the possibility that they may have experienced WGDs since their divergence from Selaginella, a hypothesis that can be tested with additional lycophyte genome sequences.

The large but diploid genomes of conifers, like that of Selaginella, appear not to have experienced any WGDs (Ahuja 2005; Morse *et al.* 2009). Polyploidy is rare in extant gymnosperms and examples of polyploid conifer species are few and far between. Colchicine-induced polyploids in pines grow poorly and show low survival rates, and pine hybrids do not show increased genome sizes (Williams *et al.* 2002; Ahuja 2005); thus, it seems polyploidy may not be tolerated at least in conifers. Regarding paleopolyploidy, there is equivocal evidence for a WGD in the history of *Welwitschia mirabilis* (Welwitschiaceae) (Cui *et al.* 2006), one of the three genera in the Gnetales, and Ahuja (2005) also raises the idea of ancient polyploidy. But current data are insufficient to determine whether such WGDs occurred and if so when; as with ferns, we must await advances in sequencing and analysis methods to answer these questions.

The large genomes of homosporous ferns, coupled with their exceedingly high chromosome numbers, initially led researchers to assume most species were polyploid; however, isozyme analyses (Haufler & Soltis 1986) were not consistent with this. This suggested the possibility of numerous ancient WGDs followed by gene silencing to return the genome to a diploid genetic condition, but without the concomitant loss of physical genetic material characteristic of diploidization in most angiosperm lineages (Haufler 1987). Efforts to document polyploidy events in the history of the *C. richardii* genome identified large numbers of duplicate genes (Nakazato *et al.* 2006), but chromosomal rearrangements prevented conclusive recognition of a WGD. More recent efforts using *Ks* values calculated for paralogs identified from expressed sequence tag (EST) collections from *C. richardii* and *Adiantum capillus-veneris* (Polypodiaceae) have demonstrated that the two share a WGD, but likely only a single WGD, dated at approximately 180 million years ago (Barker 2009; Barker & Wolf 2010). WGDs, it appears, do not explain the large fern or gymnosperm genomes.

8.2.2 Whole genome duplications in angiosperms

In contrast to the apparent low frequency of WGDs among non-flowering land plants, such duplications have occurred often during the diversification of flowering plants (e.g., Bowers *et al.* 2003; Blanc & Wolfe 2004; Cui *et al.* 2006). Early evidence led to conflicting conclusions regarding the number of duplications in the history of the Arabidopsis genome (e.g., Arabidopsis Genome Initiative 2000; Vision *et al.* 2000; Bowers *et al.* 2003). Bowers *et al.* (2003) identified three duplication events: (1) α, which they dated after the divergence of the Arabidopsis lineage from other eurosids; (2) β, which occurred before the diversification of the eudicots but after the monocot-eudicot divergence; and (3) γ, which they dated prior to the monocot-eudicot split (Figure 8.1; Bowers *et al.* 2003). Subsequent analyses confirmed these earlier

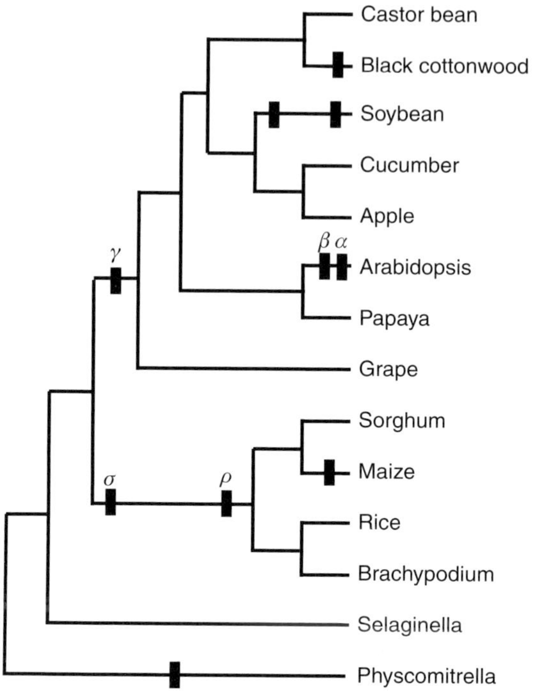

Figure 8.3 Angiosperm phylogeny showing current estimates of timing and phylogenetic distribution of whole genome duplications identified in sequenced land plant genomes. See the text for details. γ event is the paleohexaploidy that preceded divergence of euasterids and eurosids. The earlier of the soybean duplications appears to be legume-wide. The σ event may be monocot-wide.

large-scale duplications, the β duplication coincident with the origin of the eurosids or older, and the γ duplication potentially correlated with the origin or very early diversification of the angiosperms (e.g., Simillion *et al.* 2002; Adams & Wendel 2005; De Bodt *et al.* 2005; Cui *et al.* 2006).

An analysis of homologous regions in the grape (*Vitis vinifera*, Vitaceae), Arabidopsis, and black cottonwood genomes helped to clarify the phylogenetic position and timing of WGDs in the history of those species. Jaillon *et al.* (2007) concluded that all three species shared a hexaploid ancestor and that this paleotriplication is shared by all eurosids (Figure 8.3). Velasco *et al.* (2007), based on dating of *Ks* data from a second grape genome, suggested alternatively that all three share one duplication (γ), Arabidopsis and black cottonwood share a second (β), and that each of the three species experienced an independent duplication (α). However, subsequent analyses suggested unequal substitution rates may have led to erroneous conclusions (Tuskan *et al.* 2006; Semon & Wolfe 2007; Tang *et al.* 2008b; Paterson *et al.* 2010), and the absence of the β duplication from the genome of papaya (*C. papaya*, Caricaceae)

(Ming *et al.* 2008), a member of the same order as Arabidopsis (Brassicales), suggests that Arabidopsis and black cottonwood could not share the β duplication (Figure 8.3). Subsequent analyses have shown the triplicate genome structure in papaya (Ming *et al.* 2008), castor bean (*R. communis*, Euphorbiaceae; Chan *et al.* 2010), cucumber (*C. sativus*, Cucurbitaceae; Huang *et al.* 2009), apple (*Malus* × *domestica*, Rosaceae; Velasco *et al.* 2010), and soybean (Schmutz *et al.* 2010) confirming that the most recent common ancestor of the eurosid clade was hexaploid (Figure 8.3; Van de Peer *et al.* 2009a). Further evidence from tomato (*Solanum lycopersicum*, Solanaceae) BAC sequence data (Tang *et al.* 2008b), coffee (*Coffea arabica*, Rubiaceae) genomic sequence (Cenci *et al.* 2010), and Asteraceae ESTs (Barker *et al.* 2008) indicate that the paleohexaploid event predates the eurosid-euasterid split (Figure 8.3). Thus at a minimum, core eudicots (with the possible exception of the earliest diverging lineages such as Gunnerales) have a paleohexaploid ancestor. It is plausible the γ event is considerably older but it is not yet clear whether this event is shared by monocots or basal angiosperms, the genomes of which also show evidence of paleopolyploidy (Cui *et al.* 2006; Jaillon *et al.* 2007; Tang *et al.* 2008b; Van de Peer *et al.* 2009a).

The paleohexaploid event is generally referred to as a single event; however it is not clear whether the hexaploid genome arose by a single triplication event or a combination of duplications and hybridizations. Some evidence favors a WGD followed by hybridization with a diploid to form a triploid, which then underwent a second WGD to form a hexaploid (Lyons *et al.* 2008), however this is by no means clear. Pinning down the timing of this major event in genome evolution is key to understanding the role of WGDs in angiosperm diversification and success; if there is no WGD shared by all (or nearly all) angiosperms, then the hypothesis that paleopolyploidy played a major role in promoting the radiation and success of the flowering plants would not be supported. Rather, if the hexaploid event occurred in the eudicot lineage following the divergence of the monocot lineage, one might hypothesize that it was pivotal in the enormous success of the core eudicot clade, which includes the vast majority of angiosperms. Additional genome sequence data from genomes such as those of the basal eudicot columbine (*Aquilegia coerulea*, Ranunculaceae) and basal angiosperms may help define the evolutionary window within which this potentially transformative event occurred.

The evidence of Jaillon *et al.* (2007) and Tang *et al.* (2008b) indicates that Arabidopsis experienced two additional WGDs (α and β duplications) after its lineage diverged from the grape and black cottonwood lineages (Figure 8.3). Further evidence from the genome of papaya, a member of the same order as Arabidopsis (Brassicales) indicates that these duplications occurred after those two lineages diverged, as evidence for them has not been found in the papaya genome (Ming *et al.* 2008). Based on published (to date) genomes of core eudicot angiosperm species, WGDs subsequent to the γ event have been detected in black cottonwood (a single duplication; Jaillon *et al.* 2007; Tang *et al.* 2008b), soybean (one duplication early in legume evolution, a second more recent in the soybean lineage; Schmutz *et al.* 2010), and apple (a single duplication; Velasco *et al.* 2010) but have not been identified in grape, papaya, castor bean or cucumber (Figure 8.3; Tang *et al.* 2008b; Van de Peer *et al.* 2009a). Additional evidence shows both shared and independent ancient and more recent duplications in *Medicago truncatula* (Fabaceae), lettuce (Asteraceae), cotton (Malvaceae), tomato (Solanaceae), and potato (Solanaceae) (Van de Peer *et al.* 2009a).

WGDs have also figured importantly in the ancient evolutionary history of monocots, particularly the economically important cereal grasses. Studies have identified a WGD that predated the diversification of the grasses (although the exact timing appears controversial), and it has been suggested that this duplication may have promoted the diversification of the cereal-grass lineage (Goff *et al.* 2002; Paterson *et al.* 2004, 2009; Salse *et al.* 2008; Schnable *et al.* 2009; Thiel *et al.* 2009; Devos 2010). Tang *et al.* (2010) identified an additional, more ancient duplication (σ duplication), which they suggest occurred early in the diversification of the monocots (Figure 8.3). Cui *et al.* (2006) found two duplications in the history of *Acorus americanus*, a monocot belonging to the lineage that is the sister-group of all other monocots. The divergence of this lineage would clearly predate the diversification of the grasses; therefore *Acorus* might be expected to share the σ duplication with cereal grasses, but not the ρ duplication. As with eudicots, data from additional monocots are needed to characterize duplications in the monocots. Nonetheless, it is clear that polyploidy is a significant phenomenon in angiosperm evolution in a way that is not characteristic of the evolution of other embryophyte lineages; the frequency of occurrence over the course of angiosperm history suggests that it has played a role in evolution and diversification from the level of speciation perhaps to the level of diversification of major clades of flowering plants.

8.2.3 Impact of whole genome duplications on plant evolution

The aftermath of WGDs may include phenomena such as substantial genome reorganization due to improper pairing at meiosis, preferential gene loss and retention favoring some gene families, elimination of large and small noncoding genomic regions, and sub- and neofunctionalization of duplicate genes (e.g., Lynch & Force 2000; Blanc & Wolfe 2004; Paterson *et al.* 2004; Bowers *et al.* 2005; Maere *et al.* 2005; Bikard *et al.* 2009; Hufton & Panopoulou 2009; Soltis *et al.* 2009; Van de Peer *et al.* 2009b). These genome changes may in turn lead to reproductive isolation, speciation, and diversification of plant lineages (Wendel 2000; Wood *et al.* 2009). Several paleopolyploidy events are suggested to have coincided with key events such as the origin of angiosperms or their rapid early diversification (e.g., Vision *et al.* 2000; De Bodt *et al.* 2005; Maere *et al.* 2005; Cui *et al.* 2006; Van de Peer *et al.* 2009b) or the diversification of some of the most speciose groups of angiosperms, including Fabaceae, Poaceae, and Asteraceae (e.g., Paterson *et al.* 2004; Soltis *et al.* 2009). WGDs have also been correlated with lineage-specific events, such as the origin of the pome fruit in the Pyreae, the tribe of Rosaceae that includes apple (Velasco *et al.* 2010). Molecular dating puts a number of WGDs at around 60–70 million years ago, which raises the possibility that they were in some fashion related to the environmental events and mass extinction of the Cretaceous-Tertiary boundary. It has been suggested that duplicated genomes were better able to survive and diversify following this event because of an increased ability to tolerate a rapidly and cataclysmically changing condition (Fawcett *et al.* 2009; Soltis & Burleigh 2009; Van de Peer *et al.* 2009b). When these catastrophic events in evolutionary history cause massive extinctions, polyploids may be better positioned than diploids to survive and exploit reduced competition due to redundancy, broader environmental tolerance, increased heterozygosity, and other factors associated with polyploidy. Thus, WGDs may be successful only rarely and in specific ecological situations, but when they are successful they may allow for dramatic increase in species diversity and complexity.

8.3 Transposable elements

More so even than WGDs, accumulation of repetitive elements, particularly TEs, is implicated as the major factor that distinguishes large and small plant genomes (e.g., Flavell *et al.* 1974; Kubis *et al.* 1998; Kumar & Bennetzen 1999; Cavalier-Smith 2005; Gaut & Ross-Ibarra 2008). More than half of all plant DNA is transposon DNA (SanMiguel *et al.* 1998; Bennetzen 2005), and new elements continue to be discovered as detection algorithms are improved and more sequence variants are identified (e.g., International Rice Genome Sequencing Project 2005; Ming *et al.* 2008; Huang *et al.* 2009; Schnable *et al.* 2009). Major families of TEs are found in algal genomes (Merchant *et al.* 2007), and among sequenced land plant genomes, TEs account for anywhere from 15% of the genome of Arabidopsis (Arabidopsis Genome Initiative 2000) to 85% in the genome of maize (Schnable *et al.* 2009; Table 8.1). In the relatively large genome of bread wheat (*Triticum aestivum*, Poaceae; 17 000 Mb; Vitte & Bennetzen 2006), TEs are estimated to represent 90% of the genome. In the sequenced moss and lycophyte genomes, they comprise 40% (Selaginella, Banks *et al.* 2011) and 50% (Physcomitrella, Rensing *et al.* 2008) of the genome, respectively (Table 8.1).

The discovery that variation in TE abundance, even among closely related species (e.g., Tikhonov *et al.* 1999) or haplotypes (e.g., Fu & Dooner 2002), is the major factor that underlies different genome sizes rather than differences in gene content, proved part of the solution to the puzzling "C-value paradox": the lack of correlation between genome size and organismal complexity that confounded researchers who expected such a correlation based on the assumption that more complex organisms required more genes and thus a larger genome (Thomas 1971). The abundance of transposons in large genomes, with the attendant difficulties they create in sequencing and assembling such a genome, is a principal reason that sequencing projects have focused on smaller genomes.

8.3.1 Retrotransposons

In most land plant genomes sequenced to date, retrotransposons (type I elements), which transpose via transcription of an RNA molecule, make up the bulk of the transposable-element fraction of the genome; in apple, papaya, and soybean, retrotransposons make up about 40% of the genome (Ming *et al.* 2008; Schmutz *et al.* 2010; Velasco *et al.* 2010), and other estimates range from <10% in Arabidopsis, the genome of which has a relatively small proportion of TEs (Arabidopsis Genome Initiative 2000), to 80% in wheat (Kumar & Bennetzen 1999). Because retrotransposons generate an RNA intermediate, which is then reverse transcribed to form a new DNA molecule that will integrate into the genome, these type I elements are capable of dramatic amplification in the genome, thereby contributing to increasing genome size.

Among retrotransposons, long terminal repeat (LTR) retrotransposons are the most common and make up the largest percentage of retroelements, and frequently of TEs, in plant genomes. Their relative abundance in the genome varies but is often high; this group of retrotransposons is estimated to account for 70% of the maize genome (Baucom *et al.* 2009; Schnable *et al.* 2009), 50% of the Physcomitrella genome (Rensing *et al.* 2008), 35% of the Selaginella genome (Banks *et al.* 2011), and down to 10% of the TE-poor cucumber genome (Huang *et al.* 2009). Larger genomes, at least among angiosperms, tend to have a greater percentage of these elements, thus Bennetzen (2005) noted that angiosperms genomes larger than 2000 Mb were at least 50% LTR-retrotransposons. Expansion and reduction in the number of these elements alone appears to account for much of the variation in angiosperm genome size. LTR-retrotransposons are abundant and diverse in gymnosperms, and both gymnosperm-specific elements and elements that are found in other land plant lineages have been identified, as well as elements restricted to specific gymnosperm taxa (i.e., pine; Friesen *et al.* 2001; Morse *et al.* 2009; Kovach *et al.* 2010; Hao *et al.* 2011). Kovach *et al.* (2010) show that the genomes of conifers contain older and more diverse retroelements than those of angiosperms, and Morse *et al.* (2009) identified a novel element in pine (*Gymny*) that occupies genomic space equivalent to the entire Arabidopsis genome.

The most common LTR-retrotransposon elements are *copia*- and *gypsy*-like repeats, which generally comprise 50–70% of the LTR-retrotransposons in land plant genomes (e.g., Huang *et al.* 2009; Rensing *et al.* 2008; Banks *et al.* 2011; Vogel *et al.* 2010). The *copia* and *gypsy* families are large and diverse, each including a wide range of element types. These elements range in size to over 10 Kb, and may be present in 10 000–1 000 000 copies (Bennetzen 2000). Other type I elements, such as short and long interspersed nuclear elements (SINEs and LINEs), have not been annotated in all angiosperm genomes, although these elements, which predominate in mammalian genomes, are likely to be present (Baucom *et al.* 2009). In general, these non-LTR-retrotransposon type I elements make up a small percentage of the genome, although, for instance, LINEs appear to be overrepresented in the genomes of lily (*Lilium speciosum*, Liliaceae; Leeton & Smyth 1993) and grape (Jaillon *et al.* 2007; see also Baucom *et al.* 2009), and Hao *et al.* (2011) found that they were abundant in sequence data from three conifers. SINE families have been identified in mosses, gymnosperms, and angiosperms but not in the genome of Selaginella (Banks *et al.* 2011), raising the possibility that these elements were lost in the lycophyte lineage. Many families appear to be taxonomically limited (Yagi *et al.* 2011), although some (i.e., the Au SINE family; Yagi *et al.* 2011) have been found in both angiosperms and gymnosperms suggesting broad distribution.

8.3.2 DNA elements

DNA (type II) elements, which transpose directly, typically make up a smaller percentage of the mobile element genome fraction. In rice there are more than twice as many type II elements as type I, but because type I elements are longer they still make up a greater proportion of the genome (International Rice Genome Sequencing Project 2005). In contrast, in larger genomes such as those of maize and soybean, type I elements are more abundant than type II (75% vs. 5% in maize; 40% vs. 15% in soybean) (Messing *et al.* 2004; Baucom *et al.* 2009; Schnable *et al.* 2009; Schmutz *et al.* 2010). Because type II elements excise and directly reinsert in the genome they tend not to increase in number at the rate of retrotransposons. DNA elements such as CACTA-like, hAT, and *mutator*-like elements (MULEs and other members of the *Mutator* family), miniature inverted repeat TEs (MITEs), and *helitrons* (rolling-circle transposons) have been found in most sequenced angiosperm genomes (e.g., Arabidopsis Genome Initiative 2000; International Rice Genome Sequencing Project 2005) and they may vary greatly in abundance. Although MITEs are considerably smaller than CACTA elements, the former are more abundant in rice and therefore occupy more genome space; in other grass species, MITEs have not proliferated as extensively and the larger CACTA elements represent a larger fraction of the genome (Paterson *et al.* 2009; Devos 2010). The distribution of DNA transposons also varies among land plant lineages. In conifers, Kovach *et al.* (2010) identified transposons such as hAT and *Mutator*-type elements, as well as *helitrons*; although, Hao *et al.* (2011) found that DNA elements were rare in conifers. Although *helitrons*, MITEs, hAT elements, and a novel family of type II elements were found in the Selaginella genome, CACTA elements and MULEs, along with other DNA elements, appear to be absent (Banks *et al.* 2011). Likewise, the genome of Physcomitrella has only a single family of *helitrons* whereas most species have several (Selaginella has four, for instance; Rensing *et al.* 2008; Banks *et al.* 2011). These taxonomically distinct element distributions suggest lineage-specific genome dynamics related to TE activity.

8.3.3 Transposable elements and genome size

Differences in TE abundance, particularly LTR-retrotransposon abundance, largely drive differences in land plant genome size. The genome fraction of type II elements does not differ as extensively among species, and unlike LTR-retrotransposon quantity, is not correlated with genome size (Devos 2010). The impact of TE expansion on genome size depends not only on the extent of amplification of a given family but also on the size of the elements. Thus, MITEs may be present in large numbers but individually are so small (around 100–500 bp) that there may be little effect on overall genome size (e.g., Naito *et al.* 2006). In contrast, activity in a group of LTR-retrotransposons, which range in size from about 2 to >10 Kb (Bennetzen 2000), can have a dramatic effect on genome size. In five species of cereal grasses and legumes, Vitte and Bennetzen (2006) estimated copy numbers of the five most abundant LTR-retrotransposons to range from about 100–300 in the 470 Mb genome of *Lotus japonicus* (Fabaceae) to 14 000–170 000 in the 6100 Mb genome of *Triticum monococcum* (Poaceae), a diploid relative of wheat.

LTR-retrotransposon accumulation can be very rapid, occurring in bursts over a relatively short time span, possibly due to stress or intrinsic factors that result in the activation and amplification of these elements (McClintock 1984; Wessler *et al.* 1995; Friesen *et al.* 2001; Grandbastien *et al.* 2005; Baucom *et al.* 2009; Devos 2010). These periods of intense activity, along with the large size of these elements, are responsible for their dominant role in modulating genome size in plants. The maize genome is thought to have doubled in size within a few million years due to LTR-retrotransposon proliferation (SanMiguel & Bennetzen 1998). The greater size of the sorghum genome relative to the rice genome is largely due to the expanded complement of LTR-retrotransposons in the former (Paterson *et al.* 2009), and Piegu *et al.* (2006) showed dramatic and rapid expansion of LTR-retrotransposon families in the genome of the rice relative *Oryza australiensis* in the past 3–4 million years, leading to a doubling of genome size. In three diploid species of *Gossypium* (the genus to which cotton belongs), which vary almost threefold in genome size, significant differences were found in the expansion of retrotransposon families (Hawkins *et al.* 2006). Devos (2010) compared TE abundance and diversity in four grasses (rice, sorghum, maize, and Brachypodium) and found that short bursts of activity of relatively few elements can account for the differences in genome sizes. The genome of Physcomitrella also shows bursts of activity within the past 10 million years (Rensing *et al.* 2008).

In contrast, TEs may be relatively inactive in gymnosperms. Friesen *et al.* (2001), surveying LTR-retrotransposons in gymnosperms, concluded that their diversity and abundance are likely due to rapid radiation; however, these bursts may have occurred deep in the past, as Kovach *et al.* (2010)

suggest that only two conifer-specific LTR-retrotransposon elements may have been active in the past 140 million years. Transposon activity is ongoing: crop varieties differ in TE insertion patterns, suggesting very recent activity since these species were domesticated within the past 10 000 years (Vicient *et al.* 2001; Ellis *et al.* 1998; Gribbon *et al.* 1999; Vitte & Panaud 2005), and four rice cultivars with high MITE copy numbers showed 50–65 new insertions per progeny (Naito *et al.* 2006).

Although it is well accepted that genome expansion in plant lineages is mediated by the amplification of TE families and by genome duplication, until the turn of the century it was less clear whether and how genomic material is eliminated, prompting Bennetzen and Kellogg (1997) to question whether plant genomes "have a one-way ticket to genome obesity." All plant genomes are not exceedingly large, and theory suggests that large genomes are not adaptive (Knight *et al.* 2005; Gaut & Ross-Ibarra 2008), therefore it seemed clear that mechanisms must exist to reduce genome size in the face of ongoing TE proliferation. More recently, it has been demonstrated that genome expansion is countered by loss of DNA, and that species differ not only in the number of genome duplications and their rate of TE replication, but also in the efficiency with which their genomes purge nongenic material (Petrov *et al.* 2000; Petrov 2002; Hawkins *et al.* 2006; Hawkins *et al.* 2009). One process by which chromosomal material is removed is intrastrand recombination between the LTRs of an LTR-retrotransposon. This results in the production of a solo LTR (a single LTR flanked by the target site duplication characteristic of transposons) via the loss of one LTR and the accompanying internal sequence of the retrotransposon (Figure 8.4). Recombination events between homologous LTR-retrotransposons at different loci in the genome can also cause fragmentation of elements and loss of genomic DNA (Figure 8.4), although duplications or chromosomal rearrangements may also result from this process. In rice, processes leading to the loss of retroelements may account for the deletion of almost 200 Mb of sequence in 8 million years (Bennetzen *et al.* 2005; Ma *et al.* 2004). In Brachypodium 17.5 Mb of sequence is estimated to have been lost this way (Vogel *et al.* 2010). Illegitimate recombination and double-stranded break repair are also thought to cause many small genomic deletions including internal deletions within TEs (e.g., Petrov *et al.* 1996, 2000; Kirik *et al.* 2000; Shirasu *et al.* 2000; Devos *et al.* 2002; Petrov 2002; Ma *et al.* 2004; Bennetzen *et al.* 2005; Vitte & Panaud 2005; Piegu *et al.* 2006; Hu *et al.* 2011). The most common deletion size in the rice and Arabidopsis genomes is 1–2 bp; in rice these small deletions do not appear to have a profound impact on genome size, but in Arabidopsis, these and other small deletions appear to account for most of the 65 Mb size difference between the genomes of *A. thaliana* and *Arabidopsis lyrata* (Devos *et al.* 2002; Ma *et al.* 2004; Bennetzen 2005; Hu *et al.* 2011).

Fragmentation and elimination of TEs, particularly retrotransposons, can occur rapidly after insertion. A variety of studies have shown that most intact retrotransposons are young, at most a few million years old, suggesting rapid

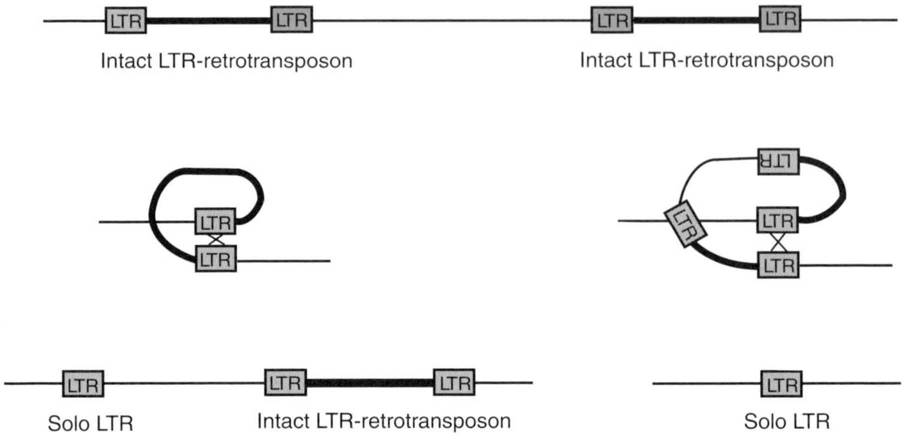

Figure 8.4 Elimination of DNA and formation of solo long terminal repeat (LTR). The top line shows a strand of DNA (thin line) with two intact LTR-retrotransposons, each consisting of two LTRs and internal sequence (thick line). On the left hand side, recombination occurs by pairing of the two LTRs of a single element, eliminating the internal DNA of the element, and leaving a solo LTR. On the right hand side, recombination occurs between the LTRs of two different elements, resulting in the deletion of the internal DNA and one LTR of each element as well as the DNA separating the two elements. A single LTR results.

elimination and alteration of inserted elements (SanMiguel *et al.* 1998; Ma *et al.* 2004; Vitte & Panaud 2005 and references therein; Naito *et al.* 2006; Devos 2010). Bennetzen *et al.* (2005) calculated that most intact elements are less than 3 million years old, with older elements showing signs of recombination, deletion, or other events leading to loss of DNA. In Brachypodium, 13 groups of type I elements were found to be less than 20 000 years old, and 66 groups were less than 100,000 years old (Vogel *et al.* 2010). Rates of deterioration can vary even among closely related species: Hu *et al.* (2011) calculated that the average age of LTR-retrotransposons in the genome of *A. lyrata* is about one-third that of those in *A. thaliana*. In fact, these processes that remove DNA can make it difficult to recognize ancient transposons, since they are likely to have undergone some type of recombination or deletion that disrupts their structure, as well as nucleotide substitutions. Genomes may appear to be overrepresented in young LTR-retrotransposons for this reason (Ma *et al.* 2004; Garfinkel 2005).

8.3.4 Dynamics of TE amplification and removal

The TE-mediated processes that contribute to genome size, such as LTR-retrotransposon proliferation and unequal recombination and illegitimate recombination, occur at different rates in different taxa, thus contributing to differences in genome size and also to lack of synteny among species in genomic regions rich in TEs. Studies show that not all TE families are equally amplified in large genomes, and that the specific families that are amplified vary, even among closely related species or cultivars (e.g., Vitte & Panaud 2005; Hawkins *et al.* 2006; Piegu *et al.* 2006; Vitte & Bennetzen 2006; Ammiraju *et al.* 2007; Devos 2010). In maize, the 20 most numerous LTR-retrotransposon families (of over 400) contribute about 70% of the sequence of the maize genome (Baucom *et al.* 2009; Schnable *et al.* 2009). Hawkins *et al.* (2006) showed that a 16-fold increase in one type of *gypsy*-like LTR-retrotransposon occurred in *Gossypium exiguum* (Malvaceae) in comparison with *Gossypium kirkii* within the past 10 million years. Likewise, studies have documented differential expansion of *gypsy*- and *copia*-like elements in species of *Oryza* and *Vicia* (Fabaceae), in some cases leading to a doubling of genome size within the past 3 million years (Hill *et al.* 2005; Piegu *et al.* 2006; Zuccolo *et al.* 2007). Hu *et al.* (2011) suggest that expansion of specific clades of LTR-retrotransposons after the divergence of *A. lyrata* and *A. thaliana* is largely responsible for the difference in genome size between these two species. Morse *et al.* (2009) suggest that variation in genome size among pine species is due primarily to amplification of TE families; however, the pine genome appears to contain a high percentage of novel elements, and the specific families involved in modulating genome size in pine have not yet been identified (Morse *et al.* 2009; Kovach *et al.* 2010).

It has been suggested that LTR-retrotransposon families may be able to proliferate in some plant genomes but not others because some genomes may be less efficient at silencing TEs, which therefore remain active, or because some LTR-retrotransposons with restricted taxonomic distribution have exceptional amplification ability (Naito *et al.* 2006; Vitte & Bennetzen 2006; Baucom *et al.* 2009). Naito *et al.* (2006) suggest that silencing of TEs is impaired in rice cultivars with high TE copy numbers, consistent with the first hypothesis, whereas Baucom *et al.* (2009) found in an analysis of the rice genome that rare mutations may enable elements to escape host recognition and undergo a burst of amplification (Baucom *et al.* 2009; see also Devos 2010), consistent with the second. It is likely that these possibilities are not mutually exclusive and that these and potentially other factors all contribute to the differential expansion of TE families in different genomes.

Processes of TE-mediated DNA removal also vary among different genomes. Whereas most genomes have multiple *helitron* families, Rensing *et al.* (2008) suggest the Physcomitrella genome has only one because it has purged elements that have not been recently active through DNA excision. The greater proportion of solo LTRs versus intact elements in rice compared to Arabidopsis suggests that the rice genome loses DNA through unequal recombination at a higher rate than the Arabidopsis genome (Ma *et al.* 2004; Bennetzen *et al.* 2005); maize has an even lower percentage of solo LTRs (Devos *et al.* 2002). Deletions associated with double-strand break repair appear to be larger and to occur more frequently in Arabidopsis than tobacco (Kirik *et al.* 2000). Vitte and Bennetzen (2006) show that the rate of unequal and illegitimate recombination, and resulting DNA loss, differs among angiosperm species including grasses, legumes, and Arabidopsis; this was not correlated with genome size or phylogenetic relatedness, and therefore they conclude the phenomenon is quite labile. Morse *et al.* (2009) suggest that processes of chromosomal rearrangement, which may be mediated by transposon activity and lead to DNA loss, may vary among pine subgenera, and may account for genome size variation among the subgenera. Nonetheless, in spite of the large amounts of DNA that can be lost via transposon activity, several studies suggest that the dominant factor underlying variation in genome size is not DNA loss but is expansion of LTR-retrotransposon families (e.g., SanMiguel & Bennetzen 1998; Ma *et al.* 2004; Bennetzen *et al.* 2005; Vitte & Panaud 2005; Hawkins *et al.* 2006; Piegu *et al.* 2006; Vitte & Bennetzen 2006; Devos 2010). For instance, in spite of the loss of almost 200 Mb of genetic material, the rice genome on balance has grown slightly in the past few hundred thousand years (Ma & Bennetzen 2004).

8.3.5 Distribution of transposable elements in plant genomes

The distribution of TEs in plant genomes, and their resultant effect on genome structure, is variable, depending both on element type and on plant species. In the small genome of Arabidopsis, with a relatively low percentage of TEs, elements are generally clustered in centromeric and other gene-poor regions (Arabidopsis Genome Initiative 2000; Bennetzen 2000; Bennetzen *et al.* 2005). Small grass genomes, such as rice, Brachypodium, and sorghum, are similar in having the majority of TEs, particularly LTR-retrotransposons, spatially isolated from regions with a high density of protein-coding genes (International Rice Genome Sequencing Project 2005; Paterson *et al.* 2009; Devos 2010; Vogel *et al.* 2010). In larger genomes such as maize and wheat, TEs are found in centromeric areas, but are also found throughout the genome in repeat-rich regions that alternate with gene-rich regions (Liu *et al.* 2007; Schnable *et al.* 2009; Devos 2010). Even within the gene-rich chromosomal regions there are subregions with high gene density and subregions with high TE density (Vitte & Panaud 2005). Evidence from dating of LTR-retrotransposon sequences in sorghum and maize suggests that these elements insert at a similar rate in all regions of the genome, but that they are eliminated more rapidly from gene-rich regions (Liu *et al.* 2007; Paterson *et al.* 2009; Devos 2010). However, some elements may also be targeted preferentially to repeat-rich regions (e.g., LTR-retrotransposons; Baucom *et al.* 2009; Tsukahara *et al.* 2009) or to gene-rich regions (i.e., type II elements such as MITEs and *helitrons*; e.g., Bennetzen 2000; Paterson *et al.* 2009; Schnable *et al.* 2009). In rice, longer elements such as LTR-retrotransposons are found in gene-poor centromeric regions, whereas shorter elements such as LINEs tend to be more abundant in gene-rich regions (International Rice Genome Sequencing Project 2005) and in maize, *gypsy*-like LTR-retrotransposons are found in gene-poor regions but *copia*-like elements insert preferentially near genes (Baucom *et al.* 2009; Paterson *et al.* 2009;

Schnable *et al.* 2009; Devos 2010). Morse *et al.* (2009) found that most pine-specific *Gymny* (*gypsy*-like) LTR-retroelements are located in gene-poor regions. Naito *et al.* (2006) show in rice that most MITE insertions within genes are rapidly eliminated, and insertions within 5 kb of genes are also eliminated albeit more slowly; they suggest that such insertions interfere with gene function and reduce fitness. Hu *et al.* (2011) note that there are fewer TEs located near genes in the genome of *A. thaliana* compared to *A. lyrata*, suggesting either that they are purged more efficiently in the former species, or that the genome of *A. lyrata* is better able to tolerate their effects in proximity to genes.

LTR-retrotransposons in repeat-rich regions of large genomes such as those of maize, wheat, and soybean are generally high-copy elements and are often found nested successively within each other or tandemly arrayed in blocks (SanMiguel *et al.* 1996; Rostoks *et al.* 2002; Bennetzen *et al.* 2005; Liu *et al.* 2007; Baucom *et al.* 2009; Devos 2010; Schmutz *et al.* 2010); in smaller genomes like those of rice and Arabidopsis LTR-retrotransposons are more often present as single elements and are more evenly distributed (Arabidopsis Genome Initiative 2000; Bennetzen *et al.* 2005). In the relatively small genome of Physcomitrella, 14% of LTR-retrotransposons are nested inside another, but all but one of these are no longer intact (Rensing *et al.* 2008), suggesting such nested transposition is selected against in this genome. This phenomenon and its distribution among plant taxa have not been explored sufficiently to determine if it plays an important role in genome function and evolution.

8.3.6 Impact of transposable elements on genome structure

Mobile element activity can have an enormous impact on genome structure, and in particular LTR-retrotransposons are implicated as important architects of plant genome structure and function. In general, TE density is inversely correlated with rates of recombination; through repressing recombination in the heterochromatic regions in which they are abundant, TEs restrict recombination to gene-rich regions (Arabidopsis Genome Initiative 2000). In grass genomes, retroelement abundance is lowest in regions far from the centromeres; in these regions, recombination and transcription are highest (Paterson *et al.* 2009). Thus, TE density has a direct effect on genome structure and function by modulating recombination and transcription. TEs also promote chromosomal rearrangements including inversions, deletions, translocations, or segmental duplications through a variety of means. Excision of DNA elements leaves sticky ends, which may be repaired or which may fuse with other sticky ends resulting in chromosomal rearrangements; these rearrangements lead to meiotic incompatibility with other members of the species and have been implicated as mechanisms of speciation in plants (Bennetzen 2000). Illegitimate or unequal recombination or events such as unsuccessful transpositions can lead to dramatic differences in genome structure between closely related species, as shown for *A. thaliana* (five chromosomes) and *A. lyrata* (eight chromosomes) (Hu *et al.* 2011).

Differential amplification as well as differential transposition and insertion of TEs can disrupt synteny and colinearity among genomes of closely related species or cultivars. Several studies have shown that even when gene-rich regions show extensive colinearity, TE-rich regions show little correspondence and can differ substantially even among different maize lines (e.g., SanMiguel *et al.* 1996; Bennetzen *et al.* 1998; Fu & Dooner 2002; Vogel *et al.* 2010). Devos (2010) suggests that retrotransposon activity is a more significant factor than recombination in disrupting colinearity. In addition, the hitchhiking of genes along with transposing mobile elements also causes a break down in synteny and colinearity among closely related species (Ma *et al.* 2004; Bennetzen 2005; Bennetzen *et al.* 2005; Bowers *et al.* 2005; Tang *et al.* 2008a). *A. thaliana* and *A. lyrata* diverged about 10 million years ago; excluding major chromosomal rearrangements, 10% of their genomes is no longer syntenic (Hu *et al.* 2011), largely the result of gene translocation by mobile elements.

8.3.7 Impact on gene diversity, expression, and function

Mobile elements can generate new alleles and affect the structure and function of protein-coding genes by inserting into or near a gene sequence. It is likely that TE insertions into genes are generally deleterious and therefore selected against (Bennetzen 2000); however, insertion into introns and promoters may be tolerated. For instance the fact that MITEs generally do not insert into exons may explain their ability to proliferate (Naito *et al.* 2006). Introns in maize are larger than those of sorghum or rice due to the insertion of TEs in maize introns (Yu *et al.* 2002; Schnable *et al.* 2009), and insertion of LINEs has been implicated in large intron size in the grape genome (Jaillon *et al.* 2007). If an element inserts into the promoter of a gene, as MITES often do, it can not only affect the size of the gene, but transcriptional functions of the element may become integrated into the regulation of the gene (Wessler *et al.* 1995; Bennetzen 2000). The result can be altered spatial and temporal expression patterns, as well as potentially new mechanisms of epigenetic regulation for the affected gene (Bennetzen 2000). The presence of transposon remnants, particularly footprints of excised type II elements, in numerous plant promoters suggests this may be a common means by which transcription regulation has been modified during plant evolution (Wessler *et al.* 1995; Bennetzen 2000). An analysis of 13 genes with insertions within an intron or exon or within 200 bp showed that expression of two was reduced and one was increased, demonstrating that TE activity can have a significant impact on gene activity (Naito *et al.* 2006).

Other studies have suggested that in general, TE insertions near genes may have little or no effect on the functioning of those genes because the elements are rapidly epigenetically silenced (Kashkush *et al.* 2003; Lippman *et al.* 2004; Vitte & Bennetzen 2006; Lisch 2009). In general only a small proportion of TEs are transcribed, appearing in EST and other transcriptome datasets (e.g.,

Parchman *et al.* 2010). In Arabidopsis, only 4% of identified TEs had corresponding ESTs (Arabidopsis Genome Initiative 2000). In lodgepole pine (*Pinus contorta*), 6% of ESTs were retrotransposon transcript (Parchman *et al.* 2010); most EST collections have had a lower percent of retrotransposon sequence but they are still a common component of these libraries, suggesting that at least at a low level these elements remain active and may therefore influence the activity of nearby genes as well as proliferate in the genome (e.g., Vicient *et al.* 2001). Expression can also be altered when a transposing element carries a protein-coding gene or fragments to a new chromosomal environment, although some evidence suggests that plant gene activity is not necessarily seriously disrupted upon transposition (Ilic *et al.* 2003).

It has also been suggested that mobile elements, including LTR-retrotransposons, *helitrons*, MULEs (Pack-MULEs), and possibly CACTA and other DNA elements, may increase gene diversity by creating chimeric genes (Bennetzen 2000; Jiang *et al.* 2004; Gupta *et al.* 2005; Lai *et al.* 2005; Morgante *et al.* 2005; Paterson *et al.* 2009; Devos 2010; Vogel *et al.* 2010). "Exon shuffling" can occur when a TE carries a gene fragment with it to a new location. The resulting insertion can alter the structure of a gene; for instance, splice sites may be disrupted so that introns can potentially become functional exons, and functions from multiple genes can be encoded in a single chimeric transcript. In fact, it has been suggested that introns originated as a mechanism for creating functional linkages at the borders of fragments from different genes (Gilbert 1979, cited in Bennetzen 2000). In the maize genome, 425 LTR-retrotransposons were found to carry fragments from 180 genes, although none carried fragments of multiple genes; over 1500 DNA transposons, including *helitrons* and Pack-MULEs, were also found carrying protein-coding gene fragments (Baucom *et al.* 2009; Schnable *et al.* 2009). Thousands of Pack-MULE elements carrying genomic fragments of 45–1000 bp with both introns and exons have been annotated in rice (Jiang *et al.* 2004), where these elements make up 1.5% of the genome (Hanada *et al.* 2009). In about a quarter of these cases, elements were found to contain fragments of multiple genes (Jiang *et al.* 2004). *Helitrons* may also include sequence fragments of several genes; just more than 2% of the maize genome consists of these elements, 60% of which include at least one gene fragment (Yang & Bennetzen 2009). It should be noted, though, that the fragments carried by these mobile elements are usually considerably smaller than a protein-coding gene, and although some are transcribed, it has not been shown that they produce functional proteins. Nonetheless, many expressed Pack-MULEs and *helitrons* are under purifying selection, suggesting that they may produce functional products and that these elements are capable of creating new, transcribed chimeric sequences (Morgante *et al.* 2005; Devos 2010).

In theory, repeated transposition of elements containing fragments of multiple genes can lead to amplification of the chimeric gene in the genome. There is evidence, though, that Pack-MULEs and *helitrons* may be rapidly silenced, preventing their amplification in the genome (Bennetzen 2005). Whereas LTR-retrotransposon elements may be replicated over 10 000 times in a genome, individual Pack-MULE and *helitron* elements are generally found in low copy numbers (Bennetzen 2005; Bennetzen *et al.* 2005). It has been suggested that these elements may silence themselves epigenetically, preventing production of chimeric proteins that might have a dominant negative interaction with the normal proteins from which the component fragments were borrowed; further, such epigenetic silencing of TEs containing gene fragments may result in silencing of the genes from which the elements co-opted fragments (Bennetzen 2005; Morgante *et al.* 2005). Thus the functioning of these gene-carrying TEs may have important consequences for gene and genome structure and function; however, their role in creating novel functional genes requires additional investigation.

8.4 Gene family expansions

Although WGDs and TE dynamics are considered the biggest contributors to genome size, expansion of gene families in land plant lineages has also played a part in genome growth. Differential gene family diversification has also doubtless contributed to adaptation and diversification of plant lineages by providing genomes with unique gene complements through processes of gene and segmental duplication. By and large, the genomes of diploid land plants have similar numbers of genes, and furthermore have the same types of genes, that is, more-or-less the same major gene families are found in all land plant genomes (e.g., Tuskan *et al.* 2006; Floyd & Bowman 2007; Rensing *et al.* 2008; Paterson *et al.* 2009; Banks *et al.* 2011). However, the complement of genes in each gene family can vary considerably, with some families having diversified extensively in some plant lineages while being highly conserved in others. Lineage-specific gene family expansion is correlated with the origin of phenotypic novelty, and thus may facilitate adaptation to new environments as well as increased complexity (Lespinet *et al.* 2002; Koonin *et al.* 2004; Vogel & Chothia 2006).

8.4.1 Land plant gene diversification

A comparison of sequenced land plant genomes shows that the genome of their common ancestor already possessed most of the gene families known to be important in directing the development of the angiosperms, in spite of the dramatic morphological and physiological differences that distinguish the major groups of embryophytes (Banks *et al.* 2011); in fact orthologs of genes involved in basic cellular processes such as translation and metabolism, as well as genes involved in physiological processes such as drought resistance and dormancy, have been found in algal genomes (Vandepoele & Van de Peer 2005; Merchant *et al.* 2007). Thus orthologs of genes involved in the development of vascular tissue are present in the genome of Physcomitrella in spite of the fact that mosses do not have true vascular tissue (Banks *et al.* 2011). Land plant evolution appears to have progressed by way of sequential and independent expansion of developmental and regulatory gene families in each land plant lineage as a result of duplication, with subsequent diversification providing new functions (Banks *et al.* 2011). Twenty-seven gene families were found only in the sequenced vascular plant genomes (lycophyte and angiosperms) but not moss; these include genes involved in meristem regulation, hormone signaling, and flowering. An additional 43 gene families were found only in angiosperms, including genes involved in these same functions as well as light signaling (Banks *et al.* 2011).

Gene family expansions in moss have occurred at a lower rate than in the genomes of angiosperms (Rensing *et al.* 2008). Some gene families, such as families of signaling factors, disease resistance genes, and TFs, show independent amplification in the moss, lycophyte, and angiosperm lineages (Rensing *et al.* 2008; Banks *et al.* 2011). Other families, such as histidine kinases and response regulators, parts of two-component signaling pathways, are expanded only in moss, suggesting functional differences, in this case in signaling modes, in these lineages. Physcomitrella represents the land plant lineage with the smallest degree of adaptation to land, nonetheless, many elements of the genome correlate with its terrestrial habitat and adaptations to increased sun exposure, decreased water reliability, etc. (Rensing *et al.* 2008). The vegetative structures of Physcomitrella are desiccation-tolerant, and the genome of Physcomitrella contains orthologs of genes involved in desiccation-tolerance in Arabidopsis. The moss genome includes cytochrome P450 genes that are involved in the formation of cutin and these are absent from green algal genomes, consistent with the presence of a waxy protective cuticle in land plants, however the moss genome lacks orthologs of P450s involved in the formation of lignin. Likewise, genes involved in many hormone signaling pathways are found in Physcomitrella and angiosperms but not algal genomes, perhaps related to increasing organismal complexity (Rensing *et al.* 2008). The genome of the lycophyte Selaginella appears to lack orthologs of genes required in angiosperms for the regulation of processes such as developmental phase change and the establishment of leaf polarity, suggesting that these processes are regulated differently in lycophytes, perhaps related to the apparent independent origin of leaves in that group (Banks *et al.* 2011). Three families of genes involved in secondary metabolism were analyzed in Selaginella, including P450s and genes involved in the formation of volatiles, and all show independent diversification within the lycophyte lineage relative to moss and angiosperms. The expansion and increase in complexity of these gene families within the lycophyte lineage suggests that lycophytes may have a unique array of secondary compounds in comparison with seed plants (Banks *et al.* 2011).

TFs show considerably more diversification in plant lineages than in other eukaryotic groups (Riechmann *et al.* 2000; Shiu *et al.* 2005). Diversification of TFs is thought to drive increasing organismal complexity, and analyses of land plant genomes show independent diversification of TF families in the moss, lycophyte, and angiosperm genomes (Floyd & Bowman 2007; Richardt *et al.* 2007; Rensing *et al.* 2008; Lang *et al.* 2010; Banks *et al.* 2011). In a comparison of transcription associated proteins (TAPs, including both TFs and indirect transcription regulators (TRs)) in the genomes of algae, Physcomitrella, Selaginella, and angiosperms, the total number of TAPs was correlated with the number of different cell types—one measure of complexity—supporting this hypothesis (Richardt *et al.* 2007; Lang *et al.* 2010). In addition, there are more and larger TAP gene families in land plants relative to algae, with two apparent bursts of diversification occurring with the colonization of land and with the origin of the angiosperms or seed plants (Lang *et al.* 2010). The increase in number and size of these families along with the increase in complexity of land plants further suggests a relationship between these two factors; Arabidopsis and rice, both flowering plants, were found to have similar TF inventories (Goff *et al.* 2002).

8.4.2 Angiosperm gene diversification

Comparisons of sequenced angiosperm genomes have shown that the distribution of genes into broadly defined gene families, as well as into functional categories, is largely similar, but that family and subfamily membership is more variable when categories are more narrowly defined. Ming *et al.* (2008) compared gene family membership in five sequenced angiosperm genomes (papaya, Arabidopsis, black cottonwood, rice, and grape) and found that the genome of papaya has a reduced number of genes across the board, consistent with its small gene complement. Likewise, black cottonwood has more protein-coding genes than Arabidopsis but the distribution of functional protein domains is similar (Tuskan *et al.* 2006), and gene families present in the sorghum genome are largely similar in size and number to members in the Arabidopsis, rice, and black cottonwood genomes (Paterson *et al.* 2009). Apple and soybean were found to have a relatively high total number of TFs, but the proportional distribution of these genes into different TF families was largely similar in all genomes analyzed (Schmutz *et al.* 2010; Velasco *et al.* 2010).

These studies suggest a relatively uniform expansion of gene families in genomes with a larger gene complement. Nonetheless, studies within broad categories have found differences in gene distribution among plant genomes. In a comparison of Arabidopsis and soybean, the latter was shown to have a greater proportion of TCP and HD-Zip TFs, whereas Arabidopsis has an excess of HSF proteins (Schmutz *et al.* 2010). About 250 gene groups, including subfamilies of the AP2/ERF and MYB TF families as well as cytochrome P450 family, appear to be legume-specific, having been found only in the genomes of soybean and *M. truncatula*; some of these are related to legume-specific functions such as nodulation (Schmutz *et al.* 2010). The papaya genome shows expansion of the MADS-box and TCP TF families (Ming *et al.* 2008); however, the increased number of MADS-box genes is due to dramatic amplification of only about half the subfamilies, with the other half actually having fewer representatives or being entirely missing in papaya (Ming *et al.* 2008). As several of the families with differential representation in plant genomes have been implicated in plant development, it has been suggested that the lineage-specific expansion of particular gene subfamilies reflects differential regulation of developmental processes in these different species (Schmutz *et al.* 2010).

Comparisons of protein-coding genes in the genome of the monocots rice and sorghum and the eudicot Arabidopsis show some potentially clade-specific gene family expansions related to differences between these two major plant groups (International Rice Genome Sequencing Project 2005; Paterson *et al.* 2009). Grass genomes for instance contain genes encoding prolamins, seed storage proteins not found in eudicots, and differences are seen in gene complements related to cell wall structure and synthesis, reflecting the difference in cell walls characteristic of grasses and eudicots (International Rice Genome Sequencing Project 2005; Paterson *et al.* 2009). As is typical, though, it was not possible to assign a family or function to a large percentage of the proteins in all three genomes (International Rice Genome Sequencing Project 2005; Paterson *et al.* 2009) so the analysis is incomplete. Characterization of these apparently novel genes is required to fully understand the differences in gene content among genomes and what it means for plant structure, function, and evolution.

Analyses of the gene inventories in plant genomes has uncovered unique or diversified gene groups related to species-specific features and functions (e.g., Paterson *et al.* 2010). The apple genome includes about 2–7 times as many genes putatively involved in the metabolism of sorbitol, a key carbohydrate involved in fruit development, as other sequenced species (Velasco *et al.* 2010). In the black cottonwood genome, Tuskan *et al.* (2005) found expansion of genes related to the perennial woody habit of these trees, for instance genes involved in the formation of secondary xylem and lignin, as well as tannins and secondary metabolites related to stress and biotic interactions. The genome of papaya, a plant that occupies middle ground between a woody tree and an herb in that it produces wood but relatively little, has an intermediate number of genes related to lignin and cellulose biosynthesis in comparison to Arabidopsis and cucumber (which lack wood) and black cottonwood (a woody tree) (Ming *et al.* 2008; Huang *et al.* 2009). Papaya does have a large number of genes related to starch metabolism and production of volatile compounds, potentially related to fruit development and dispersal in this species (Ming *et al.* 2008). Not surprisingly, the grape genome is over-represented in genes related to the characteristics that are desirable for wine such as terpenoids, which provide aromas to wine. Likewise, the genome of castor bean contains genes specifically related to the synthesis of ricin and related compounds, and the fatty acids that make this a valuable oilseed crop (Chan *et al.* 2010). In addition, genes have been identified in both cucumber and papaya in the biosynthetic pathways that regulate production of taxon-specific secondary compounds such as cucurbitacin in cucumber (Ming *et al.* 2008; Huang *et al.* 2009).

Sorghum is highly drought tolerant, and sequencing of the genome suggests this may be due to diversification of cytochrome P450 genes, which function in stress response and which are 50% more abundant in the sorghum genome than the rice genome. The sorghum genome contains six copies of miRNA169g, which is upregulated in rice in response to drought and is thought to target (NF-Y) B TFs, which function in drought tolerance in Arabidopsis and maize (Paterson *et al.* 2009). Cucumber is a vine that climbs using tendrils, the growth of which is regulated by gibberellic acid and which involves cell wall modification; in the cucumber genome, Huang *et al.* (2009) identified three clades of GA biosynthesis genes not found in other sequenced genomes, as well as amplification of a subfamily of cell wall expansins, which they postulate are related to the development of the tendril (Huang *et al.* 2009). Some unique features do not find ready candidate genes in sequenced genomes, however: cucumber and papaya are dioecious but expanded or unique gene families that might be related to this condition have not been identified (Ming *et al.* 2008; Huang *et al.* 2009).

Several gene families appear to have undergone repeated diversification in multiple plant lineages, including genes involved in hormone synthesis and regulation, F-box proteins involved in ubiquitin-mediated protein degradation, and disease resistance genes. These are all large gene families important in species-specific regulation and responses, thus it is not surprising that they should be highly variable and subject to rapid amplification and diversification. Disease resistance genes appear to be abundant in angiosperm genomes and to have independently diversified in different lineages (Goff *et al.* 2002; Tuskan *et al.* 2006; Jaillon *et al.* 2007; Velasco *et al.* 2007; Huang *et al.* 2009; Chan *et al.* 2010; Schmutz *et al.* 2010). Nucleotide-binding site leucine-rich repeat (NBS-LRR) disease resistance genes are more abundant in the genomes of apple, black cottonwood, grape, and soybean, members of the eurosid II clade of angiosperms, than in Arabidopsis, a eurosid I clade member, suggesting that expansion of this group may predate diversification of the eurosid II clade (Schmutz *et al.* 2010; Velasco *et al.* 2010). However, the genome of Arabidopsis has more of the subfamily of Toll/interleukin region (TIR)-NBS-LRR genes than the genomes of the eurosid II taxa, and these genes are absent from monocot genomes (Goff *et al.* 2002; Velasco *et al.* 2010). Thus at one level (NBS-LRR), a pattern of increased diversification in eurosid II is seen, but at another level (TIR-NBS-LRR), increased diversification in eurosid I (or at least Arabidopsis) is seen. The papaya and cucumber genomes have relatively few NBS-LRR genes: 55 and about 60, respectively, in contrast to the 200 in Arabidopsis, 400 in black cottonwood, and 600 in rice (Ming *et al.* 2008; Huang *et al.* 2009). Papaya has predominately NBS-LRR, whereas the largest fraction in Arabidopsis is TIR-NBS-LRR and in black cottonwood is coiled-coil (CC)-NBS-LRR. In cucumber, the LOX (lipoxygenase) gene family, which also plays a role in disease resistance, is expanded, perhaps providing a complementary means of defense to the relatively depauperate NBS-LRR family. Among the grasses, sorghum, which has a genome twice the size of rice, has half as many NBS-LRR genes (211 in sorghum and 410 in rice). All the sorghum NBS-LRR genes are of the CC-NBS-LRR family; only two sorghum genes have TIR domains and these lack NBS domains (Paterson *et al.* 2009). It is clear that these families of genes amplify and diversify rapidly in different plant lineages creating significant differences even among closely related species, suggesting tremendous evolutionary flexibility in disease resistance as well as regulation of development and metabolic processes.

8.5 Tandem gene duplications

Tandem duplications—two or more genes repeated sequentially on a chromosome—are common features of eukaryotic genomes including those of plants from algae to angiosperms, representing about 15% of plant genomes (Merchant *et al.* 2007; Gaut & Ross-Ibarra 2008; Vogel *et al.* 2010). They have been implicated as an important factor in lineage-specific gene family expansions, as well as in genome size and overall gene number increase (Arabidopsis Genome Initiative 2000; Schmidt 2002). One in 700 Arabidopsis seeds is estimated to contain a novel copy-number variant (Gaut & Ross-Ibarra 2008). Whereas WGDs preserve proportions of gene types and therefore, theoretically the stoichiometry of gene products, tandem duplications affect only a few genes and therefore alter the relative proportions of genes and gene products in functional pathways and networks. In addition, tandem duplicates tend to affect genes toward the ends of biosynthetic pathways, suggesting that they persist more effectively in positions that do not affect entire pathways (Gaut & Ross-Ibarra 2008). Tandem arrays have been implicated in novel phenotypes and adaptation including submersion tolerance in rice (Xu *et al.* 2006), diversification of secondary metabolites in Arabidopsis (Kliebenstein *et al.* 2001), and development of the tendril in cucumber (Huang *et al.* 2009).

At the time of publication of the Arabidopsis genome (Arabidopsis Genome Initiative 2000), about 17% of the genome was estimated to consist of tandem repeats, indicating a high degree of potential redundancy. A similar percentage (14%) was found in the rice genome, however more detailed analyses showed that duplications may be interrupted by other genes, therefore searching for repeats with an enlarged window increased the percent to almost 30% of the rice genome (International Rice Genome Sequencing Project 2005). Most of the larger arrays contain genes encoding protein kinase domains, suggesting widespread duplication and diversification of these regulators (International Rice Genome Sequencing Project 2005). More recent analyses suggest that about 25% of the Arabidopsis genome consists of tandem arrays of two or more genes, with almost 60% of the arrays consisting of just two genes. The black cottonwood genome contains 3.5 times as many tandem arrays as Arabidopsis, consistent with the larger size of the black cottonwood genome, but these arrays occupy only about 15% of the genome (Tuskan *et al.* 2006). This suggests that the black cottonwood genome may be more stable, undergoing rearrangements more slowly, consistent with its longer life span; support for this hypothesis is provided by the fewer chromosomal rearrangements and slower rate of nucleotide substitution relative to Arabidopsis (Tuskan *et al.* 2006).

Tandem duplication is also implicated in the expansion of large and dynamic gene families such as the NBS-LRR, F-box, and cytochrome P450 gene families in many plant genomes (International Rice Genome Sequencing Project 2005; Schmutz *et al.* 2010; Vogel *et al.* 2010; Hu *et al.* 2011). NBS-LRR genes are highly clustered in papaya, cucumber, and other genomes, suggesting amplification by tandem duplication in those species (Ming *et al.* 2008; Huang *et al.* 2009). Although tandem duplications are found in the moss genome, they are much less common, including only about 1% of the protein-coding sequence, as opposed to approximately 10–15% in Arabidopsis, black cottonwood, and rice (Lang *et al.* 2010; Rensing *et al.* 2008). These tandem arrays are overrepresented for genes involved in photosynthesis and carbon metabolism (Rensing *et al.* 2008).

8.6 Fern and gymnosperm genomes

Due to their large size, it has not yet been possible to fully sequence a genome of a fern or gymnosperm, so that many attributes of these genomes cannot be accurately evaluated at this point. Nonetheless, some distinctive features are apparent. Fern genomes are without a doubt extremely large, and recent analyses suggest that ferns experienced a single WGD in their evolutionary history (Barker 2009; Barker & Wolf 2010). However, many angiosperm species experienced more WGDs but have smaller genomes and fewer chromosomes, so clearly other factors are at work in the fern genome to maintain such a large genome. Current hypotheses suggest that the fern genome is less dynamic than those of angiosperms, such that after duplications, whether whole-genome or smaller-scale, genes are silenced but not lost, as postulated by Haufler (1987; see also Barker 2009; Barker & Wolf 2010). Consistent with this, Rabinowicz *et al.* (2005) estimated a protein-coding gene count of 42 300 in *C. richardii*, which would represent a relatively low density of genes in the genome. It is intriguing to note that ferns are alone among land plants in showing a correlation between genome size and chromosome number (Nakazato 2009), possibly a byproduct of the apparent requirement for bivalent pairing, another distinctive trait of ferns. These factors may contribute to the distinctive dynamics of the fern genome. It is further worth noting that although the average chromosome number of homosporous ferns is a stunningly high 57, three times the average angiosperm haploid number of 16, the average for heterosporous ferns is a modest 14. Whether there is any significance to the restriction of high chromosome numbers to the homosporous ferns will be determined when sequencing and analytical techniques allow for the sequencing of these large genomes.

Gymnosperms also have large genomes, suggesting a scenario similar to that in ferns, but their modest chromosome numbers suggest a different phenomenon is at work. Large genomes generally have relatively low complexity, defined as the amount of unique information in a genome and generally measured in terms of repetitive versus single-copy or unique sequence content. Because conifers have very large genomes, it would be reasonable to predict that they have a very high percentage of TEs and therefore low complexity. However, based on measurements from methylation filtration, Wakamiya *et al.* (1993) estimated that the 22 000 Mb genome of Eastern white pine (*Pinus strobus*) was 14% single-copy sequence, a much higher fraction than sequenced plant genomes. Kinlaw and Neale (1997), based on reassociation kinetics and Southern analyses, suggested that conifer genomes may have a higher-than-expected complexity due to greater diversification of protein-coding gene families in their genomes. Analysis of SAGE tags by Lorenz and Dean (2002) is consistent with this hypothesis. Kovach *et al.* (2010) also note that low-complexity regions of conifer genomes are rich not only in genes, but also in low-copy retroelements, so that the unexpectedly high complexity of these genomes may be due to abundance of these low-copy elements as well as to expanded gene families. Sequencing in conifers suggests these genomes do contain a large percentage of repetitive elements, many of which have not been identified (Morse *et al.* 2009; Kovach *et al.* 2010), therefore a clearer picture of genome complexity as well as gene family diversity awaits a conifer genome sequence.

Rabinowicz *et al.* (2005) estimated the gene number in loblolly pine to be 224 000 and suggested this enormous number is due to tandem gene duplications that are common in pine (Krutovsky *et al.* 2004). Kovach *et al.* (2010) estimated that 1% of the sequence of ten BACs from loblolly pine was protein-coding and speculated that these genomes may be rich in pseudogenes; based on genome size (21 600 Mb) and an estimated average gene size of 1700 bp based on their data, this would suggest approximately 127 000 genes—about half that of Rabinowicz *et al.* (2005) but still considerably higher than genomes sequenced to date. Large angiosperm genomes are thought to be large due to increased TE complement rather than more protein-coding genes (although see Rabinowicz *et al.* 2005), and this has also been suggested for gymnosperm genomes (Morse *et al.* 2009). However, the calculations of Rabinowicz *et al.* (2005) and those based on Kovach *et al.* (2010) suggest a gene count significantly higher than those obtained for moss, lycophyte, and angiosperm genomes, suggesting either that pine genomes evolve and function quite differently from those of other plant species and that genetic control of organismal processes may differ in gymnosperms, or that these calculations are inaccurate for reasons that themselves will turn out to be extremely interesting.

8.7 Arabidopsis genome

A. thaliana is by far the most intensely studied member of the plant kingdom, and it is not surprising that its genome has also been more extensively analyzed than most others. It is fascinating that the history of the Arabidopsis genome is thought to include five WGDs, yet the genome is considerably smaller than most of the other plant genomes sequenced to date. The genus *Arabidopsis* has nine species (O'Kane & Al-Shehbaz 2003; Beck *et al.* 2007); the phylogeny is not well resolved, however, *A. thaliana* and *A. lyrata*, both with sequenced genomes, are thought to have diverged about 10 million years ago (Beilstein *et al.* 2010). The genome of *A. lyrata* is estimated at 207 Mb, which is more characteristic of the genus than the reduced 125 Mb genome of *A. thaliana*. In addition, the *A. lyrata* genome is predicted to include approximately 32 700 protein-coding genes, in contrast to the 27 000 of *A. thaliana*. Significant rearrangements are detectable in the genome of *A. thaliana* relative to other species of *Arabidopsis* and other Brassicaceae, the most conspicuous being a reduction from eight chromosomes to five via major chromosomal shuffling that included three fusions with the loss of three centromere regions as well as nine to ten other major rearrangements (Kuittinen *et al.* 2004; Yogeeswaran *et al.* 2005; Lysak *et al.* 2006; Tang *et al.* 2008a; Hu *et al.* 2011). The loss of three centromere regions accounts for some of the difference in size between the genomes of *A. thaliana* and *A. lyrata* (about 10%) but Hu *et al.* (2011) suggest that "hundreds of thousands" of deletions, as well as insertions and rearrangements, across the genomes are responsible for the differences between them. Although 90% of the two genomes are syntenic, with extensive colinearity, 50% of the *A. lyrata* genome appears to be absent from the *A. thaliana* genome, and 25% of the *A. thaliana* genome from the *A. lyrata* genome. In addition, the *A. lyrata* genome includes a higher percentage of TEs, with evidence of more recent activity, suggesting that purging of TEs, which may also result in reduction of genome size, has been more efficient in *A. thaliana*. The result is a smaller genome in *A. thaliana*, with a higher percentage of the sequence being protein coding. The combined evidence hints at selection for a reduced genome size in *A. thaliana*, which Hu *et al.* (2011) suggest may be related to the shorter life span and selfing habit of *A. thaliana* in contrast to the outcrossing *A. lyrata*.

Comparisons have also been made between the genomes of two of the most widely used accessions of *A. thaliana*, Colombia and Landsberg *erecta*; these provide crucial information on the different genetic backgrounds for plant biologists using these lines for their research. These two strains diverged around 200 000 years ago, and show extensive polymorphisms consistent with the rapid rate of evolution in this short-lived annual (Tang *et al.* 2008b; Paterson *et al.* 2010). The Arabidopsis Genome Initiative (2000), comparing a 90 Mb portion of the two genomes, found over 25 000 single nucleotide polymorphisms (SNPs) and 15 000 indels ranging in size from 2 bp to over 38 Kb, with only 5% of these larger than 50 bp. Comparing gene-rich regions of the genomes of the two accessions, Ziolkowski *et al.* (2009) identified 8500 indels that were >100 bp; in addition to disrupting colinearity, the indels affected about 1200 protein-coding genes. A large proportion of these indels appear to be related to transposon activity, and in many cases genes appeared to be deleted; however, a large portion of the "missing" genes were found elsewhere in the genome, possibly by the action of transposons (Arabidopsis Genome Initiative 2000). Ziolkowski *et al.* (2009) showed that *gypsy*- and *copia*-like LTR-retrotransposons have been active, particularly the latter, as have *helitrons* and MULEs. Because *helitrons* and MULEs have been implicated in translocating and combining fragments of plant genes, they suggest this process is important in *A. thaliana* and may underlie its plasticity (Ziolkowski *et al.* 2009).

8.8 Domestication

Most sequenced genomes are from plants that are of economic importance, and many have been under intense human selection for 10 000 years or more. Domestication involves a number of different processes, but two common phenomena are a genetic bottleneck, which can result in a reduction of genetic diversity in the crop, and directed selection, which can further reduce genetic diversity and also lead to novel phenotypes. Because many of the features of the genomes of sequenced species have been under human selection, which may be considerably stronger than natural selection, there may have been rapid accumulation of features related to the traits prized by humans. For instance the elevated number of genes in the grape genome that are related to desirable wine attributes may be a result of human selection (Jaillon *et al.* 2007).

The process by which various crop plants became domesticated over the past 10 000+ years involved intense selection for specific traits that improve yield and make harvest easier and more efficient. Cultivated species by and large differ from their wild relatives by a suite of traits known as the "domestication syndrome" (Harlan *et al.* 1973). Among these are a sturdier plant, synchronized flowering time, larger edible parts (generally fruits or seeds), seed retention, and increased apical dominance with decreased branching. Recent studies combining archeological and molecular data suggest that for most plant species, domestication took place over several thousand years even though models suggest traits such as these can become fixed within a few generations given strong selection (Eyre-Walker *et al.* 1998; Allaby *et al.* 2008; Brown *et al.* 2009; Purugganan & Fuller 2009; Tang *et al.* 2009). Although the early steps of domestication probably varied from species to species and from environment to environment, the first step may have occurred when nomads, who visited the same areas on an annual basis, took advantage of plants that grew from seed tossed away with garbage in previous years. In other cases people may have burned patches of land to promote the growth of desirable food species and help eliminate undesirable species. Eventually it is thought that people began collecting seed and actively sowing it for a more secure harvest. Finally, early farmers began to collect seed specifically from plants that had desirable traits such as increased fruit size or a more convenient growth form (Doebley *et al.* 2006; Tang *et al.* 2009).

As a result of the collection of seed from relatively few individuals with desirable traits, most domesticated species probably experienced a genetic bottleneck in their history, which would have resulted in a decrease in genetic diversity in comparison with their wild relatives. The estimated degree of reduction in diversity varies, with some crop species showing nearly 60% and others showing little or no decrease (Wright *et al.* 2005; Doebley *et al.* 2006; Kilian *et al.* 2007; Aguilar-Melendez *et al.* 2009). However, reduction in variability is not uniform across the genome. Rather, genes that contribute to the phenotypes that are selected show reduced allelic diversity relative both to the genes of wild progenitors and to other genes in the domesticated plant's genome. Even within genes selection may not be uniform; analysis of *tb1* alleles showed a 60% loss of variability in the coding region, but a 98% loss in the 5′ UTR (Wang & Dooner 2006; Doebley *et al.* 2006); this may reflect the importance of changes in gene regulation in the domestication syndrome. Even so, selection may not be confined to the particular target locus. Recent studies have documented selective sweeps in which genomic regions linked to a target gene also experience intense selection and therefore reduced allelic diversity (Clark *et al.* 2004; Palaisa *et al.* 2004; Olsen *et al.* 2006; Tian *et al.* 2009). In some cases these alleles that are inadvertently swept along may be deleterious, but they are maintained due to tight linkage with the selected locus (Lu *et al.* 2006; Tang *et al.* 2009). In cultivation, plants with these detrimental alleles can survive, however it is likely that in wild populations they would not and the allele would be eliminated (Doebley *et al.* 2006, Purugganan & Fuller 2009).

QTL studies suggest that intense selection on only a small number of genes was responsible for the emergence of the traits of the domestication syndrome (Paterson 2002; Doebley 2004). In contrast, the results of selection screens have suggested that several hundred or more genes may show evidence of selection in maize and sunflower (Wright *et al.* 2005; Chapman *et al.* 2008). To date few genes directly implicated in domestication traits have been identified; those that have been identified are mostly TFs, and exhibit altered regulation in domesticated versus wild species (e.g., Doebley *et al.* 1997; Frary *et al.* 2000; Doebley *et al.* 2006; Li *et al.* 2006; Simons *et al.* 2006; Purugganan & Fuller 2009). In contrast, many more of the genes that specify the differences among varieties of individual crops have been identified. Most of these encode enzymes, and the mutant alleles, rather than showing changes in regulation, are altered in the coding sequence such that the function of the protein is altered or abolished entirely (reviewed in Purugganan & Fuller 2009; Tang *et al.* 2009). The contrasting types of gene and genome differences found on the one hand between crops and their wild relatives and on the other hand among varieties of a single crop suggest that the domestication process involves changes in transcription regulation, whereas the diversification of different cultivars and crop varieties involves changes in specific steps in pathways that affect desirable traits such as color and taste (e.g., Doebley *et al.* 2006; Purugganan & Fuller 2009).

8.9 Future directions

The relatively low cost of high-throughput next generation sequencing and the accumulation of genomic sequence data are allowing the expansion of the pool of sequencing targets from single lines of high value species such as crop plants to a wide variety of species of interest for comparative genomic and population-level biological questions. As more and more genome sequences become available, the burdensome computational analyses potentially become simplified; once a reference genome has been published, for instance the genome of the Colombia accession of *A. thaliana* (Arabidopsis Genome Initiative 2000), it becomes computationally much simpler to sequence genomes of other Arabidopsis accessions or closely related species, as the original sequence can be used as a scaffold on which new data can be organized. Thus an initiative headed by the Joint Genome Institute and initiated in 2008 is sequencing the genomes of 1001 Arabidopsis accessions to identify genomic differences associated with phenotypic variation among these lines. As of this writing the collaboration had finished 249 genomes with 90 more underway (http://www.1001genomes.org/index.html). In addition, a sequenced genome can be used as a reference to facilitate assembly of a relatively closely related species, for instance the use of the *A. thaliana* genome to assign *A. lyrata* scaffolds to linkage groups (Hu *et al.* 2011, or see Zhu & Buell 2007). Currently the computational aspects of genome analysis are by far the most time consuming and cumbersome; each additional genome sequence makes this aspect easier and quicker.

To date genome sequencing has focused on small genomes due to the technological and logistical limitations related to sequencing genomes rich in repetitive sequence. This skew in sequenced genomes toward small genome sizes may be distorting our understanding of genome structure and evolution. The sequencing of the maize genome, which is rich in repetitive sequence, opens the door for successful sequencing of these large genomes, and technological improvements in sequencing also hint that we may be able to undertake these previously unthinkable genomes. Perhaps the biggest gap in our ability to understand plant genome evolution is the lack of a genome sequence for a fern or a gymnosperm. Our ability to trace the history of gene families, to understand the dynamics of large genomes, to answer questions such as "how many genes does a pine have?" and "why are fern genomes so big?" are hampered by our lack of a fully sequenced fern or gymnosperm genome. Questions related to the origin of seed plants and of angiosperms, as well as the evolution and regulation of key adaptive features of land plants, will also be more easily and thoroughly addressed with a more phylogenetically complete set of plant genomes. As the cost of sequencing drops, it will also become more feasible to sequence additional mosses and lycophytes, as well as a greater diversity of angiosperms including basal angiosperms, basal eudicots, and a wider phylogenetic spread of eudicots (for instance currently almost all published angiosperm genomes are eurosids). A comparison among currently available sequenced angiosperm genomes shows that every new genome has new information to offer. Clearly at the very least we need multiple representatives from each major land plant lineage; no single species can speak for its entire lineage, and furthermore we can only identify adaptation-specific genome features by comparison among relatively closely related species.

As technology and analysis continue to bring down the cost and time involved in sequencing, we can expect not only genome sequences of many additional species of land plants, but also new applications of this technology to ecological, developmental, evolutionary, physiological, and other questions. The lower cost of genome sequencing is making "population genomics" available for the first time: the ability to sequence entire genomes on the population level to understand variation in structure and function at this level. Knowledge of genome dynamics within populations can give us unprecedented insight into processes of adaptation and speciation as well as information on variation in coding and noncoding regions of genes, TE diversity and activity, and copy-number variants within populations. For instance, plants with large genomes have been shown to have lower rates of photosynthesis, growth, and speciation, and to be less well adapted to extreme environments, however the connection among genome size, phenotype, and fitness is unclear (Gaut & Ross-Ibarra 2008), and the role of TEs in these relationships needs examination. It is therefore important to understand the effect of selection on TE diversity and abundance on fitness, which requires both a genomic and a population-level perspective (Gaut & Ross-Ibarra 2008). Population-level questions and genomic analyses are finding common ground in questions related to the processes involved in the evolution of genome features (Flowers & Purugganan 2008), and hopefully there will be more cross-talk between these two previously separate fields.

The sequencing of a number of land plant genomes, ranging from mosses to numerous angiosperms, has shed light on the complexities of genome structure and evolution, and in particular the astonishingly dynamic nature of plant genomes. Our understanding of the role of WGDs, tandem duplications, and TE amplification and removal on genome structure and function has increased substantially through the comparative study of the genomes of both closely and distantly related species. However, even with whole genome sequences of chlorophyte algae, a moss, a lycophyte, and numerous angiosperms, we are not yet able to answer key questions about the molecular basis of adaptation to land. Genes and gene family expansions have been identified that may be related to features such as the formation of cutin in

Physcomitrella (Rensing *et al.* 2008), a key adaptation to land, but such findings are rare; more commonly identified are lineage-specific gene functions such as nodulation in soybean (Schmutz *et al.* 2010) or ricin synthesis in castor bean (Chan *et al.* 2010). Also very common is lineage-specific expansion of gene families related to disease resistance, stress response, synthesis of secondary compounds, and other features that doubtless are critical for the survival and diversification of those plant groups. But these differences in gene complement are not likely to reflect the significant morphological and developmental changes that distinguish the major clades of land plants. We have not identified the suite of genes that allows for the development of vasculature (and with that true leaves, shoots, and roots), a dominant sporophyte, heterospory, and endosporic development in Selaginella. Likewise, in spite of numerous fully sequenced angiosperm genomes, we do not know the molecular changes that underlie the development of the integument of the megasporangium, the carpel, and pollen. The lack of fern and gymnosperm genome sequences hampers our ability to study the evolution of these features, but in addition, it is likely that many major structural and functional differences among plant groups are due to the differential expansion and diversification, including changes in regulation, of preexisting gene families. Thus, the sequencing and comparative analyses of the nuclear genomes of charophyte algae, as well as fern and gymnosperm genomes, in combination with expression, functional, and *in silico* systems analyses, will be critical to uncovering the molecular bases of the key structural and functional features of land plants.

References

Adams, K.L. & Palmer, J.D. (2003) Evolution of mitochondrial gene content: gene loss and transfer to the nucleus. *Molecular Phylogenetics and Evolution*, **29**, 380–395.

Adams, K.L. & Wendel, J.F. (2005) Polyploidy and genome evolution in plants. *Current Opinion in Plant Biology*, **8**, 135–141.

Aguilar-Melendez, A., Morrell, P.L., Roose, M.L., *et al.* (2009) Genetic diversity and structure in semiwild and domesticated chiles (*Capsicum annuum*; Solanaceae) from Mexico. *American Journal of Botany*, **96**, 1190–1202.

Ahuja, M.R. (2005) Polyploidy in gymnosperms: revisited. *Silvae Genetica*, **54**, 59–69.

Allaby, R.G., Fuller, D.Q. & Brown, T.A. (2008) The genetic expectations of a protracted model for the origins of domesticated crops. *Proceedings of the National Academy of Sciences of the United States of America*, **105**, 13982–13986.

Ammiraju, J.S., Zuccolo, A., Yu, Y., *et al.* (2007) Evolutionary dynamics of an ancient retrotransposon family provides insights into evolution of genome size in the genus Oryza. *Plant Journal*, **52**, 342–351.

Arabidopsis Genome Initiative (2000) Analysis of the genome sequence of the flowering plant Arabidopsis thaliana. *Nature*, **408**, 796–815.

Arumuganathan, K. & Earle, E.D. (1991) Nuclear DNA content of some important plant species. *Plant Molecular Biology Reporter*, **9**, 208–218.

Bancroft, I. (2001) Duplicate and diverge: the evolution of plant genome microstructure. *Trends in Genetics*, **17**, 89–93.

Banks, J.A., Nishiyama, T., Hasebe, M., *et al.* (2011) The Selaginella genome identifies genetic changes associated with the evolution of vascular plants. *Science*, **332**, 960–963.

Barker, M.S. (2009) Evolutionary genomic analyses of ferns reveal that high chromosome numhers are a product of high retention and fewer rounds of polyploidy relative to angiosperms. *American Fern Journal*, **99**, 136–141.

Barker, M.S., Kane, N.C., Matvienko, M., *et al.* (2008) Multiple paleopolyploidizations during the evolution of the Compositae reveal parallel patterns of duplicate gene retention after millions of years. *Molecular Biology and Evolution*, **25**, 2445–2455.

Barker, M.S. & Wolf, P.G. (2010) Unfurling fern biology in the genomics age. *BioScience*, **60**, 177–185.

Baucom, R.S., Estill, J.C., Chaparro, C., *et al.* (2009) Exceptional diversity, non-random distribution, and rapid evolution of retroelements in the B73 maize genome. *PLoS Genetics*, **5**, e1000732.

Beaulieu, J.M. (2010) The right stuff: evidence for an 'optimal' genome size in a wild grass population. *New Phytologist*, **187**, 883–885.

Beck, J.B., Al-Shehbaz, I.A., O'Kane, S.L., Jr., *et al.* (2007) Further insights into the phylogeny of Arabidopsis (Brassicaceae) from nuclear Atmyb2 flanking sequence. *Molecular Phylogenetics and Evolution*, **42**, 122–130.

Becker, B. & Marin, B. (2009) Streptophyte algae and the origin of embryophytes. *Annals of Botany*, **103**, 999–1004.

Beilstein, M.A., Nagalingum, N.S., Clements, M.D., *et al.* (2010) Dated molecular phylogenies indicate a Miocene origin for Arabidopsis thaliana. *Proceedings of the National Academy of Sciences of the United States of America*, **107**, 18724–18728.

Bennett, M.D. & Smith, J.B. (1991) Nuclear DNA amounts in angiosperms. *Philosophical Transactions of the Royal Society of London B*, **334**, 309–345.

Bennetzen, J.L. (2000) Transposable element contributions to plant gene and genome evolution. *Plant Molecular Biology*, **42**, 251–269.

Bennetzen, J.L. (2005) Transposable elements, gene creation and genome rearrangement in flowering plants. *Current Opinion in Plant Biology*, **15**, 621–627.

Bennetzen, J.L. & Kellogg, E.A. (1997) Do plants have a one-way ticket to genomic obesity? *Plant Cell*, **9**, 1509–1514.

Bennetzen, J.L., Ma, J. & Devos, K.M. (2005) Mechanisms of recent genome size variation in flowering plants. *Annals of Botany*, **95**, 127–132.

Bennetzen, J.L., SanMiguel, P., Chen, M., *et al.* (1998) Grass genomes. *Proceedings of the National Academy of Sciences of the United States of America*, **95**, 1975–1978.

Bikard, D., Patel, D., Le Mette, C., *et al.* (2009) Divergent evolution of duplicate genes leads to genetic incompatibilities within *A. thaliana*. *Science*, **323**, 623–626.

Blanc, G., Barakat, A., Guyot, R., *et al.* (2000) Extensive duplication and reshuffling in the Arabidopsis genome. *Plant Cell*, **12**, 1093–1101.

Blanc, G., Duncan, G., Agarkova, I., *et al.* (2010) The *Chlorella variabilis* NC64A genome reveals adaptation to photosymbiosis, coevolution with viruses, and cryptic sex. *Plant Cell*, **22**, 2943–2955.

Blanc, G. & Wolfe, K.H. (2004) Widespread paleopolyploidy in model plant species inferred from age distributions of duplicate genes. *Plant Cell*, **16**, 1667–1678.

Bowers, J.E., Arias, M.A., Asher, R., *et al.* (2005) Comparative physical mapping links conservation of microsynteny to chromosome structure and recombination in grasses. *Proceedings of the National Academy of Sciences of the United States of America*, **102**, 13206–13211.

Bowers, J.E., Chapman, B.A., Rong, J., *et al.* (2003) Unravelling angiosperm genome evolution by phylogenetic analysis of chromosomal duplication events. *Nature*, **422**, 433–438.

Brown, T.A., Jones, M.K., Powell, W., *et al.* (2009) The complex origins of domesticated crops in the Fertile Crescent. *Trends in Ecology and Evolution*, **24**, 103–109.

Cavalier-Smith, T. (2005) Economy, speed and size matter: evolutionary forces driving nuclear genome miniaturization and expansion. *Annals of Botany*, **95**, 147–175.

Cenci, A., Combes, M.C. & Lashermes, P. (2010) Comparative sequence analyses indicate that Coffea (Asterids) and Vitis (Rosids) derive from the same paleo-hexaploid ancestral genome. *Molecular Genetics and Genomics*, **283**, 493–501.

Chan, A.P., Crabtree, J., Zhao, Q., *et al.* (2010) Draft genome sequence of the oilseed species Ricinus communis. *Nature Biotechnology*, **28**, 951–956.

Chapman, M. A., Pashley, C.H., Wenzler, J., *et al.* (2008) A genomic scan for selection reveals candidates for genes involved in the evolution of cultivated sunflower (*Helianthus annuus*). *Plant Cell*, **20**, 2931–2945.

Clark, R.M., Linton, E., Messing, J., *et al.* (2004) Pattern of diversity in the genomic region near the maize domestication gene tb1. *Proceedings of the National Academy of Sciences of the United States of America*, **101**, 700–707.

Cui, L., Wall, P.K., Leebens-Mack, J.H., *et al.* (2006) Widespread genome duplications throughout the history of flowering plants. *Genome Research*, **16**, 738–749.

Davies, B.J., O'Brien, I.E.W. & Murray, B.G. (1997) Karyotypes, chromosome bands and genome size variation in New Zealand endemic gymnosperms. *Plant Systematics and Evolution*, **208**, 169–185.

De Bodt, S., Maere, S. & Van de Peer, Y. (2005) Genome duplication and the origin of angiosperms. *Trends in Ecology and Evolution*, **20**, 591–597.

Derelle, E., Ferraz, C., Rombauts, S., *et al.* (2006) Genome analysis of the smallest free-living eukaryote *Ostreococcus tauri* unveils many unique features. *Proceedings of the National Academy of Sciences of the United States of America*, **103**, 11647–11652.

Deutsch, M. & Long, M. (1999) Intron-exon structures of eukaryotic model organisms. *Nucleic Acids Research*, **27**, 3219–3228.

Devos, K.M. (2010) Grass genome organization and evolution. *Current Opinion in Plant Biology*, **13**, 139–145.

Devos, K.M., Brown, J.K. & Bennetzen, J.L. (2002) Genome size reduction through illegitimate recombination counteracts genome expansion in Arabidopsis. *Genome Research*, **12**, 1075–1079.

Doebley, J. (2004) The genetics of maize evolution. *Annual Review of Genetics*, **38**, 37–59.

Doebley, J., Stec, A. & Hubbard, L. (1997) The evolution of apical dominance in maize. *Nature*, **386**, 485–488.

Doebley, J.F., Gaut, B.S. & Smith, B.D. (2006) The molecular genetics of crop domestication. *Cell*, **127**, 1309–1321.

Ellegren, H. (2002) Mismatch repair and mutational bias in microsatellite DNA. *Trends in Genetics*, **18**, 552.

Ellis, T.H., Poyser, S.J., Knox, M.R., *et al.* (1998) Polymorphism of insertion sites of Ty1-copia class retrotransposons and its use for linkage and diversity analysis in pea. *Molecular & General Genetics*, **260**, 9–19.

Eyre-Walker, A., Gaut, R.L., Hilton, H., *et al.* (1998) Investigation of the bottleneck leading to the domestication of maize. *Proceedings of the National Academy of Sciences of the United States of America*, **95**, 4441–4446.

Fawcett, J.A., Maere, S. & Van de Peer, Y. (2009) Plants with double genomes might have had a better chance to survive the Cretaceous-Tertiary extinction event. *Proceedings of the National Academy of Sciences of the United States of America*, **106**, 5737–5742.

Flavell, R.B., Bennett, M.D., Smith, J.B., *et al.* (1974) Genome size and the proportion of repeated nucleotide sequence DNA in plants. *Biochemical Genetics*, **12**, 257–269.

Flowers, J.M. & Purugganan, M.D. (2008) The evolution of plant genomes: scaling up from a population perspective. *Current Opinion in Genetics and Development*, **18**, 565–570.

Floyd, S.K. & Bowman, J.L. (2007) The ancestral developmental tool kit of land plants. *International Journal of Plant Sciences*, **168**, 1–35.

Frary, A., Nesbitt, T.C., Grandillo, S., *et al.* (2000) fw2.2: a quantitative trait locus key to the evolution of tomato fruit size. *Science*, **289**, 85–88.

Friesen, N., Brandes, A. & Heslop-Harrison, J.S. (2001) Diversity, origin, and distribution of retrotransposons (gypsy and copia) in conifers. *Molecular Biology and Evolution*, **18**, 1176–1188.

Fu, H. & Dooner, H. K. (2002) Intraspecific violation of genetic colinearity and its implications in maize. *Proceedings of the National Academy of Sciences of the United States of America*, **99**, 9573–9578.

Garfinkel, D.J. (2005) Genome evolution mediated by Ty elements in Saccharomyces. *Cytogenetic and Genome Research*, **110**, 63–69.

Gaut, B.S. & Ross-Ibarra, J. (2008) Selection on major components of angiosperm genomes. *Science*, **320**, 484–486.

Goff, S.A., Ricke, D., Lan, T.H., *et al.* (2002) A draft sequence of the rice genome (*Oryza sativa* L. ssp. *japonica*). *Science*, **296**, 92–100.

Grandbastien, M.A., Audeon, C., Bonnivard, E., *et al.* (2005) Stress activation and genomic impact of *Tnt1* retrotransposons in Solanaceae. *Cytogenetic and Genome Research*, **110**, 229–241.

Grant, D., Cregan, P. & Shoemaker, R.C. (2000) Genome organization in dicots: genome duplication in Arabidopsis and synteny between soybean and Arabidopsis. *Proceedings of the National Academy of Sciences of the United States of America*, **97**, 4168–4173.

Greilhuber, J., Borsch, T., Muller, K., *et al.* (2006) Smallest angiosperm genomes found in Lentibulariaceae, with chromosomes of bacterial size. *Plant Biology*, **8**, 770–777.

Gribbon, B.M., Pearce, S.R., Kalendar, R., *et al.* (1999) Phylogeny and transpositional activity of *Ty1-copia* group retrotransposons in cereal genomes. *Molecular and General Genetics*, **261**, 883–891.

Gupta, S., Gallavotti, A., Stryker, G.A., *et al.* (2005) A novel class of *Helitron*-related transposable elements in maize contain portions of multiple pseudogenes. *Plant Molecular Biology*, **57**, 115–127.

Hall, J. & Delwiche, C. (2007) In the shadow of giants; systematics of the charophyte green algae. In: *Unraveling the Algae: the Past, Present, and Future of Algal Systematics* (eds J. Brodie & J. Lewis). pp. 155–170. CRC Press, Boca Raton, FL.

Hanada, K., Vallejo, V., Nobuta, K., *et al.* (2009) The functional role of pack-MULEs in rice inferred from purifying selection and expression profile. *Plant Cell*, **21**, 25–38.

Hao, D., Yang, L. & Xiao, P. (2011) The first insight into the *Taxus* genome via fosmid library construction and end sequencing. *Molecular Genetics and Genomics*, **285**, 197–205.

Harlan, J.R., De Wet, J.M.J. & Price, E.G. (1973) Comparative evolution of cereals. *Evolution*, **27**, 311–325.

Haufler, C.H. (1987) Electrophoresis is modifying our concepts of evolution in homosporous pteridophytes. *American Journal of Botany*, **74**, 953–966.
Haufler, C.H. & Soltis, D.E. (1986) Genetic evidence suggests that homosporous ferns with high chromosome numbers are diploid. *Proceedings of the National Academy of Sciences of the United States of America*, **83**, 4389–4393.
Hawkins, J.S., Kim, H., Nason, J.D., *et al.* (2006) Differential lineage-specific amplification of transposable elements is responsible for genome size variation in *Gossypium*. *Genome Research*, **16**, 1252–1261.
Hawkins, J.S., Proulx, S.R., Rapp, R.A., *et al.* (2009) Rapid DNA loss as a counterbalance to genome expansion through retrotransposon proliferation in plants. *Proceedings of the National Academy of Sciences of the United States of America*, **106**, 17811–17816.
Hill, P., Burford, D., Martin, D.M., *et al.* (2005) Retrotransposon populations of *Vicia* species with varying genome size. *Molecular Genetics and Genomics*, **273**, 371–381.
Hu, T.T., Pattyn, P., Bakker, E.G., *et al.* (2011) The *Arabidopsis lyrata* genome sequence and the basis of rapid genome size change. *Nature Genetics*, **43**, 476–481.
Huang, S., Li, R., Zhang, Z., *et al.* (2009) The genome of the cucumber, *Cucumis sativus* L. *Nature Genetics*, **41**, 1275–1281.
Hufton, A.L. & Panopoulou, G. (2009) Polyploidy and genome restructuring: a variety of outcomes. *Current Opinion in Genetics & Development*, **19**, 600–606.
Ilic, K., SanMiguel, P.J. & Bennetzen, J.L. (2003) A complex history of rearrangement in an orthologous region of the maize, sorghum, and rice genomes. *Proceedings of the National Academy of Sciences of the United States of America*, **100**, 12265–12270.
Jaillon, O., Aury, J.M., Noel, B., *et al.* (2007) The grapevine genome sequence suggests ancestral hexaploidization in major angiosperm phyla. *Nature*, **449**, 463–467.
Jiang, N., Bao, Z., Zhang, X., *et al.* (2004) Pack-MULE transposable elements mediate gene evolution in plants. *Nature*, **431**, 569–573.
Kapraun, D.F. (2007) Nuclear DNA content estimates in green algal lineages: chlorophyta and streptophyta. *Annals of Botany*, **99**, 677–701.
Kashkush, K., Feldman, M. & Levy, A.A. (2003) Transcriptional activation of retrotransposons alters the expression of adjacent genes in wheat. *Nature Genetics*, **33**, 102–106.
Kellogg, E.A. (2003) Genome evolution: it's all relative. *Nature*, **422**, 383–384.
Khandelwal, S. (1990) Chromosome evolution in the genus *Ophioglossum*. *Botanical Journal of the Linnean Society*, **102**, 205–217.
Kilian, B., Ozkan, H., Walther, A., *et al.* (2007) Molecular diversity at 18 loci in 321 wild and 92 domesticate lines reveal no reduction of nucleotide diversity during *Triticum monococcum* (Einkorn) domestication: implications for the origin of agriculture. *Molecular Biology and Evolution*, **24**, 2657–2668.
Kinlaw, C.S. & Neale, D.B. (1997) Complex gene families in pine genomes. *Trends in Plant Science*, **2**, 356–359.
Kirik, A., Salomon, S. & Puchta, H. (2000) Species-specific double-strand break repair and genome evolution in plants. *EMBO Journal*, **19**, 5562–5566.
Kliebenstein, D.J., Lambrix, V.M., Reichelt, M., *et al.* (2001) Gene duplication in the diversification of secondary metabolism: tandem 2-oxoglutarate-dependent dioxygenases control glucosinolate biosynthesis in Arabidopsis. *Plant Cell*, **13**, 681–693.
Knight, C.A., Molinari, N.A. & Petrov, D.A. (2005) The large genome constraint hypothesis: evolution, ecology and phenotype. *Annals of Botany*, **95**, 177–190.

Koonin, E.V., Fedorova, N.D., Jackson, J.D., *et al.* (2004) A comprehensive evolutionary classification of proteins encoded in complete eukaryotic genomes. *Genome Biology*, **5**: R7.

Kovach, A., Wegrzyn, J.L., Parra, G., *et al.* (2010) The *Pinus taeda* genome is characterized by diverse and highly diverged repetitive sequences. *BMC Genomics*, **11**, 420.

Krutovsky, K.V., Troggio, M., Brown, G.R., *et al.* (2004) Comparative mapping in the Pinaceae. *Genetics*, **168**, 447–461.

Kubis, S., Schmidt, T. & Heslop-Harrison, J. (1998) Repetitive DNA eements as a major component of plant genomes. *Annals of Botany*, **82**, 45–55.

Kuittinen, H., de Haan, A.A., Vogl, C., *et al.* (2004) Comparing the linkage maps of the close relatives *Arabidopsis lyrata* and *A. thaliana*. *Genetics*, **168**, 1575–1584.

Kumar, A. & Bennetzen, J.L. (1999) Plant retrotransposons. *Annual Review of Genetics*, **33**, 479–532.

Lagercrantz, U. & Lydiate, D.J. (1996) Comparative genome mapping in *Brassica*. *Genetics*, **144**, 1903–1910.

Lai, J., Li, Y., Messing, J., *et al.* (2005) Gene movement by *Helitron* transposons contributes to the haplotype variability of maize. *Proceedings of the National Academy of Sciences of the United States of America*, **102**, 9068–9073.

Lang, D., Weiche, B., Timmerhaus, G., *et al.* (2010) Genome-wide phylogenetic comparative analysis of plant transcriptional regulation: a timeline of loss, gain, expansion, and correlation with complexity. *Genome Biology and Evolution*, **2**, 488–503.

Lang, D., Zimmer, A.D., Rensing, S.A. & Reski, R. (2008) Exploring plant biodiversity: the Physcomitrella genome and beyond. *Trends in Plant Science*, **13**, 542–549.

Leeton, P.R. & Smyth, D.R. (1993) An abundant LINE-like element amplified in the genome of Lilium speciosum. *Molecular and General Genetics*, **237**, 97–104.

Leitch, I.J. & Bennett, M.D. (1997) Polyploidy in angiosperms. *Trends in Plant Science*, **2**, 470–476.

Leitch, I.J., Hanson, L., Winfield, M., *et al.* (2001) Nuclear DNA C- values complete familial representation in gymnosperms. *Annals of Botany*, **88**, 843–849.

Lespinet, O., Wolf, Y.I., Koonin, E.V., *et al.* (2002) The role of lineage-specific gene family expansion in the evolution of eukaryotes. *Genome Research*, **12**, 1048–1059.

Li, C., Zhou, A. & Sang, T. (2006) Rice domestication by reducing shattering. *Science*, **311**, 1936–1939.

Lippman, Z., Gendrel, A.V., Black, M., *et al.* (2004) Role of transposable elements in heterochromatin and epigenetic control. *Nature*, **430**, 471–476.

Lisch, D. (2009) Epigenetic regulation of transposable elements in plants. *Annual Review of Plant Biology*, **60**, 43–66.

Little, D.P., Moran, R.C., Brenner, E.D., *et al.* (2007) Nuclear genome size in Selaginella. *Genome*, **50**, 351–356.

Liu, R., Vitte, C., Ma, J., *et al.* (2007) A GeneTrek analysis of the maize genome. *Proceedings of the National Academy of Sciences of the United States of America*, **104**, 11844–11849.

Lorenz, W.W. & Dean, J.F. (2002) SAGE profiling and demonstration of differential gene expression along the axial developmental gradient of lignifying xylem in loblolly pine (Pinus taeda). *Tree Physiology*, **22**, 301–310.

Lu, J., Tang, T., Tang, H., *et al.* (2006) The accumulation of deleterious mutations in rice genomes: a hypothesis on the cost of domestication. *Trends in Genetics*, **22**, 126–131.

Lynch, M. & Conery, J.S. (2000) The evolutionary fate and consequences of duplicate genes. *Science*, **290**, 1151–1155.

Lynch, M. & Force, A.G. (2000) The origin of interspecific genomic incompatibility via gene duplication. *American Naturalist*, **156**, 590–605.

Lyons, E., Pedersen, B., Kane, J., *et al.* (2008) The value of nonmodel genomes and an example using SynMap within CoGe to dissect the hexaploidy that predates the rosids. *Tropical Plant Biology*, **1**, 181–190.

Lysak, M.A., Berr, A., Pecinka, A., *et al.* (2006) Mechanisms of chromosome number reduction in *Arabidopsis thaliana* and related Brassicaceae species. *Proceedings of the National Academy of Sciences of the United States of America*, **103**, 5224–5229.

Ma, J. & Bennetzen, J.L. (2004) Rapid recent growth and divergence of rice nuclear genomes. *Proceedings of the National Academy of Sciences of the United States of America*, **101**, 12404–12410.

Ma, J., Devos, K.M. & Bennetzen, J.L. (2004) Analyses of LTR-retrotransposon structures reveal recent and rapid genomic DNA loss in rice. *Genome Research*, **14**, 860–869.

Maere, S., De Bodt, S., Raes, J., *et al.* (2005) Modeling gene and genome duplications in eukaryotes. *Proceedings of the National Academy of Sciences of the United States of America*, **102**, 5454–5459.

International Rice Genome Sequencing Project (2005) The map-based sequence of the rice genome. *Nature*, **436**, 793–800.

McClintock, B. (1984) The significance of responses of the genome to challenge. *Science*, **226**, 792–801.

Merchant, S.S., Prochnik, S.E., Vallon, O., *et al.* (2007) The Chlamydomonas genome reveals the evolution of key animal and plant functions. *Science*, **318**, 245–250.

Messing, J., Bharti, A.K., Karlowski, W.M., *et al.* (2004) Sequence composition and genome organization of maize. *Proceedings of the National Academy of Sciences of the United States of America*, **101**, 14349–14354.

Ming, R., Hou, S., Feng, Y., *et al.* (2008) The draft genome of the transgenic tropical fruit tree papaya (*Carica papaya* Linnaeus). *Nature*, **452**, 991–996.

Morgante, M., Brunner, S., Pea, G., *et al.* (2005) Gene duplication and exon shuffling by *helitron*-like transposons generate intraspecies diversity in maize. *Nature Genetics*, **37**, 997–1002.

Morgante, M., Hanafey, M. & Powell, W. (2002) Microsatellites are preferentially associated with nonrepetitive DNA in plant genomes. *Nature Genetics*, **30**, 194–200.

Morse, A.M., Peterson, D.G., Islam-Faridi, M.N., *et al.* (2009) Evolution of genome size and complexity in *Pinus*. *PLoS ONE*, **4**, e4332.

Murray, B.G., Leitch, I.J. & Bennett, M.D. (2010) Gymnosperm DNA C-values database (release 4.0, December 2010).

Naito, K., Cho, E., Yang, G., *et al.* (2006) Dramatic amplification of a rice transposable element during recent domestication. *Proceedings of the National Academy of Sciences of the United States of America*, **103**, 17620–17625.

Nakazato, T. (2009) Fern genome structure and evolution. *American Fern Journal*, **99**, 134–136.

Nakazato, T., Jung, M.K., Housworth, E.A., *et al.* (2006) Genetic map-based analysis of genome structure in the homosporous fern *Ceratopteris richardii*. *Genetics*, **173**, 1585–1597.

O'Kane, S.L. & Al-Shehbaz, I.A. (2003) Phylogenetic position and generic limits of *Arabidopsis* (Brassicaceae) based on sequences of nuclear ribosomal DNA. *Annals of the Missouri Botanical Garden*, **90**, 603–612.

Obermayer, R., Leitch, I.J., Hanson, L., *et al.* (2002) Nuclear DNA C-values in 30 species double the familial representation in pteridophytes. *Annals of Botany*, **90**, 209–217.

Ohno, S. (1970) *Evolution by Gene Duplication*. Springer, New York.

Ohri, D. & Khoshoo, T.N. (1986) Genome size in gymnosperms. *Plant Systematics and Evolution*, **153**, 119–132.

Olsen, K.M., Caicedo, A.L., Polato, N., *et al.* (2006) Selection under domestication: evidence for a sweep in the rice waxy genomic region. *Genetics*, **173**, 975–983.

Ouyang, S., Zhu, W., Hamilton, J., *et al.* (2007) The TIGR Rice Genome Annotation Resource: improvements and new features. *Nucleic Acids Research*, **35**, D883–D887.

Palaisa, K., Morgante, M., Tingey, S., *et al.* (2004) Long-range patterns of diversity and linkage disequilibrium surrounding the maize Y1 gene are indicative of an asymmetric selective sweep. *Proceedings of the National Academy of Sciences of the United States of America*, **101**, 9885–9890.

Parchman, T.L., Geist, K.S., Grahnen, J.A., *et al.* (2010) Transcriptome sequencing in an ecologically important tree species: assembly, annotation, and marker discovery. *BMC Genomics*, **11**, 180.

Paterson, A.H. (2002) What has QTL mapping taught us about plant domestication? *New Phytologist*, **154**, 591–608.

Paterson, A.H., Bowers, J.E., Bruggmann, R., *et al.* (2009) The *Sorghum bicolor* genome and the diversification of grasses. *Nature*, **457**, 551–556.

Paterson, A.H., Bowers, J.E. & Chapman, B.A. (2004) Ancient polyploidization predating divergence of the cereals, and its consequences for comparative genomics. *Proceedings of the National Academy of Sciences of the United States of America*, **101**, 9903–9908.

Paterson, A.H., Freeling, M., Tang, H., *et al.* (2010) Insights from the comparison of plant genome sequences. *Annual Review of Plant Biology*, **61**, 349–372.

Pellicer, J., Fay, M.F. & Leitch, I.J. (2010) The largest eukaryotic genome of them all? *Botanical Journal of the Linnean Society*, **164**, 10–15.

Petrov, D.A. (2002) DNA loss and evolution of genome size in Drosophila. *Genetica*, **115**, 81–91.

Petrov, D.A., Lozovskaya, E.R. & Hartl, D.L. (1996) High intrinsic rate of DNA loss in Drosophila. *Nature*, **384**, 346–349.

Petrov, D.A., Sangster, T.A., Johnston, J.S., *et al.* (2000) Evidence for DNA loss as a determinant of genome size. *Science*, **287**, 1060–1062.

Piegu, B., Guyot, R., Picault, N., *et al.* (2006) Doubling genome size without polyploidization: dynamics of retrotransposition-driven genomic expansions in Oryza australiensis, a wild relative of rice. *Genome Research*, **16**, 1262–1269.

Prochnik, S.E., Umen, J., Nedelcu, A.M., *et al.* (2010) Genomic analysis of organismal complexity in the multicellular green alga *Volvox carteri*. *Science*, **329**, 223–226.

Purugganan, M.D. & Fuller, D.Q. (2009) The nature of selection during plant domestication. *Nature*, **457**, 843–848.

Qiu, Y.-L. (2008) Phylogeny and evolution of charophytic algae and land plants. *Journal of Systematics and Evolution*, **46**, 287–306.

Rabinowicz, P.D., Citek, R., Budiman, M.A., *et al.* (2005) Differential methylation of genes and repeats in land plants. *Genome Research*, **15**, 1431–1440.

Rensing, S.A., Lang, D., Zimmer, A.D., *et al.* (2008) The Physcomitrella genome reveals evolutionary insights into the conquest of land by plants. *Science*, **319**, 64–69.

Richardt, S., Lang, D., Reski, R., *et al.* (2007) PlanTAPDB, a phylogeny-based resource of plant transcription-associated proteins. *Plant Physiology*, **143**, 1452–1466.

Riechmann, J.L., Heard, J., Martin, G., *et al.* (2000) Arabidopsis transcription factors: genome-wide comparative analysis among eukaryotes. *Science*, **290**, 2105–2110.

Rostoks, N., Park, Y.J., Ramakrishna, W., *et al.* (2002) Genomic sequencing reveals gene content, genomic organization, and recombination relationships in barley. *Functional & Integrative Genomics*, **2**, 51–59.

Salse, J., Bolot, S., Throude, M., *et al.* (2008) Identification and characterization of shared duplications between rice and wheat provide new insight into grass genome evolution. *Plant Cell*, **20**, 11–24.

SanMiguel, P. & Bennetzen, J. (1998) Evidence that a recent increase in maize genome size was caused by the massive amplification of intergene retrotransposons. *Annals of Botany*, **82** (Suppl. A), 37–44.

SanMiguel, P., Gaut, B.S., Tikhonov, A., *et al.* (1998) The paleontology of intergene retrotransposons of maize. *Nature Genetics*, **20**, 43–45.

SanMiguel, P., Tikhonov, A., Jin, Y.K., *et al.* (1996) Nested retrotransposons in the intergenic regions of the maize genome. *Science*, **274**, 765–768.

Schmidt, R. (2002) Plant genome evolution: lessons from comparative genomics at the DNA level. *Plant Molecular Biology*, **48**, 21–37.

Schmutz, J., Cannon, S.B., Schlueter, J., *et al.* (2010) Genome sequence of the palaeopolyploid soybean. *Nature*, **463**, 178–183.

Schnable, P.S., Ware, D., Fulton, R.S., *et al.* (2009) The B73 maize genome: complexity, diversity, and dynamics. *Science*, **326**, 1112–1115.

Semon, M. & Wolfe, K.H. (2007) Consequences of genome duplication. *Current Opinion in Genetics & Development*, **17**, 505–512.

Shahmuradov, I.A., Akbarova, Y.Y., Solovyev, V.V., *et al.* (2003) Abundance of plastid DNA insertions in nuclear genomes of rice and Arabidopsis. *Plant Molecular Biology*, **52**, 923–934.

Shirasu, K., Schulman, A.H., Lahaye, T., *et al.* (2000) A contiguous 66-kb barley DNA sequence provides evidence for reversible genome expansion. *Genome Research*, **10**, 908–915.

Shiu, S.H., Shih, M.C. & Li, W.H. (2005) Transcription factor families have much higher expansion rates in plants than in animals. *Plant Physiology*, **139**, 18–26.

Simillion, C., Vandepoele, K., Van Montagu, M.C., *et al.* (2002) The hidden duplication past of Arabidopsis thaliana. *Proceedings of the National Academy of Sciences of The United States of America*, **99**, 13627–13632.

Simons, K.J., Fellers, J.P., Trick, H.N., *et al.* (2006) Molecular characterization of the major wheat domestication gene Q. *Genetics*, **172**, 547–555.

Smarda, P., Horova, L., Bures, P., *et al.* (2010) Stabilizing selection on genome size in a population of Festuca pallens under conditions of intensive intraspecific competition. *New Phytologist*, **187**, 1195–1204.

Soltis, D. & Soltis, P. (1999) Polyploidy: recurrent formation and genome evolution. *Trends in Ecology and Evolution*, **14**, 348–352.

Soltis, D.E., Albert, V.A., Leebens-Mack, J., *et al.* (2009) Polyploidy and angiosperm diversification. *American Journal of Botany*, **96**, 336–348.

Soltis, D.E. & Burleigh, J.G. (2009) Surviving the K-T mass extinction: new perspectives of polyploidization in angiosperms. *Proceedings of the National Academy of Sciences of the United States of America*, **106**, 5455–5456.
Tanaka, T., Antonio, B.A., Kikuchi, S., *et al.* (2008) The Rice Annotation Project Database (RAP-DB): 2008 update. *Nucleic Acids Research*, **36**, D1028–D1033.
Tang, H., Bowers, J.E., Wang, X., *et al.* (2008a) Synteny and collinearity in plant genomes. *Science*, **320**, 486–488.
Tang, H., Bowers, J.E., Wang, X., *et al.* (2010) Angiosperm genome comparisons reveal early polyploidy in the monocot lineage. *Proceedings of the National Academy of Sciences of the United States of America*, **107**, 472–477.
Tang, H., Sezen, U. & Paterson, A.H. (2009) Domestication and plant genomes. *Current Opinion in Plant Biology*, **13**, 1–7.
Tang, H., Wang, X., Bowers, J.E., *et al.* (2008b) Unraveling ancient hexaploidy through multiply-aligned angiosperm gene maps. *Genome Research*, **18**, 1944–1954.
Thiel, T., Graner, A., Waugh, R., *et al.* (2009) Evidence and evolutionary analysis of ancient whole-genome duplication in barley predating the divergence from rice. *BMC Evolutionary Biology*, **9**, 209.
Thomas, C.A., Jr. (1971) The genetic organization of chromosomes. *Annual Review of Genetics*, **5**, 237–256.
Tian, F., Stevens, N.M. & Buckler, E.S.T. (2009) Tracking footprints of maize domestication and evidence for a massive selective sweep on chromosome 10. *Proceedings of the National Academy of Sciences of the United States of America*, **106** (Suppl. 1), 9979–9986.
Tikhonov, A.P., SanMiguel, P.J., Nakajima, Y., *et al.* (1999) Colinearity and its exceptions in orthologous adh regions of maize and sorghum. *Proceedings of the National Academy of Sciences of the United States of America*, **96**, 7409–7414.
Tirichine, L. & Bowler, C. (2011) Decoding algal genomes: tracing back the history of photosynthetic life on Earth. *Plant Journal*, **66**, 45–57.
Tsukahara, S., Kobayashi, A., Kawabe, A., *et al.* (2009) Bursts of retrotransposition reproduced in Arabidopsis. *Nature*, **461**, 423–426.
Tuskan, G.A., Difazio, S., Jansson, S., *et al.* (2006) The genome of black cottonwood, Populus trichocarpa (Torr. & Gray). *Science*, **313**, 1596–1604.
Van de Peer, Y., Fawcett, J.A., Proost, S., *et al.* (2009a) The flowering world: a tale of duplications. *Trends in Plant Science*, **14**, 680–688.
Van de Peer, Y., Maere, S. & Meyer, A. (2009b) The evolutionary significance of ancient genome duplications. *Nature Reviews Genetics*, **10**, 725–732.
Vandepoele, K. & Van de Peer, Y. (2005) Exploring the plant transcriptome through phylogenetic profiling. *Plant Physiology*, **137**, 31–42.
Velasco, R., Zharkikh, A., Affourtit, J., *et al.* (2010) The genome of the domesticated apple (Malus × domestica Borkh.). *Nature Genetics*, **42**, 833–839.
Velasco, R., Zharkikh, A., Troggio, M., *et al.* (2007) A high quality draft consensus sequence of the genome of a heterozygous grapevine variety. *PLoS ONE*, **2**, e1326.
Vicient, C.M., Jaaskelainen, M.J., Kalendar, R., *et al.* (2001) Active retrotransposons are a common feature of grass genomes. *Plant Physiology*, **125**, 1283–1292.
Vision, T.J., Brown, D.G. & Tanksley, S.D. (2000) The origins of genomic duplications in Arabidopsis. *Science*, **290**, 2114–2117.
Vitte, C. & Bennetzen, J.L. (2006) Analysis of retrotransposon structural diversity uncovers properties and propensities in angiosperm genome evolution. *Proceedings of the National Academy of Sciences of the United States of America*, **103**, 17638–17643.

Vitte, C. & Panaud, O. (2005) LTR retrotransposons and flowering plant genome size: emergence of the increase/decrease model. *Cytogenetic and Genome Research*, **110**, 91–107.

Vogel, C. & Chothia, C. (2006) Protein family expansions and biological complexity. *PLoS Computational Biology*, **2**(5): e48.

Vogel, J.P. Garvin, D.F., Mockler, T.C., *et al.* (2010) Genome sequencing and analysis of the model grass *Brachypodium distachyon*. *Nature*, **463**, 763–768.

Wakamiya, I., Newton, R.J., Johnston, J.S., *et al.* (1993) Genome size and environmental factors in the genus Pinus. *American Journal of Botany*, **80**, 1235–1241.

Wang, Q. & Dooner, H.K. (2006) Remarkable variation in maize genome structure inferred from haplotype diversity at the bz locus. *Proceedings of the National Academy of Sciences of the United States of America*, **103**, 17644–17649.

Wang, W., Tanurdzic, M., Luo, M., *et al.* (2005) Construction of a bacterial artificial chromosome library from the spikemoss Selaginella moellendorffii: a new resource for plant comparative genomics. *BMC Plant Biology*, **5**, 10.

Wendel, J.F. (2000) Genome evolution in polyploids. *Plant Molecular Biology*, **42**, 225–249.

Wessler, S.R., Bureau, T.E. & White, S.E. (1995) LTR-retrotransposons and MITEs: important players in the evolution of plant genomes. *Current Opinion in Genetics & Development*, **5**, 814–821.

Williams, C.G., Joyner, K.L., Auckland, L.D., *et al.* (2002) Genomic consequences of interspecific *Pinus* spp. hybridization. *Biological Journal of the Linnean Society*, **75**, 503–508.

Wood, T.E., Takebayashi, N., Barker, M.S., *et al.* (2009) The frequency of polyploid speciation in vascular plants. *Proceedings of the National Academy of Sciences of the United States of America*, **106**, 13875–13879.

Wright, S.I., Bi, I.V., Schroeder, S.G., *et al.* (2005) The effects of artificial selection on the maize genome. *Science*, **308**, 1310–1314.

Xu, K., Xu, X., Fukao, T., *et al.* (2006) Sub1A is an ethylene-response-factor-like gene that confers submergence tolerance to rice. *Nature*, **442**, 705–708.

Yagi, E., Akita, T. & Kawahara, T. (2011) A novel Au SINE sequence found in a gymnosperm. *Genes & Genetic Systems*, **86**, 19–25.

Yang, L. & Bennetzen, J.L. (2009) Distribution, diversity, evolution, and survival of Helitrons in the maize genome. *Proceedings of the National Academy of Sciences of the United States of America*, **106**, 19922–19927.

Yogeeswaran, K., Frary, A., York, T.L., *et al.* (2005) Comparative genome analyses of Arabidopsis spp.: inferring chromosomal rearrangement events in the evolutionary history of A. thaliana. *Genome Research*, **15**, 505–515.

Yu, J., Hu, S., Wang, J., *et al.* (2002) A draft sequence of the rice genome (*Oryza sativa* L. ssp. indica). *Science*, **296**, 79–92.

Zhu, W. & Buell, C.R. (2007) Improvement of whole-genome annotation of cereals through comparative analyses. *Genome Research*, **17**, 299–310.

Ziolkowski, P.A., Koczyk, G., Galganski, L., *et al.* (2009) Genome sequence comparison of Col and Ler lines reveals the dynamic nature of Arabidopsis chromosomes. *Nucleic Acids Research*, **37**, 3189–3201.

Zuccolo, A., Sebastian, A., Talag, J., *et al.* (2007) Transposable element distribution, abundance and role in genome size variation in the genus *Oryza*. *BMC Evolutionary Biology*, **7**, 152.

Annual Plant Reviews (2013) **45**, 277–320
doi: 10.1002/9781118305881.ch9
http://onlinelibrary.wiley.com

Chapter 9

DEVELOPMENT AND THE EVOLUTION OF PLANT FORM

Barbara A. Ambrose[1] and Cristina Ferrándiz[2]

[1] *The New York Botanical Garden, Bronx, NY, USA*
[2] *Instituto de Biología Molecular y Celular de Plantas (IBMCP), Consejo Superior de Investigaciones Cientificas-Universidad Politécnica de Valencia, 46022 Valencia, Spain*

Abstract: Development in land plants is the process whereby a single cell becomes a large multicellular organism. Developmental networks specify every organ, tissue, or cell type from egg to juvenile to a reproducing adult. Therefore, it would follow that changes in developmental networks result in changes in form, and understanding these developmental networks and their changes would provide great insight into the evolution of plant form. Only recently has the field of evolutionary developmental biology (evo-devo) been recognized as an important evolutionary mechanism; however, this field is still in its nascent stages in plants. Much of the plant evo-devo studies have focused on the flower and to some extent on the leaf. However, to understand the evolution of plant form, these ideas and studies need to be extended across the land plants. We review what we know about the contribution of development to the evolution of plant form and discuss avenues of future research.

Keywords: plant development; evo-devo; leaf development; ABC model; flower development; co-option; modularity; MADS-box genes.

9.1 Introduction

The evolution of plant form has been studied for centuries using morphology, development, and comparative embryology. These studies have included both paleobotanical evidence as well as focusing on extant lineages, and these have been important for generating hypotheses on the evolution of plant morphological features. Phylogenetic analyses using molecular characters have refined some of these hypotheses, and more recently developmental genetics has provided some compelling data to further evaluate the evolution

Annual Plant Reviews Volume 45: The Evolution of Plant Form, First Edition.
Edited by Barbara A. Ambrose and Michael Purugganan.

of particular plant morphological features, revealing some general concepts on the role of development in evolution (e.g., Ambrose *et al.* 2000; Nickrent *et al.* 2000; Carroll 2008).

Research in developmental genetics is focused on unraveling the gene regulatory networks that are necessary to build a particular morphological feature (reviewed in Carroll 2008). The field of evolutionary developmental biology (also known as evo-devo) is concerned with the molecular genetics of these developmental pathways and how changes in the gene regulatory networks underlie changes in form (Wagner *et al.* 2000; Carroll 2008). The study of these gene regulatory networks has revealed some general concepts about the logic of these developmental pathways and shown how developmental features play a role in the evolution of organism. Unlike population genetics, evolutionary developmental biology can provide an explanation for the evolution of body plans and the origin of morphological novelties (Wagner *et al.* 2000). In addition, evo-devo studies have been important in contributing molecular data, and together with the morphological and paleontological data have been the basis for discussions about the homology of structures.

9.1.1 A brief historical overview of evolutionary developmental biology

The role of development in the evolution of form arose with the explosion of developmental genetic studies in the fruit fly, *Drosophila melanogaster* (*Drosophila*) (reviewed in Carroll 2008). In the 1980s, geneticists began characterizing homeotic mutations in *Drosophila*. Homeosis, the transformation of one organ type into the identity of another organ, was defined by William Bateson in the late 1800s. Geneticists began to identify the molecular basis of these mutations and found that mutations in similar types of genes also resulted in homeotic mutations. These genes were referred to as *Homeobox* (*Hox*) genes, and ultimately eight linked *Hox* genes were identified in *Drosophila* that affected the specification of segments in the embryo. In addition, the order of the *Hox* genes on the chromosome paralleled the order of the function of these *Hox* genes in the development of the embryo from "head" to "tail."

The real beginning of evo-devo studies occurred when *Hox* clusters were identified throughout the animal kingdom and similar patterning functions could be ascribed to these *Hox* clusters in morphologically diverse animals. For example, *Pax6*, is the mouse *Hox* homolog of *eyeless* from *Drosophila* and both are important for eye field selector function in both animals. Eyes evolved independently at least three times in animals and the eyes of mice are morphologically distinct from eyes in *Drosophila*, with a lens and retina as opposed to the ommatidia common to insects. Amazingly, ommatidia form when the mouse *Pax6* gene is introduced into *Drosophila* (Halder *et al.* 1995). Although all eyes are not homologous, developmental genetic studies demonstrated that there was a common genetic component that was deployed independently in the evolution of similar structures, a phenomenon termed deep homology (Shubin *et al.* 1997; Shubin *et al.* 2009).

The beginnings of plant evolutionary developmental biology also began when homeotic mutants in flowering plants were characterized (Bowman *et al.* 1989; Coen & Meyerowitz 1991). Homeotic mutants such as *apetala3* in *Arabidopsis thaliana* (*Arabidopsis*) have flowers whose petals are replaced by sepals and stamens replaced by carpels. Geneticists began to analyze the molecular basis for these homeotic mutations and found that mutations in similar types of genes, namely, MADS-box genes, resulted in many of these floral homeotic mutations (Sommer *et al.* 1990; Yanofsky *et al.* 1990; Jack *et al.* 1992). Orthologs of MADS-box genes were identified across the flowering plants and mutations in these resulted in similar homeotic mutations in morphologically diverse flowers (see Section 9.2.1 for more detail). These early studies spawned the isolation of hundreds of floral MADS-box genes across the flowering plants and the beginning of modern plant evo-devo.

9.1.2 General concepts in evolutionary developmental biology

Developmental pathways that underlie form are composed of gene regulatory networks (e.g., Davidson & Erwin 2006; Peter & Davidson 2011), and evo-devo studies have shown that many components of gene regulatory networks encode transcription factors or signaling molecules. Once the genes and gene regulatory network underlying a particular morphology have been elucidated, a comparative approach can be used to try and understand how changes in the gene regulatory network can account for phenotypic changes. Evolution of gene regulatory networks can occur by duplication and diversification of genes, and by mutations in the coding sequence or mutations in the *cis*-regulatory elements of these loci. These evo-devo studies have allowed us to understand how development can act as a mechanism in the evolution of form.

Evo-devo studies have highlighted the role of large gene families (e.g., MADS-box genes) in the evolution of developmental processes, particularly large transcription factor families. The duplication and diversification of gene families provides the raw material for evolutionary processes, and gene duplication is an important source of genetic variation (Bowman *et al.* 2007; Conant & Wolfe 2008; Romanel *et al.* 2009). Following duplication, the gene and its copy can be functionally redundant and constraints on one of the copies can be relaxed. One copy can then accumulate mutations and take on a part of the original gene function (subfunctionalization), take on an entirely new function (neofunctionalization), or accumulate mutations until it is nonfunctional (becoming a pseudogene) (Ohno 1970).

The evolution of developmental networks can occur by changes in the coding sequence, and/or the *cis*-regulatory sequences of a gene in the network. Changes in the coding sequences of genes will affect the interaction of proteins in the network and the expression of downstream genes. Although duplication and diversification of genes and mutations in developmental genes can result in changes in the gene regulatory network, it has been argued that mutations in *cis*-regulatory elements are the main mechanism underlying the evolution of form (Carroll 2008).

Research in animal and plant evolutionary developmental biology has revealed some general concepts that arise from evo-devo, including modularity, co-option, developmental constraint, redundancy, evolvability, canalization, and developmental plasticity (Arthur 2002, 2004; Jenner & Wills 2007). All of these general concepts provide a framework to describe how developmental pathways can evolve and provide insight into how plant form has evolved.

For example, modularity is a property of a developmental system so that components of that system can be separated and modified independent of the rest of the system (Bolker 2000; Schlosser & Wagner 2004). One illustrative case is that of insects, which are composed of serial segments that can be multiplied or modified to have legs or wings, independent of other segments and the rest of the organism. It has been argued that modularity is an important property of developmental systems because it reduces the constraints and allows the developmental system to explore additional morphospace independent of the entire organism (Bolker 2000; Schlosser & Wagner 2004). Modularity is not only seen in body plan construction but also in the organization of protein domains and the *cis*-regulatory elements of a developmental gene (Riechmann *et al.* 2000; Carroll 2008; Charoensawan *et al.* 2010). Developmental networks are also modular in organization, and it is this modularity that gives the network robustness and evolvability. Another concept that has been revealed in evo-devo studies is co-option. Co-option is the recruitment of a gene (or pathway) to a new use, or the recruitment of a gene independently in evolution to a similar function (Arthur 2002).

Evolutionary developmental biology studies usually begin with the characterization of a gene involved in development and subsequently a more detailed analysis of that gene in a developmental network. Once the function of the gene or gene regulatory network has been elucidated, a comparative approach is used to try and understand which changes in the gene or gene regulatory network can account for phenotypic changes.

9.2 Plant evolutionary developmental biology

There are two areas of plant research that best illustrate how evolutionary developmental studies contribute to our knowledge of the evolution of plant form. The first is the well-studied molecular genetics of flower development, which illustrates the concepts of modularity and canalization of flower development, provides tools to address questions of organ homology, the importance of duplication, and diversification of large gene families, and provides an explanation for the origin of an important evolutionary innovation—the flower. The second area of research is the molecular genetics of leaf development, which illustrates the evolution of developmental pathways, co-option of regulatory modules, data to address questions of homology, the influence of developmental constraint, and is beginning to provide an explanation for the origin of another important plant innovation—the leaf. We will also briefly discuss the development of some plant structures that have received somewhat less attention in plant molecular genetics including gametophytes and spores. Finally, the development of functional model systems across the land plants is crucial to the future of plant evo-devo.

9.2.1 The evolution and development of the flower

9.2.1.1 How to build a flower

Most flowers in angiosperms follow a common basic plan, which consists of a standard arrangement of four concentric whorls. The two outer whorls or perianth are composed of sterile organs, most commonly sepals and petals, and the two inner whorls comprise the male and female reproductive organs.

Two decades ago, a genetic model for specifying floral organ identity was proposed based on mutant analyses in the two eudicot model species *Arabidopsis thaliana* and *Antirrhinum majus*. In both *Arabidopsis* and *Antirrhinum*, mutants had been described that showed transformations of one class of floral organs into another (Bowman *et al.* 1989; Schwarz-Sommer *et al.* 1990). These transformations were classified as *a*, *b*, or *c* types and generally affected two adjacent whorls. Class *a* mutants showed transformations of the two outer whorls, in which sepals took carpel identity and petals had stamen identity. Class *b* mutants developed sepals instead of petals and carpels instead of stamens. Finally, class *c* mutants showed transformations of stamens into petals and carpels into sepals and, in addition, they lost floral determinacy forming new flowers from the center of the floral meristem that repeated this pattern of homeotic transformations—lost determinacy.

Haughn and Sommerville made the first attempt to explain the phenotypes of these floral homeotic mutants with some simple logical rules that involved three classes of genes (A, B, C) which in different combinations of on/off states could specify the four types of floral organs (Haughn & Somerville 1988). A few years later, when the first homeotic genes started to be molecularly identified and Schwartz-Sommer *et al.* had already started to rework these "*abc*" rules (Schwarz-Sommer *et al.* 1990), Coen and Meyerowitz published the seminal "War of the whorls" paper, where the ABC model of floral development was clearly delineated (Coen & Meyerowitz 1991).

The ABC model was drawn from the three classes of homeotic mutants explained earlier and proposed that three gene functions, A, B, and C, defined by these mutant classes, would be responsible for the specification of floral organ identity. A, B, and C functions should be expressed in two adjacent whorls, overlapping with each other. Thus, A-function alone would specify sepals in the outer whorl, A + B would define petals in whorl 2, B + C stamens in whorl 3, and C alone carpels in the center of the flower. Mutual repression of A and C functions also had to be included in the model to explain *a* and *c* phenotypes, as well as an additional role of C-function in preventing indeterminate floral meristem growth (Figure 9.1a). In addition to the elegance of this simple yet powerful model, an added value was that

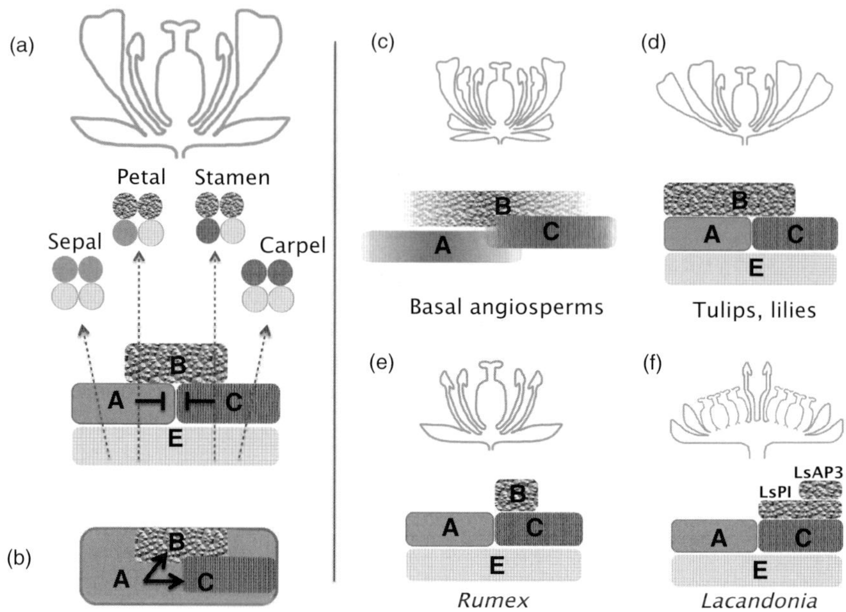

Figure 9.1 Models of genetic control of floral organ identity. (a) The ABCE model. Floral quartets of MADS factors direct the specification of sepals (A + E), petals (A + B + E), stamens (B + C + E), and carpels (C + E). (b) The (A) BC model according to Causier *et al.* 2010. The newly defined (A)-function would provide a floral context, required for B- and C-function activities. In this model, B-function alone promotes petal identity, B + C stamen identity and C-alone carpel identity, while the absence of B and C in the floral context provided by (A) results in sepal development. (c–f) Different versions of the ABCE model explain diverse floral morphologies (c) The "fading borders" hypothesis explains floral organs of intermediate morphology in basal angiosperms like *Amborella* or *Persea.* (d–f) The "sliding boundaries" hypothesis explains petaloid tepals in several nongrass monocots (d), the sepaloid perianth in Rumex acetosa (e), or the unusual inside-out morphology of *Lacandonia schismatica* (f).

it was inferred from the phenotypes of equivalent mutants in two relatively distant species of quite different floral morphology, and therefore strongly indicated that it could represent a general mechanism for flower development in angiosperms.

Concurrent with the proposal of the model, the identification of genetic functions to which ABC roles could be assigned strongly supported the model and provided a wealth of molecular data (Bradley *et al.* 1993; Goto & Meyerowitz 1994; Jack *et al.* 1992; Jofuku *et al.* 1994; Mandel *et al.* 1992; Sommer *et al.* 1990; Tröbner *et al.* 1992). All the identified ABC genes, except the *Arabidopsis* class A *APETALA2* gene, were closely related in sequence, and belonged to the MADS-box family of transcription factors. Moreover, MADS-box proteins were shown to interact in protein complexes, providing a molecular clue to the mechanisms that explained the combinatorial nature of the model (Davies *et al.* 1996b; Egea-Cortines *et al.* 1999). In addition, genes with equivalent functions in different species showed interspecific homology, reinforcing the idea of the general applicability of the model throughout the angiosperms.

Still, the original ABC model did not provide answers for some essential questions. For example, ectopic expression of ABC genes was shown to transform one type of floral organ into another, as predicted by the model, but was not sufficient to transform leaves into floral organs (Mizukami & Ma 1992; Davies *et al.* 1996a; Krizek & Meyerowitz 1996). These results were in conflict with an old but attractive theory of the German philosopher and writer Goethe, who proposed that all lateral organs shared a leaf-like developmental plan (Goethe 1790). This necessitated an extension of the basic model with an additional E-function, specific to the floral context, which was proposed and eventually assigned to a subclade of MADS-box genes, the *SEPALLATA* (*SEP*) subfamily, expressed in all floral whorls (Figure 9.1a).

Sepallata mutants showed flowers where only leaf-like organs developed in a whorled but indeterminate pattern, strongly supporting the idea of a common ground plan for leaves and floral organs (Pelaz *et al.* 2000; Ditta *et al.* 2004). E-function (SEP) proteins were subsequently shown to act as molecular "glue" in high order MADS-box complexes (Immink *et al.* 2009a). Thus, the ABC model was reformulated into what we know today as the ABCE or quartet model: whorls 1, 2, 3, and 4 contain, respectively, tetramers of A-A-E-E, A-B-B-E, B-B-C-E, or C-C-E-E MADS-box proteins, which bind to DNA of target genes to direct floral organ specific development (Figure 9.1a). Several excellent reviews have addressed the molecular details and the historical development of the quartet model (Causier *et al.* 2010; Immink *et al.* 2009b; Krizek & Fletcher 2005; Theissen & Melzer 2007).

The regulation of temporal and spatial expression of ABCE functions was shown to be crucial for establishing floral organ identity. The genetic pathways from floral meristem specification to the establishment of ABCE functions have been studied and it was shown how they were based on feed-forward loops and a final feedback loop that reinforced and conferred robustness to the floral program. In *Arabidopsis*, LEAFY (LFY), a plant-specific transcription factor with a major role in floral meristem specification, is also a key player in the upregulation of ABCE genes (Weigel *et al.* 1992). *LFY* is uniformly expressed in the young floral meristem, but achieves temporal and spatial regulation of floral organ identity genes through the action of different cofactors (Parcy *et al.* 1998). *LFY* appears to be sufficient to activate *APETALA1* (*AP1*), a MADS-box gene with a dual role in floral meristem specification and A-function.

Once activated by LFY, AP1 upregulates *SEP3*. SEP3 then forms a protein complex with AP1, switching its activity from floral specification to A-function. Subsequently, SEP3 is also able to act as a LFY cofactor to upregulate the other *SEP* genes, and the B- and C-function genes, although SEP3 is not as important for LFY activity as the other LFY cofactors (reviewed in Liu *et al.* 2009). LFY directly interacts with UNUSUAL FLORAL ORGANS (UFO), an F-box protein, to activate the B-class gene *APETALA3* (*AP3*) in the *UFO* domain of expression, which coincides with the B-class expression domain (Lee *et al.* 1997; Chae *et al.* 2008). Activation of the C-class gene *AGAMOUS* (*AG*) in the center of the floral meristem requires the concerted action of LFY and WUSCHEL (WUS), a homeodomain protein with a key role in stem cell maintenance of all aerial meristems. Finally, in a negative feedback loop, AG terminates floral meristem growth by repressing *WUS* expression in the center of the floral meristem (Lenhard *et al.* 2001; Lohmann *et al.* 2001).

9.2.1.2 The evolution of flower developmental pathways

Conceptually, the quartet model could be considered a paradigmatic example of an evolutionary framework. It was based on a modular structure, with combinatorial activities that resulted in different outcomes, and these activities were largely controlled by regulation of expression patterns. Moreover, since it was proposed from independent studies in two relatively distant species, conservation across species was strongly supported. These features of the model provided powerful mechanisms to explain evolutionary innovation, immediately suggesting predictions that could explain morphological novelties. Accordingly, from the early years when the model was initially proposed and MADS-box genes were identified as the main players involved, intensive research comprising an ever growing diversity of species has shown how the model shows both conservation and divergence and has provided a body of evidence to study the evolution of organ identity. Still, we remain limited by the comparatively small number of comprehensively studied species, especially in evolutionary nodes where innovations appeared and functional studies are challenging.

The elusive A-function. From molecular phylogenies and comparative development studies, it is clear now that the best-supported part of the model refers to the specification of stamens and carpels by B and C functions. The A-function, however, remains controversial. *Arabidopsis* is the only species so far where clear loss-of-function A-class mutants have been described. *Arabidopsis ap1* mutants show a partial loss of floral meristem identity and also some loss of A-function, with sepals transformed into bract-like organs and petals that do not form (Bowman *et al.* 1993). In the other A-class mutant, *ap2*, sepals are transformed into carpels and petals into stamens (Bowman *et al.* 1989). The other few mutants in *AP1*-like genes characterized in different species (for example, *squamosa* in Antirrhinum, *proliferating inflorescence meristems* in legumes or *macrocalyx* in tomato; Huijser *et al.* 1992; Taylor *et al.* 2002; Vrebalov *et al.* 2002) do not show these dual defects, being mainly affected in the floral meristem specification function. Likewise, mutants in *AP2* orthologs like *lip1* and *lip2* show only mild defects in sepal and petal development but not misregulation of C-class genes.

For these reasons, many authors have favored a variant of the ABCE model, an earlier version known as the BC model (Litt 2007; Schwarz-Sommer *et al.* 1990), or its more recent reformulation, the (A)BC model (Causier *et al.* 2010).

This alternative to the quartet model proposes that the (A) function would comprise both *AP1*-like and *SEP* genes and would not have a major role in perianth organ specification. (A)-class main roles then would be to specify the floral meristem, establishing a floral context where the default identity for floral organs would be sepals, and subsequently to regulate the expression pattern of B- and C-class genes. Once established the floral program, only B and C functions would be required for the specification of floral organ identity—carpels (C), stamens (B + C), and petals (B), with sepals being defined by the absence of B and C activities (Figure 9.1b).

The debate on quartet versus (A)BC model is hampered by the scarcity of mutants in different species, which makes it difficult to distinguish the general from the particular. But even after revisiting the original A-function, postulated as required for sepal and petal identity and to restrict C-function expression to the inner whorls, the quartet model in which A-function would require *AP1*-like genes still deserves some credit. The evolution of a dual perianth composed of sepals and petals appears to be fixed only in core eudicots, coincidental in evolutionary time with a whole genome duplication that originated the euAP1 clade (Litt & Irish 2003; see Section "Duplication and diversification of large gene families"). Moreover, while the *in vivo* formation of MADS tetramers has not been definitively proven yet, indirect evidence obtained *in vitro* or in yeast strongly supports the quartet ensemble, being one of these AP1/SEP/AP3/PI (Honma & Goto 2001; Ferrario *et al.* 2003). Also, the simultaneous ectopic expression of *AP1* and the B-class genes *PI* and *AP3* has been shown to convert cauline leaves into petals, without further development of floral meristems (Pelaz *et al.* 2001), and, for instance, the *ap1*-like mutant in *Medicago truncatula* (*mtpim*) shows conversions of sepals into leaf-like organs, casting some shadows on the idea that sepals could be the "default" floral organ state (Benlloch *et al.* 2006).

It is thus likely that *AP1*-like genes are part of the floral organ identity toolkit, while not so much in charge of the spatial repression of the C-function. This role in *Arabidopsis* would be mainly fulfilled by *AP2*, in turn repressed in the center of the flower by the microRNA miR172 or other cadastral regulators. In other species, *AP2*-like genes have not been assigned equivalent functions. Notably, however, a similar cadastral function is performed in *Petunia hybrida* or *A. majus* by a parallel module composed of NF-YA transcription factors regulated by a different microRNA, miR169 (Cartolano *et al.* 2007). The parallel AP2/miR172 and NF-YA/miR169 pathway constitutes an example of variation in molecular networks, which is not translated into structural differences in the flowers. It could also be a consequence of parallel redundant networks, which differ in relative importance in different species. This is suggested by the weak *ap2*-like phenotype of loss-of-function mutants in *LIP1* and *LIP2*, two AP2-related genes from *Antirrhinum* (Keck *et al.* 2003), or the common role in C-function repression performed by similar transcriptional cofactors in *Arabidopsis* (*LEUNIG*) and *Antirrhinum* (*STYLOSA*) (Conner & Liu 2000; Navarro *et al.* 2004)

The B function works with borders: from fading to sliding. The debate affecting the specific role of putative A-function genes in floral organ specification does not touch the widely accepted universality of B and C-function roles in specifying petals, stamens, and carpels. B-class and C-class genes have been identified from a huge number of species and corresponding mutants have been characterized for several of them, which in general exhibit the expected homeotic defects predicted by the quartet model (reviewed in Irish & Sussex 1992; Krizek & Fletcher 2005; Kater *et al.* 2006).

Most common differences in floral architecture affect perianth organ identity. While core eudicots appear to have fixed the whorled arrangement of sepals–petals–stamens–carpels, considerably more variation occurs in basal angiosperms and monocots. Basal angiosperms generally show a perianth where no clear distinction can be made between sepals and petals. Perianth organs are often arranged in spiral phyllotaxis and show variable merosity. In some species, all perianth organs are petaloid, but in others, outer tepals are morphologically different from inner tepals, and intergrading organs of intermediate morphology are observed. Also, in some basal angiosperms like *Asimina* species, clearly different outer sepals and inner petals develop.

In all basal angiosperms, homologs to ABCE genes are found and, in general, expression patterns are consistent with predictions of floral organ identity by the quartet model. *AG*-like gene expression is found only in reproductive organs, likely reflecting this restricted expression pattern as ancestral. However, MADS-box genes related to the B-class genes *AP3* or *PI* are expressed in broader domains, encompassing the whole of the perianth and frequently also the carpels in many basal angiosperms (Kim *et al.* 2005). The presence of B-class genes in the perianth is consistent with a modified version of the (A)BC model where broad B-gene expression explains the petaloid identity of these organs in most basal angiosperms. Also, in *Asimina* species, B-class gene expression is restricted to the petal domain, suggesting that a reduction in the ancestral broad domain of B-class expression could have originated the *Asimina* dual perianth (Kim *et al.* 2005). To explain gradual transitions in floral organ identity found in many basal angiosperms, the "fading borders" hypothesis has been proposed. This model suggests that gradients of expression of organ identity genes, and especially of B-genes, could generate domains with different relative proportions of overlapping activities that led to organs of intermediate morphology (Figure 9.1c). Precise information of expression patterns of B-genes in different basal angiosperm species is limited, but studies in *Amborella* or *Persea* support this scenario (Soltis *et al.* 2007a, 2007b; Theissen & Melzer 2007).

Many nongrass monocot species, such as lilies or tulips, have whorled tepaloid perianths. In general, although with a few exceptions, these species express B-class genes related to *AP3* and *PI* in whorls 1, 2, and 3 (Kanno *et al.* 2003), supporting the prediction of van Tunen *et al.* (1993) of a "sliding border" for the B-class genes, expanding to the whorl 1 and therefore conferring petaloid identity to these organs (Figure 9.1d).

In fact, from these and other studies in basal angiosperms and monocots, the petaloid character of the perianth together with a broad and may be undefined expression pattern of B-class genes could be inferred as ancestral. Dual perianth organ identity would have evolved independently many times in different clades of angiosperms, possibly through the refinement of B-class gene expression domains. From the perspective of the quartet model, authors like Theissen and Melzer (2007) have suggested that the establishment of positive autoregulatory loops through changes in promoters or increasingly restricted protein complex formation could have contributed to define the borders of B- and C-function activities.

B and C genes are already present in gymnosperms, so they predate the origin of angiosperm flower, suggesting an ancestral role already in the specification of male and female reproductive structures. Interestingly, B-type proteins from gymnosperms tested so far have been shown to form only homodimers (Wang *et al.* 2010). In contrast, in basal angiosperms and monocots, homodimerization, facultative, and obligate heterodimerization of B-class proteins have been shown to occur (Su *et al.* 2008). Finally, in core eudicots, B-proteins form obligate heterodimers, and only these heterodimers are able to provide B-function, both maintaining its own expression and acting on other downstream targets (reviewed in Theissen & Melzer 2007; Immink *et al.* 2009a). This trend toward obligate heterodimerization possibly coupled to autoregulation, could have thus provided a mechanism for restricting and sharpening the borders of B-activity domain to specify defined whorls or petals and stamens.

Variations in the quartet model and the evolution of novel morphologies. Novel floral morphologies not only necessarily included in these evolutionary trends but also caused by changes in floral organ identity can be interpreted using the ABCE model. For example, the eudicot *Rumex acetosa*, has a two whorled perianth composed only of sepaloid organs. Accordingly, only the stamens express B-class genes, suggesting that an inward shift of their expression has originated this new morphology (Figure 9.1e; Ainsworth *et al.* 2005).

Similarly, shifts in B-class gene expression appear to underlie the highly unusual floral morphology of the monocot *Lacandonia schismatica*. In *Lacandonia*, small flowers form with a tepaloid perianth of sepaloid morphology and an inverted arrangement of the reproductive organs, where numerous carpels surround a central whorl of stamens. Studies on B- and C-function genes in this species have shown how the B-class gene *LsPI* is expressed broadly in young floral meristems to restrict later to the domain that will give rise to stamens and carpels, overlapping with the C-class gene *LsAG*. In contrast, the expression of the other B-class gene, *LsAP3*, is restricted to the central zone of the meristems where stamens will develop. Thus, simultaneous expression of *LsAP3*, *LsPI*, and *LsAG* is only achieved in the central whorl of stamens, while in the surrounding carpels only *LsPI* and *LsAG* are found, and all of them are absent in the tepaloid organs (Figure 9.1f; Alvarez-Buylla *et al.* 2010).

An additional example of morphological innovation that can be explained by modifications in the quartet model is provided by species of the basal eudicot genus *Aquilegia*. Most *Aquilegia* species have a perianth composed of two morphologically different types of petaloid organs and an additional whorl of a novel type of organ, the staminodium, placed between stamens and carpels. *Aquilegia* has one *PI*-like gene but three *AP3*-like genes. Detailed studies of B-class gene function have shown how the three *AP3*-like genes are expressed in different spatial and temporal patterns, including petals, stamens, and staminodia, and that each organ type expresses a specific combination of B-type genes. Moreover, downregulation of *AqvPI* caused transformation of both stamens and staminodia into carpels, indicating that staminodia are probably derived from stamens, and suggesting that gene duplication and subsequent subfunctionalization of *AP3* genes could have originated this novel organ identity program (Kramer *et al.* 2007).

As we have seen, variations in the ABC architecture are able to explain floral morphological diversity. However, the same ABC system has also been shown to specify very different looking organs. Grass flowers differ significantly from their eudicot relatives in the identity of the sterile perianth organs, which is composed of grass-specific organs known as glumes, lemma, palea, and lodicules. The glumes were considered as bracts, while homology of lemma, palea, and lodicules to other angiosperm floral organs has been the subject of debate. The lodicule is a small glandular-like organ, placed between the palea/lemma and the stamens. At anthesis, lodicules expand to separate the lemma and palea and thus expose the anthers to the wind. Functional studies of flowers of grasses, such as rice or maize, have shown how mutations in B-class genes lead to homeotic transformations of lodicules into palea-like organs and stamens into carpels, strongly supporting the homology of lodicules and petals (Ambrose *et al.* 2000; Nagasawa *et al.* 2003; Xiao *et al.* 2003; Whipple *et al.* 2004). The petaloid nature of lodicules represents a remarkable example of a novel type of organ derived from a preexisting identity program. The unique morphology of lodicules would then be more likely the result of a change in the set of downstream targets of B-type genes throughout the course of grass evolution (Whipple *et al.* 2007).

Duplication and diversification of large gene families. Gene duplication is a common phenomenon in plants which, when followed by neo- and/or subfunctionalization, can lead to large gene families and gene networks, enabling functional diversification and the subsequent evolution of morphological novelties. Given the central role of MADS-box genes in floral development and evolution, considerable efforts have been directed to identify a vast number of members of this family across the land plants and to reconstruct their phylogeny. These studies have revealed the highly complicated structure of the family where relationships of orthology and paralogy give rise to multiple cases of neo- and subfunctionalization, gene co-option, and even convergent evolution (Jaramillo & Kramer 2007). Still, general patterns on MADS phylogenies correlate diversification of the MADS family with major steps in floral evolution. These studies have shown how key MADS-box genes related to floral organ specification could have undergone simultaneous duplication, probably through polyploidization, just before or coincident with the origin of the angiosperms (Irish & Litt 2005; Zahn *et al.* 2005). Moreover, the patterns of duplication of ABCE genes (*AP1, AP3/PI, AG,* and *SEP* subfamilies) are related and basically coincident with the origin of major clades within angiosperms, therefore suggesting a role for these duplications in angiosperm diversification.

For instance, in gymnosperms only one copy of a B-class homolog is present, whereas at least two copies (related to *AP3* and *PI*) are found in angiosperms (Kim *et al.* 2004). Then, two clades of *AP3*-like genes are found only in core eudicots (euAP3 and TM6) indicating a second event of duplication before the origin of this clade (Kramer *et al.* 1998). Likewise, angiosperms possess two lineages of C-like genes, one comprising *AG*-like genes related to C-function and another involved in ovule identity specification originated from a duplication predating angiosperm origin (Kramer *et al.* 2004). *AG*-like genes have also undergone more duplication events before and within the monocot and the dicot clades. These duplications have resulted in paralogs where changes mainly in expression patterns but also in protein function have given rise to functional diversification (Hasebe *et al.* 1998; Davies *et al.* 1999; Causier *et al.* 2005; Airoldi *et al.* 2010). Likewise, for E-function genes we found a similar scenario. Two lineages of *SEP* genes (*SEP1/2/4* and *SEP3*) are thought to be present in the common ancestor of angiosperms while no clear homologs of *SEP* genes have been found in gymnosperms, suggesting that SEP genes were lost in the ancestor of extant gymnosperms (Zahn *et al.* 2005; Theissen & Melzer 2007). The essential role of E-class genes in the specification of all floral organs, as key components of the floral quartets, suggests that the origin and diversification of the *SEP*-clade might have played a key role in the origin of the angiosperm flower (Wang *et al.* 2010).

Notably, the clade where the putative A-function *AP1*-like genes belong (euAP1) appears at the base of the eudicots as the result of a duplication in the more ancient *AP1/FUL* lineage (Litt & Irish 2003). The origin of the euAP1 clade coincides with that of the euAP3 lineage (see Section "The elusive A-function") and with an important morphological innovation—the fixation of the whorled structure of the eudicot flower, with a dual perianth of clearly distinct sepals and petals. Although as previously discussed, the role of *AP1*-like genes in the specification of perianth organs is still debateable, the temporal concurrence of euAP1/euAP3 clades, sepals, and petals suggests a role for these eudicot gene clades in this morphological innovation.

9.2.1.3 The origin of the flower

Although floral organ identity is well studied and the quartet model constitutes a powerful working framework, the evolutionary origin of the flower remains an open question. The fossil record is still far from comprehensive, and we have not yet identified fossils unequivocally representing stem-group angiosperms. Therefore, we lack a clear picture of how the ancestral parents of extant flowers looked like, what was the temporal sequence of the morphological novelties that appeared in the first flowers, and what were the molecular changes responsible for these innovations. In spite of these uncertainties, Goethe's theory of leaves and floral organs sharing a common ground plan is nowadays accepted and strongly supported by the *sepallata* mutant phenotypes (Goethe 1790; Pelaz *et al.* 2000; Ditta *et al.* 2004). Thus, flowers are viewed as "compressed" branches and floral reproductive organs (stamens and carpels) derive from leaves bearing reproductive structures (sporophylls) along the axis of these branches.

Three major evolutionary steps can be proposed to transform these ancestral reproductive branches into "modern" flowers. Since most reproductive branches in gymnosperms are unisexual, the first step could be the origin of bisexuality. In addition, reproductive branches in gymnosperms are usually indeterminate, so determinacy would have to be acquired. Finally, sterile perianth organs should have originated. Although there is no definitive data on the order or even the nature of these steps, this sequence appears to be a plausible scenario.

Several recent publications have discussed current theories on the origin of bisexuality based on paleobotanic data and the rapidly increasing amount of molecular data (Frohlich & Chase 2007; Specht & Bartlett 2009; Melzer *et al.* 2010). Briefly, three different hypotheses are currently debated, which basically differ with respect to the nature (male or female) of the ancestral branch of sporophylls.

The Mostly Male Theory (MMT; Frohlich & Parker 2000) proposes that flowers are derived from male strobili of a gymnosperm-like ancestor. In the male strobili of this ancestor, the male sporophylls near the apex would have first become bisexual by developing ectopic ovules, and subsequently, these bisexual sporophylls would have lost male sporangia and become functionally female. Additional evolutionary steps toward the flower would be the loss of the residual female strobili, the transformation of sporophylls in the male/female strobili by closure of the apical female sporophylls to form carpels and the basal male sporophylls into stamens (Figure 9.2).

The MMT was proposed mainly on the basis of evolutionary studies on *LEAFY* gene function. *LFY* plays a central role in the specification of floral meristems in angiosperms and, subsequently, in the upregulation of floral organ identity genes. Although in angiosperm species only one *LFY* gene is found, in gymnosperms two paralogs are present, one *LFY* ortholog, and *NEEDLY* (*NLY*), for which no direct ortholog has been found in angiosperms. Initial expression studies in certain gymnosperms showed that each paralog was expressed predominantly in male (*LFY*) or female cones (*NLY*). Thus, the absence of *NLY* in angiosperms pointed to the female strobilus having been lost from the common ancestor to flowering plants and therefore to the male origin of the ancestral flower (Mouradov *et al.* 1998). Several independent observations supported this theory. For example, the *Arabidopsis lfy* mutant is

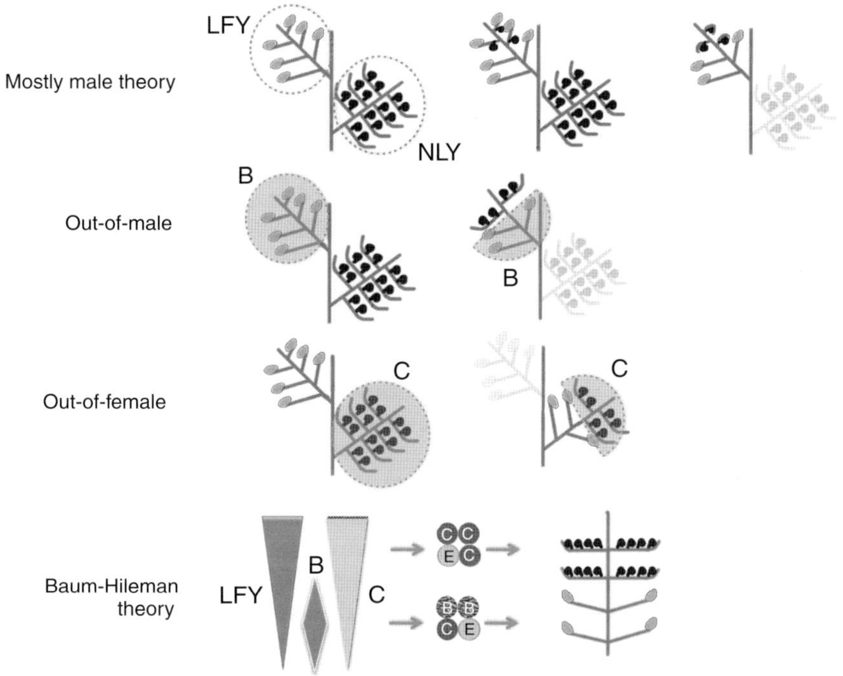

Figure 9.2 Hypotheses for the origin of the flower. The Mostly Male Theory proposes that male sporophylls expressing *LFY* homologs and female sporophylls expressing *NLY* homologs coexisted in the angiosperm ancestor. As the first step, ectopic ovules formed on male sporophylls. Subsequently, these sporophylls lost microsporangia and closed around the ovule to form the carpel, and female sporophylls were also lost. The out-of-male hypothesis also proposes that bisexual flowers originated from male strobili. Basipetal movement of B-function MADS-box gene expression would have left distal female sporophylls, which later became carpels. The out-of-female hypothesis is conceptually similar, but in this case, B-type MADS-box gene expression would be restricted to the apical part of a previously female strobilus, leaving male structures at the base. According to the hypothesis of Baum and Hileman (2006), a temporal increase of LFY levels in reproductive axes occurred in the ancestor of the flowering plants. This change affected differentially the accumulation of LFY targets, B- and C- MADS-box genes, generating high concentrations of C-function-rich MADS-box complexes close to the distal part, and higher proportions of B-function-rich MADS-box complexes toward the base. The different composition of these complexes would direct the patterning of the strobilus into apical female and basal male reproductive structures, and these later became carpels and stamens, respectively.

able to produce carpels but no stamens, suggesting that *LFY* is only essential for male specification. In addition, it has been described that ectopic ovules form in different single mutants of several angiosperm species and even in the leaves of a gymnosperm like *Ginkgo*, suggesting that ectopic ovules are easily produced. However, recent conflicting data showing coexpression of *LFY* and *NLY* in both female and male cones from other gymnosperms (Carlsbecker *et al.* 2004; Vazquez-Lobo *et al.* 2007) have weakened this theory.

Two other sister hypotheses on flower origin, the out-of-male (OOM)/out-of-female (OOF) theories, both proposed by Theissen *et al.* (2002), are based on the observation that some extant conifers possess bisexual cones. These theories incorporate molecular data and the ABCE model of floral organ identity in an attempt to explain how bisexuality of the flower was originated by temporary or spatial shifts in patterns of expression of genes determining the male/female identity of reproductive organs, downstream of *LFY/NLY* (Figure 9.2). The most important difference of OOM/OOF and MMT is that the former do not involve extensive loss of female developmental programs and therefore, postulate that carpels have female origin instead of the ectopic formation of ovules on male organs proposed by MMT. The OOM hypothesis proposes that in an ancestral male strobilus, the expression of male-identity B-class MADS-box genes could have been downregulated in the distal part to leave female organs in these apical positions (Theissen *et al.* 2002) by an unknown signal of hormonal or other nature. Conversely, the OOF theory postulates that B-class genes could have been upregulated in the basal portion of a female cone, likewise producing a bisexual axis. Some recent studies on pollen versus seed cones in extant conifers show most spontaneous bisexual cones appear to be modified seed cones rather than modified pollen cones, thus favoring the out-of-female hypothesis (Rudall *et al.* 2011).

A further theory has been proposed by Baum and Hileman (2006; Baum and Hileman Theory (BHT)), which can be considered a variant of the OOM hypothesis, but with much more molecular detail. BHT incorporates increasing knowledge of molecular mechanisms of flower identity specification in extant species to formulate a highly speculative model that, in addition to the origin of bisexuality, goes further in an attempt to explain the acquisition of floral determinacy and even perianth origin. The BHT takes into account the central role of *LFY* in the upregulation of B- and C-function genes and the quartet model of floral organ identity. According to this hypothesis, LFY protein levels would increase with time in male reproductive branches, as has been observed in extant angiosperm inflorescences (Blázquez *et al.* 1997, 1998; Schmid *et al.* 2003). The BHT incorporates two more assumptions: that this tendency of temporal LFY accumulation would have been reinforced during evolution, and also that C-function proteins should be able to accumulate at much higher quantitative levels than B-function proteins.

Increasing levels of LFY together with this higher capacity of C-function accumulation would create a distal maxima of C-function. This apical excess of C-function would then sequester all available SEP proteins, preventing the formation of B-SEP complexes and thus resulting in the development of megasporophylls at the apex of the strobilus (Figure 9.2). The BHT also suggests that a change in *WUSCHEL cis*-regulatory sequences during evolution could have occasioned *WUS* repression by the apical excess of C-function, resulting in floral determinacy. This hypothesis also speculates further on the origin of the sterile perianth, involving a newly acquired essential role of WUS as a LFY cofactor to spatially define the activation of C-function genes.

The final requirement of WUS for C-function activation, together with the central spatial pattern of *WUS* in the meristem, would be its exclusion from the outer/basal region of the ancestral flower. This would result in the loss of stamen identity in these regions and the development of sterile organs of petaloid appearance and therefore implying the andropetaloid origin of these organs. An attractive aspect of the BHT is that it makes a huge number of predictions that potentially could be experimentally tested. However, as already discussed, functional studies in representative species of most interesting clades, including gymnosperms and basal angiosperms, are currently out of reach, and we will have to wait before more conclusive evidence is gathered.

9.2.2 The evolution and development of leaves

9.2.2.1 What is a leaf?

Leaves are one of the most conspicuous features of land plants and arise as lateral determinate organs from an indeterminate apical meristem. The leaves of most vascular plants are all lateral organs that are vascularized and have determinate growth (Gifford & Foster 1989), with a broad lamina with adaxial–abaxial asymmetry, which is well suited for photosynthesis and respiration. Although leaves can be easily recognized by their position and structure, the leaves that are found throughout the land plants are not considered homologous (Bower 1935).

9.2.2.2 Leaf evolution and homology across the land plants

The evolution and homology of leaves has been a longstanding question among botanists. Leaves in the bryophytes are found only in the gametophyte generation and are not considered homologous to the leaves found in any of the vascular plants (Domin 1931). Vascular plant leaves are found only in the sporophyte generation and are classified as either microphylls or megaphylls, the former found in lycophytes and the latter in the euphyllophytes (Figure 9.3). Microphylls are defined by their generally small size, single unbranched vascular strand, and lack of a leaf gap. Megaphylls are generally large, have complex venation, and are associated with a leaf gap. However, there are exceptions to these definitions, some lycophyte microphylls have branched vasculature and some fossil lycophyte microphylls were large (Gifford & Foster 1989). In spermatophytes, exceptions to the megaphyll definition also exist, such as the needles found in pines. Microphylls and megaphylls are not considered homologous, and megaphylls are proposed to have arisen anywhere from two to seven times independently in the euphyllophytes. There are several hypotheses regarding the evolution of megaphylls, the most popular being the telome theory, which proposes that megaphylls evolved through a transformation of branches (Wilson 1953; Kenrick & Crane 1997).

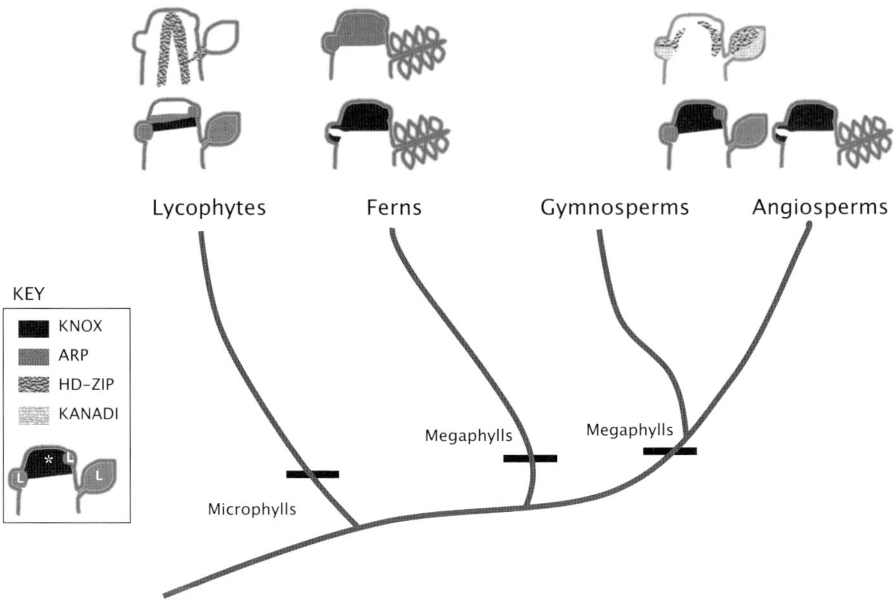

Figure 9.3 Leaf evolution and expression of leaf development genes. Illustration depicting phylogenetic relationships of vascular land plants: lycophytes, ferns, and seed plants (angiosperms and gymnosperms). Microphylls evolved once in the lycophyte lineage. Hypotheses regarding megaphyll evolution include these structures evolving anywhere from two to seven times. Diagrams of vegetative shoots with shoot apical meristems (*) and leaf primordia (L) and leaves (L) summarizing the known expression patterns of Class I *KNOX*, *ARP*, Class III *HD-Zip*, and *KAN* leaf development genes in vascular plants. In angiosperms, expression patterns are known for some leaf development genes in species with simple leaves and species with complex leaves.

Paleobotanical studies have been important in constructing hypotheses of leaf evolution in vascular plants. Fossils of vascular plants from the Silurian and Devonian have highly branched structures that lack laminate leaves and Paleozoic fossils indicate that laminate leaves evolved independently in the euphyllophyte clades (Gifford & Foster 1989). Although there is a wide range of vascular plant leaf morphology, these morphologies exist in all euphyllophyte clades indicating a high level of convergence in leaf development (Boyce & Knoll 2002). Analyses of fossil leaf morphology and vasculature show that spermatophytes and monilophytes underwent similar leaf developmental trajectories. Boyce and Knoll (2002) suggest that this indicates modifications of a common developmental network and that this shared developmental network constrained the early evolutionary direction of leaf development and the number of leaf morphologies that could be attained. Comparative development of angiosperm leaves provides some support for these hypotheses, however, molecular genetic support across the euphyllophytes is still lacking.

9.2.2.3 Genes and pathways necessary for simple leaf development in angiosperms

Molecular genetic studies in angiosperms have elucidated a large part of the regulatory network necessary for the development of a simple angiosperm leaf. The development of lateral organs involves the specification of organ position on the flanks of the meristem, establishment of boundaries between the determinate organ and indeterminate meristem, and specification of the cell types of the lateral organ. This would appear to be a logical sequence of events but there is in fact a lot of feedback between the different processes composing the leaf regulatory network. In many angiosperms, the KNOX-ARP (Class I **KN**OTTED1-like HOMEOB**OX** (KNOX) and **A**SYMMETRIC LEAVES1/**R**OUGH SHEATH2/**P**HANTASTICA (ARP MYB domain proteins) regulatory module is important for setting the domains between an indeterminate shoot apex and a determinate lateral organ (Figure 9.3; Vollbrecht *et al.* 1991; Lincoln *et al.* 1994; Long *et al.* 1996; Waites *et al.* 1998; Timmermans *et al.* 1999; Tsiantis *et al.* 1999; Byrne *et al.* 2000; Reiser *et al.* 2000; Semiarti *et al.* 2001). And in many angiosperms, the interaction between Class III homeodomain leucine zipper (HD-ZIPs), KANADI (KAN), and YABBY (YAB) proteins are important for vascular development, meristem identity, and adaxial–abaxial asymmetry in lateral organs (Baima *et al.* 1995; McConnell & Barton 1998; Siegfried *et al.* 1999; Eshed *et al.* 2001; McConnell *et al.* 2001; Emery *et al.* 2003; Eshed *et al.* 2004; Prigge *et al.* 2005; Byrne 2006). Simple leaf development studies have focused on the KNOX/ARP and HD-ZIP/KAN/YAB modules. We will review the developmental network involving these two genetic modules mainly from work done in the angiosperm *Arabidopsis thaliana* unless otherwise specified.

Lateral determinate organs are produced on the flanks of the indeterminate meristem, and determinate primordia initiate at sites of auxin maxima (Reinhardt *et al.* 2000; Benkova *et al.* 2003; Reinhardt *et al.* 2003). Auxin influx and efflux carriers (e.g., PIN-FORMED1 (PIN1)) direct auxin flow in the shoot (Reinhardt *et al.* 2000; Reinhardt *et al.* 2003; Bainbridge *et al.* 2008), and plants unable to synthesize auxin or those treated with auxin transport inhibitors have defects in lateral organ formation (Okada *et al.* 1991). In addition, *pin1* mutants, that are unable to create auxin maxima, have defects in lateral organ formation (Okada *et al.* 1991; Galweiler *et al.* 1998).

PIN1 and KNOX homeodomain transcription factors are expressed in a complementary pattern in the shoot (Heisler *et al.* 2005). Generally, KNOX proteins are expressed in the vegetative and floral meristems and downregulated in leaf primordia and floral organs (Vollbrecht *et al.* 1991; Kerstetter *et al.* 1994; Long *et al.* 1996; Reiser *et al.* 2000). The Class I KNOX mutant *shootmeristemless* (*stm*) germinates with no apical meristem and although adventitious branches may form, the meristems of these branches soon terminates as well

(Barton & Poethig 1993; Long *et al.* 1996). Class I KNOX proteins are required for the maintenance of indeterminacy and prevention of cell differentiation in the shoot apical meristem. ARP and KNOX are also expressed in complementary patterns in the shoot. The ARP protein, ASYMMETRIC LEAVES1 (AS1) is expressed in leaf primordia and developing leaves (Figure 9.3; Byrne *et al.* 2000, 2002). A loss of ARP function results in KNOX expression in the leaves and the resulting plant resembles a KNOX overexpression phenotype. The ARP protein AS1 forms a heterodimer with the LOB domain protein, AS2, which binds to the *cis*-regulatory region of Class I KNOX genes and represses KNOX transcription (Guo *et al.* 2008). In addition, Class I KNOX negatively regulate *AS1* (Byrne *et al.* 2000).

The complementary expression domains and antagonism of ARP and KNOX proteins have been proposed as an important developmental module to specify determinate leaves from the flanks of the indeterminate SAM. In addition, auxin and the ARP protein, AS1, have been shown to act together to repress KNOX expression in the leaves (Hay *et al.* 2006). The auxin defective and *as1* double mutants produce leaves with more indeterminate features than either single mutant and KNOX genes are ectopically expressed in these indeterminate regions of the double mutant.

The boundary between the indeterminate apical meristem and determinate lateral organs is finely tuned by three NAC-domain transcription factors, CUP-SHAPED COTYLEDON1–3 (CUC1, CUC2, CUC3) (Aida *et al.* 1997; Takada *et al.* 2001; Vroemen *et al.* 2003). *CUCs* are expressed in the embryonic shoot apical meristem and in the boundary region between the apical meristem and lateral determinate organs (Aida *et al.* 1999; Ishida *et al.* 2000; Takada *et al.* 2001; Vroemen *et al.* 2003; Hibara *et al.* 2006). Plants with mutations in any two of the three *Arabidopsis CUC* genes have fused lateral organs while plants overexpressing *CUC1* have meristems forming on cotyledons (Aida *et al.* 1997; Takada *et al.* 2001; Vroemen *et al.* 2003; Hibara *et al.* 2006). The *35S:CUC1* plants have ectopic *STM* expression in the cotyledons while a loss of CUC function results in the loss of *STM* expression and therefore a loss of meristem formation during embryogenesis (Aida *et al.* 1999; Takada *et al.* 2001). In addition, a loss of STM results in an expansion of *CUC* gene expression and a plant with no apical meristem and fused cotyledons (Aida *et al.* 1999). Therefore, *CUC* genes promote KNOX expression during embryogenesis and KNOX expression represses *CUC* expression in the meristem. CUCs have redundant roles in SAM formation during embryogenesis and lateral organ boundary formation and maintenance throughout the plant lifecycle (Aida *et al.* 1997, 1999; Ishida *et al.* 2000; Takada *et al.* 2001; Vroemen *et al.* 2003; Hibara *et al.* 2006). Given the relationship between KNOX and auxin and KNOX and CUC, it is not surprising that there is an interaction between CUCs and auxin. Plants mutant for the auxin efflux protein, *pin1*, have expanded *CUC2* expression suggesting that auxin represses *CUC* expression to the boundary of the initiating primordia (Vernoux *et al.* 2000).

After leaf initiation, the angiosperm leaf soon acquires polarity, which is important for proper cell type specification in this lateral organ. The Class

III HD-ZIP, KAN (GARP-type transcription factors), and YABBY proteins are important for meristem identity and adaxial–abaxial asymmetry in lateral organs (Figure 9.3; Siegfried *et al.* 1999; Eshed *et al.* 2001; Kerstetter *et al.* 2001; McConnell *et al.* 2001; Emery *et al.* 2003; Prigge *et al.* 2005). HD-ZIPs are expressed in the SAM, vasculature, and adaxial region of lateral organs and are necessary for the initiation of the SAM and axillary meristems, specification of adaxial identity in lateral determinate organs such as leaves and floral organs, and determination of vascular tissue patterning (Baima *et al.* 1995; McConnell & Barton 1998; McConnell *et al.* 2001; Emery *et al.* 2003; Prigge *et al.* 2005; Byrne 2006). *KANs* are expressed in the abaxial region of lateral organs and are necessary to specify abaxial identity and vascular patterning (Eshed *et al.* 2001; Kerstetter *et al.* 2001; Emery *et al.* 2003; Eshed *et al.* 2004). *YABBYs* are expressed in the abaxial region of lateral organs and are necessary for abaxial identity in lateral organs (Sawa *et al.* 1999; Siegfried *et al.* 1999; Sarojam *et al.* 2010).

A loss of HD-ZIP function results in peg-shaped cotyledons with abaxial epidermal features while a loss of KAN or YABBY proteins results in radial lateral organs that are adaxialized (Siegfried *et al.* 1999; Eshed *et al.* 2001; Emery *et al.* 2003; Eshed *et al.* 2004; Sarojam *et al.* 2010). *KANs* are ectopically expressed in the HD-ZIP loss of function mutants and HD-ZIPs are ectopically expressed in the *KAN* loss of function mutants (Eshed *et al.* 2001). HD-ZIP proteins and KAN proteins are mutually antagonistic (Eshed *et al.* 2001; Kerstetter *et al.* 2001; Reinhart *et al.* 2002; Rhoades *et al.* 2002; Emery *et al.* 2003). Mutations that result in abaxialized lateral organs also have meristems that cease to function early in development (Siegfried *et al.* 1999; Eshed *et al.* 2001; Emery *et al.* 2003). Micro RNAs 165/166 are important for repressing HD-ZIP activity and restricting its action to the adaxial side of the leaf (McConnell *et al.* 2001; Rhoades *et al.* 2002; Emery *et al.* 2003). In model angiosperm species, the juxtaposition of abaxial and adaxial identity in leaves has been proposed to be necessary for laminar outgrowth (Waites & Hudson 1995; McConnell *et al.* 2001; Rhoades *et al.* 2002; Emery *et al.* 2003; Eshed *et al.* 2004).

Molecular genetic studies have been important to further elucidate the developmental network necessary to specify the leaves. Genes involved in auxin signaling along with *KANs* promote abaxial identity (Pekker *et al.* 2005). Proteins that are important for leaf determinacy also play a role in the axial patterning of the leaves. The ARP protein AS1 along with the LOB domain protein AS2 promotes adaxial identity by positively affecting HD-ZIP expression (Lin *et al.* 2003, 2005). In addition, AS1 and AS2 repress the activity of *YABBY* genes in the adaxial region (Iwakawa *et al.* 2007).

9.2.2.4 Genes and pathways necessary for compound leaf development in angiosperms

Comparative molecular genetic studies have shown that the leaf regulatory network that specifies the leaf primordia from the shoot apical meristem is similar to the regulatory network that species leaflets from the rachis of a compound or dissected leaf (Figure 9.4). Differences in KNOX and

Figure 9.4 Co-option of a regulatory network. Illustration of shoots that produce simple leaves (left) and shoots that produce complex leaves (right). Leaf primordia and leaflet primordia are indicated by gray circles. Development of determinate leaves from an indeterminate meristem is mediated in part by a regulatory network involving KNOX, ARP, auxin, PIN1, and CUCs. KNOX are integral for maintaining an indeterminate meristem (*). Auxin maxima are found at leaf primordia initiation sites and auxin maxima are mediated in part by PIN1. The boundary of the leaf primordia are finely tuned by CUC proteins whose expression may be regulated by auxin. CUCs and KNOX are part of a positive regulatory loop. KNOX expression is known to be reactivated in complex leaf primordia and the expression and function of PIN1 and CUCs in leaflet primordia of complex leaves is similar to that found in simple leaf primordia indicating a co-option of this regulatory network in the development of complex leaves.

ARP expression patterns and regulation have provided a molecular genetic explanation for the morphological differences between simple versus compound leaves in some angiosperm species (Kessler & Sinha 2004; Hay and Tsiantis 2006; Kellogg 2006; Kimura *et al.* 2008). In simple-leaved angiosperms, KNOX proteins are expressed in the SAM, downregulated in incipient leaf primordia, and not expressed in mature leaves (Vollbrecht *et al.* 1991; Lincoln *et al.* 1994; Nishimura *et al.* 1999).

However, in angiosperms with compound leaves (e.g., tomato), KNOX proteins are expressed in the incipient leaf primordia but not in mature leaves (Janssen *et al.* 1998). In tomato, overexpression of KNOX produces young leaves that are coiled like a fern fiddlehead (circinate vernation) and mature leaves that are highly dissected (Hareven *et al.* 1996; Chen *et al.* 1997; Janssen *et al.* 1998). This coiled morphology in tomato KNOX overexpression plants is hypothesized to be due to faster cell growth on the abaxial compared to the adaxial side of the leaf. In addition, the natural variation in the level of leaf dissection found in various tomato cultivars is due to differences in KNOX expression and its interactions (Kimura *et al.* 2008).

Developmental genetic studies in *Cardamine hirsuta* have nicely illustrated how changes in the leaf developmental network result in changes in leaf morphology (Hay & Tsiantis 2006). *C. hirsuta* is closely related to *Arabidopsis* but has compound leaves. Similar to what is found in other compound leaved angiosperm species, *C. hirsuta KNOX* expression is reactivated in the leaflet primordia (Bharathan *et al.* 2002; Hay & Tsiantis 2006). In addition, an increase in KNOX expression in *C. hirsuta* results in more complex leaves while a reduction in KNOX expression gives simple leaves (Hay & Tsiantis 2006). Hay and Tsiantis (2006) also found that the difference in *C. hirsuta* KNOX expression in leaves compared to *Arabidopsis* KNOX expression are due to changes in the *cis*-regulatory region of *C. hirsuta* KNOX.

Studies on the regulation of these leaf developmental networks are important for understanding how changes in this regulatory network can result in different leaf morphologies. Analyses of angiosperm KNOX sequences have shown that a conserved noncoding sequence in the 5′ regulatory region termed the K-box is found in angiosperms with simple and complex leaves (Uchida *et al.* 2007). Deletion of the K-box in simple leaved species results in an *as1* (ARP) mutant phenotype. These results show that the K-box is necessary for the persistent repression of KNOX in leaves.

Although KNOX gene expression is reactivated in the leaflet primordia of many compound leaved angiosperms, the compound leaved *Pisum sativum* (pea) is an exception (Hofer *et al.* 1997; Bharathan *et al.* 2002). The pea leaf mutant *unifoliata* has a simple leaf and the mutation mapped to an ortholog of the meristem identity gene *LEAFY* (*LFY*) (Hofer *et al.* 1997). This likely represents a co-option of a protein necessary for flower development to leaf development in pea.

Not only are Class I KNOX proteins reactivated in the leaflet primordia of most compound leaved angiosperms, but research so far indicates that the entire leaf developmental network is redeployed for leaflet formation in compound leaved species (Figure 9.4). For example, auxin maxima are detected at the sites of leaflet primordia initiation of several compound-leaved angiosperm species (Barkoulas *et al.* 2008; DeMason & Polowick 2009; Koenig *et al.* 2009). In addition, in *C. hirsuta* the accumulation of auxin coincides with a reduction in Class I KNOX expression in developing leaflet primordia similar to what is found in developing leaf primordia on the flanks of the SAM (Barkoulas *et al.* 2008). The NAC transcription factors also play a role in defining the boundary between the leaflets and rachis of the compound leaf similar to their role in defining the boundary between the leaf primordia and SAM (Blein *et al.* 2008, 2009). Surprisingly, pea LFY interacts in the developmental network with auxin and NAC transcription factors to form compound leaves similar to the role played by KNOX with auxin and NAC transcription factors in forming compound leaves in other angiosperms (Blein *et al.* 2008, 2010). The developmental genetic research on compound leaves illustrates how the simple leaf developmental network has been redeployed to specify leaflets in compound leaved species.

9.2.2.5 Leaf genes and pathways across the land plants

The identification of the leaf regulatory networks throughout the land plants will be important for understanding when these networks evolved and to elucidate the evolution of land plant leaves. KNOX homologs have been isolated throughout the land plants (Bharathan *et al.* 2002; Harrison *et al.* 2005). Functional analyses of KNOX genes in *Physcomitrella patens* indicate that KNOX proteins are essential for proper development of the diploid sporophytes but not for the development of the leafy gametophyte (Sakakibara *et al.* 2008). These results support the idea that the leaves of bryophytes are not homologous to the leaves of vascular plants (Domin 1931). HD-ZIP and KAN transcription factors have been identified in bryophytes (*P. patens*), lycophytes (*Selaginella* spp.), and seed plants, and are likely to be found in ferns (Sakakibara *et al.* 2001; Floyd *et al.* 2006; Prigge & Clark 2006; Floyd & Bowman 2007). YABBY transcription factors have only been identified thus far in seed plants and it has been suggested that the evolution of the YABBY protein family was important for the evolution of the seed plant leaf (Floyd & Bowman 2007; Sarojam *et al.* 2010). KNOX expression patterns have been assessed in three fern species with compound leaves (*Ceratopteris richardii, Osmunda regalis*, and *Anogramma chaeophylla*) and found to be expressed in the apical meristem, leaf primordia, and leaf margins (Figure 9.3; Bharathan *et al.* 2002; Harrison *et al.* 2005; Sano *et al.* 2005). Unlike angiosperms, it appears that KNOX genes are not downregulated in incipient leaf primordia in ferns. ARP expression has only been studied in *O. regalis* and found in the apical meristem and leaf primordia (Harrison *et al.* 2005).

9.2.2.6 Leaf genes and leaf homology

Studies of KNOX, ARP, and HD-ZIP expression patterns have been used to provide evidence for various hypotheses on microphyll and megaphyll origins and megaphyll homology (Harrison *et al.* 2005; Floyd & Bowman 2006). The leaves of vascular plants are all lateral organs that are vascularized, have determinate growth, and have adaxial–abaxial polarity (Gifford & Foster 1989). The KNOX-ARP and HD-ZIP-KAN interactions are important for the specification of these defining leaf characteristics. KNOX proteins are expressed in the apical region in the lycophyte *Selaginella kraussiana* but not in its leaf primordia (Figure 9.3; Harrison *et al.* 2005). ARP homologs were also isolated from *S. kraussiana* and are expressed in the apical region, leaf primordia, and developing leaves.

The expression of KNOX and ARP genes in the lycophyte *S. kraussiana* has been interpreted as support for a common genetic mechanism specifying leaf determinacy in microphylls and megaphylls (Harrison *et al.* 2005). Although, it has been argued that the KNOX-ARP interaction is important for the determinate growth of leaves, others have suggested that KNOX genes are not important for the determinacy of a leaf but whether a leaf is simple or compound (Harrison *et al.* 2005; Floyd & Bowman 2006). A thorough investigation of KNOX and ARP expression patterns in ferns and gymnosperms with simple and compound leaves will provide further data to compare to lycophyte and seed plant expression patterns.

The HD-ZIPs and KANADIs are important for leaf development, vascular patterning, and adaxial–abaxial polarity in leaves of seed plants (Floyd & Bowman 2007). HD-ZIP homologs have been isolated from across the streptophytes, and phylogenetic analyses indicate that one HD-ZIP protein was present in the common ancestor of all embryophytes (Floyd & Bowman 2004; Floyd *et al.* 2006; Prigge & Clark 2006). Two Class III HD-ZIP proteins each have been isolated from the lycophytes *Selaginella moellendorffii, S. kraussiana,* and *C. richardii*, and three from *Psilotum nudum* (Floyd & Bowman 2004, 2006, 2007; Floyd *et al.* 2006; Prigge & Clark 2006). *In situ* hybridization in *S. kraussiana* showed that one HD-ZIP gene is strongly expressed in the apical cells and the provascular tissue, whereas the other HD-ZIP gene is weakly expressed in the apical cells, adaxial region of the developing microphyll where the ligule will form, and the provascular tissue (Figure 9.3; Floyd & Bowman 2006). This suggests that HD-ZIP proteins have a role in apical meristem function and vascular development. A comparison of HD-ZIP expression patterns in seed plants and the lycophyte *S. kraussiana* has been interpreted as support for the independent origin of megaphylls and microphylls (Floyd & Bowman 2006). HD-ZIP and KAN expression patterns have not been assessed in any ferns but are integral to our understanding of megaphyll evolution and development. These comparative studies are crucial to address questions of homology and may illustrate how regulatory networks underlying development constrain the evolution of form.

9.3 Future directions

9.3.1 Morphological features

We have learned a lot about the developmental pathways necessary to build flowers and leaves. Many of the evo-devo studies in flowers and leaves have contributed important molecular data to the discussion on organ homology. The gene regulatory networks have also been well studied for flowering time, meristem maintenance, and root development in angiosperms. Yet much remains to be discovered in plant evolutionary developmental biology, not only in elucidating the gene regulatory networks important for the development of these plant structures but also how these networks evolve. Further studies are needed to understand the developmental genetics of the alternation of generations, gametophyte development, vasculature, spores, and seeds. To understand the gene regulatory networks underlying these features we will need to perform plant evolutionary developmental studies across the land plants. We will briefly discuss two examples that require more developmental genetic studies (alternation of generations and sporangia) and one (meristems) that warrants a comparative approach across the land plants. The comparative approach is a critical aspect of evo-devo studies and to understand the evolution of morphological features across the land plants we will need to develop model organisms across the land plants.

9.3.2 Alternation of generations

One of the most dramatic changes in body plan is the alternation of generations exhibited by all land plants. All land plants alternate between two multicellular forms—the haploid gametophyte and the diploid sporophyte—and during land plant evolution there has also been a reduction in one phase and an elaboration in the other phase. In bryophytes, the gametophyte is the dominant phase while in vascular plants the sporophyte is the dominant phase of the life cycle.

Several theories, including the antithetic theory and homologous theory, have been proposed to explain the evolution of the alternation of generation in land plants (Gifford & Foster 1989). The antithetic or interpolation theory proposes that a delay in meiosis of the zygote and subsequent rounds of mitosis produced the multicellular sporophyte. Although the homologous or transformation theory proposes that the sporophyte and gametophyte generations evolved different morphologies from an ancestor with an isomorphic life cycle. Current data strongly support the antithetic theory (Bower 1972; Kenrick & Crane 1997). Phylogenetic analyses indicate that the closest relatives to the land plants are charophytes (Karol *et al.* 2001), which have a dominant multicellular haploid phase and a diploid phase represented by a unicellular zygote. The antithetic theory proposes that land plants evolved from an algal ancestor with a charophyte-type life cycle and that the multicellular diploid generation evolved by a delay in meiosis and an increase in mitotic divisions (Bower 1972; Kenrick & Crane 1997; Graham & Wilcox 2000).

One of the most intriguing aspects of the alternation of generations is that two morphologically distinct forms are generated from the same genome. Meiosis and syngamy are the turning points for the generation of the gametophyte and sporophyte phases, respectively. In fact, regardless of ploidy the gametophyte generation follows meiosis and the sporophyte generation follows syngamy. However, other switches besides meiosis and syngamy clearly exist. For example, apogamy and apospory occur in ferns and bryophytes, where the alternate generation is formed without syngamy or meiosis, respectively (Bower 1935; Niklas & Kutschera 2010). New resources and comparative analyses are beginning to shed some light on the developmental genetics of this dramatic change in body plan and the evolution of the two generations during land plant evolution.

A single plant genome is able to generate two morphologically distinct plant forms by differential gene expression. This has recently been demonstrated in *Arabidopsis* with large-scale expression techniques. Several studies found that the transcriptomes from *Arabidopsis* male and female gametophytes were unique from sporophyte transcriptomes (Pina *et al.* 2005; Yu *et al.* 2005). For example, in *Arabidopsis*, the pollen transcriptome is smaller than any sporophyte transcriptome but it is unique and has more selectively expressed and enriched genes compared to any other sporophyte transcriptome (Pina *et al.* 2005). However, we still do not know the developmental mechanism for how large-scale differential gene expression is generated in the two generations. It is clear that not only is activation of gene expression important but so is repression of gene expression. Gene silencing mechanisms are known to be important for the regulation of developmental processes in plants and animals, including regulation by small RNAs, DNA methylation, and histone modification (Baulcombe 2004; Loidl 2004; Chan *et al.* 2005). We are now beginning to uncover the molecular basis of gene silencing mechanisms that may be involved in the alternation of generations.

Histone modification likely plays a large role in differential gene expression between the gametophyte and sporophyte phases. Different heterochromatin banding patterns of chromosomes were noted between the gametophyte and sporophyte phases in the liverworts (Newton 1987). Heterochromatin banding is associated with methylation of histones. The Polycomb recruiting complex 2 (PRC2 or Polycomb group protein (PcG complex)) mediates the methylation of lysine 27 on histone 3 (K3K27me3) (Hennig *et al.* 2005; Hennig & Derkacheva 2009). PRC2 complexes are found in animals and plants. PcG complexes are modular in organization and several different PcG complexes have been found in *A. thaliana*. These *A. thaliana* PcG complexes are composed of two WD-40 proteins (FIE and MSI1), one SET domain protein (CLF, SWN, or MEA), and one zinc finger protein (EMF2, VRN2, or FIS2). We know a lot about the role of PcG complexes in *A. thaliana* development, however we have little data from across the land plants. However, recent research in *P. patens* has uncovered roles for PpFIE and PpCLF in moss development (Mosquna *et al.* 2009; Okano *et al.* 2009). Mutations in PpFIE or PpCLF result in gametophore meristems that overproliferate and do not reach the reproductive stage and produce gametes. However, PpFIE and PpCLF mutants produce sporophyte structures that look like sporophytes and express sporophyte specific genes. These results suggest that both PpCLF and PpFIE are necessary to repress apogamy in the gametophyte of *P. patens*.

The antithetic theory proposes a delay in meiosis and an increase in mitotic divisions to explain the evolution of the multicellular sporophytic phase. Therefore, analyses of cell cycle genes could provide some molecular data to further assess hypotheses on the alternation of generations. Not only is the PcG complex found in plants and animals but some genes involved in the cell cycle are also conserved across the eukaryotes. One conserved negative regulator of the cell cycle is Retinoblastoma (Ebel *et al.* 2004), and functional analyses of the *A. thaliana* RETINOBLASTOMA RELATED (RBR) protein has shown that it is required for differentiation of the male and female gametophytes (Johnston *et al.* 2008, 2010; Borghi *et al.* 2010). Functional analyses in *A. thaliana* have shown that RBR is important to derepress expression of genes involved in the PRC2 complex as well as genes involved in DNA methylation. Further research will provide great insights into the connection between gene silencing and the cell cycle and perhaps those developmental pathways involved in the alternation of generations.

9.3.3 Gametophytes

We know a lot more about the molecular genetics of sporophyte development compared to gametophyte development. This is mainly due to the fact that a majority of developmental genetic studies have been performed in angiosperms that have a dominant sporophyte phase and a much smaller gametophyte phase. However, we still have a surprising amount of information about the transcriptomes and development of the female gametophyte (embryo sac) and the male gametophyte (pollen) of *Arabidopsis* (Pina *et al.* 2005; Yu *et al.* 2005; Yang *et al.* 2010). Mutant screens for gametophyte development are helping to elucidate the function of gametophyte expressed genes as well as novel genes (Pagnussat *et al.* 2005; Boavida *et al.* 2009).

Overall, transcription factors are underrepresented in the pollen transcriptome. However, Type I MADS-box and Type II MIKC* MADS-box genes are transcription factors that are overrepresented in the pollen transcriptome (Pina *et al.* 2005). Mutations in Type II MIKC* MADS-box genes have reduced pollen germination that results in reduced fertility (Verelst *et al.* 2007; Adamczyk and Fernandez 2009). In addition, Type II MIKC* MADS-box genes have been shown to be part of a pollen developmental network that is composed of other MADS-box genes, particularly Type I MADS-box genes (Verelst *et al.* 2007; Adamczyk & Fernandez 2009). Type I MADS-have recently been shown to have important functions in female gametophyte development in *Arabidopsis* (Portereiko *et al.* 2006; Yoo *et al.* 2006; Bemer *et al.* 2008; Colombo *et al.* 2008; Kang *et al.* 2008; Steffen *et al.* 2008; Ambrose 2010; Masiero *et al.* 2011). Interestingly, of all the MADS-box genes present in the *P. patens* genome, more than half are Type II MIKC* MADS-box genes while there are only two Type I MADS-box genes (Rensing *et al.* 2008). Comparative functional analyses in *Arabidopsis* and *P. patens* will reveal the roles of these particular types of MADS-box genes in gametophyte development and evolution. Further developmental genetic studies in *P. patens*, which has a conspicuous gametophytic phase, will see leaps and bounds in our knowledge about the evolution and development of the gametophyte.

9.3.4 Sporangia and spores

Sporophytes produce sporangia and meiosis occurs in the sporangia to produce haploid spores (Gifford & Foster 1989). Sporangia and spores have been well studied in the angiosperms, however, sporangia position, morphology, and development are diverse throughout the land plants (Kenrick & Crane 1997). In the lycophytes, the sporangia are affiliated with the microphyll and one hypothesis for the evolution of microphylls is through sporangia sterilization. Developmental studies of sporangia revealed the differences between leptosporangiate and eusporangiate species more than a century ago (Gifford & Foster 1989). The development of sporangia is well studied across the land plants and is an important character for taxonomic purposes. Therefore the framework for understanding the molecular genetic basis for sporangia development is well established and needs to be extended across the land plants.

The spores, and sporopollenin wall surrounding the spores, are homologous across the land plants (Blackmore & Barnes 1987). The seed plants produce microspores or pollen grains, and developmental genetic pathways underlying *Arabidopsis* sporopollenin and microspore development are beginning to be understood (Ma 2005; Ariizumi & Toriyama 2011). For example, the gene *SPOROCYTELESS* in *Arabidopsis* is important for the initiation of sporogenesis and is regulated by the MADS-box gene *AGAMOUS* (Yang *et al.* 1999; Ito *et al.* 2004). However, a comparative approach is needed to understand the evolution of spore developmental genetics in the land plants (Brown & Lemmon 2011).

Homosporous land plants produce spores that are all the same size and these produce gametophytes with both antheridium (male) and archegonium (female) (Gifford & Foster 1989). Heterosporous land plants have two different sporangia that produce spores of two different sizes; the microspores will produce male gametophytes while megaspores will produce female gametophytes. Heterospory is found in distinct lineages and heterospory is thought to have been important for the evolution of the seed (Bateman & DiMichele 1994). It is necessary to study the evolutionary genetics of heterospory to understand the transition from heterospory to seed, and homospory to understand the transition from homospory to heterospory. To understand these transitions, developmental genetic studies will need to be performed across the land plants.

9.3.5 Meristems

The shoot apical meristem is found at the apex of land plant sporophyte shoots. The organization of the meristem has been described by one of seven different types (Wardlaw 1968), but even though the cellular organization may appear distinct, all meristems perform the same function. The apical meristem gives rise to all adult organs of the plant, and the plant must also maintain this totipotent group of cells. However the apical meristem of seed plants is composed of a pool of cells spanning several cell layers while the apical meristem of ferns is represented by a single apical cell with two or three cutting faces (Wardlaw 1968; Gifford & Foster 1989; Steeves & Sussex 1989). In lycophytes, members of the Selaginellaceae have one or two apical cells or a group of cells while members of the Lycopodiaceae and Isoetaceae have an apical meristem composed of a pool of cells similar to the apical meristem found in seed plants (Gifford & Foster 1989). Although, these meristems appear to differ, many are organized similarly with one quiescent cell or a pool of quiescent cells surrounded by small cytoplasmically dense rapidly dividing cells (McAlpin & White 1974; Stevenson 1976; Gifford & Foster 1989). Therefore, it is not surprising that homologs of genes necessary for meristem maintenance are found throughout the land plants (Floyd & Bowman 2007).

Meristems must balance the proliferation of cells to make lateral organs and maintain a set of pluripotent cells. The gene regulatory network needed to maintain meristems cells in an undifferentiated state has been well studied in angiosperms (Ha *et al.* 2010). Molecular genetic studies in *A. thaliana* have shown that a KNOX regulatory network and a CLAVATA-WUSCHEL regulatory network are both important for maintaining the shoot apical meristem. There has been little comparative work on the CLAVATA-WUSCHEL pathway outside of angiosperms. There has, however, been some comparative research on the role of KNOX genes in meristem development.

In *P. patens*, the gametophyte develops by apical tip growth and gives rise to the leafy plant and the sporophyte meristem is determinate and gives rise to the sporophyte (Lee *et al.* 2005; Vidali *et al.* 2007). Functional analyses of KNOX orthologs in *P. patens* indicate that KNOX proteins are important for the proper development of the sporophytes but have no role in the indeterminate haploid gametophyte (Sakakibara *et al.* 2008). Sakakibara *et al.* (2008) concluded that the KNOX pathway for specifying indeterminate meristems arose *de novo* in the angiosperms. KNOX genes function in both bryophyte and angiosperm sporophytes but the meristem in bryophytes is determinate while the angiosperm meristem is indeterminate. It is possible that the bryophyte intercalary meristem constrains the developmental potential of the sporophyte. An analysis of the KNOX developmental network and its regulation is necessary to better understand the role of KNOX proteins in the determinate sporophyte meristem of bryophytes. In *C. richardii*, KNOX orthologs are expressed in the sporophyte meristem and the margins of developing leaves but are not detected in the gametophyte meristem (Sano *et al.* 2005). KNOX orthologs are expressed similarly in the *C. richardii* meristem despite the meristem appearing structurally different from angiosperm meristems. These results also support the idea that KNOX orthologs are dispensable for gametophyte meristem development (Sano *et al.* 2005; Sakakibara *et al.* 2008).

Axillary meristems are formed from the flanks of the shoot apical meristem during the generation of lateral organs in gymnosperms and angiosperms (Gifford & Foster 1989). Axillary meristems contribute to the plasticity of form and the generation of novel forms, contributing to diverse architecture of seed plants. Although branching may sometimes occur in ferns, axillary meristems are not formed consistently and regularly in the ferns as they are in the seed plants. The genes that are responsible for the generation of meristems in the axils of lateral organs are not well known. However, it is clear that this pathway did not become well established until late in land plant evolution. Meristems are found not only in the shoot apex but also in the root tip and in vascular cambium, and although distinct regulatory networks function in the shoot, root, and vascular meristems, recent work has shown that some of the genes in these networks are homologous (Fiers *et al.* 2007; Hirakawa *et al.* 2010). In fact, recent work on unraveling the gene regulatory network present in vascular cambium has nicely illustrated the power of the candidate gene approach (Hirakawa *et al.* 2010).

However, there are some interesting aspects of meristems outside the flowering plants that warrant further study. One interesting structure is the rhizophore, which is found in lycophytes and arises from the angle meristem where the shoot branches (Webster 1992). The rhizophore is particularly intriguing because it can develop into a shoot or a root depending on the environmental and developmental conditions (Worsdell 1910; Imaichi & Kato 1989). In fact the identity of the rhizophore has been variously interpreted. It has been considered as either a root or shoot because it shares morphological features with each. The rhizophore has also been interpreted as a novel organ. The developmental plasticity of this meristem is unique among plant meristems and understanding the molecular genetic network underlying rhizophore development will illuminate mechanisms of stem cell maintenance and cell fate specification in plant meristems.

9.3.6 Development of model organisms

Much of what we know about the developmental genetics of these plant morphological features comes from studies in angiosperm species. A lot of focus on flower development and evolution reflects the availability of genetic and genomic tools in the angiosperms. The candidate gene approach to understand plant morphological diversity within the angiosperms and to some extent across the land plants has been very powerful and will remain very useful for lineages with large genomes and long life cycles. The candidate gene approach has also been useful for providing more data in the assessment of homology. Transcriptomes are being generated across the land plants, which will facilitate the candidate gene approach for evo-devo studies across the land plants (http://www.onekp.com/project.html; http://www.jgi.doe.gov/genome-projects). This may be particularly useful in gymnosperms and ferns, whose genomes are prohibitively large for current whole genome sequencing efforts. In addition, transcriptome and whole genome sequences have been generated for the lycophytes *S. moellendorffii* and *S. apoda* (Banks *et al.* 2011; D. Stevenson, personal communication). A comparison of plant genomes will allow a comparison of major plant transcription factor families and signaling proteins (expansion and contraction of plant families) that allow hypotheses to be proposed on the evolution of major features during land plant evolution (Floyd & Bowman 2007).

However, to understand the development of morphological features that are found outside the angiosperms and to understand the evolution of developmental pathways, we need to develop additional plant model systems. Model organisms are essential for unraveling the developmental genetics of particular morphological features and are also important for understanding how developmental pathways have evolved. For a species to be effectively used as a model, it needs to have certain characteristics and community resources: it must be easily and quickly grown, and it must possess natural variation, a genetic map, reliable expression protocols, reverse and forward genetic resources, transcriptome sequences, and genome sequences (Abzhanov *et al.* 2008). Major model systems have been developed across the animal kingdom, however, except for the bryophyte, *P. patens*, major plant model systems are restricted to angiosperms (Abzhanov *et al.* 2008). To fully understand land plant evolution and development, major model systems must be developed across the backbone of the land plant phylogenetic tree. To develop a major model system it is necessary to generate forward and reverse genetic resources. Homologous recombination in *P. patens* has been an important aspect of developing this bryophyte into a major model system (Schaefer & Zryd 1997). Genomes and functional tools are being developed for additional bryophytes, which are important resources for the development of a model organism. Sequencing of the liverwort genome, *Marchantia polymorpha* is currently underway and transformation of this species has also been documented (http://www.jgi.doe.gov/genome-projects; Ishizaki *et al.* 2008).

9.4 Conclusions

Evolutionary developmental biology studies in animals and plants have both demonstrated that they share some general concepts that show the logic of developmental pathways and how development influences evolution. Plant evolutionary developmental biology has provided many insights into the evolution and development of leaves and flowers. However, there is still much more work to be done on these plant morphological features as well as additional morphological features such as the alternation of generations, spores, and meristems. To better understand the evolution and development of plant form we need to expand our studies across the land plants and to do this we need to develop additional model organisms.

References

Abzhanov, A., Extavour, C.G., Groover, A. *et al.* (2008) Are we there yet? Tracking the development of new model systems. *Trends in Genetics*, **24**, 353–360.

Adamczyk, B.J. & Fernandez, D.E. (2009) MIKC* MADS domain heterodimers are required for pollen maturation and tube growth in Arabidopsis. *Plant Physiol*, **149**, 1713–1723.

Aida, M., Ishida, T., Fukaki, H. *et al.* (1997) Genes involved in organ separation in Arabidopsis: an analysis of the cup-shaped cotyledon mutant. *Plant Cell*, **9**, 841–857.

Aida, M., Ishida, T. & Tasaka, M. (1999) Shoot apical meristem and cotyledon formation during Arabidopsis embryogenesis: interaction among the CUP-SHAPED COTYLEDON and SHOOT MERISTEMLESS genes. *Development*, **126**, 1563–1570.

Ainsworth, C., Rahman, A., Parker, J. *et al.* (2005) Intersex inflorescences of Rumex acetosa demonstrate that sex determination is unique to each flower. *The New Phytologist*, **165**, 711–720.

Airoldi, C.A., Bergonzi, S. & Davies, B. (2010) Single amino acid change alters the ability to specify male or female organ identity. *Proceedings of the National Academy of Sciences of the United States of America*, **107**, 18898–18902.

Alvarez-Buylla, E.R., Ambrose, B.A., Flores-Sandoval, E. *et al.* (2010) 'B-function expression in the flower center underlies the homeotic phenotype of Lacandonia schismatica (Triuridaceae). *Plant Cell*, **22**, 3543–3559.

Ambrose, B.A. (2010) MADS-box genes in plant evolution and development. *International Journal of Plant Developmental Biology*, **4**, 30–37.

Ambrose, B.A., Lerner, D.R., Ciceri, P. *et al.* (2000) Molecular and genetic analyses of the silky1 gene reveal conservation in floral organ specification between eudicots and monocots. *Molecular Cell*, **5**, 569–579.

Ariizumi, T. & Toriyama, K. (2011) Genetic regulation of sporopollenin synthesis and pollen exine development. *Annual Reviews in Plant Biology*, **62**, 437–460.

Arthur, W. (2002) The emerging conceptual framework of evolutionary developmental biology. *Nature*, **415**, 757–764.

Arthur, W. (2004) The effect of development on the direction of evolution: toward a twenty-first century consensus. *Evolution and Development*, **6**, 282–288.

Baima, S., Nobili, F., Sessa, G. *et al.* (1995) The expression of the Athb-8 homeobox gene is restricted to provascular cells in Arabidopsis thaliana. *Development*, **121**, 4171–4182.

Bainbridge, K., Guyomarc'h, S., Bayer, E. *et al.* (2008) Auxin influx carriers stabilize phyllotactic patterning. *Genes and Development*, **22**, 810–823.

Banks, J.A., Nishiyama, T., Hasebe, M. *et al.* (2011) The Selaginella genome identifies genetic changes associated with the evolution of vascular plants. *Science*, **332**, 960–963.

Barkoulas, M., Hay, A., Kougioumoutzi, E. *et al.* (2008) A developmental framework for dissected leaf formation in the Arabidopsis relative Cardamine hirsuta. *Nature Genetics*, **40**, 1136–1141.

Barton, M.K. & Poethig, R.S. (1993) Formation of the shoot apical meristem in *Arabidopsis thaliana*: an analysis of development in the wild type and in the *shoot meristemless* mutant. *Development*, **119**, 823–831.

Bateman, R.M. & Dimichele, W.A. (1994) Heterospory: the most iterative key innovation in the evolutionary history of the plant kingdom. *Biological Reviews*, **69**, 345–417.

Baulcombe, D. (2004) RNA silencing in plants. *Nature*, **431**, 356–363.

Baum, D.A. and Hileman, L.C. (2006). A developmental genetic model for the origin of the flower. In: *Flowering and Its Manipulation*, (ed. C. Ainsworth), Blackwell Publishing Ltd., Oxford.

Bemer, M., Wolters-Arts, M., Grossniklaus, U. *et al.* (2008) The MADS domain protein DIANA acts together with AGAMOUS-LIKE80 to specify the central cell in Arabidopsis ovules. *Plant Cell*, **20**, 2088–2101.

Benkova, E., Michniewicz, M., Sauer, M. *et al.* (2003) Local, efflux-dependent auxin gradients as a common module for plant organ formation. *Cell*, **115**, 591–602.

Benlloch, R., d'Erfurth, I., Ferrandiz, C. *et al.* (2006) Isolation of mtpim proves Tnt1 a useful reverse genetics tool in Medicago truncatula and uncovers new aspects of AP1-like functions in legumes. *Plant Physiology*, **142**, 972–983.

Berger, Y., Harpaz-Saad, S., Brand, A. *et al.* (2009) The NAC-domain transcription factor GOBLET specifies leaflet boundaries in compound tomato leaves. *Development*, **136**, 823–832.

Bharathan, G., Goliber, T.E., Moore, C. *et al.* (2002) Homologies in leaf form inferred from KNOXI gene expression during development. *Science*, **296**, 1858–1860.

Blackmore, S. & Barnes, S.H. (1987) Embryophyte spore walls: origin, development, and homologies. *Cladistics*, **3**, 185–195.
Blázquez, M.A., Green, R., Nilsson, O. *et al.* (1998) Gibberellins promote flowering of Arabidopsis by activating the *LEAFY* promoter. *Plant Cell*, **10**, 791–800.
Blázquez, M.A., Soowal, L.N., Lee, I. *et al.* (1997) LEAFY expression and flower initiation in Arabidopsis, *Development*, **124** (19), 3835–3844.
Blein, T., Hasson, A. & Laufs, P. (2010) Leaf development: what it needs to be complex. *Current Opinion in Plant Biology*, **13**, 75–82.
Blein, T., Pulido, A., Vialette-Guiraud, A. *et al.* (2008) A conserved molecular framework for compound leaf development. *Science*, **322**, 1835–1839.
Boavida, L.C., Shuai, B., Yu, H.J. *et al.* (2009) A collection of Ds insertional mutants associated with defects in male gametophyte development and function in Arabidopsis thaliana. *Genetics*, **181**, 1369–1385.
Bolker, J.A. (2000) Modularity in development and why it matter to evo-devo. *American Zoologist*, **40**, 770–776.
Borghi, L., Gutzat, R., Futterer, J. *et al.* (2010) Arabidopsis RETINOBLASTOMA-RELATED is required for stem cell maintenance, cell differentiation, and lateral organ production. *Plant Cell*, **22**, 1792–1811.
Bower, F.O. (1935) *Primitive Land Plants, Also Known as the Archegoniatae*, Macmillan & Co., Ltd., London.
Bower, F.O. (1972) *The Origin of a Land Flora: A Theory Based upon theFacts of Alternation*. Hafner Publishing Co., New York.
Bowman, J.L., Alvarez, J., Weigel, D. *et al.* (1993) Control of flower development in *Arabidopsis thaliana* by *APETALA1* and interacting genes. *Development*, **119**, 721–743.
Bowman, J.L., Floyd, S.K. & Sakakibara, K. (2007) Green genes-comparative genomics of the green branch of life. *Cell*, **129**, 229–234.
Bowman, J.L., Smyth, D.R. & Meyerowitz, E.M. (1989) Genes directing flower development in Arabidopsis. *Plant Cell*, **1**, 37–52.
Boyce, C.K. & Knoll, A.H. (2002) Evolution of developmental potential and the multiple independent origins of leaves in Paleozoic vascular plants. *Paleobiology*, **28**, 70–100.
Bradley, D., Carpenter, R., Sommer, H. *et al.* (1993) Complementary floral homeotic phenotypes result from opposite orientations of a transposon at the *plena* locus of Antirrhinum. *Cell*, **72**, 85–95.
Brown, R.C. & Lemmon, B.E. (2011) Spores before sporophytes: hypothesizing the origin of sporogenesis at the algal-plant transition. *New Phytologist*, **190**, 875–881.
Byrne, M.E. (2006) Shoot meristem function and leaf polarity: the role of class III HD-ZIP genes. *PLoS Genetics*, **2**, e89.
Byrne, M.E., Barley, R., Curtis, M. *et al.* (2000) Asymmetric leaves1 mediates leaf patterning and stem cell function in Arabidopsis. *Nature*, **408**, 967–971.
Byrne, M.E., Simorowski, J. & Martienssen, R.A. (2002) ASYMMETRIC LEAVES1 reveals knox gene redundancy in Arabidopsis. *Development*, **129**, 1957–1965.
Carlsbecker, A., Tandre, K., Johanson, U. *et al.* (2004) The MADS-box gene DAL1 is a potential mediator of the juvenile-to-adult transition in Norway spruce (Picea abies). *The Plant Journal* **40**, 546–557.
Carroll, S.B. (2008) Evo-devo and an expanding evolutionary synthesis: a genetic theory of morphological evolution. *Cell*, **134**, 25–36.

Cartolano, M., Castillo, R., Efremova, N. *et al.* (2007) A conserved microRNA module exerts homeotic control over Petunia hybrida and Antirrhinum majus floral organ identity. *Nature Genetics*, **39**, 901–905.

Causier, B., Castillo, R., Zhou, J. *et al.* (2005) Evolution in action: following function in duplicated floral homeotic genes. *Current Biology*, **15**, 1508–1512.

Causier, B., Schwarz-Sommer, Z. & Davies, B. (2010) Floral organ identity: 20 years of ABCs. *Seminars in Cell & Developmental Biology*, **21**, 73–79.

Chae, E., Tan, Q.K.-G., Hill, T.A. *et al.* (2008) An Arabidopsis F-box protein acts as a transcriptional co-factor to regulate floral development. *Development*, **135**, 1235–1245.

Chan, S.W., Henderson, I.R. & Jacobsen, S.E. (2005) Gardening the genome: DNA methylation in Arabidopsis thaliana. *Nature Reviews in Genetics*, **6**, 351–360.

Charoensawan, V., Wilson, D. & Teichmann, S.A. (2010) Lineage-specific expansion of DNA-binding transcription factor families. *Trends in Genetics*, **26**, 388–393.

Chen, J.J., Janssen, B.J., Williams, A. *et al.* (1997) A gene fusion at a homeobox locus: alterations in leaf shape and implications for morphological evolution. *Plant Cell*, **9**, 1289–1304.

Coen, E.S. & Meyerowitz, E.M. (1991) The war of the whorls: genetic interactions controlling flower development. *Nature*, **353**, 31–37.

Colombo, M., Masiero, S., Vanzulli, S. *et al.* (2008) AGL23, a type I MADS-box gene that controls female gametophyte and embryo development in Arabidopsis. *Plant Journal*, **54**, 1037–1048.

Conant, G.C. & Wolfe, K.H. (2008) Turning a hobby into a job: how duplicated genes find new functions. *Nature Reviews in Genetics*, **9**, 938–950.

Conner, J. & Liu, Z. (2000) LEUNIG, a putative transcriptional corepressor that regulates AGAMOUS expression during flower development. *Proceedings of the National Academy of Sciences of the United States of America*, **97**, 12902–12907.

Davidson, E.H. & Erwin, D.H. (2006) Gene regulatory networks and the evolution of animal body plans. *Science*, **311**, 796–800.

Davies, B., Di Rosa, A., Eneva, T.H. *et al.* (1996a) Alteration of tobacco floral organ identity by expression of combinations of Antirrhinum MADS-box genes. *Plant Journal*. **10**, 663–677.

Davies, B., Egea-Cortines, M., de Andrade Silva, E. *et al.* (1996b) Multiple interactions amongst floral homeotic MADS box proteins. *EMBO Journal*, **15**, 4330–4343.

Davies, B., Motte, P., Keck, E. *et al.* (1999) PLENA and FARINELLI: redundancy and regulatory interactions between two Antirrhinum MADS-box factors controlling flower development. *EMBO Journal*, **18**, 4023–4034.

Demason, D.A. & Polowick, P.L. (2009) Patterns of DR5: GUS expression in organs of pea (Pisum sativum). *International Journal of Plant Sciences*, **170**, 1–11.

Ditta, G., Pinyopich, A., Robles, P. *et al.* (2004) The SEP4 gene of Arabidopsis thaliana functions in floral organ and meristem identity. *Current Biology*, **14**, 1935–1940.

Domin, K. (1931) Phylogenetic evolution of the phyllome. *American Journal of Botany*, **18**, 237–242.

Ebel, C., Mariconti, L. & Gruissem, W. (2004) Plant retinoblastoma homologues control nuclear proliferation in the female gametophyte. *Nature*, **429**, 776–780.

Egea-Cortines, M., Saedler, H. & Sommer, H. (1999) Ternary complex formation between the MADS-box proteins SQUAMOSA, DEFICIENS and GLOBOSA is involved in the control of floral architecture in Antirrhinum majus. *EMBO Journal*, **18**, 5370–5379.

Emery, J.F., Floyd, S.K., Alvarez, J. *et al.* (2003) Radial patterning of Arabidopsis shoots by class III HD-ZIP and KANADI genes. *Current Biology*, **13**, 1768–1774.

Eshed, Y., Baum, S.F., Perea, J.V. *et al.* (2001) Establishment of polarity in lateral organs of plants. *Current Biology*, **11**, 1251–1260.

Eshed, Y., Izhaki, A., Baum, S.F. *et al.* (2004) Asymmetric leaf development and blade expansion in Arabidopsis are mediated by KANADI and YABBY activities. *Development*, **131**, 2997–3006.

Ferrario, S., Immink, R.G., Shchennikova, A. *et al.* (2003) The MADS box gene FBP2 is required for SEPALLATA function in petunia. *Plant Cell*, **15**, 914–925.

Fiers, M., Ku, K.L. & Liu, C.M. (2007) CLE peptide ligands and their roles in establishing meristems. *Current Opinion in Plant Biology*, **10**, 39–43.

Floyd, S.K. & Bowman, J.L. (2004) Gene regulation: ancient microRNA target sequences in plants. *Nature*, **428**, 485–486.

Floyd, S.K. & Bowman, J.L. (2006) Distinct developmental mechanisms reflect the independent origins of leaves in vascular plants. *Current Biology*, **16**, 1911–1917.

Floyd, S.K. & Bowman, J.L. (2007) The ancestral developmental toolkit of land plants. *International Journal of Plant Science*, **168**, 1–35.

Floyd, S.K., Zalewski, C.S. & Bowman, J.L. (2006) Evolution of class III homeodomain-leucine zipper genes in streptophytes. *Genetics*, **173**, 373–388.

Frohlich, M.W. & Parker, D.S. (2000). The mostly male theory of flower evolu- tionary origins: from genes to fossils. *Systematic Botany* **25**, 155–170.

Frohlich, M.W. & Chase, M.W. (2007) After a dozen years of progress the origin of angiosperms is still a great mystery. *Nature*, **450**, 1184–1189.

Galweiler, L., Guan, C., Muller, A. *et al.* (1998) Regulation of polar auxin transport by AtPIN1 in Arabidopsis vascular tissue. *Science*, **282**, 2226–2230.

Gifford, E.M. & Foster, A.S. (1989) *Morphology and Evolution of Vascular Plants*, W.H. Freeman and Co., New York.

Goethe, J.W.V. (1790) *Versuch die Metamorphose der Pflanzen zu erklären*, Gotha: C.W, Ettinger.

Goto, K. & Meyerowitz, E.M. (1994) Function and regulation of the *Arabidopsis* floral homeotic gene *PISTILLATA*. *Genes and Development*, **8**, 1548–1560.

Graham, L.K. & Wilcox, L.W. (2000) The origin of alternation of generations in land plants: a focus on matrotrophy and hexose transport. *Philosophical Transactions of the Royal Society B: Biological Sciences*, **355**, 757–766; discussion 766–757.

Guo, M., Thomas, J., Collins, G. *et al.* (2008) Direct repression of KNOX loci by the ASYMMETRIC LEAVES1 complex of Arabidopsis. *Plant Cell*, **20**, 48–58.

Ha, C.M., Jun, J.H. & Fletcher, J.C. (2010) Shoot apical meristem form and function. *Current Topics in Developmental Biology*, **91**, 103–140.

Halder, G., Callaerts, P. & Gehring, W.J. (1995) Induction of ectopic eyes by targeted expression of the eyeless gene in Drosophila. *Science*, **267**, 1788–1792.

Hareven, D., Gutfinger, T., Parnis, A. *et al.* (1996) The making of a compound leaf: genetic manipulation of leaf architecture in tomato. *Cell*, **84**, 735–744.

Harrison, C.J., Corley, S.B., Moylan, E.C. *et al.* (2005) Independent recruitment of a conserved developmental mechanism during leaf evolution. *Nature*, **434**, 509–514.

Hasebe, M., Wen, C.K., Kato, M. *et al.* (1998) Characterization of MADS homeotic genes in the fern Ceratopteris richardii. *Proceedings of the National Academy of Sciences of the United States of America*, **95**, 6222–6227.

Hay, A., Barkoulas, M. & Tsiantis, M. (2006) ASYMMETRIC LEAVES1 and auxin activities converge to repress BREVIPEDICELLUS expression and promote leaf development in Arabidopsis. *Development*, **133**, 3955–3961.

Hay, A. & Tsiantis, M. (2006) The genetic basis for differences in leaf form between Arabidopsis thaliana and its wild relative Cardamine hirsuta. *Nature Genetics*, **38**, 942–947.

Haughn, G.W. & Somerville, C.R. (1988) Genetic control of morphogenesis in *Arabidopsis*. *Developmental Genetics*, **9**, 73–89.

Heisler, M.G., Ohno, C., Das, P. *et al.* (2005) Patterns of auxin transport and gene expression during primordium development revealed by live imaging of the Arabidopsis inflorescence meristem. *Current Biology*, **15**, 1899–1911.

Hennig, L., Bouveret, R. & Gruissem, W. (2005) MSI1-like proteins: an escort service for chromatin assembly and remodeling complexes. *Trends in Cell Biology*, **15**, 295–302.

Hennig, L. & Derkacheva, M. (2009) Diversity of Polycomb group complexes in plants: same rules, different players? *Trends in Genetics*, **25**, 414–423.

Hibara, K., Karim, M.R., Takada, S. *et al.* (2006) Arabidopsis CUP-SHAPED COTYLEDON3 regulates postembryonic shoot meristem and organ boundary formation. *Plant Cell*, **18**, 2946–2957.

Hirakawa, Y., Kondo, Y. & Fukuda, H. (2010) TDIF peptide signaling regulates vascular stem cell proliferation via the WOX4 homeobox gene in Arabidopsis. *Plant Cell*, **22**, 2618–2629.

Hofer, J., Turner, L., Hellens, R. *et al.* (1997) UNIFOLIATA regulates leaf and flower morphogenesis in pea. *Current Biology*, **7**, 581–587.

Honma, T. & Goto, K. (2001) Complexes of MADS-box proteins are sufficient to convert leaves into floral organs. *Nature*, **409**, 525–529.

Huijser, P., Klein, J., Lönnig, W.-E. *et al.* (1992) Bracteomania, an inflorescence anomaly, is caused by the loss of function of the MADS-box gene *squamosa* in *Antirrhinum*. *EMBO Journal*, **11**, 1239–1249.

Imaichi, R. & Kato, M. (1989) Developmental anatomy of the shoot apical cell, rhizophore and root of *Selaginella uncinata*. *Botanical Magazine Tokyo*, **102**, 369–380.

Immink, R.G., Tonaco, I.A., de Folter, S. *et al.* (2009a) SEPALLATA3: the 'glue' for MADS box transcription factor complex formation. *Genome Biology*, **10**, R24.

Immink, R.G.H., Kaufmann, K. & Angenent, G.C. (2009b) The 'ABC' of MADS domain protein behaviour and interactions. *Seminars in Cell & Developmental Biology*. **22**, 87–92

Irish, V.F. & Litt, A. (2005) Flower development and evolution: gene duplication, diversification and redeployment. *Current Opinion in Genetics & Development*, **15**, 454–460.

Irish, V.F. & Sussex, I.M. (1992) A fate map of the *Arabidopsis* embryonic shoot apical meristem. *Development*, **115**, 745–753.

Ishida, T., Aida, M., Takada, S. *et al.* (2000) Involvement of CUP-SHAPED COTYLEDON genes in gynoecium and ovule development in Arabidopsis thaliana. *Plant and Cell Physiology*, **41**, 60–67.

Ishizaki, K., Chiyoda, S., Yamato, K.T. *et al.* (2008) Agrobacterium-mediated transformation of the haploid liverwort Marchantia polymorpha L., an emerging model for plant biology. *Plant and Cell Physiology*, **49**, 1084–1091.

Ito, T., Wellmer, F., Yu, H. *et al.* (2004) The homeotic protein AGAMOUS controls microsporogenesis by regulation of SPOROCYTELESS. *Nature*, **430**, 356–360.

Iwakawa, H., Iwasaki, M., Kojima, S. *et al.* (2007) Expression of the ASYMMETRIC LEAVES2 gene in the adaxial domain of Arabidopsis leaves represses cell proliferation in this domain and is critical for the development of properly expanded leaves. *Plant Journal*, **51**, 173–184.

Jack, T., Brockman, L.L. & Meyerowitz, E.M. (1992) The homeotic gene APETALA3 of Arabidopsis thaliana encodes a MADS box and is expressed in petals and stamens. *Cell*, **68**, 683–697.

Janssen, B.J., Lund, L. & Sinha, N. (1998) Overexpression of a homeobox gene, LeT6, reveals indeterminate features in the tomato compound leaf. *Plant Physiology*, **117**, 771–786.

Jaramillo, M. & Kramer, E. (2007) The role of developmental genetics in understanding homology and morphological evolution in plants. *International Journal of Plant Science*, **168**, 61–72.

Jenner, R.A. & Wills, M.A. (2007) The choice of model organisms in evo-devo. *Nature Reviews in Genetics*, **8**, 311–319.

Jofuku, K.D., den Boer, B.G.W., Van Montagu, M. *et al.* (1994) Control of Arabidopsis flower and seed development by the homeotic gene *APETALA2*. *Plant Cell*, **6**, 1211–1225.

Johnston, A.J., Kirioukhova, O., Barrell, P.J. *et al.* (2010) Dosage-sensitive function of retinoblastoma related and convergent epigenetic control are required during the Arabidopsis life cycle. *PLoS Genetics*, **6**, e1000988.

Johnston, A.J., Matveeva, E., Kirioukhova, O. *et al.* (2008) A dynamic reciprocal RBR-PRC2 regulatory circuit controls Arabidopsis gametophyte development. *Current Biology*, **18**, 1680–1686.

Kanno, A., Saeki, H., Kameya, T. *et al.* (2003) Heterotopic expression of class B floral homeotic genes supports a modified ABC model for tulip (Tulipa gesneriana). *Plant Molecular Biology*, **52**, 831–841.

Karol, K.G., Mccourt, R.M., Cimino, M.T. *et al.* (2001) The closest living relatives of land plants. *Science*, **294**, 2351–2353.

Kater, M.M., Dreni, L. & Colombo, L. (2006) Functional conservation of MADS-box factors controlling floral organ identity in rice and Arabidopsis. *Journal of Experimental Botany*, **57**, 3433–3444.

Keck, E., McSteen, P., Carpenter, R. *et al.* (2003) Separation of genetic functions controlling organ identity in flowers. *EMBO Journal*, **22**, 1058–1066.

Kellogg, E.A. (2006) The difference between simple and complex leaves. *Nature Genetics*, **38**, 865–866.

Kenrick, P. & Crane, P.R. (1997) *The Origin and Early Diversification of Land Plants: A Cladistic Study*, Smithsonian Institution Press, Washington, DC.

Kerstetter, R., Vollbrecht, E., Lowe, B. *et al.* (1994) Sequence analysis and expression patterns divide the maize knotted1-like homeobox genes into two classes. *Plant Cell*, **6**, 1877–1887.

Kerstetter, R.A., Bollman, K., Taylor, R.A. *et al.* (2001) KANADI regulates organ polarity in Arabidopsis. *Nature*, **411**, 706–709.

Kessler, S. & Sinha, N. (2004) Shaping up: the genetic control of leaf shape. *Current Opinion in Plant Biology*, **7**, 65–72.

Kim, S., Koh, J., Yoo, M.J. *et al.* (2005) Expression of floral MADS-box genes in basal angiosperms: implications for the evolution of floral regulators. *The Plant Journal*, **43**, 724–744.

Kim, S., Yoo, M.J., Albert, V.A. *et al.* (2004) Phylogeny and diversification of B-function MADS-box genes in angiosperms: evolutionary and functional implications of a 260-million-year-old duplication. *American Journal of Botany*, **91**, 2102–2118.

Kimura, S., Koenig, D., Kang, J. *et al.* (2008) Natural variation in leaf morphology results from mutation of a novel KNOX gene. *Current Biology*, **18**, 672–677.

Koenig, D., Bayer, E., Kang, J. *et al.* (2009) Auxin patterns Solanum lycopersicum leaf morphogenesis. *Development*, **136**, 2997–3006.

Kramer, E.M., Dorit, R.L. & Irish, V.F. (1998) Molecular evolution of genes controlling petal and stamen development: duplication and divergence within the APETALA3 and PISTILLATA MADS-box gene lineages. *Genetics*, **149**, 765–783.

Kramer, E.M., Holappa, L., Gould, B. *et al.* (2007) Elaboration of B gene function to include the identity of novel floral organs in the lower eudicot Aquilegia. *Plant Cell*, **19**, 750–766.

Kramer, E.M., Jaramillo, M.A. & Di Stilio, V.S. (2004) Patterns of gene duplication and functional evolution during the diversification of the AGAMOUS subfamily of MADS box genes in angiosperms. *Genetics*, **166**, 1011–1023.

Krizek, B.A. & Fletcher, J.C. (2005) Molecular mechanisms of flower development: an armchair guide. *Nature Reviews in Genetics*, **6**, 688–698.

Krizek, B.A. & Meyerowitz, E.M. (1996) The *Arabidopsis* homeotic genes *APETALA3* and *PISTILLATA* are sufficient to provide the B class organ identity function. *Development*, **122**, 11–22.

Kang, I.H., Steffen, J.G., Portereiko, M.F. *et al.* (2008) The AGL62 MADS domain protein regulates cellularization during endosperm development in Arabidopsis. *Plant Cell*, **20**, 635–647.

Lee, K.J., Sakata, Y., Mau, S.L. *et al.* (2005) Arabinogalactan proteins are required for apical cell extension in the moss Physcomitrella patens. *Plant Cell*, **17**, 3051–3065.

Lee, I., Wolfe, D.S., Nilsson, O. *et al.* (1997) A *LEAFY* co-regulator encoded by *UNUSUAL FLORAL ORGANS*. *Current Biology*, **7**, 95–104.

Lenhard, M., Bohnert, A., Jurgens, G. *et al.* (2001) Termination of stem cell maintenance in Arabidopsis floral meristems by interactions between WUSCHEL and AGAMOUS. *Cell*, **105**, 805–814.

Li, H., Xu, L., Wang, H. *et al.* (2005) The Putative RNA-dependent RNA polymerase RDR6 acts synergistically with ASYMMETRIC LEAVES1 and 2 to repress BREVIPEDICELLUS and MicroRNA165/166 in Arabidopsis leaf development. *Plant Cell*, **17**, 2157–2171.

Lin, W.C., Shuai, B. & Springer, P.S. (2003) The Arabidopsis LATERAL ORGAN BOUNDARIES-domain gene ASYMMETRIC LEAVES2 functions in the repression of KNOX gene expression and in adaxial-abaxial patterning. *Plant Cell*, **15**, 2241–2252.

Lincoln, C., Long, J., Yamaguchi, J. *et al.* (1994) A knotted1-like homeobox gene in Arabidopsis is expressed in the vegetative meristem and dramatically alters leaf morphology when overexpressed in transgenic plants. *Plant Cell*, **6**, 1859–1876.

Litt, A. (2007) An Evaluation of A-Function: evidence from the APETALA1 and APETALA2 Gene Lineages. *International Journal of Plant Sciences*, **168**, 73–91.

Litt, A. & Irish, V.F. (2003) Duplication and diversification in the APETALA1/FRUITFULL floral homeotic gene lineage: implications for the evolution of floral development. *Genetics*, **165**, 821–833.

Liu, C., Xi, W., Shen, L. *et al.* (2009) Regulation of floral patterning by flowering time genes. *Developmental Cell*, **16**, 711–722.

Lohmann, J.U., Hong, R.L., Hobe, M. *et al.* (2001) A molecular link between stem cell regulation and floral patterning in Arabidopsis. *Cell*, **105**, 793–803.

Loidl, P. (2004) A plant dialect of the histone language. *Trends in Plant Science*, **9**, 84–90.

Long, J.A., Moan, E.I., Medford, J.I. *et al.* (1996) A member of the KNOTTED class of homeodomain proteins encoded by the STM gene of Arabidopsis. *Nature*, **379**, 66–69.

Ma, H. (2005) Molecular genetic analyses of microsporogenesis and microgametogenesis in flowering plants. *Annual Reviews in Plant Biology*, **56**, 393–434.

Mandel, M.A., Gustafson-Brown, C., Savidge, B. *et al.* (1992) Molecular characterization of the *Arabidopsis* floral homeotic gene *APETALA1*. *Nature*, **360**, 273–277.

Masiero, S., Colombo, L., Grini, P.E. *et al.* (2011) The emerging importance of type I MADS box transcription factors for plant reproduction. *Plant Cell*, **23**, 865–872.

Mcalpin, B.W. & White, R.A. (1974) Shoot organization in the filicales: the promeristem. *American Journal of Botany*, **61**, 562–579.

Mcconnell, J.R. & Barton, M.K. (1998) Leaf polarity and meristem formation in Arabidopsis. *Development*, **125**, 2935–2942.

Mcconnell, J.R., Emery, J., Eshed, Y. *et al.* (2001) Role of PHABULOSA and PHAVOLUTA in determining radial patterning in shoots. *Nature*, **411**, 709–713.

Melzer, R., Wang, Y.Q. & Theissen, G. (2010) The naked and the dead: the ABCs of gymnosperm reproduction and the origin of the angiosperm flower. *Seminars in Cell & Developmental Biology*, **21**, 118–128.

Mizukami, Y. & Ma, H. (1992) Ectopic expression of the floral homeotic gene *AGAMOUS* in transgenic Arabidopsis plants alters floral organ identity. *Cell*, **71**, 119–131.

Mosquna, A., Katz, A., Decker, E.L. *et al.* (2009) Regulation of stem cell maintenance by the Polycomb protein FIE has been conserved during land plant evolution. *Development*, **136**, 2433–2444.

Mouradov, A., Glassick, T., Hamdork, B. *et al.* (1998) NEEDLY, a *Pinus radiata* ortholog of FLORICAULA/LEAFY genes, expressed in both reproductive and vegetative meristems. *Proceedings of the National Academy of Sciences of the United States of America*, **95**, 6537–6542.

Nagasawa, N., Miyoshi, M., Sano, Y. *et al.* (2003) SUPERWOMAN1 and DROOPING LEAF genes control floral organ identity in rice. *Development*, **130**, 705–718.

Navarro, C., Efremova, N., Golz, J.F. *et al.* (2004) Molecular and genetic interactions between STYLOSA and GRAMINIFOLIA in the control of Antirrhinum vegetative and reproductive development. *Development*, **131**, 3649–3659.

Newton, M.E. (1987) Developmental Aspects of Chromatin Condensation in Liverworts. *The Bryologist*, **90**, 376–382.

Nickrent, D.L., Parkinson, C.L., Palmer, J.D. *et al.* (2000) Multigene phylogeny of land plants with special reference to bryophytes and the earliest land plants. *Molecular Biology and Evolution*, **17**, 1885–1895.

Niklas, K.J. & Kutschera, U. (2010) The evolution of the land plant life cycle. *New Phytol*, **185**, 27–41.

Nishimura, A., Tamaoki, M., Sato, Y. *et al.* (1999) The expression of tobacco knotted1-type class 1 homeobox genes correspond to regions predicted by the cytohistological zonation model. *Plant Journal*, **18**, 337–347.

Ohno, S. (1970) *Evolution by Gene Duplication*, Springer-Verlag, Berlin, New York.

Okada, K., Ueda, J., Komaki, M.K. *et al.* (1991) Requirement of the auxin polar transport system in early stages of Arabidopsis floral bud formation. *The Plant Cell*, **3**, 677–684.

Okano, Y., Aono, N., Hiwatashi, Y. *et al.* (2009) A polycomb repressive complex 2 gene regulates apogamy and gives evolutionary insights into early land plant evolution. *Prroceedings of the National Academy of Sciences of the United States of America*, **106**, 16321–16326.

Pagnussat, G.C., Yu, H.J., Ngo, Q.A. *et al.* (2005) Genetic and molecular identification of genes required for female gametophyte development and function in Arabidopsis. *Development*, **132**, 603–614.

Parcy, F., Nilsson, O., Busch, M.A. *et al.* (1998) A genetic framework for floral patterning. *Nature*, **395**, 561–566.

Pekker, I., Alvarez, J.P. & Eshed, Y. (2005) Auxin response factors mediate Arabidopsis organ asymmetry via modulation of KANADI activity. *Plant Cell*, **17**, 2899–2910.

Pelaz, S., Ditta, G.S., Baumann, E. *et al.* (2000) B and C floral organ identity functions require SEPALLATA MADS-box genes. *Nature*, **405**, 200–203.

Pelaz, S., Tapia-Lopez, R., Alvarez-Buylla, E.R. *et al.* (2001) Conversion of leaves into petals in Arabidopsis. *Current Biology*, **11**, 182–184.

Peter, I.S. & Davidson, E.H. (2011) Evolution of gene regulatory networks controlling body plan development. *Cell*, **144**, 970–985.

Pina, C., Pinto, F., Feijo, J.A. *et al.* (2005) Gene family analysis of the Arabidopsis pollen transcriptome reveals biological implications for cell growth, division control, and gene expression regulation. *Plant Physiology*, **138**, 744–756.

Portereiko, M.F., Lloyd, A., Steffen, J.G. *et al.* (2006) AGL80 is required for central cell and endosperm development in Arabidopsis. *Plant Cell*, **18**, 1862–1872.

Prigge, M.J. & Clark, S.E. (2006) Evolution of the class III HD-Zip gene family in land plants. *Evolution and Development*, **8**, 350–361.

Prigge, M.J., Otsuga, D., Alonso, J.M. *et al.* (2005) Class III homeodomain-leucine zipper gene family members have overlapping, antagonistic, and distinct roles in Arabidopsis development. *Plant Cell*, **17**, 61–76.

Reinhardt, D., Mandel, T. & Kuhlemeier, C. (2000) Auxin regulates the initiation and radial position of plant lateral organs. *Plant Cell*, **12**, 507–518.

Reinhardt, D., Pesce, E.R., Stieger, P. *et al.* (2003) Regulation of phyllotaxis by polar auxin transport. *Nature*, **426**, 255–260.

Reinhart, B.J., Weinstein, E.G., Rhoades, M.W. *et al.* (2002) MicroRNAs in plants. *Genes and Development*, **16**, 1616–1626.

Reiser, L., Sanchez-Baracaldo, P. & Hake, S. (2000) Knots in the family tree: evolutionary relationships and functions of knox homeobox genes. *Plant Molecular Biology*, **42**, 151–166.

Rensing, S.A., Lang, D., Zimmer, A.D. *et al.* (2008) The Physcomitrella genome reveals evolutionary insights into the conquest of land by plants. *Science*, **319**, 64–69.

Rhoades, M.W., Reinhart, B.J., Lim, L.P. *et al.* (2002) Prediction of plant microRNA targets. *Cell*, **110**, 513–520.

Riechmann, J.L., Heard, J., Martin, G. *et al.* (2000) Arabidopsis transcription factors: genome-wide comparative analysis among eukaryotes. *Science*, **290**, 2105–2110.

Romanel, E.A., Schrago, C.G., Counago, R.M. *et al.* (2009) Evolution of the B3 DNA binding superfamily: new insights into REM family gene diversification. *PLoS One*, **4**, e5791.

Rudall, P.J., Hilton, J., Vergara-Silva, F. *et al.* (2011) Recurrent abnormalities in conifer cones and the evolutionary origins of flower-like structures. *Trends in Plant Science*, **16** 151–159.

Sakakibara, K., Nishiyama, T., Deguchi, H. *et al.* (2008) Class 1 KNOX genes are not involved in shoot development in the moss Physcomitrella patens but do function in sporophyte development. *Evolution and Development*, **10**, 555–566.

Sakakibara, K., Nishiyama, T., Kato, M. *et al.* (2001) Isolation of homeodomain-leucine zipper genes from the moss Physcomitrella patens and the evolution of homeodomain-leucine zipper genes in land plants. *Molecular Biology and Evolution*, **18**, 491–502.

Sano, R., Juarez, C.M., Hass, B. *et al.* (2005) KNOX homeobox genes potentially have similar function in both diploid unicellular and multicellular meristems, but not in haploid meristems. *Evolution & Development*, **7**, 69–78.

Sarojam, R., Sappl, P.G., Goldshmidt, A. *et al.* (2010) Differentiating Arabidopsis shoots from leaves by combined YABBY activities. *Plant Cell*, **22**, 2113–2130.

Sawa, S., Watanabe, K., Goto, K. *et al.* (1999) FILAMENTOUS FLOWER, a meristem and organ identity gene of Arabidopsis, encodes a protein with a zinc finger and HMG-related domains. *Genes and Development*, **13**, 1079–1088.

Schaefer, D.G. & Zryd, J.P. (1997) Efficient gene targeting in the moss Physcomitrella patens. *Plant Journal*, **11**, 1195–1206.

Schlosser, G. & Wagner, G.P. (2004) *Modularity in Development and Evolution*, University of Chicago Press, Chicago, IL.

Schmid, M., Uhlenhaut, N.H., Godard, F. *et al.* (2003) Dissection of floral induction pathways using global expression analysis. *Development*, **130**, 6001–6012.

Schwarz-Sommer, Z., Huijser, P., Nacken, W. *et al.* (1990) Genetic control of flower development: homeotic genes in *Antirrhinum majus*. *Science*, **250**, 931–936.

Semiarti, E., Ueno, Y., Tsukaya, H. *et al.* (2001) The ASYMMETRIC LEAVES2 gene of Arabidopsis thaliana regulates formation of a symmetric lamina, establishment of venation and repression of meristem-related homeobox genes in leaves. *Development*, **128**, 1771–1783.

Shubin, N., Tabin, C. & Carroll, S. (1997) Fossils, genes and the evolution of animal limbs. *Nature*, **388**, 639–648.

Shubin, N., Tabin, C. & Carroll, S. (2009) Deep homology and the origins of evolutionary novelty. *Nature*, **457**, 818–823.

Siegfried, K.R., Eshed, Y., Baum, S.F. *et al.* (1999) Members of the YABBY gene family specify abaxial cell fate in Arabidopsis. *Development*, **126**, 4117–4128.

Soltis, D.E., Chanderbali, A.S., Kim, S. *et al.* (2007a) The ABC model and its applicability to basal angiosperms. *Annals of Botany*, **100**, 155–163.

Soltis, D.E., Ma, H., Frohlich, M.W. *et al.* (2007b) The floral genome: an evolutionary history of gene duplication and shifting patterns of gene expression. *Trends in Plant Science*, **12**, 358–367.

Sommer, H., Beltran, J.P., Huijser, P. *et al.* (1990) Deficiens, a homeotic gene involved in the control of flower morphogenesis in Antirrhinum majus: the protein shows homology to transcription factors. *EMBO Journal*, **9**, 605–613.

Specht, C.D. & Bartlett, M.E. (2009) Flower evolution: the origin and subsequent diversification of the angiosperm flower. *Annual Reviews in Ecology, Evolution and Systematics*, **40**, 217–243.

Steeves, T.A. & Sussex, I.M. (1989) *Patterns in Plant Development*, Cambridge University Press, Cambridge; New York.

Steffen, J.G., Kang, I.H., Portereiko, M.F. *et al.* (2008) AGL61 interacts with AGL80 and is required for central cell development in Arabidopsis. *Plant Physiology*, **148**, 259–268.

Stevenson, D.W. (1976) The Cytohistological and Cytohistochemical Zonation of the Shoot Apex of Botrychium multifidum. *American Journal of Botany*, **63**, 852–856.

Su, K., Zhao, S., Shan, H. *et al.* (2008) The MIK region rather than the C-terminal domain of AP3-like class B floral homeotic proteins determines functional specificity in the development and evolution of petals. *The New Phytologist*, **178**, 544–558.

Takada, S., Hibara, K., Ishida, T. *et al.* (2001) The CUP-SHAPED COTYLEDON1 gene of Arabidopsis regulates shoot apical meristem formation. *Development*, **128**, 1127–1135.

Taylor, S.A., Hofer, J.M.I., Murfet, I.C. *et al.* (2002) PROLIFERATING INFLORESCENCE MERISTEM, a MADS-box gene that regulates floral meristem identity in pea. *Plant Physiology*, **129**, 1150–1159.

Theissen, G., Becker, A., Winter, K.-U. *et al.* (2002). How the land plants learned their floral ABCs: the role of MADS- box genes in the evolutionary origin of flowers. In: *Developmental Genetics and Plant Evolution* (eds Q.B.C. Cronk R.M. Bateman & J.A. Hawkins), pp. 173–205, Taylor & Francis, London.

Theissen, G. & Melzer, R. (2007) Molecular mechanisms underlying origin and diversification of the angiosperm flower. *Annals of Botany*, **100**, 603–619.

Timmermans, M.C., Hudson, A., Becraft, P.W. *et al.* (1999) ROUGH SHEATH2: a Myb protein that represses knox homeobox genes in maize lateral organ primordia. *Science*, **284**, 151–153.

Tröbner, W., Ramirez, L., Motte, P. *et al.* (1992) *GLOBOSA*: a homeotic gene which interacts with *DEFICIENS* in the control of *Antirrhinum* floral organogenesis. *EMBO Journal*, **11**, 4693–4704.

Tsiantis, M., Schneeberger, R., Golz, J.F. *et al.* (1999) The maize rough sheath2 gene and leaf development programs in monocot and dicot plants. *Science*, **284**, 154–156.

Uchida, N., Townsley, B., Chung, K.H. *et al.* (2007) Regulation of SHOOT MERISTEMLESS genes via an upstream-conserved noncoding sequence coordinates leaf development. *Proceedings of the National Academy of Sciences of the United States of America*, **104**, 15953–15958.

van Tunen, A.J., Eikelboom, W. & Angenent, G.C. (1993). Floral organogenesis in *Tulipa. Flowering Newsletters*, **16**, 33–38.

Vazquez-Lobo, A., Carlsbecker, A., Vergara-Silva, F. *et al.* (2007) Characterization of the expression patterns of LEAFY/FLORICAULA and NEEDLY orthologs in female and male cones of the conifer genera Picea, Podocarpus, and Taxus: implications for current evo-devo hypotheses for gymnosperms. *Evolution & Development*, **9**, 446–459.

Verelst, W., Twell, D., De Folter, S. *et al.* (2007) MADS-complexes regulate transcriptome dynamics during pollen maturation. *Genome Biology*, **8**, R249.

Vernoux, T., Kronenberger, J., Grandjean, O. *et al.* (2000) PIN-FORMED 1 regulates cell fate at the periphery of the shoot apical meristem. *Development*, **127**, 5157–5165.

Vidali, L., Augustine, R.C., Kleinman, K.P. *et al.* (2007) Profilin is essential for tip growth in the moss Physcomitrella patens. *Plant Cell*, **19**, 3705–3722.

Vollbrecht, E., Veit, B., Sinha, N. *et al.* (1991) The developmental gene Knotted-1 is a member of a maize homeobox gene family. *Nature*, **350**, 241–243.

Vrebalov, J., Ruezinsky, D., Padmanabhan, V. *et al.* (2002) A MADS-box gene necessary for fruit ripening at the tomato ripening-inhibitor (rin) locus. *Science*, **296**, 343–346.

Vroemen, C.W., Mordhorst, A.P., Albrecht, C. *et al.* (2003) The CUP-SHAPED COTYLEDON3 gene is required for boundary and shoot meristem formation in Arabidopsis. *Plant Cell*, **15**, 1563–1577.

Wagner, G.P., Chiu, C.H. & Laubichler, M. (2000) Developmental evolution as a mechanistic science: the inference from developmental mechanisms to evolutionary processes. *American Zoologist*, **40**, 819–831.

Waites, R. & Hudson, A. (1995) *phantastica:* a gene required for dorsiventrality of leaves in *Antirrhiunum majus. Development*, **121**, 2143–2154.

Waites, R., Selvadurai, H.R., Oliver, I.R. *et al.* (1998) The PHANTASTICA gene encodes a MYB transcription factor involved in growth and dorsoventrality of lateral organs in Antirrhinum. *Cell*, **93**, 779–789.

Wang, Y.-Q., Melzer, R. & Theißen, G. (2010) Molecular interactions of orthologues of floral homeotic proteins from the gymnosperm Gnetum gnemon provide a clue to the evolutionary origin of 'floral quartets'. *The Plant Journal*, **64**, 177–190.

Wardlaw, C.W. (1968) *Morphogenesis in Plants: A Contemporary Study*, Methuen, London.

Webster, T.J. (1992) Developmental problems in Selaginella (Selaginellaceae) in an evolutionary context. *Annals of the Missouri Botanical Garden*, **79**, 632–647.

Weigel, D., Alvarez, J., Smyth, D.R. *et al.* (1992) LEAFY controls floral meristem identity in Arabidopsis, *Cell* **69**, 843–859.

Whipple, C.J., Ciceri, P., Padilla, C.M. *et al.* (2004) Conservation of B-class floral homeotic gene function between maize and Arabidopsis, *Development* **131**, 6083–6091.

Whipple, C.J., Zanis, M.J., Kellogg, E.A. *et al.* (2007) Conservation of B class gene expression in the second whorl of a basal grass and outgroups links the origin of lodicules and petals, *Proceedings of the National Academy of Sciences of the United States of America*, **104**, 1081–1086.

Wilson, C.L. (1953) The telome theory. *Botanical Review*, **19**, 417–437.

Worsdell, W.C. (1910) The rhizophore of *Selaginella. New Phytologist*, **9**, 242–253.

Xiao, H., Wang, Y., Liu, D. *et al.* (2003) Functional analysis of the rice AP3 homologue OsMADS16 by RNA interference, *Plant Molecular Biology* **52**, 957–966.

Yang, W.C., Shi, D.Q. & Chen, Y.H. (2010) Female gametophyte development in flowering plants. *Annual Reviews in Plant Biology*, **61**, 89–108.

Yang, W.C., Ye, D., Xu, J. *et al.* (1999) The SPOROCYTELESS gene of Arabidopsis is required for initiation of sporogenesis and encodes a novel nuclear protein. *Genes & Development*, **13**, 2108–2117.

Yanofsky, M.F., Ma, H., Bowman, J.L. *et al.* (1990) The protein encoded by the Arabidopsis homeotic gene agamous resembles transcription factors. *Nature*, **346**, 35–39.

Yoo, S.K., Lee, J.S. & Ahn, J.H. (2006) Overexpression of AGAMOUS-LIKE 28 (AGL28) promotes flowering by upregulating expression of floral promoters within the autonomous pathway. *Biochemichal & Biophysical Research Communications*, **348**, 929–936.

Yu, H.J., Hogan, P. & Sundaresan, V. (2005) Analysis of the female gametophyte transcriptome of Arabidopsis by comparative expression profiling. *Plant Physiology*, **139**, 1853–1869.

Zahn, L.M., Kong, H., Leebens-Mack, J.H. *et al.* (2005) The evolution of the SEPALLATA subfamily of MADS-box genes: a preangiosperm origin with multiple duplications throughout angiosperm history. *Genetics*, **169**, 2209–2223.

Annual Plant Reviews (2013) **45**, 321–356
doi: 10.1002/9781118305881.ch10
http://onlinelibrary.wiley.com

Annual Plant Reviews

Chapter 10

DEVELOPMENT IN THE WILD: PHENOTYPIC PLASTICITY

Kathleen Donohue

Biology Department, Duke University, Durham, NC, USA

Abstract: Development of organisms in the wild occurs in ecologically variable environments. Phenotypic plasticity occurs when a single genotype alters its phenotype in response to its environment. Some traits are more plastic than others, and whether a particular trait evolves plasticity depends on the degree of environmental variation experienced by that trait, the quality of environment-dependent natural selection on that trait, and the strength of genetic correlations between the trait expressed in different environments. Thus, identifying the molecular basis of pleiotropy of traits expressed in different environments should be an important agenda in studies of development and plasticity. The strength of pleiotropy across environments can be influenced by environment-dependent gene expression and signal transduction and by the structure of genetic pathways. The degree of environmental dependence of many of these molecular processes, however, is only beginning to be elucidated. Phenotypic plasticity influences adaptation, niche breadth, and ecological ranges and has the potential to influence the evolution of reproductive isolation. Understanding the genetic and ecological mechanisms of plasticity, therefore, will enhance our knowledge of the genetic basis of adaptation and the evolution of diversity.

Keywords: adaptation; epigenetics; gene expression; parental effects; phenology; phytochrome; plasticity.

10.1 Development in the wild is phenotypic plasticity

Plant form is the product of a developmental sequence in which events that occur at early life stages shape events at later life stages. In the wild, this temporal sequence of development occurs within a temporal sequence of seasons and changing ecological conditions. The moist and sunny conditions that a new germinant experiences in early spring do not persist: the canopy closes,

Annual Plant Reviews Volume 45: The Evolution of Plant Form, First Edition.
Edited by Barbara A. Ambrose and Michael Purugganan.

the days grow longer, rain showers stop, insects hatch out, fungal spores mobilize, and soil microbes grow, die, replace each other, and alter the soil. Development in the wild is not merely the unfolding of a programmed protocol encoded by DNA sequences. It is a constant dialog between the ecological environment and the developmental processes that determine morphology, physiology, and behavior. It comes down to the interaction between the environment and genes.

Within the lifetime of a given plant, the environment inhabited by the plant changes, caused by processes completely independent of the plant itself: the weather, for example, cycles of water flow, or the life cycles of other organisms. These changes may be predictable to different degrees. Leaf-out always occurs in spring, but the precise date is unknown. Water levels in ponds rise predictably, but the precise node on the plant at which it ceases rising cannot be predicted with accuracy. Successful accommodation of these environmental vagaries requires responding to environmental inputs on the spot.

The growing plant itself induces further changes in the environment it experiences subsequently. Leaves it produces later in life may shade those produced earlier unless they are well placed. The act of growing upward early in life can increase the light available to the plant at later life stages. Nutrients harvested from the soil during early growth may no longer be available later in life. The formation and elongation of branches can determine the subsequent floral display and exposure to pollinators and dispersers. Simply by growing and developing, the plant alters the environment it experiences, and it must respond to that altered environment appropriately in order to maximize its fitness.

Environments change across generations as well. This, too, affects plant form, physiology, and behavior. The persistence and prevalence of any given allele associated with a developmental response to the environment depends on how advantageous it has been across its evolutionary history, and across all environments to which it has been exposed. An allele that has an advantage across several generations, and through all the environmental fluctuations that occurred during those generations, will persist with greater success than an allele that excels in one generation but fails completely in the next. Thus plant responses to environments that vary across generations, even if they are fairly constant within the lifetime of a single plant, may determine the extant forms that we see today.

Environmental change—within and across generations—is therefore inevitable. Inevitably, it has shaped the evolution of plant form. Accommodating those environmental changes as they occur within a plant's lifetime, or as they have occurred in previous generations, is an incessant challenge for plants. To understand plant form is to understand how the genotype of a plant interacts with the ecological environment it experiences to effect development. To understand the evolution of plant form is to understand the adaptive value, the genetic mechanisms, and the genetic constraints of these interactions between genotype and environment.

10.1.1 Why are some traits more plastic than others?

The study of phenotypic plasticity emerged from the study of systematics and local adaptation. It became clear from studies in systematics that distinct-looking plants collected from different locations did not necessarily represent different taxa; plants from seeds of different apparent taxa grown in a common garden looked similar, and plants from seeds of a single species looked very different when grown under different conditions (Turresson 1925; Hall 1932). To make sense of plant diversity, it became necessary to control for environmental effects on plant phenotypes (Gregor 1931; Gregor *et al.* 1936). Taxonomists aimed to identify traits that indicated stable differences among taxa across a broad range of conditions. Much taxonomic work of the early twentieth century was attuned to identifying traits that exhibited such stable differences among taxa and that therefore represented diagnostic differences between them. The classic work of Clausen *et al.* (1940, 1947), which solidified the study of ecological genetics and phenotypic plasticity in plants, was motivated by the intention to determine whether morphologically distinct entities remain morphologically distinct when one controls the effects of the environment in which plants grow naturally.

Deliberate controls of growth conditions exposed the enormous plasticity in plant traits, making plasticity itself a focus of study. Yet some traits are not plastic. Structures such as flowers, for example, are notoriously invariant and are used to distinguish species and higher taxa. Certain aspects of organ identity determination are also very predictable across taxa, such as the placement of reproductive structures along branches. Taxonomic categories are intentionally based on such traits that remain distinct under a variety of environmental conditions. Although many taxonomic categories based on morphology have been revised by genetic data, in several cases morphological distinctions do reflect taxonomic categories. In that sense, understanding why certain traits are not plastic and remain stably different across taxa could contribute to our understanding of the basis of morphological distinctions among defined taxa. Is it actually the case, then, that nonplastic traits differ more across taxa than plastic ones, or is that trend simply the byproduct of an outdated mode of assigning species or a statistical consequence of plastic traits exhibiting more variance within a given taxon? If nonplastic traits are better diagnostic characters of taxonomic differences, does that pattern have biological significance? To evaluate that question requires considering the adaptive and genetic mechanisms whereby some traits evolve to be plastic and others do not.

While many aspects of plant structure exhibit a high degree of phenotypic plasticity, it is just as important to recognize that some aspects do not. Indeed, it is this contrast that requires explanation. In order to understand why some traits are more plastic than others, it is necessary to understand the adaptive value and genetic constraints on the evolution of phenotypic plasticity of particular traits.

10.1.2 Manifestations of phenotypic plasticity in plants

The function that describes how a phenotype of any given genotype changes in response to the environment is called the "reaction norm" (or "norm of reaction"). Plants exhibit plasticity in phenology, physiology, and morphology. These different forms of plasticity can have different effects on subsequent exposure to environmental conditions, have different degrees of reversibility, and occur at different time and spatial scales.

10.1.2.1 Phenology and habitat selection through plasticity

Plasticity in phenology, or the timing of events in a life cycle, is the major determinant of the overall life cycle that a plant exhibits. Phenology determines generation time—the length of time between reproductive episodes and the number of reproductive episodes it completes within a given growing season. It determines the relative timing of developmental events with respect to each other—for example, when reproduction occurs with respect to the size of the plant. And crucially, it determines the conditions during the seasonal cycle that subsequent life stages are exposed to.

Phenology determines life-history expression. For example, many plants are not induced to flower until they experience a prolonged cold period (vernalization). In *Campanulastrum americanum*, the life stage that receives this cold period determines which life history is expressed: if seeds are dispersed early enough in autumn to germinate, the seedlings will experience cold and flower the following spring and become annuals; if seeds are dispersed later and germinate in spring, the plant will not experience that cold treatment until the next winter, so they will be biennials (Galloway 2001). In *Arabidopsis thaliana*, some genotypes do not have this cold requirement in order to flower (Ratcliffe 1965), and such genotypes are sometimes able to exhibit a rapid-cycling life history with multiple generations per year compared to the cold-requiring genotypes that exhibit a winter annual life cycle (Thomspon 1994; Donohue 2009). In this case, these phenological interactions across life stages are genotype-specific.

Adaptive phenology times developmental events such that the life cycle that is expressed is suitable for the seasonal conditions it experiences. For example, seeds frequently remain dormant in hot summer conditions in which seedlings would not survive, germinating only under the more favorable conditions of autumn (Baskin & Baskin 1998). Flowering is frequently postponed until after winter (Chouard 1960), thereby preventing exposure of sensitive reproductive meristems to freezing, coordinating flower display with the presence of pollinators, and providing more favorable temperatures for embryonic development. The matching of life stage to the appropriate season for that stage is critical to the survival and reproduction of plants.

This ability of phenology to "match" life stages to the appropriate season makes obvious one more general consequence of phenology: it determines the environment that the plant experiences following that response. A seed that is capable of germinating beneath a heavy canopy will produce a seedling that has to contend with that canopy in order to survive. Thus, the adaptive value of postgermination traits will depend on the germination behavior of the seed (Donohue *et al.* 2010). In this example, the seedling would need to accommodate the low light levels through changes in physiology or through a pronounced shade-avoidance response via elongation. A seed that germinated in full sun would result in a seedling that required neither, but would instead need very different physiology and morphology appropriate for water conservation and heat stress. In this manner, phenology is a sort of habitat selection, such that the environmental conditions that elicit the developmental transition are those that the following life stage will experience (Donohue 2005). While perhaps most obvious in phenological life-stage transitions, this dynamic of habitat selection applies to other forms of phenotypic plasticity as well. For example, by elongating, a plant increases its light supply; by developing roots in particular directions it increases its water supply to its tissues; and by altering branching patterns, clonal plants are able to forage and increase their nutrient supply (reviewed in Bazzaz 1991).

Plants therefore often respond to environmental conditions in ways that alter their subsequent exposure to them. In this sense, plants can exhibit habitat selection, and they do so through phenotypic plasticity. To understand variation in plant form and behavior, it is necessary to understand not only how plants respond to environmental factors they experience, but also how they came to experience those particular environmental conditions in the first place. In fact, as will be discussed, these two processes of habitat selection and plasticity to that habitat are expected to coevolve to determine plant form, physiology, and behavior.

10.1.2.2 Reversible and irreversible plasticity—physiology and morphology

Plants are capable of responding to environmental conditions extremely rapidly, and these responses are frequently reversible. The reversible closure of stomata is a rapid response to loss of turgor pressure as a consequence of low water availability, for example. Photosynthesis rates respond to light and temperature. Very rapid and reversible plasticity is frequently referred to as "lability" (Bradshaw 1965; Schlichting & Pigliucci 1998).

In contrast, morphological plasticity is far more likely to be irreversible, as a structure once developed is unlikely to be resorbed. Structures may be excised, however, making early morphological plasticity, if not reversible, at least impermanent. Shade leaves made early in life, for example, may drop and be replaced by sun leaves.

The modular structure of plants enables different modules to respond to environmental conditions somewhat independently, depending on the degree of physiological integration across modules. Within-plant variation in the morphology of individual structures is likely the result of independent responses of particular modules to the particular environments they are exposed to (Winn 1996). A familiar example is the presence of shade leaves lower in the canopy of the plant, while sun leaves occur near the top and periphery of the canopy (Rubio de Casas *et al*. 2007). Heterophylly in aquatic plants is another example, with narrow, elongate leaves being present at nodes below the water surface and broad leaves above it (Cook & Johnson 1968). Localized responses to microenvironmental stimuli are therefore possible and common in plants.

Plant architecture can then be considered to be determined by the relations of modular structures to each other. In addition to the particular environment-dependent morphology of a given organ, plant architecture is determined by the environment-dependent determination of which organs develop, and where. An undifferentiated meristem may develop into a leaf, branch, or reproductive organ. Frequently, organ identity is determined by environmental conditions. In many rosette plants, for example, a developing node can produce a leaf that contributes to the rosette, or it can make the irreversible transition to reproduction. Whether another leaf is produced or whether the meristem becomes reproductive is determined by environmental inputs into a genetic pathway that represses flowering. In many plant species, such as *A. thaliana*, cold temperature or long photoperiod act in pathways that repress genes that in turn repress flowering; once these environmental stimuli are perceived, the flowering pathway is activated and the meristem becomes reproductive (Simpson & Dean 2002; Michaels *et al*. 2005; Baurle & Dean 2006; Dennis 2007). Likewise, light cues of low red : far-red can elicit similar inductions of floral developmental pathways. Plants sense a developing vegetative canopy by perceiving the low red : far-red light that results from the canopy (Ballaré *et al*. 1990). Suppression of branching and allocation of meristems to reproduction instead is a classical component of the shade-avoidance response—a major reorganization of plant architecture in response to vegetation shade (Smith & Whitelam 1997). In short, plasticity of organ identity is a major determinant of plant architecture.

The modular organization of plants enables plastic responses to be localized. However, some environmental stimuli elicit more systemic responses. Inducible resistance to herbivores and pathogens can be very localized, causing necrosis of a small number of cells to prevent the spread of the consumer. Alternatively, herbivore and pathogens can also elicit systemic responses that cause phenotypic changes throughout the shoot in the form of the production of secondary compounds, lignin, or trichomes (reviewed in Howe & Jander 2008).

In summary, phenotypic plasticity can be reversible or irreversible. Thus, the temporal scales of environmental change and responses to such changes are relevant to interpreting the functional consequences and adaptive value of plasticity. Phenotypic plasticity can be localized or systemic, and the spatial extent of the plastic response determines the morphological heterogeneity of a given individual. The modular organization of plants enables localized responses to environmental stimuli. In this manner, similar organs on a plant can have a different morphology depending on the conditions in which they developed. Moreover, the identity of organs can be influenced by environmental conditions. Thus, the plant architecture—as determined by the spatial relationship among organs—variation in structure of organs on any given individual, and physiology itself are highly dependent on how genes interact with the environment.

10.2 Why are some traits more plastic than others? The evolution of phenotypic plasticity

If organisms could express the optimum phenotype in each environment they encounter, with no cost or compromise, it would be to their advantage to do so. Organisms do not, however, exhibit unlimited phenotypic plasticity. The evolution of phenotypic plasticity depends on the adaptive value of expressed phenotypes within each environment that is experienced, costs of being able to exhibit plasticity, and constraints on abilities to attain optimal phenotypes in any given environment. To understand why any given trait exhibits the observed degree of phenotypic plasticity—and to determine why some traits are more plastic than others—requires knowing each of those elements that determine the evolution of plasticity of any given trait. More specifically, the adaptive value of phenotypic plasticity depends on environment-dependent selection on phenotypes and the costs of being plastic. Constraints on the evolution of plasticity involve lack of genetic variance for a trait within any given environment and genetic covariance between the traits expressed in different environments that hinder trait evolution, as discussed in Section 10.2.1 (Figure 10.1).

10.2.1 The adaptive value of plasticity: scales of environmental variation

The adaptive value of phenotypic plasticity depends on the scale of environmental variation experienced by the alleles governing phenotypic expression (Figure 10.1, upper). Theoretical treatments of the evolution of plasticity have been formulated with different scales of environmental variation: organisms that experience only one environment during their life, but which environment it experiences is uncertain (coarse-grained—constant, uncertain environment); environmental variation across generations, such that progeny phenotypes are determined by responses to parental environments (parental effects); and within-generation environmental heterogeneity such that any given individual experiences different environments (fine-grained—within-generation).

Adaptive value

Frequency of environments
Strength of selection in each environment
Quality of environment (hard selection)
"Costs" of being plastic

Genetic constraints

Genetic variation for each trait in each environment (diagonals)
Genetic correlations between traits within each environments (dark grey)
Genetic correlations between traits expressed in different environments (light green))

	$Trait_1Env_1$	$Trait_2Env_1$	$Trait_1Env_2$	$Trait_2Env_2$
$Trait_1Env_1$	$T_1E_1^2$	(T_2E_1,T_1E_1)	(T_1E_2,T_1E_1)	(T_2E_2,T_1E_1)
$Trait_2Env_1$	(T_1E_1,T_2E_1)	$T_2E_1^2$	(T_1E_2,T_2E_1)	(T_2E_2,T_2E_1)
$Trait_1Env_2$	(T_1E_1,T_1E_2)	(T_2E_1,T_1E_2)	$T_1E_2^2$	(T_2E_2,T_1E_2)
$Trait_2Env_2$	(T_1E_1,T_2E_2)	(T_2E_1,T_2E_2)	(T_1E_2,T_2E_2)	$T_2E_2^2$

Figure 10.1 Summary of factors influencing the adaptive value of plasticity (upper) and genetic constraints on the evolution of plasticity (lower). The genetic matrix shows the genetic correlations that influence the evolution of plasticity for two traits in two environments. Diagonal elements are genetic variances for each trait in each environment. Dark grey elements are genetic covariances between traits within a given environment. Light grey elements are genetic covariances between the same and different traits in different environments.

10.2.1.1 Constant, uncertain environment

The original models of the evolution of phenotypic plasticity were developed to model a situation in which individuals of a given genotype are distributed across a heterogeneous environment and experience each environment with a given probability (Falconer 1952, 1990; Via & Lande 1985). The optimum phenotype was considered to differ in the different environments. Barring any constraints, the optimum level of plasticity would be determined by the optimum phenotype within each environment; the organism should always express the optimum phenotype in any given environment.

Constraints do occur, however, and organisms usually are not able to express the optimum phenotype in all the environments they experience because of physiological and genetic constraints. Which phenotype or set of environment-dependent phenotypes they should express depends on the nature of those constraints and on the nature of the environments they experience, specifically the following (Figure 10.1, upper): the probability of experiencing each environment; the strength of natural selection on the phenotypes in each environment; and additional costs of the ability to alter one's phenotype that are independent of selection on the expressed phenotypes per se (Via & Lande 1985; Van Tienderen 1991).

If a genotype experiences only one environment, there is no adaptive advantage to phenotypic plasticity, and the constraints and costs of plasticity would favor a specialist to that one environment. Likewise, if a genotype experiences one environment much more frequently than another, all else being equal, its evolved phenotype should be closer to the optimum in the more frequent environment, presuming it cannot attain the optimum in both. If, however, natural selection is much stronger in one environment than another, all else being equal, the evolved phenotype would more quickly attain the optimum in the environment with stronger selection. The outcome, however, also depends on whether natural selection is hard or soft: if one environment is so harsh that even highly adapted individuals produce very few offspring, whereas a small phenotypic adjustment might increase fitness greatly in a less harsh environment, then phenotypes would evolve toward the optimum in the less harsh environment, in which they can attain a proportionally higher relative fitness advantage. The overall outcome therefore depends on the balance between the frequency of the different environments that are experienced, the strength of selection in each environment, and the quality of each environment (Figure 10.1, upper; Via & Lande 1985; Van Tienderen 1991; Gavrilets & Scheiner 1993; Scheiner 1993; DeWitt & Scheiner 2004).

If a cost of being plastic exists, a plastic individual will have lower fitness than a nonplastic individual even if it expresses the exact same phenotype in any given environment (Van Tienderen 1991). Such costs may be caused by the requirement to maintain sensory systems and signal transduction mechanisms that would not be maintained in nonplastic individuals who do not respond to environmental conditions (DeWitt *et al.* 1998). The magnitude of that cost influences the adaptive value of plasticity directly; if large, it can favor a nonplastic specialist to one environment, or a nonplastic generalist with an intermediate suboptimal phenotype, depending on the frequency of environments experienced and the strength of selection in each environment as discussed previously.

In summary, the frequency of environments experienced by a genotype, the strength of selection within and quality of each environment, and intrinsic costs of being plastic determine the adaptive value of phenotypic plasticity and the optimal norm of reaction (Figure 10.1, upper).

10.2.1.2 Across-generation plasticity—parental effects

Parental environmental effects are associated with cross-generational plasticity whereby the phenotype of offspring depends on the environment experienced by the parents (Roach & Wulff 1987; Kirkpatrick & Lande 1989; Mousseau & Fox 1998; Räsänen & Kruuk 2007). The adaptive value of such plasticity therefore depends on how accurately the environment that induces the parental effect predicts the environment experienced by progeny (Donohue & Schmitt 1998). If the environment remains the same, then the parental effect is more likely to be adaptive. For example, some plants produce larger seeds in the shade than they do in the sun. This parental effect would be adaptive if larger seeds perform better in the shade, and if those seeds actually experience shady conditions. If those seeds are just as likely to experience sunny conditions, then the parental effect would not necessarily be adaptive. Whether an adaptive phenotype is produced by the parent therefore depends on whether the environment that produces the phenotype persists long enough that the individual that carries that phenotype actually experiences that environment. Likewise, if the environment is not the same across generations, but it is nevertheless predictable from that experienced by the parent, then parental effects can be adaptive.

Parental effects differ from standard phenotypic plasticity in that its adaptive value depends not only on selection on the progeny phenotype in the progeny environment but also selection on parents in parental environments (Falconer 1965; Kirkpatrick & Lande 1989). In this sense, the total adaptive value of parental effects will be determined by environment-dependent selection in both generations.

10.2.1.3 Within-generation environmental variation

The predictability of the environment between generations is a crucial determinant of the adaptive value of parental effects; this is because the production of adaptive phenotypes depends on the rate at which the environment changes relative to the time necessary to produce a phenotype as a response to the environment. The same is true for temporal variation in environments experienced within a single generation by any given individual. The original formulations of the evolution of plasticity, which describe plasticity to environments that remain constant within a given generation but which differ across generations, do not directly address the issue of the pace of environmental change compared to the pace of responding to those changes. When environments change within the lifetime of a single individual, the adaptive value of plasticity depends on how quickly phenotypes can be produced that are favored in the environments that produce them; if a plant inhabiting a sunny environment begins to produce sun leaves, but by the time the leaves are produced the environment has grown shady, then responding to the sunny conditions would not be adaptive. Thus lag times of the production of phenotypes become important determinants of fitness in environments with temporal variation within a given generation (DeWitt & Scheiner 2004; Engelmann & Schlichting 2005; Gabriel *et al.* 2005).

Equally important is the correspondence between the environmental factor that imposes selection and the environmental factor that induces phenotypic plasticity. Plasticity is far more likely to be adaptive when these two environmental factors are identical, and it will be less likely to be adaptive as the eliciting cue becomes more unreliable as a predictor of natural selection. This is the case for environmental variation at all scales.

10.2.1.4 Determinants of environmental variation—plasticity and habitat choice

The degree and scale of environmental variation experienced by an organism fundamentally influences the adaptive value of phenotypic plasticity. What determines the variability of environments experienced by organisms? Obviously, extrinsic environmental factors can vary independently of the organisms that inhabit them. However, organisms can mediate the degree of environmental variation they expose themselves to by modifications of their behavior (Bazzaz 1991; Donohue 2003). Such behaviors then influence the adaptive value of plasticity because they determine how variable or constant is the environment that organisms experience.

Habitat selection frequently buffers environmental variation, such that organisms that inhabit an extrinsically variable environment can nonetheless experience environmental constancy. As such, habitat choice decreases the adaptive value of plasticity by alleviating selection imposed by environments other than the chosen environment (Whitlock 1996). As discussed in Section 10.1.2.1, plants have many behaviors of phenology and developmental morphology that act in the same manner as habitat choice, in that they determine the environment experienced by the plant subsequently. Therefore, the adaptive value of plasticity depends on the extent to which plants mediate the degree of environmental variation they experience via such behaviors.

It should be noted that the adaptive value of plasticity depends on the degree of variation in the environmental factor that is responsible for natural selection—not on environmental variation in any general sense. Different structures on plants are likely to be targets of selection by different environmental factors. The adaptive value of leaf shape, for example, might depend on light availability whereas the adaptive value of flower shape depends on pollinator availability. If one factor varies and the other does not, then plasticity would be adaptive only in the trait that experiences variable natural selection.

Particular organs might also experience different degrees of environmental variation with respect to specific environmental factors that exert selection. For instance, leaves, which are distributed throughout the plant, may experience great heterogeneity of temperature conditions as a consequence. In contrast, flowers that are always positioned near the apex and which are enclosed in tissue with thermoregulatory properties (via pigments; Stiles *et al.* 2007; Lacey *et al.* 2010) may experience much more constant temperature conditions.

The adaptive value of plasticity is therefore likely to differ for different traits. First, intrinsic costs of being plastic might differ across traits, especially if different traits require different sensory systems and transduction pathways in order to express plasticity. Even if those costs are similar, the adaptive value of plasticity is likely to differ across traits because different environmental factors—with different degrees of variation—impose selection on different traits. Moreover, different traits may experience different degrees of environmental variation even of the same environmental factor. The degree of variation in selection experienced by different traits is determined not only by extrinsic factors but also by properties of the plants that moderate the degree of environmental variation that is actually experienced. To explain why some traits are more plastic than others, one needs to identify environmental factors that impose selection on the trait and characterize the degree to which selection on that trait varies—given that organisms themselves modify their exposure to environmental variation.

10.2.2 Genetic constraints on the evolution of plasticity

If no constraints or costs of plasticity occur, the optimum level of plasticity would be determined by the optimum phenotype within each environment. The challenge comes when genetic correlations among traits restrict the ability to evolve those optimum phenotypes (Figure 10.1, lower). As with any set of correlated characters under selection, selection on one trait can constrain adaptive evolution of correlated traits. In the case of phenotypic plasticity, one must consider not only genetic variation for each trait (Figure 10.1, lower; diagonals) and correlations among multiple traits expressed in any given environment (Figure 10.1, lower, dark grey), as in standard quantitative-genetic models of trait evolution, but also the correlations among traits expressed in different environments (Figure 10.1, lower, light grey; Falconer 1952; 1965; Via & Lande 1985; Van Tienderen & Koelwijn 1994).

For example, if genotypes that are the tallest in one environment are also the tallest in the other environment, then height is positively genetically correlated across environments (e.g., T_1E_1 and T_1E_2 in Figure 10.1 are positively correlated). If selection varies across environments, favoring tall individuals in the first environment but short in the second, tall genotypes would have high relative fitness in the first environment but low relative fitness in the second.

With such across-environment genetic correlations, the question then becomes whether a specialist genotype evolves (which expresses a phenotype that is optimal in one environment but maladaptive in the other), a plastic generalist evolves (which expresses suboptimal but different phenotypes in each environment), or a nonplastic generalist evolves (which expresses a constant but suboptimal phenotype in both environments; Bradshaw 1965; Via & Lande 1985; Whitlock 1996). The outcome depends on how adaptive plasticity is, as discussed previously, and on the strength of the genetic correlations (within and) across environments.

10.3 The genetic basis of phenotypic plasticity and genetic constraints on plasticity

Just as the adaptive value of plasticity is likely to differ for different traits, the genetic constraints on plasticity are also bound to differ across traits. The genetic basis of environment-specific expression of particular traits therefore fundamentally influences the evolution of trait-specific plasticity.

Not all plant traits are plastic enough to express optimal phenotypes in all the environments they experience. Two explanations are (1) they cannot express the trait quickly enough to match the environment before it changes again and (2) genetic correlations across environments prevent optimal phenotypes from being attained in all environments. The second explanation pertains to all temporal scales of phenotypic plasticity, both within and across generations. Genetic correlations across environments, as with any genetic correlation, can be caused by pleiotropy, close physical linkage, or linkage disequilibrium, with pleiotropy being the most persistent constraint. Thus understanding persistent genetic constraints on the evolution of plasticity requires understanding pleiotropy across environments. To understand the constraints of pleiotropy in particular requires knowledge of the genetic basis of phenotypic plasticity: which genes and pathways are involved in environment-dependent phenotypic expression.

10.3.1 Molecular mechanisms of phenotypic plasticity: gene–environment interactions

Environment-dependent phenotypes can be caused by environment-dependent gene expression, environment-dependent gene function, or both (West-Eberhardt 2003; Bossdorf *et al.* 2008). If certain genes are transcribed and translated only in some environmental conditions, or if gene products are stable or bioactive only under some conditions, then particular features will be manifest only under some conditions, leading to environment-dependent phenotypes. QTL studies conducted under multiple environments have clearly shown that genes associated with trait variation differ across environments, indicating some independent genetic basis of trait expression in different environments (e.g., Dubcovsky *et al.* 1994; Ungerer *et al.* 2003; Lacaze *et al.* 2009; reviewed in Mitchell-Olds 1995). Such QTLs could have a number of different molecular-genetic mechanisms responsible for these environment-dependent genetic associations.

10.3.1.1 Environment-dependent gene expression via transcriptional regulation

Environment-dependent gene expression is ubiquitous. Genome-wide expression studies have documented that entire genetic pathways are variously expressed according to the environment that the organism was exposed to. Thus, environment-dependent gene expression is without doubt fundamental to phenotypic plasticity (reviewed in Aubin-Horth & Renn 2009), and can be regulated at the level of transcription or post transcription. There are numerous mechanisms that can regulate gene expression in response to the environment, and here we highlight a few as illustrative examples.

Transcriptional regulation occurs via the recruitment of specific transcription factors and through chromatin remodeling (Russo *et al.* 1996; Jaenisch & Bird 2003; Gendrel & Colot 2005; Grant-Downton & Dickinson 2005; Berger 2007; Mattick *et al.* 2009). Chromatin remodeling can determine how accessible DNA regions are to transcription. One example is via methylation, which alters chromatin structure, and is associated with transcriptional regulation (Bender 2004). Methylation typically converts cytosine to 5-methylcytosine, and when methyl groups are added to nucleotides at the regulatory regions of genes, those genes are less transcriptionally active; the more highly methylated those regions are, the less transcriptionally active is the gene.

Chromatin remodeling also occurs through posttranslational changes in the histones (Grant-Downton & Dickinson 2005; Berger 2007). As the structure of the histone molecules are altered, the manner in which the specific DNA region it associates with is also altered, changing the accessibility of that particular sequence to transcription. Histones are modified in a variety of ways, including acetylation, phosphorylation, methylation, ubiquitylation, sumoylation, and isomerization. Certain modifications have been thought to be predictably associated with certain outcomes: for example, acetylation of lysine is frequently associated with gene activation and deacetylation with the repression of gene expression. However, more recently it has been proposed that it is the combinations of modifications that determine specific responses in gene expression (Berger 2007).

Chromatin remodeling has been shown to be involved in phenotypic plasticity. One especially well-characterized example is cold (vernalization)-dependent flowering regulation (Simpson & Dean 2002; Henderson & Dean 2004; Michaels *et al.* 2005; Shindo *et al.* 2006, Sung & Amasino 2006). The transition from vegetative to reproductive development is regulated by genes whose expression depends on whether plants have received a period of cold. In *A. thaliana, Flowering Locus C* (*FLC*) represses genes that initiate the transition to reproduction. *FLC* expression is impeded by a cold treatment. Genes associated with chromatin remodeling via histone modifications have been implicated. During a cold treatment, products of genes in the *VRN* and *VIN* families are recruited to the *FLC* site, and resulting histone deacetylation and DNA methylation causes the stable repression of *FLC*, which enables flowering (Chandler *et al.* 1996; Michaels & Amasino 1999; Bastow *et al.* 2004; Sheldon *et al.* 2008). In this manner, reproductive development occurs in response to a prolonged cold temperature.

In *A. thaliana, FLC* repression is irreversible, and this leads to a winter-annual life history in which any developing meristems produced after vernalization can be reproductive. In contrast, in the perennial *Arabis alpina*, the homolog of *FLC* (*PEP1*) is reversibly repressed (Wang *et al.* 2009): therefore, meristems produced while *FLC* is repressed become reproductive, whereas meristems produced after *FLC* repression has been reversed can remain non-reproductive, and it is these meristems that permit reproduction the following year. Hence, the cyclical repression of *FLC* enables a perennial as opposed to annual life cycle. This case demonstrates that transcriptional regulation via chromatin remodeling can be environment-dependent, and that it can have various degrees of persistence.

The persistence of chromatin remodeling is a topic of intense research at present. Alteration of chromatin can be inherited to various degrees across mitotic replication events. When this occurs, gene expression changes can be transmitted to daughter cell lineages even if the original initiating environmental trigger of gene-expression change is no longer present. Methylation patterns, for example, are sometimes stably transmitted across mitotic replication events, and they can even be inherited across meiotic replication—that is, from parents to offspring, as occurs in imprinting. Histone modifications can also be transmitted across mitotic replication events. Moreover, both methylation and histone modifications are reversible (Takeda & Paszkowski 2006). The mechanisms that regulate chromatin remodeling, the stability of its inheritance across mitotic (or meiotic) replication, and its reversibility are not yet known.

The persistence of chromatin remodeling has important implications for phenotypic plasticity. First, the persistence of changes in gene expression across mitotic divisions without further environmental inputs could determine the difference between irreversible versus modulated plasticity. For example, the transition from vegetative to reproductive development may require only a single environmental triggering event (e.g., vernalization) that alters the developmental trajectory irreversibly through stable changes in gene expression. In contrast, during shade avoidance, each node is able to respond to light conditions throughout development; a pulse of vegetation shade early in development does not induce an endlessly elongating plant even after it has escaped overhead shade (Ballaré *et al.* 1990). The regulation of these different sorts of responses may depend on the mechanisms of the stability of changes in gene expression.

Second, how persistent are changes in gene expression via chromatin remodeling across different environmental conditions? Different environments can alter this persistence, both by regulating the stability of chromatin remodeling across mitotic divisions or by regulating the reversal of chromatin changes. The role of the environment in each of these processes is not known. Nonetheless, if chromatin modifications persist across many environmental conditions, plasticity could be impeded. Thus the environment-dependence of chromatin changes is of direct relevance to phenotypic plasticity, but its prevalence or its mechanisms are as yet unknown.

10.3.1.2 Environment-dependent gene expression via posttranscriptional, pretranslational regulation

After a gene is transcribed, the production of the corresponding gene product depends on the processing and translation of the mRNA. RNA can be processed in ways that affect the RNA sequence that is transcribed as well as the probability that it is transcribed (West-Eberhard 2003).

Splicing of exons determines the RNA sequence that is transcribed, and many genes have different splicing variants. Alternative splicing results in different gene products, depending on how exons are joined together. Exons can be dropped completely (exon skipping), splicing can occur at different locations within a given exon, or one of two alternate exons is retained in a given RNA sequence. Splicing is regulated by the interaction between protein complexes (repressors or activators of splicing) and specific binding sequences on the RNA that prevent splicing ("silencers") or promote it ("enhancers"). Whether alternative splicing is an important mechanism for phenotypic plasticity is still unclear as mechanisms and propensity for environment-dependent alternative splicing are not yet characterized.

RNA stability or degradation is a major determinant of whether mRNA becomes transcribed and how much protein is produced from a given mRNA (Russo *et al.* 1996; Lorkovic 2009). The more stable an mRNA is, the more it can be translated and the more protein is produced from it. mRNA is variously stabilized by poly-adenylation of the 3′ end to protect against 3′ exonucleases, and capping of the 5′ end protects mRNA from 5′ exonucleases and also regulates nuclear export. Degradation of mRNA occurs when either the 3′ poly(A)-tail or the 5′ cap is disrupted. mRNA can exhibit various degrees of stability, in part via variation in the length and protection of the poly(A)-tail.

mRNA can be actively degraded via interactions with other small or microRNA (miRNA) sequences (Matzke *et al.* 2009). Small-interfering RNAs are double-stranded RNAs and form a complex with an endonuclease that cleaves RNA. The interfering RNAs bind to complementary regions on the mRNA, enabling cleavage and subsequent degradation of the mRNA strand. miRNAs are single-stranded RNAs that are small, complementary sequences that bind to mRNA. This binding prevents translation and also accelerates removal of the poly(A)-tail, thereby accelerating mRNA degradation. miRNAs have also been implicated in the regulation of transcription (Grant-Downton & Dickinson 2005; Vauchere 2006; Kanazawa 2008). These interfering RNAs are transcribed from DNA sequences, and therefore their production might be regulated by various mechanisms of transcriptional control, as discussed previously. As such, their production can be environment-dependent and contribute to phenotypic plasticity.

For mRNA to be transcribed, it needs to be transported from the nucleus to ribosomes. This process is another checkpoint that determines whether a gene that is transcribed will be translated into a protein product. Binding to the ribosome itself is yet another point of regulation, and it is influenced by the mRNA sequence itself, antisense RNA (interference) binding, or the binding of other proteins.

Posttranscriptional regulation therefore includes alternative splicing, which determines the mRNA sequence itself, stability of the mRNA strand, which can be actively degraded in a regulated manner, transport of the mRNA to the ribosome, and binding to the ribosome. Some of these processes depend on the production of other gene products, and as such have the potential to be regulated in an environment-dependent manner and therefore involved in phenotypic plasticity. For most of them, however, their environment-dependence has not been well elucidated, especially in regard to ecologically relevant environmental variation.

10.3.1.3 Environment-dependent gene function

One may be left with the impression that environment-dependent gene expression accounts for phenotypic plasticity. Yet, at some point, some gene products must respond directly to environmental factors to initiate changes in gene expression. If particular transcription factors or histones regulate the expression of signal transduction pathways, how does the environment—the signal itself—regulate the activity of those initiating molecules? How also might the environment alter the activity of other gene products in the transduction pathway?

In the case of vernalization, discussed previously, histones are deacetylated under cold conditions, enabling them to repress *FLC* expression. The precise mechanism whereby cold accomplishes this is not known. Cold also acts as a stimulus for other plant developmental responses, such as germination. As with *FLC*, key repressors of germination (e.g., *SPATULA*) are repressed after a cold treatment (Oh *et al.* 2004; Penfield *et al.* 2005), but it is not yet known whether this occurs through histone modification.

As a more general response to cold, kinetic activity of many enzymes is affected by temperature. Allelic variants of metabolic enzymes differ in kinetic activity (binding affinities affecting Km) at low temperature as well as heat stability at high temperature, and the efficiency with which particular reactions occur would be a function of temperature. This could result in a continuous modulation of phenotype, or if the downstream response requires a threshold level of substrate, it could result in a discontinuous switch of phenotype (Pigliucci 2001).

Other facets of enzymatic reactions are also environmentally dependent. One mechanism may be associated with the presence of a substrate. If this substrate is supplied by the external environment, then the reaction will be initiated only under conditions in which the substrate is present, and the resulting phenotype will be environment-dependent. For example, the production of some coloration requires the uptake of certain elements from the environment; if the elements are not available, the coloration of the organism will be affected. Immune reactions are similar in that they are initiated only when the elicitor is present (reviewed in Howe & Jander 2008). Elicitors, such as insect herbivores or pathogens, can initiate complex reactions such as the production of secondary compounds, trichomes or hairs, or other modifications of leaf form.

Finally, certain molecules act as direct environmental sensors. The phytochrome photoreceptors offer a well-studied example and will be discussed in some detail here (Figure 10.2). Phytochrome genes exist as a small gene family (Mathews & Sharrock 1997), and they regulate diverse plant responses,

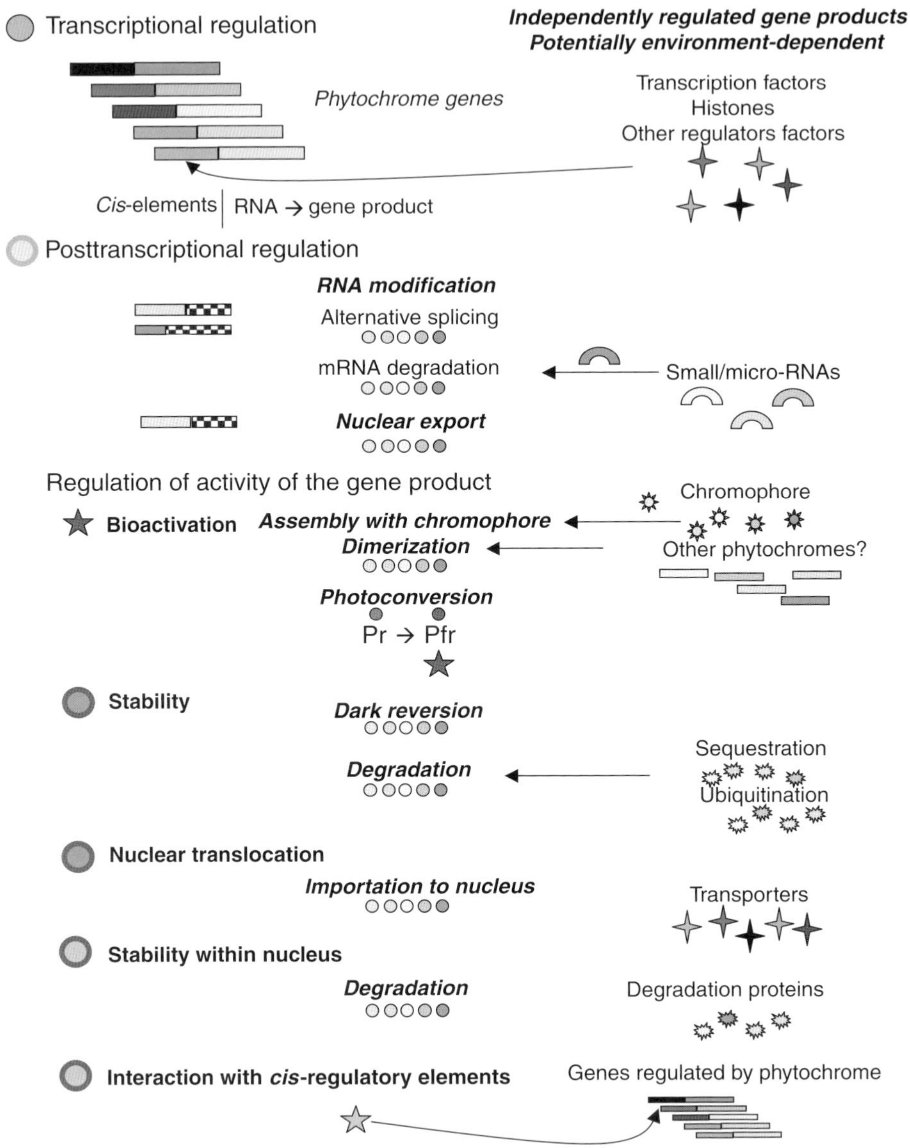

Figure 10.2 Phytochrome-regulated phenotypic expression. Phytochrome genes, arisen from gene duplication events, differ in their environment-dependent expression and environment-dependent gene products. In the simplest case, different environments, represented by different colors, enable the expression of particular phytochrome genes (e.g., dark green), and additional environmental conditions enable the activity of the gene product (e.g., light green). In this manner, different phytochromes can elicit a phenotype only under certain combinations or sequences of environments. The gene expression component actually consists of transcriptional regulation and posttranscriptional regulation.

including germination, de-etiolation, shade-avoidance responses, and flowering initiation (Casal *et al.* 2003). Phytochromes exist in two isomeric forms: a red-absorbing form (Pr) and a far-red-absorbing form (Pfr). The two are interconvertible, such that the Pfr is converted to the Pr when it is exposed to red light, and vice versa (Braslavsky *et al.* 1997, Kendrick & Spruit 1997). The two forms exist in a dynamic photoequilibrium, and the ratio of the two forms depends on the ratio of ambient red : far-red light. This ratio is an accurate indicator of competition for light, since competing leaves absorb red light and lower the red : far-red ratio (Ballaré *et al.* 1990). The Pfr is the bioactive form that is transported to the nucleus, initiating the signal transduction.

Once a "bioactive" molecule (Pfr in the case of phytochromes) is produced in response to an environmental cue and initiates signal transduction, several points of regulation involving that gene product can subsequently occur. The phytochromes again illustrate the complexity of this process (Moller *et al.* 2002; Nagy & Schafter 2002; Schafer & Bowler 2002). After the phytochrome apoprotein has been transcribed, it is assembled with a chromophore, which enables it to respond to light. Active phytochrome is in the form of dimers (Hennig & Schafter 2001; Sharrock & Clack 2004). Heterodimers have been observed in a yeast two-hybrid system, so it is possible that some phytochrome function may require the expression of more than one phytochrome gene, although this remains unknown.

Once the phytochrome has been converted to the bioactive far-red absorbing isomer, it must remain stable, and its stability and activity depends on light and temperature conditions. Phytochromes revert from the bioactive Pfr to the inactive Pr form in the dark, and the rate of this reversion depends on

Figure 10.2 (*Continued*) Various steps of gene expression can be environment-dependent and can depend on other gene products whose expression and activity might be environment-dependent. Regulation of the activity of the gene product also comprises many steps, which can be environment-dependent and can depend on other gene products whose expression and activity are environment-dependent. Colored dots beneath a process indicate that the process may be environment-dependent. In the hypothetical example of phytochromes, transcription occurs and mRNA is transcribed under "dark green" conditions. Hypothetical degrading microRNAs are formed under all but the "light-green" conditions, and alternate splice types that are not formed under light-green conditions are not formed, leaving only "light-green" stable mRNAs available for translation. That gene product combines with other gene products (the chromophore and perhaps other phytochrome apoproteins for heterodimers). Bioactivation of the assembled gene product occurs under "red" conditions, and that gene product is stable and translocated to the nucleus only under "gray" (plus "red") conditions. Within the nucleus, the bioactive gene product is stable only under "purple" conditions. It interacts with genes that are capable of being transcribed under "purple" conditions. These genes continue the signal transduction pathway. In this manner, biological processes occur only under particular combinations and sequences of environmental conditions. The sequence of the environments under which the process takes place is indicated by the colored symbols on the left. (For a color version of this figure, see Plate 10.1.)

temperature (Hennig *et al.* 1999; Eichenberg *et al.* 2000). Moreover, the degradation of phytochrome protein occurs in light, the rate depending on light wavelength and fluence rate (Higgs & Colbert 1994; Clough & Vierstra 1997; Hennig *et al.* 2000). In light, phytochrome protein is first aggregated and then degraded through ubiquitination, both processes being highly regulated.

For phytochrome signal transduction to proceed, bioactive Pfr phytochrome must be translocated into the nucleus, and this occurs in a light-dependent manner (Kircher *et al.* 1999; Nagy *et al.* 2001). Retention- and G-proteins, which bind phytochromes within the cytosol, appear to be involved in this process. Phytochromes also have sites for phosphorylation, and phosphorylated phytochromes are less active; phosphorylation, which occurs in a light-dependent manner, could impede nuclear translocation or its binding with DNA or other proteins (Yeh & Lagarias 1998; Hennig *et al.* 2000). Nuclear translocation has also been shown to be dependent on specific gene products (Huq *et al.* 2003; Genoud *et al.* 2008; Chen *et al.* 2010). Once in the nucleus, phytochrome can also be degraded, and phosphorylation may play a role in tagging phytochrome for degradation. In addition, phytochrome-degrading factors, such as *COP1*, may be excluded from the nucleus in a light-dependent manner, providing another point of regulation (Moller *et al.* 2002).

The example of the phytochromes illustrates the various points of regulation that can occur even at the protein level, after translation has been completed. In this case, the environment directly converts the protein into a particular bioactive isomeric form. However, the persistence of that bioactive isomer, the stability or degradation of the bioactive protein in the cytosol, the translocation of the protein to the nucleus, and the degradation of the protein within the nucleus all determine whether the initial signal—the conversion into a bioactive form by light—will result in changes in gene expression. This entire process is regulated by other proteins, whose allelic sensitivities and synthesis via gene expression are likely to be comparably complex and even environment-dependent.

10.3.1.4 Gene expression and bioactivity of gene products combined

Environment-dependent phenotypic expression is the result of environment-dependent gene expression and environment-dependent bioactivity of gene products. It is the combination of both processes that enables a particular outcome, and for that reason, the outcome can occur only under a particular combination of environmental conditions that permit both processes to occur (Figure 10.2).

The precise environmental regulation of biological processes requires the combination of environment-dependent gene expression and environment-dependent bioactivity of gene products. One mechanism that achieves this is gene duplication. If duplicated genes are all involved in a similar process, then gene duplication creates different versions of genes, each of which is independently regulated and each of which produces a gene product with potentially different environment-dependent bioactivity. As such, gene duplication offers a mechanism whereby a particular pathway is expressed only under particular combinations or sequences of environmental stimuli: the first enabling gene expression and the following enabling activity of the gene product. In this manner a process might proceed, for example, under short, cool days or long, warm days, but not under long, cool days.

Again, the phytochromes offer an example. The small gene family of phytochromes has arisen through a series of gene duplication events (Mathews & Sharrock. 1997). All phytochromes absorb red and far-red light, but they differ in other aspects, and as a consequence have diversified in their function. It has been shown that different phytochromes contribute to the same developmental processes, but under different conditions. For example, some phytochromes enable flowering and germination under warm conditions and others enable the same process under cool conditions (Halliday & Whitelam 2003; Halliday *et al.* 2003; Heschel *et al.* 2008). Whether other preconditions are required for the contribution of specific phytochromes has not been thoroughly tested. Because different phytochromes contribute to the same process in an environment-dependent manner, each phytochrome might promote the process under different sets of environmental conditions. First, phytochromes differ in the conditions or tissues in which they are expressed (Somers & Quail 1995; Goosey *et al.* 1997; Sharrock & Clack 2002). Phytochrome mRNAs also differ in the degree to which they are stable in light (Hennig *et al.* 2000). Thus, both transcription and posttranscriptional regulation via mRNA degradation regulate the production of phytochrome protein, and they do so differently for different phytochromes. Phytochromes also differ in the rate of dark reversion and in their nuclear translocation (Hennig *et al.* 1999; Kircher *et al.* 1999; Eichenberg *et al.* 2000), with PhyA, for example, being transported in red or far-red light, and PhyB being transported only in red light. Once in the nucleus, different phytochromes are likely to have different affinities for other regulatory compounds. Thus, both environment-dependent gene expression and environment-dependent protein activity differ among the phytochromes.

Gene duplication offers a clear scenario in which a particular process might proceed only under certain combinations of environmental cues that enable first the transcription, then the translation, and then the activity of the gene product. However, the numerous checkpoints of signal transduction via interactions among gene products, each of whose production was potentially regulated by environmental factors, offers the same sort of combinatorial regulation of biological processes (Figure 10.2).

10.3.2 How does the molecular mechanism of plasticity translate to genetic constraints on plasticity?

The evolution of phenotypic plasticity depends not simply on what the molecular mechanism of plasticity is, but more specifically on how those genes, their variation, and their interactions determine the magnitude of genetic correlations across environments. If across-environment correlations are the major constraint on the evolution of plasticity, then identifying causes of those correlations is central to understanding variation in plasticity across taxa and among traits within taxa. This requires interpreting molecular-genetic mechanisms in terms of quantitative-genetic correlations.

Plasticity can evolve when genetic correlations across environments are weak (Figure 10.1, lower). Weak genetic correlations across environments occur under three scenarios. The first is when genetic variance for the trait is absent in both environments; in this trivial case the trait cannot evolve in either environment. The second is when genetic variation for the trait is absent in one environment but not the other; in this case, depletion of genetic variation expressed in one environment does not constrain evolution in the other environment, and the phenotype can evolve to be different in each environment. The third is when genetic variation for the trait is expressed in both environments, but different genotypes respond differently to the environments—there is genetic variation for plasticity itself such that the rank order of genotypes changes across environments. Weak genetic correlations indicate that the traits have independent genetic bases. While genetic correlations can be due to linkage disequilibrium, close physical linkage, or pleiotropy, the latter is the most persistent constraint on evolution in the long term. Weak genetic correlations can occur only when pleiotropy is not strong. Therefore, it is of interest to ask whether certain molecular mechanisms of phenotypic expression are more likely than others to exhibit across-environment pleiotropy.

One question is whether genes with particular properties are more likely to influence a trait under all environmental conditions and are others more likely to influence a trait only under some conditions? Do those different categories of genes differ in particular molecular characteristics? For example, do members of one group tend to be transcription factors, histones, miRNAs? Do they have characteristic cis-regulatory regions, enabling the gene to be expressed in only certain environmental conditions? Do they have distinctive environmental sensitivities with respect to kinetic activity or stability? Are they in specific positions of genetic pathways, for example, upstream, close to environmental inputs, or far downstream?

Likewise, what aspects of genetic pathway structure itself promote pleiotropy across environments? If a given trait is produced from a single pathway, is it more likely to be expressed in an identical manner in all environments than a trait that could be produced by several different pathways—each of which might be activated by different environmental conditions? Does the branching architecture of the pathway influence the constancy of trait expression in different environments? Thus, the intense interest in the molecular-genetic basis of pleiotropy (Hahn *et al.* 2004; Promislow 2005; Hansen 2006; He & Zhang 2006) is highly relevant to the evolution of phenotypic plasticity.

Finally, since evolution requires genetic variation, it is relevant to ask which genes in these pathways exhibit natural variation and can therefore contribute to the evolution of environment-dependent phenotypes. Are variable genes in pathway positions that would affect the expression of the trait in all environments, or does variation occur in pathway positions that would affect phenotypic expression only in one environment?

This last question depends not only on where genes controlling environment-specific phenotypes are distributed in genetic pathways but also on where genetic variation is distributed in genetic pathways (Wagner 2000, 2008). Some recent research has addressed this second issue. Genes with greater pathway connectivity directly interact with more genes than those with less connectivity, so a change in those genes would affect more pathways. Such highly connected genes, therefore, are expected to be more adaptively constrained and thus less likely to harbor genetic variation. Genes in upstream and branch-point positions in pathways tend to have higher connectivity than genes downstream in a pathway, so such genes are expected to be less likely to be variable within natural populations and more likely to exhibit evidence of positive or purifying natural selection. Some evidence for such patterns exist: for example, an upstream gene in the anthocyanin pathway in *Ipomoea spp.* showed a signature of past selection whereas downstream genes did not (Rausher *et al.* 2008). However, functionally significant variation has been detected in the major flowering pathways in *A. thaliana*, even in major regulatory and integrator positions (Johanson *et al.* 2000; Caicedo *et al.* 2004; Le Corre 2005; Schwartz *et al.* 2009). It is too soon to characterize general patterns of the distribution of genetic variation in genetic pathways, but such information will be important for characterizing genetic constraints on evolution in general and the evolution of plasticity in particular.

In short, understanding constraints on the evolution of phenotypic plasticity requires understanding mechanisms of pleiotropy across environments. Across-environment pleiotropy may be influenced by gene expression regulation, environmental sensitivities of gene products, and genetic pathway structure and redundancy. In addition, characterizing genetic variation of genes in different pathway positions is necessary, and then determining whether these positions are also associated with environment-specific as opposed to environmentally constant trait expression. The environmental constancy of some traits compared to others may be due to a genetic architecture that is less conducive to environment-specific expression as much as a lack of adaptivity of plasticity.

10.4 Phenotypic plasticity and local adaptation

Phenotypic plasticity has been implicated in adaptation to novel environments. It has even been described as an intermediate step in the evolution of specialists to novel environments, contributing to overall biodiversity. The

Plastic traits (avoided for taxonomic classification)	Nonplastic traits (used for taxonomic classification)
Evolutionary prerequisites	***Evolutionary prerequisites***
Variable environments *(extrinsic variation, variation induced by development/behavior)*	Stable environments *(stable extrinsic environment habitat selection, homeostatic mechanisms)*
Low costs of plasticity	High costs of plasticity
Weak pleiotropy across environments	Strong pleiotropy across environments
Ecological implications	***Ecological implications***
[Contribute to generalist strategy OR Adaptation to temporally variable but spatially stable environments]	Contribute to specialist strategy
Wider ecological niche	Narrower ecological niche/local adaptation
Wider geographic range	Narrower geographic range/local adaptation
[Weak ecological isolation OR Adaptation to new environments and subsequent isolation (genetic assimilation)]	Contribute to ecological isolation Contribute to reproductive isolation?

Figure 10.3 Summary of evolutionary and ecological consequences of plastic and nonplastic traits.

evolutionary reversibility of phenotypic plasticity is of great interest, that is, whether plastic genotypes evolve into specialists or vice versa, as it pertains to the more general question of whether specialization is phylogenetically terminal and evolutionarily fatal. Moreover, plasticity, by influencing niche breadth, may play an important part in the evolution of species niche breadth, geographic range, and consequent ecological isolation (Figure 10.3).

10.4.1 Plasticity, niche width, and ecological isolation

Phenotypic plasticity is usually viewed as a generalist strategy that enables organisms to inhabit diverse environments, and as such it is implicated in the evolution of niche breadth. Wider niches are often associated with broader geographic ranges, so phenotypic plasticity is thought to be involved in determining species ranges as well. Particular case studies support this hypothesis.

In *Polygonum*, for example, species that exhibit phenotypic plasticity to soil edaphic conditions inhabit a wider range of soil types than specialist, non-plastic species (Sultan *et al.* 2001). These differences in niche breadth predict differences in geographic distribution, with the more plastic species having a larger ecological and geographic range.

Plasticity has also been implicated in adaptation to novel environments. It has been proposed that phenotypic plasticity enables the expression of a phenotype that is closer to the optimum in a novel environment than would be the case if plasticity were absent. While the trait need not express the actual optimum at first, a trait that is closer to the optimum would nonetheless provide a demographic advantage by slowing the rate of, or preventing, extinction long enough for the population to adapt (Baldwin 1896; Bradshaw 1965; Sultan 1987; West-Eberhardt 1989; Schlichting & Pigliucci 1998; Price *et al.* 2003).

The success of plasticity in improving fitness in novel environments depends on how similar the novel environment is to the ancestral environment that elicits plasticity—both in terms of the cues that elicit plasticity and how well those cues predict fitness in that environment. For example, in *Impatiens capensis*, plasticity to a novel overhead canopy was associated with higher fitness of a population that evolved under open-canopy conditions (Donohue *et al.* 2001). The shade-avoidance response, which evolved in response to severe competition by conspecific neighbors in an open canopy environment, was also adaptive when in a novel overhead woodland canopy environment. Thus plasticity increased fitness in the novel environment.

Phenotypic plasticity may then be an important determinant of ecological niche breadth in ways that impact the spatial distribution of species. As such, phenotypic plasticity may also influence the degree of ecological isolation among species, with plastic generalists capable of inhabiting broader ecological niches and encountering a greater number of species and populations than specialists restricted to particular habitats. As such, there is a potential role of phenotypic plasticity in the population dynamics of gene flow and reproductive isolation.

10.4.2 Phenotypic plasticity as an intermediate stage of specialization

There is increasing interest in the possibility that plasticity is an intermediate step in specialization to new environments. The first phase in this scenario is that plasticity facilitates adaptation to a novel environment, while the second phase is the constitutive expression of the adaptive phenotype in its new environment. Whether plasticity actually contributes to local adaptation—that is, specialization—to novel environments is still a matter of debate. Plasticity can apparently facilitate the colonization of new environment under some circumstances (e.g., Yeh & Price 2004); the question is under what conditions does subsequent specialization occur—that is, loss of plasticity? This process is referred to as "genetic assimilation," whereby a phenotype that was originally induced by the environment becomes expressed even when the environmental cue is absent. That is, the phenotype, formerly plastic, becomes constitutively expressed in one form (Waddington 1953; 1961; Schlichting & Pigliucci 1998; Pigliucci *et al.* 2006; Crispo 2007; Lande 2009). No convincing case of genetic assimilation has been documented in the wild, although a small number of experimental studies have documented it (Waddington 1953; Rutherford & Lindquist 1998; Queitsch *et al.* 2002). To address this issue, comparative studies could test whether canalized (nonplastic) trait expression is evolutionarily derived from plastic trait expression. If this is generally true, specialization itself could be derived, although current evidence is mixed (e.g., Pigliucci *et al.* 1999).

How generally plasticity contributes to local adaptation and specialization to new environments is not known. Studies that quantify differences in gene expression across environments, and compare it to gene expression within a given environment of different ecotypes that have locally adapted to those environments, would be highly informative. Does environment-dependent gene expression predict ecotypic differentiation in gene expression? If so, then that would be evidence that plastic gene expression can possibly contribute to the evolution of differences among populations or higher taxa.

The polarity of plastic generalist versus specialist strategies in evolutionary time is of intense interest. If nonplastic specialists are usually derived, that is, if the loss of plasticity tends to be irreversible, is specialization an evolutionary dead end? Or can specialists be rescued by the evolution of plastic trait expression toward phenotypes that are more highly adaptive in novel or former environments? The answer to this question depends, as before, on the genetic basis of environment-dependent phenotypic expression. For example, do mutations disrupt phenotypic expression in one environment but not others, leading to plasticity, or conversely, do mutations disrupt environmental regulation, leading to constitutive expression? That is, what are genetic and epigenetic mechanisms that cause plasticity and how evolutionarily reversible are changes in those mechanisms?

10.4.3 Does plasticity prevent or promote divergence?

To the extent that plasticity is in contrast to specialization, plastic organisms are expected to inhabit a wider diversity of environments than nonplastic specialists, have larger geographic ranges, less ecological isolation from other members of the species, and as a consequence, exhibit less divergence (Figure 10.3). Nonplastic specialists, in contrast, are expected to be more ecologically restricted, and populations are expected to be more geographically restricted in a heterogeneous landscape since dispersal to other appropriate patches—through other habitats—would be more difficult. Local adaptation of these nonplastic specialists would lead to ecological isolation, and further promote divergence.

Greater specialization in turn would favor mechanisms that reduce the environmental variation that is experienced, for example, via reduced dispersal, increased habitat selection, or increased mechanisms of homeostasis. Just as specialization is promoted by habitat selection, so habitat selection would be more advantageous for specialists. Thus, a feedback between habitat selection and specialization is expected, favoring ever more specialized and ecologically restricted entities. In the extreme, specialization will lead to ecological isolation, genetic divergence, and even speciation. Thus nonplastic traits may actually contribute to reproductive isolation and speciation, vindicating, to some degree, the taxonomists who troubled themselves to find stable morphological diagnostics of species-specific differences.

In contrast, proponents of genetic assimilation have at times argued that plasticity, by facilitating adaptation to novel habitats, may actually contribute to the speciation process, supposing that the novel habitat is isolated from other habitats. It is assumed that the organisms cease to experience the previous environment, and in this new environmental stability, plasticity is no longer adaptive. Because this scenario depends on a sequential and irreversible change of environment experienced by the organisms, it might be more likely to pertain to anagenesis (sequential evolution over time within a lineage) rather than cladogenesis (divergence among contemporaneous lineages), unless one makes a number of assumptions about dispersal and the scale of environmental variation experienced by the organisms. The precise ecological context of this process needs to be considered: namely the scale of environmental variation experienced by the lineage and the corresponding adaptive loss of plasticity that must be occur.

This ecological context is especially important to keep in mind in light of the extreme interest in the role that environmentally induced epigenetic changes might play in evolution and speciation. Cross-generational epigenetic inheritance—that is, an environmentally induced change in gene expression that is transmitted to offspring for several generations—has been proposed to be a major alternative to evolution by genetic inheritance, contributing to rapid adaptation, divergence, and even speciation. One pressing question regarding this possibility is whether the epigenetic mechanisms that respond to external environmental stimuli are in fact those that are likely to be inherited from parent to offspring in a stable manner. This is not known. Thus the genetic basis of plasticity is key. More fundamentally, it is possible that the intergenerational stability of epigenetic changes itself can evolve. If so, then the question is, under what circumstance would it actually be adaptive for an epigenetic change to be inherited by offspring? This is precisely the central question concerning the adaptive value of parental effects—or cross-generational plasticity. In short, the adaptive value of cross-generational inheritance of gene expression patterns depends on how predictable the offspring environment is from the parental environment. Similarly, the evolution of such cross-generational epigenetic inheritance depends on the genetic correlations in gene expression between parent and offspring environments. Indeed, the evolutionary literature on cross-generational plasticity and parental effects is likely to be directly pertinent to the evolution of epigenetic inheritance. To some degree, they are likely to be the same phenomena.

10.5 Conclusion

Some traits are more plastic than others. This special subset of nonplastic traits has been historically important in defining evolutionarily stable differences among taxa. This is not to say that plastic traits do not also differ among taxa. Certainly taxa also differ in how they respond to environmental conditions. However, it is also possible that nonplastic traits actually do play a part in the differentiation of taxa.

Why is this? Plasticity contrasts with specialization, and specialization is far more likely to contribute to ecological isolation among taxa. The loss or absence of plasticity in some trait could restrict the ecological conditions a species can successfully inhabit, isolating it from other close taxa. Some such traits, such as flowers, might be directly associated with reproduction; stable differences in reproductive compatibility could certainly contribute to reproductive isolation. It is therefore interesting to ask whether the degree of plasticity of different traits actually does influence its contribution to the evolution of differences among taxa—its contribution to reproductive isolation via their contribution to ecological isolation.

To understand the contribution of different traits to this process, it may therefore be productive to ask why some traits are more plastic than others. The reasons could be ecological or genetic. Regarding ecology, the adaptive value of plasticity of a trait depends on the frequency of selective environments experienced by that trait, and is therefore a function of variation in particular environmental factors and the strength of natural selection imposed by those factors. Not all traits experience the same degree of variation in selective environments: different environmental factors exert selection on different traits, and different traits may be differently buffered from environmental variation via phenological responses (germination cuing), physiological responses (e.g., thermoregulation), or developmental morphology (placement of organs in particular environmental conditions). Moreover, such mechanisms of environmental buffering can coevolve with specialization, leading to positive feedbacks that increasingly favor nonplasticity. Thus one goal in predicting trait-specific plasticity is to characterize environmental factors that elicit plasticity, those that exert selection on particular structures, and the many ways plants mediate the exposure of particular structures to particular environmental factors.

Regarding genetics, plasticity can evolve only when genetic correlations across environments are low. Thus understanding the genetic basis of across-environment pleiotropy is necessary for understanding variation in plasticity among traits. This includes analysis of how the environment-dependent expression and activity of particular genes as well as the genetic pathway structure of trait expression contributes to genetic correlations across environments. It also requires analysis of the genetic variation of those genes and the distribution of variable genes in genetic pathways.

Plants grow in variable environments. Their development reflects this, and is a continuous dialog between genes and environmental conditions. Perhaps what is most remarkable is that any trait is stably expressed and that such stability is in fact sometimes adaptive. Predicting which traits are stable and which developmentally plastic, which traits indicate stable differences among taxa, and which traits contribute to ecological specialization, requires consideration of the genetic basis of environment-dependent phenotypes and the ecology of environmental variation. This requires consideration not only of extrinsic causes of environmental variation but also of how organisms experience that variation, including how they mediate their own exposure to it.

References

Aubin-Horth, N. & Renn, S.C.P. (2009) Genomic reaction norms: using integrative biology to understand molecular mechanisms of phenotypic plasticity. *Molecular Ecology*, **18**, 3763–3780.

Baldwin, J.M. (1896) A new factor in evolution. *American Naturalist*, **30**, 354–451.

Ballaré, C.L., Scopel, A.L. & Sanchez, R.A. (1990) Far-red radiation reflected from adjacent leaves: an early signal of competition in plant canopies. *Science*, **247**, 329–332.

Baskin, C.C. & Baskin, J.M. (1998) *Seeds: Ecology, biogeography and evolution of dormancy and germination*. Academic Press, San Diego, CA.

Bastow, R., Mylne, J.S., Lister, C., *et al.* (2004) Vernalization requires epigenetic silencing of FLC by histone methylation. *Nature*, **427**, 164–167.

Baurle, I. & Dean, C. (2006) The timing of developmental transitions in plants. *Cell*, **125**, 655–664.

Bazzaz, F.A. (1991) Habitat selection in plants. *The American Naturalist*, **137**, S116–S130.

Bender, J. (2004) DNA methylation and epigenetics. *Annual Review of Plant Biology*, **55**, 41–68.

Berger, S.L. (2007) The complex language of chromatin regulation during transcription. *Nature*, **441**, 407–412.

Bossdorf, O., Richards, C.L. & Massimo Pigliucci, M. (2008) Epigenetics for ecologists. *Ecology Letters*, **11**, 106–115.

Bradshaw, A.D. (1965) Evolutionary significance of phenotypic plasticity in plants. *Advances in Genetics*, **13**, 115–155.

Braslavsky, S.E., Gartner, W. & Schaffner, K. (1997) Phytochrome photoconversion. *Plant, Cell and Environment*, **20**, 700–706.

Caicedo, A., Stinchcombe, J.R., Olsen, K.M., Schmitt, J. & Purugganan, M.D. (2004) Epistatic interaction between Arabidopsis FRI and FLC flowering time genes generates a latitudinal cline in a life history trait. *Proceedings of the National Academy of Science of the United States of America*, **101**, 15670–15675.

Casal, J.J., Luccioni, L.G., Oliverio, K.A. & Boccalandro, H.E. (2003) Light, phytochrome signalling and photomorphogenesis in *Arabidopsis*. *Photochemsitry and Photobiology Science*, **2**, 625–636.

Chandler, J., Wilson, A. & Dean, C. (1996) Arabidopsis mutants showing an altered response to vernalization. *The Plant Journal*, **10**, 637–644.

Chouard, P. (1960) Vernalization and its relations to dormancy. *Annual Review of Plant Physiology*, **11**, 191–238.

Clausen, J., Keck, D. & Hiesey, W.M. (1940) *Experimental studies on the nature of species. I. Effect of varied environments on western North American plants*. Carnegie Institution of Washington, Washington, DC.

Clausen, J., Keck, D.D. & Hiesey, W.M. (1947) Heredity of geographically and ecologically isolated races. *The American Naturalist*, **81**, 114–133.

Chen, M., Galvao, R.M., Li, M., *et al.* (2010) Arabidopsis HEMERA/pTAC12 initiates photomorphogenesis by phytochromes. *Cell*, **141**, 1230–1240.

Clough, R.C. & Vierstra, R.D. (1997) Phytochrome degradation. *Plant cell Environment*, **20**, 713–721.

Cook, C.D.K. (1968) Phenotypic plasticity with particular reference to three amphibious plant species. In: *Modern Methods in Plant Taxonomy*. pp. 97–111. Academic Press, London.

Crispo, E. (2007) The Baldwin effect and genetic assimilation: revisiting two mechanisms of evolutionary change mediated by phenotypic plasticity. *Evolution*, **61**, 2469–2479.

Dennis, E.S. & Peacock, W.J. (2007) Epigenetic regulation of flowering. *Current Opinion in Plant Biology*, **10**, 520–527.

DeWitt, T.J. & Scheiner, S.M. (2004) *Phenotypic Plasticity: Functional and Conceptual Approaches*. Oxford University Press, New York.

DeWitt, T.J., Sih, A. & Wilson, D.S. (1998) Costs and limits of phenotypic plasticity. *Trends in Ecology and Evolution*, **13**, 77–81.

Donohue, K. (2003) Setting the stage: plasticity as habitat selection. *International Journal of Plant Sciences*, **164**, S79–S92.

Donohue, K. (2005) Niche construction through phenological plasticity: life history dynamics and ecological consequences. *New Phytologist*, **166**, 83–92.

Donohue, K. (2009) Completing the cycle: maternal effects as the missing link in plant life cycles. *Philosophical Transactions of the Royal Society of London B, Biological Sciences*, **364**, 1059–1074.

Donohue, K., Hammond-Pyle, E., Messiqua, D., Heschel, M.S. & Schmitt, J. (2001) Adaptive divergence in plasticity in natural populations of *Impatiens capensis* and its consequences for performance in novel habitats. *Evolution*, **55**, 692–702.

Donohue, K., Rubio de Casas, R., Burghardt, L., Kovach, K. & Willis, C. (2010) Germination, post-germination adaptation, and species ecological ranges. *Annual Review of Evolution, Ecology and Systematics*, **41**, 293–319.

Donohue, K. & Schmitt, J. (1998) Maternal environmental effects: adaptive plasticity? In: *Maternal Effects as Adaptations* (eds T.A. Mousseau & C.W. Fox). pp. 137–158. Oxford University Press, Oxford.

Dubcovsky, J., Galves, A.F. & Dvork, J. (1994) Comparison of the genetic organization of the early-salt-stress response gene systme in slat-tolerant *Lophropyrum elongatum* and salt-sensitive wheat. *Theoretical and Applied Genetics*, **87**, 957–964.

Eichenberg, K., Baurle, I., Paulo, N., *et al.* (2000) *Arabidopsis* phytochromes C and E have different spectral characters from those of phytochromes A and, B. *FEBS Letters*, **470**, 107–112.

Engelmann, K.E. & Schlichting, C.D. (2005) Coarse-versus fine-grained water stress in *Arabidopsis thaliana* (Brassicaceae). *American Journal of Botany*, **92**, 101–106.

Falconer, D.S. (1952) The problem of environment and selection. *The American Naturalist*, **86**, 293–298.

Falconer, D.S. (1965) Maternal effects and selection response. In: *Genetics Today, Proceedings of the VI International Congress on Genetics* (ed. S.J. Geerts). pp. 763–774. Pergamon, Oxford.

Falconer, D.S. (1990) Selection in different environments: effects on environmental sensitivity (reaction norm) and on mean performance. *Genetical Research*, **56**, 57–70.

Gabriel, W., Luttbeg, B., Sih, A. & Tollrian, R. (2005) Environmental tolerance, heterogeneity, and the evolution of reversible plastic responses. *American Naturalist*, **166**, 339–353.

Galloway, L.F. (2001) Parental environmental effects on life history in the herbaceous plant *Campanula americana*. *Ecology*, **10**, 2781–2789.

Gavrilets, S. & Scheiner, S.M. (1993) The genetics of phenotypic plasticity. V. Evolution of reaction norm shape. *Journal of Evolutionary Biology*, **6**, 31–48.

Gendrel, A.-V. & Colot, V. (2005) Arabidopsis epigenetics: when RNA meets chromatin *Current Opinion in Plant Biology*, **8**, 142–147.

Genoud, T., Schweizer, F., Tscheuschler, A., *et al.* (2008) FHY1 mediates nuclear import of the light-activated phytochrome A photoreceptor. *PLoS Genetics*, **4**, e1000143.

Goosey, L., Palecanda, L. & Sharrock, R.A. (1997) Differential patterns of expresion of the *Arabidopsis* PHYB, PHYD, and PHYE phytochrome genes. *Plant Physiology*, **115**, 959–969.

Grant-Downton, R.T. & Dickinson, H.G. (2005) Epigenetics and its implications for plant biology: 1. The epigenetic network in plants. *Annals of Botany*, **96**, 1143–1164.

Gregor, J.W. (1931) Experimental delimitation of species. *New Phytologist*, **30**, 204–217.

Gregor, J.W., McM. Davey, V. & Lang, J.M.S. (1936) Experimental taxonomy. I: Experimental garden technique in relation to the recognition of the small taxonomic units. *New Phytologist*, **35**, 373–387.

Hahn, M.W., Conant, G.C. & Wagner, A. (2004) Molecular evolution in large genetic networks: does connectivity equal constraint? *Journal of Molecular Evolution*, **58**, 203–211.

Hall, H.M. (1932) Heredity and environment—as illustrated by transplant studies. *The Scientific Monthly*, **October**, 289–302.

Halliday, K.J., Salter, M.G., Thingnaes, E. & Whitelam, G.C. (2003) Phytochrome control of flowering is temperature sensitive and correlates with expression of the floral integrator FT. *The Plant Journal*, **33**, 875–885.

Halliday, K.J. & Whitelam, G.C. (2003) Changes in photoperiod or temperature alter the functional relationships between phytochromes and reveal roles for phyD and phyE. *Plant Physiology*, **131**, 1913–1920.

Hansen, T.F. (2006) The evolution of genetic architecture. *Annual Review of Ecology and Systematics*, **37**, 123–157.

He, X. & Zhang, J. (2006) Toward a molecular understanding of pleiotropy. *Genetics*, **174**, 1885–1891.

Henderson, I. & Dean, C. (2004) Control of Arabidopsis flowering: the chill before the bloom. *Development*, **131**, 3829–3838.

Hennig, L., Buche, C., Eichenberg, K. & Schafer, E. (1999) Dynamic properties of endogenous phytochrome A in Arabidopsis seedlings. *Plant Physiology*, **121**, 571–578.

Hennig, L., Buche, C. & Schafer, E. (2000) Degradation of phytochrome A and the high-irradiance response in *Arabidopsis*: a kinetic analysis. *Plant, Cell and Environment*, **23**, 727–734.

Hennig, L. & Schafer, E. (2001) Both subunits of the dimeric plant photoreceptor phytochrome require chromophore for stability of the far-red light-absorbing form. *Journal of Biological Chemistry*, **276**, 7913–7918.

Heschel, M.S., Butler, C.M., Barua, D., *et al.* (2008) New roles of phytochrome during germination. *International Journal of Plant Science*, **169**, 531–540.

Higgs, D.C. & Colbert, J.T. (1994) Oat phytochrome A mRNA degradation appears to occur via two distinct pathways. *Plant Cell*, **6**, 1007–1019.

Howe, G.A. & Jander, G. (2008) Plant immunity to insect herbivores. *Annual Review of Plant Biology*, **59**, 41–66.

Huq, E., Al-Sady, B. & Quail, P.H. (2003) Nuclear translocation of the photoreceptor phytochrome B is necessary for its biological function in seedling photomorphogenesis. *Plant Journal*, **35**, 660–664.

Jaenisch, R. & Bird, A. (2003) Epigenetic regulation of gene expression: how the genome integrates intrinsic and environmental signals. *Nature Genetics*, **33**, 245–254.

Johanson, U., West, J., Lister, C., *et al.* (2000) Molecular analysis of FRIGIDA, a major determinant of natural variation in Arabidopsis flowering time. Science 290,. *Science*, **290**, 344–347.

Kanazawa, A. (2008) RNA silencing manifested as visibly altered phenotypes in plants. *Plant Biotechnology*, **5**, 423–435.

Kendrick, R.E. & Spruit, C.J.P. (1977) Phototransformations of phytochromes. *Photochemistry and Photobiology*, **26**, 201–214.

Kircher, S., Kozma-Bognar, L., Kim, L., *et al.* (1999) Light quality-dependent nuclear import of the plant photoreceptors phytochrome A and B. *Plant Cell*, **11**, 1445–1456.

Kirkpatrick, M. & Lande, R. (1989) The evolution of maternal characters. *Evolution*, **43**, 485–503.

Lacaze, X., Hayes, P.M. & Korol, A. (2009) Genetics of phenotypic plasticity: QTL analysis in barley, *Hordeum vulgare*. *Heredity*, **102**, 163–173.

Lacey, E.P., Lovin, M.E., Richter, S.J. & Herington, D.A. (2010) Floral reflectance, color, and thermoregulation: What really explains geographic variation in thermal acclimation ability of ectotherms? *American Naturalist*, **175**, 335–349.

Lande, R. (2009) Adaptation to an extraordinary environment by evolution of phenotypic plasticity and genetic assimilation. *Jounal of Evolutionary Biology*, **22**, 1435–1446.

Le Corre, V. (2005) Variation at two flowering time genes within and among populations of *Arabidopsis thaliana*: comparison with markers and traits. *Molecular Ecology*, **14**, 4181–4192.

Lorkovic, Z.J. (2009) Role of plant RNA-binding proteins in development, stress response and genome organization. *Trends in Plant Science*, **14**, 229–236.

Mathews, S. & Sharrock, R.A. (1997) Phytochrome gene diversity. *Plant Cell Environment*, **20**, 666–671.

Mattick, J.S., Amaral, P.P., Dinger, M.E., Mercer, T.R. & Mehler, M.F. (2009) RNA regulation of epigenetic processes. *BioEssays: News and Reviews in Molecular, Cellular and Developmental Biology*, **31**, 51–59.

Matzke, M., Kanno, T., Daxinger, L., Huettel, B. & Matzke, A.J.M. (2009) RNA-mediated chromatin-based silencing in plants. *Current Opinion in Cell Biology*, **21**, 367–376.

Michaels, S.D. & Amasino, R.M. (1999) FLOWERING LOCUS C encodes a novel MADS domain protein that acts as a repressor of flowering. *Plant Cell*, **11**, 949–956.

Michaels, S.D., Himelblau, E., Kim, S.Y., Schomburg, F.M. & Amasino, R.M. (2005) Integration of flowering signals in winter-annual Arabidopsis. *Plant Physiology*, **137**, 49–56.

Mitchell-Olds, T. (1995) Themolecular basis of quantitative-genetic variation in natural populations. *Trends in Ecology and Evolution*, **10**, 324–327.

Moller, S.G., Ingles, P.J. & Whitelam, G.C. (2002) The cell biology of phytochrome signalling. *New Phytologist*, **145**, 553–590.

Mousseau, T.A. & Fox, C.W. eds. (1998) *Maternal Effects as Adaptations*. Oxford Univeristy Press, Oxford.

Nagy, F., Kircher, S. & Schafer, E. (2001) Intracellular trafficking of photoreceptors during light-induced signal transduciton in plants. *Journal of Cell Science*, **114**, 475–480.

Nagy, F. & Schafer, E. (2002) Phytochromes control photomorphogenesis by differentially regulated, interacting signaling pathwasy in higher plants. *Annual Review of Plant Biology*, **53**, 329–355.

Oh, E., Kim, J., Park, E., *et al.* (2004) PIL5, a phytochrome-interacting basic helix-loop-helix protein, is a key negative regulator of seed germination in *Arabidopsis thaliana*. *The Plant Cell*, **16**, 3045–3058.

Penfield, S., Josse, E.-M., Kannagara, R., *et al.* (2005) Cold and light control seed germination through the bHLH transcription factor SPATULA. *Current Biology*, **15**, 1998–2006.

Pigliucci, M. (2001) *Phenotypic Plasticity: Beyond Nature and Nurture*. John's Hopkins University Press, Baltimore.

Pigliucci, M., Cammell, K. & Schmitt, J. (1999) Evolution of phenotypic plasticity: a comparative approach in the phylogenetic neighborhood of *Arabidopsis thaliana*. *Journal of Evolutionary Biology*, **12**, 779–791.

Pigliucci, M., Murren, C.J. & Schlichting, C.D. (2006) Phenotypic plasticity and evolution by genetic assimilation. *Journal of Experimental Biology*, **209**, 2362–2367.

Price, T.D., Qvarnstrom, A. & Irwin, D.E. (2003) The role of phenotypic plasticity in driving genetic evolution. *Proceedings of the Royal Society of London B*, **270**, 1433–1440.

Promislow, D. (2005) A regulatory network analysis of phenotypic plasticity in yeast. *American Naturalist*, **165**, 515–523.

Queitsch, C., Sangster, T.A. & Lindquist, S. (2002) Hsp90 as a capacitor of phenotypic variation. *Nature*, **417**, 618–624.

Rasanen, K. & Kruuk, L.E.B. (2007) Maternal effects and evolution at ecological time scales. *Functional Ecology*, **21**, 408–421.

Ratcliffe, D. (1965) The geographical and ecological distribution of *Arabidopsis* and comments on physiological variation. *Arabidopsis Information Service*, **1S**.

Rausher, M.D., Lu, Y. & Meyerm, K. (2008) Variation in constraint versus positive selection as an explanation for evolutionary rate variation among anthocyanin genes. *Journal of Molecular Evolution*, **67**, 137–144.

Roach, D.A. & Wulff, R.D. (1987) Maternal effects in plants. *Annual Review of Ecology and Systematics*, **18**, 209–235.

Rubio de Casas, R., Vargas, P., Perez-Corona, E., *et al.* (2007) Field patterns of leaf plasticity in adults of the long-lived evergreen *Quercus coccifera*. *Annals of Botany*, **100**, 325–334.

Russo, V.E.A., Martienssen, R.A. & Riggs, A.D. (1996) *Epigenetic Mechanisms of Gene Regulation*. Cold Spring Harbor Laboratory Press, Plainview, NY.

Rutherford, S.L. & Lindquist, S. (1998) Hsp90 as a capacitor for morphological evolution. *Nature*, **396**, 336–342.

Schafer, E. & Bowler, C. (2002) Phytochrome-mediated photoperception and signal transduction in higher plants. *Embo Reports*, **3**, 1042–1048.

Scheiner, S.M. (1993) Genetics and evolution of phenotypic plasticity. *Annual Review of Ecology and Systematics*, **24**, 35–68.

Schlichting, C.D. & Pigliucci, M. (1998) *Phenotypic Evolution: A Reaction Norm Perspective*. Sinauer Associates, Inc., Sunderland, MA.

Schwartz, C., Balasubramanian, S., Warthmann, N., *et al.* (2009) Cis-regulatory changes at FLOWERING LOCUS T mediate natural variation in flowering responses of *Arabidopsis thaliana*. *Genetics*, **183**, 723–732.

Sharrock, R.A. & Clack, T. (2002) Patterns of expression and normalized levels of the five Arabidopsis phytochromes. *Plant Physiology*, **130**, 442–456.

Sharrock, R.A. & Clack, T. (2004) Heterodimerization of type II phytochromes in Arabidopsis. *Proceedings of the National Academy of Science of the United States of America*, **101**, 11500–11505.

Sheldon, C.C., Hills, M.J., Lister, C., *et al.* (2008) Resetting of FLOWERING LOCUS C expression after epigenetic repression by vernalization. *Proceedings of the National Academy of Science of the United States of America*, **105**, 2214–2219.

Shindo, C., Lister, C., Crevillen, P., Nordborg, M. & Dean, C. (2006) Variation in the epigenetic silencing of FLC contributes to natural variation in *Arabidopsis* vernalization response. *Genes and Development*, **20**, 3079–3083.

Simpson, G.G. & Dean, C. (2002) Arabidopsis, the rosetta stone of flowering time? *Science*, **296**, 285–289.

Smith, H. & Whitelam, G.C. (1997) The shade avoidance syndrome: multiple responses mediated by multiple phytochromes. *Plant, Cell and Environment*, **20**, 840–844.

Somers, D.E. & Quail, P.H. (1995) Temporal and spatial expression patterns of PHYA and PHYB genes in *Arabidopsis*. *The Plant Journal*, **7**, 413–427.

Stiles, E.A., Cech, N., Dee, S. & Lacey, E.P. (2007) The relationship between anthocyanins and thermoregulation of reproduction in *Plantago lanceolata*. *Physiologia Plantarum*, **129**, 756–765.

Sultan, S.E. (1987) Evolutionary implications of phenotypic plasticity in plants. *Evolutionary Biology*, **21**, 127–178.

Sultan, S.E. (2001) Phenotypic plasticity for fitness components in Polygonum species of contrasting ecological breadth. *Ecology*, **82**, 328–343.

Sung, S. & Amasino, R.M. (2006) Molecular genetic studies of the memory of winter. *Journal of Experimental Botany*, **57**, 3369–3377.

Takeda, S. & Paszkowski, J. (2006) DNA methylation and epigenetic inheritance during plant gametogenesis. *Chromosoma*, **115**, 27–35.

Thompson, L. (1994) The spatiotemporal effects of nitrogen and litter on the population dynamics of *Arabidopsis thaliana*. *Journal of Ecology*, **82**, 63–68.
Turesson, G. (1925) The plant species in relation to habitat and climate. *Hereditas*, **6**, 147–236.
Ungerer, M.C., Halldorsdottir, S.S., Purugganan, M.D. & Mackay, T.F.C. (2003) Genotype-environment interactions at Quantitative Trait Loci affecting inflorescence development in *Arabidopsis thaliana*. *Genetics*, **165**, 353–365.
van Tienderen, P.H. (1991) Evolution of generalists and specialists in spatially heterogenous environments. *Evolution*, **45**, 1317–1331.
van Tienderen, P.H. & Koelewijn, H.P. (1994) Selection on reaction norms, genetic correlations and constraints. *Genetical Research*, **64**, 115–125.
Vaucheret, H. (2006) Post-transcriptional small RNA pathways in plants: mechanisms and regulations. *Genes and Development*, **20**, 759–771.
Via, S. & Lande, R. (1985) Genotype-environment interaction and the evolution of phenotypic plasticity. *Evolution*, **39**, 505–522.
Waddington, C.H. (1953) Genetic assimilation of an acquired character. *Evolution*, **7**, 118–126.
Waddington, C.H. (1961) Genetic assimilation. *Advances in Genetics*, **10**, 257–290.
Wagner, A. (2000) Mutational robustness in genetic networks of yeast. *Nature Genetics*, **24**, 355–361.
Wagner, A. (2008) Robustness and evolvability: a paradox resolved. *Proceedings of the Royal Society of London B*, **275**, 91–100.
Wang, R., Farrona, S., Vincent, C., *et al.* (2009) PEP1 regulates perennial flowering in Arabis alpina. *Nature*, **459**, 423–428.
West-Eberhard, M.J. (1989) Phenotypic and the origins of diversity. *Annual Review of Ecology and Systematics*, **20**, 249–278.
West-Eberhard, M.J. (2003) *Developmental Plasticity and Evolution*. Oxford University Press, Oxford.
Whitlock, M.C. (1996) The red queen beats the jack -of-all-trades: the limitations on the evolution of phenotypic plasticity and niche breadth. *The American Naturalist*, **148**, S65–S77.
Winn, A.A. (1996) Adaptation to fine-grained environmental variation: an analysis of within-individual leaf variation in an annual plant. *Evolution*, **50**, 1111–1118.
Yeh, K.-C. & Lagarias, J.C. (1998) Eukaryotic phytochromes: light-regulated serine/threonine protein kinases with histidine kinase ancestry. *Proceedings of the National Academy of Sciences of the United States of America*, **95**, 13976–13981.
Yeh, P.J. & Price, T.D. (2004) Adaptive phenotypic plasticity and the successful colonization of a novel environment. *American Naturalist*, **164**, 531–542.

Annual Plant Reviews (2013) **45**, 357–366 http://onlinelibrary.wiley.com
doi: 10.1002/9781118305881.ch11

Chapter 11

THE EVOLUTION OF PLANT FORM: A SUMMARY PERSPECTIVE

Michael Purugganan

Department of Biology, Center for Genomics and Systems Biology, New York University, NY and New York University Abu Dhabi Institute, Abu Dhabi, United Arab Emirates, USA

Abstract: The study of the evolution of development (evo-devo) has advanced dramatically in the last two decades. Although there had been a strong interest in evo-devo for nearly a century, it was only with the advent of molecular developmental genetics that provided a strong basis for a conceptual and methodological integration of developmental studies with evolutionary biology. The study of plant evolution of development has also progressed in the last two decades beginning with the study of the molecular basis for floral diversification in the angiosperms. The confluence of studies in development, molecular genetics, phylogenetics, paleobotany, population genetics, and molecular evolution in the study of plant form is now a robust discipline, with various opportunities to further our understanding of plant morphological evolution at all levels of biological organization.

Keywords: evolution of development; MADS-box; gene family; molecular evolution

The differences in shape and form among organisms are the most visible manifestations of evolution. It is through the diversity in how different species of plants and animals look that we can often discern the magnitude of evolutionary divergence, and the study of morphological diversification has been one of the foundations for advancing evolutionary studies. For most, the study of how form evolves is how we begin to think about how evolution works. The study of how morphologies evolve has naturally focused on the study of how the process of ontogenetic development evolves, and this area of research is now known as evolutionary developmental biology (evo-devo).

Annual Plant Reviews Volume 45: The Evolution of Plant Form, First Edition.
Edited by Barbara A. Ambrose and Michael Purugganan.

Today, studies on the evolution of plant form encompass a wide array of approaches and research areas, including developmental biology, phylogenetics and systematics, population genetics, and genomics and systems biology. All these disciplines together are contributing to our knowledge of the evolution of plant form, and evo-devo has emerged from the synergies of these areas in the last few decades.

In the history of science, it is now generally accepted that the field of biology grew along distinct and largely autonomous lines throughout the last 150 years (Mayr 1985). Evolutionary biology—based on the general principles enunciated by Charles Darwin in the latter half of the nineteenth century (Darwin 1859)—grew as a discipline as it sought to understand how life evolved. This culminated in the modern synthesis of evolutionary biology in the mid-twentieth century, which arose as a confluence of the principles and practices of classical genetics, systematics, paleontology, and ecology (Mayr 1985; Gould 2002).

Developmental biology, on the other hand, progressed with its own disciplinary dynamic, focusing on the mechanisms of embryonic development to elucidate the principles underlying the formation of morphologies. Although there were early and strong interactions between developmental and evolutionary biology (epitomized, e.g., by the work of Ernst Haeckel), it is now generally regarded that these two fields worked in separate domains with very little in the way of intellectual cross-talk of any import (Mayr 1985; Gould 2002).

Despite the separate trajectories of these two biological disciplines, it was nevertheless recognized that understanding the evolution of morphology required a study of how developmental processes evolved. Leigh Van Valen, the evolutionary theorist, famously said "Evolution is the control of development by ecology," but the precise role of development in the evolutionary process remained vague. There were numerous attempts to integrate both evolutionary and developmental biology, and to seek to identify whether such an integration could yield fundamental insights into the nature of evolutionary diversification (Raff 1996; Gould 1985). It was clear that a full picture of the evolutionary process needed to come to grips with development and differentiation; after all, ontogeny was the process by which the organism's genotype is transformed into the phenotype, which is then subject to the forces of natural selection. Several evolutionary biologists did provide key conceptual ideas to the field, notably Rudy Raff (Raff 1996) and Stephen Jay Gould (Gould 1985), among others, which suggested that the integration of development and evolution would indeed prove scientifically illuminating. These attempts were mostly theoretical in nature, but allowed the elaboration of key concepts that continue to be important in the current study of the evolution of development. These include the nature of modularity of development, evolvability of traits, mechanisms of heterochrony and paedomorphosis, and the role of developmental constraints (Raff 1996; Gould 1985, 1966; Gerhart & Kirschner 1997; Wray 2000).

Despite the theoretical richness of these early evolution of development or evo-devo studies, there was still no clear intellectual synergy between the two disciplines, nor was there any large-scale movement to study this potentially transformative interdisciplinary subject. The key to further progress, however, began to be clear toward the end of the twentieth century—the key was genetics.

If evolutionary biology and development were two great intellectual strains of biology throughout the last century, the elaboration of the molecular basis for genetics is a third line of biological studies that also emerged and dominated the life sciences for the last half-century (Judson 1996). The study of the gene and molecular basis of inheritance occupied the attention of a wide range of scientists, including geneticists, physicists, chemists, microbiologists, and virologists. The molecular underpinnings of genetics and the fundamental cellular processes was at the heart of both the process of evolution and the specification of organismal development.

Genetics was a fundamental foundation for evolutionary biology, and indeed the modern synthesis of evolutionary thought was driven in part by the richness of the mathematical and conceptual framework of population genetics (Mayr 1985). The paradigm in the field—that evolution was the change in allele frequencies of genes—focused the attention of biologists on measuring genetic variation within and between populations, as well as examining the dynamics of genetic change. The advent of molecular evolution in the latter half of the twentieth century reinforced this gene-centric view of evolution, by articulating a research program centered on evolution at the protein and DNA level (Mayr 1985; Gould 2002). Nearing the close of the twentieth century, molecular evolution and population genetics as subdisciplines of evolutionary biology matured, with scientists studying genes not only to trace evolutionary histories but also to understand the diversification of genes and the phenotypes they produced.

Developmental biologists, meanwhile, were undertaking their own genetic revolution. In the 1950s, Lewis famously discovered the class of homeotic genes in *Drosophila melanogaster*, illustrating the potential power of a genetics approach to dissecting developmental processes (Lewis 1978). This was then extended by Christiane Nusslein-Volhard, Eric Wieschaus, and their colleagues and students in the 1980s, as they began to do comprehensive genetic analyses of fly development (Nusslein-Volhard & Wieschaus 1980; Nusslein-Volhard 1996), culminating in the isolation of the Drosophila homeotic genes and the discovery that many were DNA-binding transcriptional regulators (Bender *et al.* 1983). Soon after their isolation in flies, it was shown that many homeotic genes belonged to a multigene family that arose through a series of gene duplications (McGinnis *et al.* 1984a). Moreover, workers then began to find homologs of these genes in animals with disparate body plans, including vertebrates (McGinnis *et al.* 1984b). The discovery of the homeodomain gene family led to widespread interest in understanding how genes that contained a largely conserved homeodomain protein sequence could function across various branches of the animal kingdom that had diverse body plans.

It was the incorporation of molecular genetics to developmental biology that was the catalyst for modern evo-devo studies. With the gene taking center stage in the study of developmental processes, modern developmental biology could now begin to fruitfully interact with evolutionary biology. For developmental biology, genes held many of the answers to how ontogeny unfolded. For evolutionary biology, a large part of its research program centered on the study of how genes evolved. Both disciplines now had the study of the gene as a key element that united their conceptual paradigms.

The study of evo-devo further took off as developmental genetics, evolutionary genetics, molecular evolution, and phylogenetics all matured as disciplinary approaches. In animals, studies ranged from, among others, the evolution of *Hox* genes (Zhang & Nei 1996), diversification of body plans (Panganiban *et al.* 1994, 1995), sea urchin evolutionary developmental biology (Wray & Lowe 2000), and ascidian morphological diversification (Swalla & Jefferey 1996). In contrast, plant evo-devo studies had a later start, in part because it took a few more years before plant developmental genetics began a comprehensive molecular genetic analysis of developmental features. Molecular genetic studies in plants, particularly in maize, also underscored the importance of regulatory genes in various aspects of plant development, such as pigmentation (Cone *et al.* 1986; Ludwig *et al.* 1989), seed protein synthesis (Schmidt *et al.* 1987), and leaf cell differentiation (Hake *et al.* 1989). The major breakthrough in plant developmental genetics and evo-devo studies, however, came with the study of the genes underlying flower organ identity. In elegant and independent genetic studies in the model systems *Arabidopsis thaliana* and *Antirrhinum majus*, homeotic loci were identified that specified the identities of the major floral organs—the sepals, petals, stamens, and carpels (Bowman *et al.* 1991). Like their *Drosophila* counterparts, mutations in these loci resulted in interconversions in organ identity, and an elegant and predictive genetic model of floral organ development—the ABC model—was formulated that was generally applicable to species as disparate as the mustard weed and the snapdragon (Coen & Meyerowitz 1991).

Molecular isolation of these floral homeotic genes also showed that, like the *Hox* genes, they encoded DNA-binding transcription factors that were members of a large multigene family (Sommer *et al.* 1990; Yanofsky *et al.* 1990; Jack *et al.* 1992). Unlike the *Hox* genes, however, these loci did not encode homeodomain proteins, but specified members of a different class of transcription factors referred to as the MADS-box proteins (Yanofsky *et al.* 1990; Jack *et al.* 1992; Sommer *et al.* 1990). Like the animal *Hox* genes, a process of gene duplication and gene family evolution was the foundation of a key developmental system, and orthologous members of the gene family appeared to regulate homologous developmental functions across divergent flowering plant species (Purugganan *et al.* 1995; Purugganan 1997). The latter finding was bolstered by early studies of MADS-box gene function in grass species such as maize (Schmidt *et al.* 1993; Mena *et al.* 1996), which as monocots had very different floral structures from the dicotyledonous *Arabidopsis* and *Antirrhinum*, and which also together represented evolutionary lineages that spanned almost the entire ~200 myr timeframe of angiosperm diversification. Similar to the discovery of homology among *Hox* genes in animals, it also suggested a deep evolutionary conservation of developmental systems within the flowering plants.

The study of plant evo-devo initially focused on the evolution of the regulatory gene families that comprised many of the genes that were associated with plant developmental processes such as the MADS-box gene family (Purugganan *et al.* 1995; Theissen *et al.* 1996; Purugganan 1997; Martinez-Castilla & Alvarez-Buylla 2003; Nam *et al.* 2003). Since these studies, starting more than a decade ago, the field of plant evo-devo has significantly expanded and our understanding of how plant form evolves has advanced considerably. The field has been nurtured by different approaches in both developmental and evolutionary biology that have been used to inform what we now know about how plant form diversifies.

The role of phylogenetic analyses in advancing plant evo-devo studies cannot be overestimated. A robust comparative analysis in plant evo-devo arose because of rigorous studies on evolutionary relationships, at the level of regulatory gene families to the level of plant species. Early analyses of gene family diversification, particularly the MADS-box gene family, represented some of the most extensive analysis of gene family evolution in any organismal group. These molecular evolutionary studies provided the framework for our understanding of how developmental functions evolved, how to classify paralogous genes and gene groups, and examine models for developmental gene evolution. Phylogenetic analysis also led to a comprehensive picture of the history of plant evolution (Donoghue 1989; Chase *et al.* 1993; Lee *et al.* 2011). In the last two decades, better and larger gene datasets coupled with improved methods of phylogenetic analyses, have allowed systematists to provide a detailed understanding of the relationships between various plant groups (Donoghue 1989; Chase *et al.* 1993; Lee *et al.* 2011). These plant phylogenies, in turn, allowed us to examine how different traits evolved within land plants, and served as a reference that helped develop hypotheses on plant trait diversification that could be tested by molecular and developmental genetic approaches. The use of molecular phylogenies to reconstruct the history of land plant evolution was complemented by paleobotanical studies that provided a fossil context for plant diversification (Donoghue *et al.* 1989). Fossil data were crucial in unraveling some of the patterns of evolution of plant traits, including extinct taxa that provided clues to evolutionary experimentation in land plant groups.

As the studies of plant evo-devo have advanced, there remain areas that need to be addressed. One issue that has arisen in plant evo-devo studies is the strong emphasis on floral developmental evolution. In a sense, this was to be expected given the particularly strong understanding we have on the developmental genetics of floral form and floral organ specification (Friedman *et al.* 2004). There are other areas of plant evo-devo research, however, that continues to remain largely unknown and would be ripe for further analyses (Friedman *et al.* 2004). These include studies of fruit shape in angiosperms, the evolutionary developmental genetics of leaf shape, and the diversification of root architecture. There are also several major issues in plant evo-devo, such as the evolution of the meristem and the establishment of the shoot/root polarity in land plant evolution that could provide major advances in our understanding of evolutionary patterns in plant development (Friedman *et al.* 2004). In particular, comparative studies of land plant groups outside the angiosperms will be crucial, providing a molecular developmental context for how specific organs (e.g., leaves and roots) evolved multiple times, how developmental networks can diversify in macroevolution and, in general, the dynamics and constraints of the evolutionary process.

Another area of research is a comprehensive model of gene family evolution and their role in phenotypic diversification. As has been shown, a substantial number of plant genes are organized in gene families that arise from a history of gene duplication and/or polyploidization (Moore & Purugganan 2005), and many of these gene families are regulatory families that control developmental processes. The MADS-box gene family has been dissected in some detail (Purugganan *et al.* 1995; Purugganan 1997; Theissen *et al.* 1996; Martinez-Castilla & Alvarez-Buylla 2003; Nam *et al.* 2003), but continued analyses of this and other regulatory gene families are vital if we are to fully understand the nature of plant developmental evolution.

The nature of developmental gene networks and how they evolve also remains largely unknown (Cork & Purugganan 2004; Wilkins 2001). It has been postulated that changes in gene network rewiring are one of the major modes of phenotypic evolution, not only in plants but across the tree of life. Nevertheless, we still have little information on the large-scale structure of networks and how they diversify across evolutionary time. The advent of plant systems biology, however, with an emphasis on a global understanding of networks may provide one of the keys to progress in this area (Coruzzi & Gutierrez 2009; Keurentjes *et al.* 2011).

Another burgeoning area of research is the population genetics of developmental genes and the study of intraspecific developmental variation. The comparative approach has thus far been applied to species that have diverged over large spans of evolutionary time; in essence, the analysis of plant evo-devo has been largely a study of macroevolutionary patterns. The study of population genetics provides a way to examine potential developmental transformations at the beginning of the process rather than at its endpoint (Purugganan 2000; Purugganan & Suddith 1998; Sucena & Stern 2000; Hoekstra *et al.* 2006; Balhoff & Wray 2005). By examining developmental variation at its origins, we can start to understand the early evolutionary (i.e., microevolutionary) dynamics that lead to the fixation of developmental variants that underlie species differences. Moreover, there is a further extension of this approach that now examines the role of ecology in evo-devo studies, providing a bridge between ecology, evolution, and development (Gilbert & Epel 2008; Sultan 2007, 2010).

The field of plant evo-devo is also progressing with the use of more species outside of the usual plant model systems such as *A. thaliana*, *Oryza sativa*, or *Zea mays*. Genomics technologies are allowing for greater access to genomic information in various plant species, and genome sequences are available across most major land plant groups, including *Selaginella* (Banks *et al.* 2011) and *Physcomitrella* (Rensing *et al.* 2008). Moreover, expressed sequence tag (EST) sequences are also available for even more taxa, including from floral tissues (Albert *et al.* 2005). The widespread availability of genome sequences allows investigators to develop new expression analyses platforms even for nonmodel species. Molecular genetic techniques, such as virus-induced gene silencing (VIGS), also allow researchers to examine phenotypes associated with individual genes, and allow for a comparative functional developmental biology (Ruiz *et al.* 1998).

Finally, the spectacular advances in genomics technologies, including the ease and relatively low cost of large-scale sequencing, is going to provide data at an unprecedented pace and rapidly increase our understanding on the evolution of plant form particularly how genomes evolve. Indeed, this and other genomics technologies will impact all of the areas described above, and will lead to a systematic extension of the comparative developmental genetic approach not just within angiosperms but across land plants.

We are now entering a new era in the study of plant evo-devo, one in which we utilize multiple approaches that integrate both evolutionary and developmental biology. From its beginnings in the molecular comparative study of flower development, plant evo-devo study is now maturing and promises to uncover the molecular and evolutionary mechanisms that underlie the diversification of plant form. There continues to be a need to integrate developmental concepts into evolutionary thinking, but it is now clear that a complete picture of how plant morphologies evolve can only be reached from the combined vantage points of both development and evolution, and that this multidisciplinary perspective is a key to unlocking the nature of evolutionary diversification.

References

Albert, V.A., Soltis, D.E., Carlson, J., *et al.* (2005) Floral gene resources from basal angiosperms for comparative genomics research. *BMC Plant Biology*, **5**, 5–10.

Balhoff, J.P. & Wray, G.A. (2005) Evolutionary analysis of the well characterized *endo16* promoter reveals substantial variation within functional sites. *Proceedings of the National Academy of Science of the United States of America*, **102**, 8591–8596.

Banks, J.A., Nishiyama, T., Hasebe, M., *et al.* (2011) The *Selaginella* genome identifies genetic changes associated with the evolution of vascular plants. *Science*, **332**, 960–963.

Bender, W., Akam, M., Karch, F. *et al.* (1983) Molecular genetics of the *bithorax* complex in *Drosophila melanogaster*. *Science*, **221**, 23–29.

Bowman, J.L., Smyth, D.R. & Meyerowitz, E.M. (1991) Genetic interactions among floral homeotic genes of *Arabidopsis*. *Development*, **112**, 1–20.

Chase, M.W., Soltis, D.E., Olmstead, R.G., *et al.* (1993) Phylogenetics of seed plants –An analysis of nucleotide sequences from the plastid gene *rbcL*. *Annals of the Missouri Botanical Garden*, **80**, 528–580.

Coen, E. & Meyerowitz, E.M. (1991) The war of the whorls – Genetic interactions controlling flower development. *Nature*, **353**, 31–37.

Cone, K.C., Burr, F.A. & Burr, B. (1986) Molecular analysis of the maize anthocyanin regulatory locus *C1*. *Proceedings of the National Academy of Science of the United States of America*, **83**, 9631–9635.

Cork, J.M. & Purugganan, M.D. (2004) The evolution of molecular genetic pathways and networks. *Bioessays* **26**, 479–484.

Coruzzi, G. & Gutierrez, R. (2009) *Annual Plant Reviews Volume 35, Plant Systems Biology*, Blackwell Publishing Ltd., Oxford, UK.

Darwin, C. (1859) *Origin of Species*. John Murray, London.

Donoghue, M.J. (1989) Phylogenies and the analysis of evolutionary sequences, with examples from seed plants. *Evolution*, **43**, 1137–1148.

Donoghue, M.J., Doyle, J.A., Gauthier, J., *et al.* (1989) The importance of fossils in phylogeny reconstruction. *Annual Reviews of Ecology and Systematics*, **20**, 431–460.

Friedman, W.E., Moore, R.C. & Purugganan, M.D. (2004) The evolution of plant development. *American Journal of Botany*, **91**, 1726–1741.

Gerhart, J. & Kirschner, M. (1997) *Cells, Embryos and Evolution*. John Wiley & Sons, New York.

Gilbert, S.F. & Epel, D. (2008) *Ecological Developmental Biology: Integrating Epigenetics, Medicine, and Evolution*, Sinauer Press, Sunderland, MA.

Gould, S.J. (1966) Allometry and size in ontogeny and phylogeny. *Biological Reviews of the Cambridge Philosophical Society*, **41**, 587–597.

Gould, S.J. (1985) *Ontogeny and Phylogeny*. Belknap Press of Harvard University, Cambridge, MA.

Gould, S.J. (2002) *The Structure of Evolutionary Theory*. Belknap Press of Harvard University, Cambridge, MA.

Hake, S., Volbrecht, E. & Freeling, M. (1989) Cloning *Knotted*, the dominant morphological mutant in maize using *Ds2* as a transposon tag. *EMBO Journal*, **8**, 15–22.

Hoekstra, H.E., Hirschmann, R.J., Bundey, R.A., Insel, P.A. & Crossland, J.P. (2006) A single amino acid mutation contributes to adaptive beach mouse color pattern. *Science*, **313**, 101–104.

Jack, T., Brockman, L.L. & Meyerowitz, E.M. (1992) The homeotic gene *Apetala3* of *Arabidopsis thaliana* encodes a MADS box and is expressed in petals and stamens. *Cell*, **68**, 683–697.

Judson, H.F. (1996) *The Eighth Day of Creation: Makers of the Revolution in Biology, 25th Anniversary Edition*. Cold Spring Harbor Laboratory Press, Cold Spring, NY.

Keurentjes, J.J.B., Angenent, G.C., Dicke, M., *et al.* (2011) Redefining plant systems biology: from cell to ecosystem. *Trends in Plant Science*, **16**, 183–190.

Lee, E.K., Cibrian-Jaramillo, A., Kolokotronis, S.O., *et al.* (2011) A functional phylogenomics view of the seed plants. *PLoS Genetics*, **7**, e1002411.

Lewis, E. B. (1978) Gene complex controlling segmentation in *Drosophila*. *Nature*, **276**, 565–570.

Ludwig, S.R., Habera, L.F., Dellaporta, S.L., *et al.* (1989) *Lc*, a member of the maize *R* gene family responsible for tissue-specific anthocyanin production, encodes a protein similar to transcriptional activators and the contains the myc-homology region. *Proceedings of the National Academy of Science of the United States of America*, **86**, 7092–7096.

Martinez-Castilla, L.P. & Alvarez-Buylla, E.R. (2003) Adaptive evolution in the Arabidopsis MADS-box gene family inferred from its complete resolved phylogeny. *Proceedings of the National Academy of Science of the United States of America*, **101**, 13407–13412.

Mayr, E. (1985) *The Growth of Biological Thought: Diversity, Evolution, and Inheritance.* Belknap Press of Harvard University, Cambridge, MA.

McGinnis, W., Levine, M.S., Hafen, E., *et al.* (1984a) A conserved DNA sequence in homeotic genes of the *Drosophila Antennapedia* and *Bithorax* complexes. *Nature*, **308**, 428–433.

McGinnis, W., Garber, R.L., Wirz, J., *et al.* (1984b) A homologous protein coding sequence in *Drosophila* homeotic genes and its conservation in other metazoans. *Cell*, **37**, 403–408.

Mena, M., Ambrose, B.A., Meeley, R.B., *et al.* (1996) Diversification of C-function activity in maize flower development. *Science*, **274**, 1537–1540.

Moore, R.C. & Purugganan, M.D. (2005) The evolutionary dynamics of plant duplicate genes. *Current Opinion in Plant Biology*, **8**, 122–128.

Nam, J., dePamphilis, C.W., Ma, H., *et al.* (2003) Antiquity and evolution of the MADS-box gene family controlling flower development in plants. *Molecular Biology and Evolution*, **20**, 1435–1447.

Nusslein-Volhard, C. (1996) The identification of genes controlling development in flies and fishes (Nobel lecture). *Angewandte Chemie*, **35**, 2177–2188.

Nusslein-Volhard, C. & Wieschaus, E. (1980) Mutations affecting segment number and polarity in *Drosophila*. *Nature*, **287**, 795–801.

Panganiban, G., Nagy, L. & Carroll, S. (1994) The role of the *distal-less* gene in the development and evolution of insect limbs. *Developmental Biology*, **163**, 540–548.

Panganiban, G., Sebring, A., Nagy, L., *et al.* (1995) The development of crustacean limbs and the evolution of arthropods. *Science*, **270**, 1363–1366.

Purugganan, M.D. (1997) The MADS-box floral homeotic gene lineages predate the origin of seed plants: phylogenetic and molecular clock estimates. *Journal of Molecular Evolution*, **45**, 392–396.

Purugganan, M.D. (2000) The molecular population genetics of regulatory genes. *Molecular Ecology*, **9**, 1451–1461.

Purugganan, M.D. & Suddith, J.I. (1998) Molecular population genetics of the *Arabidopsis CAULIFLOWER* regulatory gene: nonneutral evolution and naturally occurring variation in floral homeotic function. *Proceedings of the National Academy of Science of the United States of America*, **95**, 8130–8134.

Purugganan, M.D., Rounsley, S.D., Schmidt, R.J., *et al.* (1995) Molecular evolution of flower development – Diversification of the plant MADS-box regulatory gene family. *Genetics*, **140**, 345–356.

Raff, R.A. (1996) *The Shape of Life: Genes, Development, and the Evolution of Animal Form.* University of Chicago Press, Chicago, IL.

Rensing, S.A., Lang, D., Zimmer, A.D., *et al.* (2008) The *Physcomitrella* genome reveals evolutionary insights into the conquest of land by plants. *Science*, **319**, 64–69.

Ruiz, M.T., Voinnet, O. & Baulcombe, D.C. (1998) Initiation and maintenance of virus-induced gene silencing. *Plant Cell*, **10**, 937–946.

Schmidt, R.J., Burr, F.A. & Burr, B. (1987) Transposon tagging and molecular analysis of the maize regulatory locus *Opaque-2*. *Science*, **238**, 960–963.

Schmidt, R.J., Veit, B., Mandel, M.A., *et al.* (1993) Identification and molecular characterization of *Zag1*, the maize homolog of the *Arabidopsis* floral homeotic gene *Agamous*. *Plant Cell*, **5**, 729–737.

Sommer, H., Beltran, J.P., Huijser, P., *et al.* (1990) *Deficiens*, a homeotic gene involved in the control of flower morphogenesis in *Antirhinnum majus* – the protein shows homology to transcription factors. *EMBO Journal*, **9**, 605–613.

Sucena, E. & Stern, D.L. (2000) Divergence of larval morphology between *Drosophila sechellia* and its sibling species caused by *cis*-regulatory evolution of *ovo/shaven-baby*. *Proceedings of the National Academy of Science of the United States of America*, **97**, 4530–4534.

Sultan, S.E. (2007) Development in context: the timely emergence of eco-devo. *Trends in Ecology and Evolution*, **22**, 575–582.

Sultan, S.E. (2010) Plant developmental responses to the environment: Eco-Devo insights. *Current Opinion in Plant Biology*, **13**, 96–101.

Swalla, B.J. & Jeffery, W.R. (1996) Requirement of the *Manx* gene for expression of chordate features in a tailless ascidian larva. *Science*, **274**, 1205–1208.

Theissen, G., Kim, J.T. & Saedler, H. (1996) Classification and phylogeny of the MADS-box multigene family suggest defined roles of MADS-box gene subfamilies in the morphological evolution of eukaryotes. *Journal of Molecular Evolution*, **43**, 484–516.

Wilkins, A.S. (2001) *The Evolution of Developmental Pathways*. Sinauer Associates, Sunderland, MA.

Wray, G.A. (2000) The evolution of embryonic patterning mechanisms in animals. *Seminars in Cell and Developmental Biology*, **11**, 385–393.

Wray, G.A. & Lowe, C.J. (2000) Developmental regulatory genes and echinoderm evolution. *Systematic Biology*, **49**, 28–51.

Yanofsky, M.F., Ma, H., Bowman, J.L., *et al.* (1990) The protein encoded by the *Arabidopsis* homeotic gene *Agamous* resembles transcription factors. *Nature*, **346**, 35–39.

Zhang, J.Z. & Nei, M. (1996) Evolution of *antennapedia*-class homeobox genes. *Genetics*, **142**, 295–303.

INDEX

Annual Plant Reviews Volume 45: The Evolution of Plant Form, First Edition.
Edited by Barbara A. Ambrose and Michael Purugganan.

$199.95 RS. 13,036.74

24